Inequalities for Differential Forms

Inequalities for Differential Forms

Ravi P. Agarwal
Florida Institute of Technology
Melbourne, FL, USA

Shusen Ding
Seattle University
Seattle, WA, USA

and

Craig Nolder
Florida State University
Tallahassee, FL, USA

 Springer

Ravi P. Agarwal
Department of Mathematical Sciences
Florida Institute of Technology
Melbourne, FL 32901
agarwal@fit.edu

Shusen Ding
Department of Mathematics
Seattle University
Seattle, WA 98122
sding@seattleu.edu

Craig Nolder
Department of Mathematics
Florida State University
Tallahassee, FL 32306
nolder@math.fsu.edu

ISBN 978-0-387-36034-8   e-ISBN 978-0-387-68417-8
DOI 10.1007/978-0-387-68417-8
Springer New York Dordrecht Heidelberg London

Library of Congress Control Number: 2009931765

Mathematics Subject Classification (2000): Primary: 26D10, 58A10, 35J60, Secondary: 26D15, 26D20, 30C65, 31B05, 46E35, 53A45

Printed on acid-free paper

Springer is part of Springer Science+Business Media (www.springer.com)

100591848X

This book is dedicated to
Sadhna Agarwal, Yuhao Ding, Raymond W. Nolder, Laura Yang

Preface

Differential forms have been widely studied and used in many fields, such as physics, general relativity, theory of elasticity, quasiconformal analysis, electromagnetism, and differential geometry. They can be used to describe various systems of partial differential equations and to express different geometrical structures on manifolds. Hence, differential forms have become invaluable tools for many fields. One of the purposes of this monograph is to present a series of estimates and inequalities for differential forms, particularly, for the forms satisfying the homogeneous A-harmonic equations, or the nonhomogeneous A-harmonic equations, or the conjugate A-harmonic equations in $\mathbf{R}^n$, $n \geq 2$. These estimates and inequalities are critical tools to investigate the properties of solutions to the nonlinear differential equations and to control oscillatory behavior in domains or on manifolds. These results can be further used to explore the global integrability of differential forms and to estimate the integrals of differential forms. Throughout this monograph we always keep in our mind that differential forms are the extensions of functions (functions are 0-forms). Hence, all results about differential forms presented in this monograph remain valid for functions defined in $\mathbf{R}^n$.

In Chapter 1, we study various versions of the Hardy–Littlewood inequalities for differential forms satisfying the conjugate A-harmonic equation. We first introduce some definitions and notation related to differential forms, which will be used in this monograph. Then, we discuss different versions of the A-harmonic equations and weight classes. From Sections 1.5, 1.6, and 1.7, we present the local and global Hardy–Littlewood inequalities with different weights in John domains and $L^s(\mu)$-averaging domains, respectively. We also give the best integrable exponents in Section 1.8. Finally, we investigate the Hardy–Littlewood inequalities with Orlicz norms.

In Chapter 2, we concentrate on the L^p-estimates for solutions of the nonhomogeneous A-harmonic equation. We also extend these estimates to the

$A_r(\Omega)$-weighted cases. We conclude Chapter 2 with the global norm comparison inequalities and some applications to the compositions of operators.

Chapters 3 and 4 treat the Poincaré inequalities and the Caccioppoli inequalities, respectively. Specifically, we present the Poincaré inequalities with L^p-norms and Orlicz norms for differential forms. We provide some estimates for Green's operator and the projection operator. As applications of the Poincaré inequalities, we also obtain some estimates for Jacobians of the Sobolev mappings. We develop both local and global Caccioppoli-type estimates with different weights in a domain or on a manifold in Chapter 4. Roughly speaking, these estimates provide upper bounds for the L^s-norm of ∇u (if u is a function) or du (if u is a form) in terms of the L^s-norm of differential form u. We also discuss Caccioppoli-type estimates with Orlicz norms.

Chapters 5 and 6 are concerned with the imbedding inequalities and the reverse Hölder inequalities, respectively. The imbedding inequalities for functions can be found in almost every book on partial differential equations; see Sections 7.7 and 7.8 in [63], for example. Hence, we only study the imbedding inequalities for differential forms in Chapter 5. We also explore the imbedding inequalities for some operators applied to differential forms and discuss various weighted cases. In Chapter 6, various versions of the reverse Hölder inequalities are established.

Chapter 7 is devoted to the integral estimates for some related operators, such as the homotopy operator, Laplace–Beltrami operator, and the gradient operator. We also develop some estimates for the compositions of operators, including the Hardy–Littlewood maximal operator and the sharp maximal operator. We know that the Jacobian of a quasiconformal mapping satisfies a stronger estimate, the reverse Hölder inequality. Then, what kind of estimates can we expect for the Jacobian of a mapping in a Sobolev class? We discuss the integrability of Jacobians in Chapter 8. Finally, in Chapter 9, we develop norm comparison theorems related to BMO-norms and Lipschitz norms. We also prove that the integrability exponents described in the Lipschitz norm comparison theorem are the best possible.

This monograph presents an up-to-date account of the advances made in the study of inequalities for differential forms and will hopefully stimulate further research in this area.

We would like to express our deep gratitude to our colleagues and friends who gave us various help and support during the preparation of this monograph. In particular, we are grateful for the valuable discussions with Professor Janet Mills, Professor Wynne Guy, and Professor John Carter. During the preparation of this monograph, Professor Yuming Xing, Professor Bing Liu, and Professor Peilin Shi generously devoted considerable time and effort in

reading various versions of the manuscript and giving us many precious and
thoughtful suggestions which led to substantial improvements in the text. We
also want to thank Ms. Damle Vaishali at Springer Verlag, New York, for her
support and cooperation.

Melbourne, Florida   *Ravi P. Agarwal*
Seattle, Washington   *Shusen Ding*
Tallahasse, Florida   *Craig A. Nolder*

August, 2008

Contents

Chapter 1
Hardy–Littlewood inequalities

In this first chapter, we discuss various versions of the Hardy–Littlewood inequality for differential forms, including the local cases, the global cases, one weight cases, and two-weight cases. We know that differential forms are generalizations of the functions, which have been widely used in many fields, including potential theory, partial differential equations, quasiconformal mappings, nonlinear analysis, electromagnetism, and control theory; see [1–19], for example. During recent years new interest has developed in the study of the L^p theory of differential forms on manifolds [20, 21]. For $p = 2$, the L^p theory has been well studied. However, in the case of $p \neq 2$, the L^p theory is yet to be fully developed. The development of the L^p theory of differential forms makes it possible to transport all notations of differential calculus in $\mathbf{R}^n$ to the field of differential forms. The outline of this chapter is first to provide background materials, such as the definitions of differential forms and A-harmonic equations, some classes of weight functions and domains, and then, introduce different versions of Hardy–Littlewood inequalities on various domains with some specific weights or norms.

1.1 Differential forms

We first introduce some preliminaries about differential forms. But, we do not try to include all basic properties and results related to differential forms in this section. We only briefly review some basic definitions and terminology related to differential forms which are needed in later chapters.

1.1.1 Basic elements

Throughout this monograph Ω is used to denote an open subset of $\mathbf{R}^n$, $n \geq 2$, and $\mathbf{R} = \mathbf{R}^1$. Balls are denoted by B and σB is the ball with the same center

R.P. Agarwal et al., *Inequalities for Differential Forms*,
DOI 10.1007/978-0-387-68417-8_1, © Springer Science+Business Media, LLC 2009

as B with $\mathrm{diam}(\sigma B) = \sigma \mathrm{diam}(B)$. We do not distinguish the balls from cubes. The n-dimensional Lebesgue measure of a set $E \subseteq \mathbf{R}^n$ is denoted by $|E|$. We call w a weight if $w \in L^1_{loc}(\mathbf{R}^n)$ and $w > 0$ a.e. For $0 < p < \infty$, we denote the weighted L^p norm of a measurable function f over E by

$$||f||_{p,E,w} = \left(\int_E |f(x)|^p w dx \right)^{1/p}$$

if the above integral exists. Let $e_1 = (1,0,\ldots,0)$, $e_2 = (0,1,\ldots,0)$, ..., $e_n = (0,0,\ldots,1)$ be the standard unit basis of $\mathbf{R}^n$. For $l = 0,1,\ldots,n$, the linear space of l-vectors, spanned by the exterior products $e_I = e_{i_1} \wedge e_{i_2} \wedge \cdots \wedge e_{i_l}$, corresponding to all ordered l-tuples $I = (i_1, i_2, \ldots, i_l)$, $1 \le i_1 < i_2 < \cdots < i_l \le n$, is denoted by $\wedge^l = \wedge^l(\mathbf{R}^n)$. The Grassman algebra

$$\wedge = \wedge(\mathbf{R}^n) = \oplus_{l=0}^n \wedge^l(\mathbf{R}^n)$$

is a graded algebra with respect to the exterior products. For $\alpha = \sum \alpha^I e_I \in \wedge$ and $\beta = \sum \beta^I e_I \in \wedge$, the inner product in $\wedge$ is given by

$$< \alpha, \beta >= \sum \alpha^I \beta^I$$

with summation over all l-tuples $I = (i_1, i_2, \ldots, i_l)$ and all integers $l = 0, 1, \ldots, n$.

A *differential l-form* ω on Ω is a Schwartz distribution on Ω with values in $\wedge^l(\mathbf{R}^n)$. We use $D'(\Omega, \wedge^l)$ to denote the space of all differential l-forms and $L^p(\Omega, \wedge^l)$ to denote the l-forms

$$\omega(x) = \sum_I \omega_I(x) dx_I = \sum \omega_{i_1 i_2 \cdots i_l}(x) dx_{i_1} \wedge dx_{i_2} \wedge \cdots \wedge dx_{i_l}$$

with $\omega_I \in L^p(\Omega, \mathbf{R})$ for all ordered l-tuples I. Thus $L^p(\Omega, \wedge^l)$ is a Banach space with norm

$$||\omega||_{p,\Omega} = \left(\int_\Omega |\omega(x)|^p dx \right)^{1/p} = \left(\int_\Omega (\sum_I |\omega_I(x)|^2)^{p/2} dx \right)^{1/p}.$$

It is easy to see that

$$\{dx_{i_1} \wedge dx_{i_2} \wedge \cdots \wedge dx_{i_l}, \ 1 \le i_1 < i_2 < \cdots < i_l \le n\}$$

is a basis of the space $\wedge^l$, and hence $\dim(\wedge^l) = \dim(\wedge^l(\mathbf{R}^n)) = \binom{n}{l}$ and

$$\dim(\wedge) = \sum_{l=0}^n \dim(\wedge^l(\mathbf{R}^n)) = \sum_{l=0}^n \binom{n}{l} = 2^n.$$

We should also notice that

$$dx_i \wedge dx_j = -dx_j \wedge dx_i, \ i \neq j$$

and

$$dx_i \wedge dx_i = 0.$$

We denote the *exterior derivative* by $d : D'(\Omega, \wedge^l) \to D'(\Omega, \wedge^{l+1})$ for $l = 0, 1, \ldots, n-1$. The *exterior differential* can be calculated as follows:

$$d\omega(x) = \sum_{k=1}^{n} \sum_{1 \leq i_1 < \cdots < i_l \leq n} \frac{\partial \omega_{i_1 i_2 \cdots i_l}(x)}{\partial x_k} dx_k \wedge dx_{i_1} \wedge dx_{i_2} \wedge \cdots \wedge dx_{i_l}.$$

Its formal adjoint operator $d^\star$ which is called the *Hodge codifferential* is defined by

$$d^\star = (-1)^{nl+1} \star d\star : D'(\Omega, \wedge^{l+1}) \to D'(\Omega, \wedge^l),$$

where $l = 0, 1, \ldots, n-1$ and $\star$ is the Hodge star operator that will be defined in Definition 1.1.1. For $\omega \in D'(\Omega, \wedge^l)$ the vector-valued differential form

$$\nabla \omega = \left(\frac{\partial \omega}{\partial x_1}, \ldots, \frac{\partial \omega}{\partial x_n} \right)$$

consists of differential forms

$$\frac{\partial \omega}{\partial x_i} \in D'(\Omega, \wedge^l),$$

where the partial differentiation is applied to the coefficients of ω. Similarly, $W^{1,p}(\Omega, \wedge^l)$ is used to denote the *Sobolev space of l-forms* which equals $L^p(\Omega, \wedge^l) \cap L^p_1(\Omega, \wedge^l)$ with norm

$$\|\omega\|_{W^{1,p}(\Omega, \wedge^l)} = diam(\Omega)^{-1} \|\omega\|_{p,\Omega} + \|\nabla \omega\|_{p,\Omega}.$$

The notations $W^{1,p}_{loc}(\Omega, \mathbf{R})$ and $W^{1,p}_{loc}(\Omega, \wedge^l)$ are self-explanatory. For $0 < p < \infty$ and a weight $w(x)$, the weighted norm of $\omega \in W^{1,p}(\Omega, \wedge^l)$ over Ω is denoted by

$$\|\omega\|_{W^{1,p}(\Omega, \wedge^l), w} = diam(\Omega)^{-1} \|\omega\|_{p,\Omega,w} + \|\nabla \omega\|_{p,\Omega,w}. \qquad (1.1.1)$$

We define the *Hodge star operator* $\star : \wedge \to \wedge$ as follows.

Definition 1.1.1. If $\omega = \alpha_{i_1 i_2 \cdots i_k}(x_1, x_2, \ldots, x_n) dx_{i_1} \wedge dx_{i_2} \wedge \cdots \wedge dx_{i_k}$, $i_1 < i_2 < \cdots < i_k$, is a differential k-form, then

$$\star \omega = sign(\pi) \alpha_{i_1 i_2 \cdots i_k}(x_1, x_2, \ldots, x_n) dx_{j_1} \wedge \cdots \wedge dx_{j_{n-k}},$$

where $\pi = (i_1, \ldots, i_k, j_1, \ldots, j_{n-k})$ is a permutation of $(1, \ldots, n)$ and $sign(\pi)$ is the signature of the permutation.

It should be noticed that Hodge star operator can be defined equivalently as follows.

Definition 1.1.2. If $\omega = \alpha_{i_1 i_2 \cdots i_k}(x_1, x_2, \ldots, x_n) dx_{i_1} \wedge dx_{i_2} \wedge \cdots \wedge dx_{i_k} = \alpha_I dx_I$, $\ i_1 < i_2 < \cdots < i_k$, is a differential k-form, then

$$\star\omega = \star\alpha_{i_1 i_2 \cdots i_k} dx_{i_1} \wedge dx_{i_2} \wedge \cdots \wedge dx_{i_k} = (-1)^{\sum(I)} \alpha_I dx_J,$$

where $I = (i_1, i_2, \ldots, i_k)$, $J = \{1, 2, \ldots, n\} - I$, and

$$\sum(I) = \frac{k(k+1)}{2} + \sum_{j=1}^{k} i_j.$$

For example,
$$\star dx_1 = (-1)^2 dx_2 \wedge dx_3 = dx_2 \wedge dx_3$$

in $\wedge^1(\mathbf{R}^3)$. Applying $\star$ to a 2-form

$$\omega = a_{12} dx_1 \wedge dx_2 + a_{13} dx_1 \wedge dx_3 + a_{23} dx_2 \wedge dx_3$$

in $\wedge^2(\mathbf{R}^3)$, we have

$$\star\omega = \star(a_{12} dx_1 \wedge dx_2 + a_{13} dx_1 \wedge dx_3 + a_{23} dx_2 \wedge dx_3)$$

$$= (-1)^{\frac{2(2+1)}{2}+1+2} a_{12} dx_3 + (-1)^{\frac{2(2+1)}{2}+1+3} a_{13} dx_2 + (-1)^{\frac{2(2+1)}{2}+2+3} a_{23} dx_1$$

$$= a_{12} dx_3 - a_{13} dx_2 + a_{23} dx_1.$$

Let $u = u^1 dx_1 + u^2 dx_2 + \cdots + u^n dx_n$ be a 1-form in $\wedge^1(\mathbf{R}^n)$. Then, the $(n-1)$-form $\star u$ can be found as

$$\star u = u^1 dx_2 \wedge dx_3 \wedge \cdots \wedge dx_n - \cdots + (-1)^{n+1} u^n dx_1 \wedge dx_2 \wedge \cdots \wedge dx_{n-1}.$$

Thus, we have

$$d(\star u) = u^1_{x_1} dx_1 \wedge dx_2 \wedge \cdots \wedge dx_n + \cdots + u^n_{x_n} dx_1 \wedge dx_2 \wedge \cdots \wedge dx_n$$

$$= (u^1_{x_1} + u^2_{x_2} + \cdots + u^n_{x_n}) dx_1 \wedge dx_2 \wedge \cdots \wedge dx_n.$$

Therefore, we find that

$$d^\star u = -(\star d\star) u = -(u^1_{x_1} + u^2_{x_2} + \cdots + u^n_{x_n}) = -\mathrm{div}\, u.$$

The Hodge star operator has the following properties:
(1) $\star$ maps k-forms to $(n-k)$-forms for $0 \le k \le n$.
(2) If $e_1, e_2, \ldots, e_n$ is the standard unit basis of $\mathbf{R}^n$ and $\alpha, \beta \in \wedge$, then

$$\star 1 = e_1 \wedge e_2 \wedge \cdots \wedge e_n \quad \text{and} \quad \alpha \wedge \star\beta = \beta \wedge \star\alpha = <\alpha, \beta> (\star 1).$$

Hence the norm of $\alpha \in \wedge$ is given by the formula

$$|\alpha|^2 = <\alpha, \alpha> = \star(\alpha \wedge \star\alpha) \in \wedge^0 = \mathbf{R}.$$

The Hodge star is an isometric isomorphism on $\wedge$ with $\star$ (from $\wedge^l$ to $\wedge^{n-l}$) and $\star\star(-1)^{l(n-l)}$ (from $\wedge^l$ to $\wedge^l$).

1.1.2 Definitions and notations

Let us recall some basic definitions, terms, and results that will be used frequently throughout this monograph. Let X and Y be the vector spaces of dimensions n and m, respectively. We denote the space of all linear transformations from X to Y by $L(X,Y)$. We write $X' = L(X,\mathbf{R})$ and $Y' = L(Y,\mathbf{R})$. For any $T \in L(X,Y)$, the dual transformation $T' \in L(Y',X')$ is defined by the formula

$$T'(g) = g \circ T \quad \text{for} \quad g \in Y' = L(Y,\mathbf{R}).$$

We say $T'(g)$ is the *pullback* of g, which is induced via the mapping $T :$ $X \rightarrow Y$. The concept of the pullback extends naturally to l-covectors,

$$T_\natural : \wedge^l(Y) \rightarrow \wedge^l(X), \ l = 1, 2, \ldots, n.$$

Specifically, for any $\omega \in \wedge^l(Y)$, we define $T_\natural\omega \in \wedge^l(X)$ by the rule

$$(T_\natural\omega)(v_1, v_2, \ldots, v_l) = \omega(Tv_1, Tv_2, \ldots, Tv_l),$$

where $v_1, v_2, \ldots, v_l \in X$.

The operation of pulling back has many nice properties. Here we list some of them:

$$T_\natural(\omega \wedge \varphi) = (T_\natural\omega) \wedge (T_\natural\varphi), \quad \omega, \varphi \in \wedge^l(Y),$$

$$(T \circ S)_\natural(\omega) = S_\natural(T_\natural\omega),$$

$$(T^{-1})_\natural = (T_\natural)^{-1} \quad \text{if } T \text{ is a linear isomorphism},$$

$$(kT)_\natural = k^l T_\natural \quad \text{on } l\text{-covectors}, \ k \in \mathbf{R}.$$

Next, we assume that $\Omega \subset \mathbf{R}^n$ and $\Omega' \subset \mathbf{R}^m$. Let $f : \Omega \rightarrow \Omega'$ be a mapping and its differential $Df(x) : \mathbf{R}^n \rightarrow \mathbf{R}^m$ be defined for almost all $x \in \Omega$. Then, the associate pullback $[Df(x)]_\natural : \wedge^l(\mathbf{R}^m) \rightarrow \wedge^l(\mathbf{R}^n)$ is well defined pointwise almost everywhere. Let ω be a differential form in Ω',

$$\omega(y) = \sum_I \omega^I(y)dy_I = \sum \omega_{i_1 i_2 \cdots i_l}(y)dy_{i_1} \wedge dy_{i_2} \wedge \cdots \wedge dy_{i_l},$$

where ω^I are functions defined at each point $y \in \Omega'$. Then, the *pullback* of ω via f, denoted by $f^\star(\omega)$, is a differential form in Ω defined pointwise almost everywhere by

$$f^\star(\omega)(x) = [Df(x)]_\natural \omega(y),$$

where $y = f(x)$ and $\omega(y)$ is the l-covector in $\wedge^l(\mathbf{R}^m)$. We should notice that

$$f^\star \circ d = d \circ f^\star$$

which is a very useful property. See [22, 23] for more properties of the pullback of differential forms.

Let V be a vector space over $\mathbf{R}$. An algebra with unit 1 generated by the elements of V over $\mathbf{R}$ satisfying the relation

$$u \wedge v = -v \wedge u$$

for arbitrary $u, v \in V$ is denoted by $\wedge(V)$ and called an *exterior algebra* of V or a *Grassmann algebra*. Here $\wedge$ stands for the product of this algebra. Let $\{e_1, e_2, \ldots, e_n\}$ be the basis for V and $\{e^1, e^2, \ldots, e^n\}$ be the basis for the dual space V'. Then, the elements $1, e^1, \ldots, e^n$ generate the algebra $\wedge(V)$ and the covectors

$$1, e^1, \ldots, e^n, e^1 \wedge e^2, \ldots, e^{n-1} \wedge e^n, \ldots, e^1 \wedge e^2 \wedge \cdots \wedge e^n$$

act as the induced basis of $\wedge(V)$. A convenient notation in exterior algebra has many applications in the study of differential forms and determinants. For example, for covectors $v^j = v_1^j e^1 + \cdots + v_n^j e^n \in V', j = 1, 2, \ldots, l$, we have

$$v^1 \wedge v^2 \wedge \cdots \wedge v^l = \sum_{1 \le i_1 < \cdots < i_l \le n} \begin{vmatrix} v_{i_1}^1 & v_{i_2}^1 & \cdots & v_{i_l}^1 \\ v_{i_1}^2 & v_{i_2}^2 & \cdots & v_{i_l}^2 \\ \vdots & \vdots & \ddots & \vdots \\ v_{i_1}^l & v_{i_2}^l & \cdots & v_{i_l}^l \end{vmatrix} e^{i_1} \wedge \cdots \wedge e^{i_l}.$$

1.1.3 Poincaré lemma

Let $\omega = \omega(x)dx_1 \wedge dx_2 \wedge \cdots \wedge dx_n$ be an n-form. After choosing an orientation of $\mathbf{R}^n$, we can define the integral by

$$\int_\Omega \omega = \int_\Omega \omega(x)dx$$

if the coefficient function ω is integrable. Let $f : \Omega \to \Omega'$ be an orientation-preserving diffeomorphism and $f^\star : L^1(\Omega', \wedge^n) \to L^1(\Omega, \wedge^n)$ be the pullback induced by f. Then, the relation

$$\int_\Omega f^\star(\omega) = \int_{\Omega'} \omega$$

plays an important role in the theory of integration on manifolds. In fact, the operation $f^\star$ is equivalent to "substitution of variables." The following Poincaré lemma is ubiquitous in the study of differential forms.

Poincaré Lemma. *Let M be a contractible differentiable manifold, and u be a differentiable k-form in M with $du = 0$. Then, u is exact, i.e., there exists a $(k-1)$-form v in M such that $dv = u$.*

It has been proved that differential forms have the following Hodge decomposition properties.

Theorem 1.1.3. *If $\omega \in L^p(\mathbf{R}^n, \wedge^k)$, $1 < p < \infty$, then there is a $(k-1)$-form α and a $(k+1)$-form β such that*

$$\omega = d\alpha + d^\star \beta$$

and $d\alpha + d^\star \beta \in L^p(\mathbf{R}^n, \wedge^k)$. Moreover, the forms $d\alpha$ and $d^\star \beta$ are unique up to a constant form and

$$\alpha \in Ker d^\star \cap L_1^p(\mathbf{R}^n, \wedge^{k-1}),$$

$$\beta \in Ker d \cap L_1^p(\mathbf{R}^n, \wedge^{k+1}).$$

Also, we have the uniform estimate

$$\|\alpha\|_{L_1^p(\mathbf{R}^n)} + \|\beta\|_{L_1^p(\mathbf{R}^n)} \leq C_p(k,n)\|\omega\|_p$$

for some constant $C_p(k,n)$ independent of ω. Here $d^\star$ is the Hodge codifferential operator (the formal adjoint of d), $k = 1, \ldots, n-1$.

See [22, 24] for the proof of the Poincaré lemma. For more operational properties and results about differential forms, see [22–26]. We conclude this section with the following generalized Hölder's inequality which will be frequently used throughout this monograph.

Theorem 1.1.4. *Let $0 < \alpha < \infty$, $0 < \beta < \infty$, and $s^{-1} = \alpha^{-1} + \beta^{-1}$. If f and g are measurable functions on $\mathbf{R}^n$, then*

$$\| fg \|_{s,E} \leq \| f \|_{\alpha,E} \cdot \| g \|_{\beta,E}$$

for any $E \subset \mathbf{R}^n$.

1.2 *A*-harmonic equations

The developments of the theory about the *A*-harmonic equation are closely
related to the theory of quasiconformal and quasiregular mappings. A series
of results about the solutions to different versions of the *A*-harmonic equation and their applications in fields such as quasiconformal mappings and
the theory of elasticity have been found recently; see [1, 2, 27–52]. In fact,
many properties of solutions to the *A*-harmonic equation are generalizations
of quasiconformal and quasiregular mappings in $\mathbf{R}^n$. Quasiconformal and
quasiregular mappings in $\mathbf{R}^n$ are natural extensions of conformal and analytic functions of one complex variable, respectively. In the two-dimensional
case these mappings were introduced by H. Grötzsch [53] in 1928 and the
higher dimensional case was first studied by M.A. Lavrent'ev [54] in 1938. Far-
reaching results were obtained also by O. Teichmüller [55] and L.V. Ahlfors
[56]. The systematic study of quasiconformal mappings in $\mathbf{R}^n$ was furnished
by F.W. Gehring [5] and J. Väisälä [57] in 1961, and the study of quasiregular mappings by Yu. G. Reshetnyak in 1966 [58]. In a highly significant
series of papers published during the period 1966–1969, Reshetnyak proved
the fundamental properties of quasiregular mappings by exploiting tools from
differential geometry, non-linear partial differential equation theory, such as
the *p-harmonic equation*

$$\operatorname{div}(\nabla u |\nabla u|^{p-2}) = 0 \tag{1.2.1}$$

for functions in $\mathbf{R}^n$, and the theory of Sobolev spaces. The *p*-harmonic equation (1.2.1) is a special case of the *A-harmonic equation*

$$\operatorname{div} A(x, \nabla u) = 0 \tag{1.2.2}$$

for functions in $\mathbf{R}^n$, where $A\colon \mathbf{R}^n \times \mathbf{R}^n \to \mathbf{R}^n$ is a mapping which satisfies
certain structural assumptions. See [59, Chapter 3] for detailed results about
A-harmonic equation of functions in $\mathbf{R}^n$. The *A*-harmonic equation has also
been studied extensively in [60–86, 51]. One of the main purposes of this
monograph is to summarize the recent developments of *A*-harmonic equations
for differential forms.

During the period 1969–1972, O. Martio, S. Rickman, and J. Väisälä [87–
90] gave a second approach to the theory of quasiregular mappings which
was based on the fact that a nonconstant quasiregular mapping is discrete
and open. On the other hand, their approach made use of tools from the theory of quasiconformal mappings, such as curve families and moduli of curve
families. The extremal length and modulus of a curve family were introduced
by L. V. Ahlfors and A. Beurling in their paper [91] on conformal invariants
in 1950. A third approach was developed by B. Bojarski and T. Iwaniec in
1983. Their methods are real analytic in nature and largely independent of
Reshetnyak's work. A fourth approach was suggested by M. Vuorinen [92],

which is a ramification of the curve family method (see [87–89]), in which conformal invariants play a central role. Each of the above four approaches yields a theory covering the whole spectrum of results of the theory of quasiregular mappings. Particularly, the fourth approach introduced by M. Vuorinen in [93–95] has been applied mainly to distortion theory. This work has been continued in [43, 96–98], where some qualitative distortion theorems were discovered. These papers also include results which are sharp as the maximal dilation K approaches 1.

1.2.1 Quasiconformal mappings

Let $f : \Omega \to \mathbf{R}^n$, $f = (f^1, \ldots, f^n)$ be a mapping of Sobolev class $W^{1,p}_{loc}(\Omega, \mathbf{R}^n)$, $1 \le p < \infty$, whose distributional differential $Df = [\partial f^i / \partial x_j] : \Omega \to GL(n)$ is locally integrable function on Ω with values in the space $GL(n)$ of all $n \times n$-matrices. A homeomorphism $f : \Omega \to \mathbf{R}^n$ of Sobolev class $W^{1,p}_{loc}(\Omega, \mathbf{R}^n)$ is said to be K-quasiconformal, $1 \le K < \infty$, if its differential matrix $Df(x)$ and the Jacobian determinant $J = J(x, f) = \det Df(x)$ satisfy

$$|Df(x)|^n \le KJ(x, f), \tag{1.2.3}$$

where $|Df(x)| = \max\{|Df(x)h| : |h| = 1\}$ denotes the norm of the Jacobi matrix $Df(x) : \mathbf{R}^n \to \mathbf{R}^n$, as a linear transformation and the constant $K \ge 1$ is independent of $x \in \Omega$. The operator norm $|Df(x)|$ stands for the magnitude of infinitesimal deformation of one-dimensional objects, the cofactors $D^\sharp f(x) \in \mathbf{R}^{n \times n}$ of the differential matrix characterize infinitesimal deformations of $(n-1)$-dimensional objects, and the n-form $J(x, f)dx = df^1 \wedge \cdots \wedge df^n$ represents an infinitesimal change of volume at the point $x \in \Omega$. Hadamard's inequality gives the bounds

$$|J(x, f)| \le |D^\sharp f(x)|^{n/(n-1)} \le |Df(x)|^n.$$

The mapping f is said to be K-quasiregular if the injectivity in the above definition is dropped. We say that f is orientation preserving (reversing) if its Jacobian determinant $J(x, f)$ is nonnegative (nonpositive) almost everywhere. A Sobolev mapping $f : \Omega \to \mathbf{R}^n$ is said to have finite distortion if (i) its Jacobian determinant is locally integrable and does not change sign and (ii) there exists a measurable function $K = K(x) \ge 1$, finite almost everywhere, such that

$$|Df(x)|^n = K(x)|J(x, f)|. \tag{1.2.3$'$}$$

Equation $(1.2.3)'$ is called the distortion equation and $K(x)$ is called the outer distortion function of f. It is well known that every mapping of finite distortion solves a nonlinear system of first-order PDEs, the so-called Beltrami system

$$D^\star f(x) Df(x) = |J(x, f)|^{\frac{2}{n}} \mathbf{G},$$

where $D^\star f(x)$ stands for the transpose of the differential matrix and $\mathbf{G} = \mathbf{G}(x)$, called the *conformal distortion tensor* of f, is acquired from the Cauchy–Green tensor by proper re-scaling, such that $\det \mathbf{G}(x) \equiv 1$ [99]. This in turn gives rise to a degenerate elliptic equation of the second order, an analog of the familiar A-harmonic equation. The matrix function $\mathbf{G} = \mathbf{G}(x) = [G_{ij}] \in \mathbf{R}^{n \times n}$ is often called the *conformal structure* or the *Riemannian metric* in Ω. This may be understood as saying that f is conformal with respect to this metric.

One difference between conformal and quasiconformal mappings is that the later need not be differentiable in the usual sense. However, by a theorem due to A. Mori, F.W. Gehring, and J. Väisälä, every quasiconformal mapping $f : \Omega \to \mathbf{R}^n$ is differentiable almost everywhere and its Jacobian determinant $J(x, f)$ is locally integrable. Moreover, quasiconformality of f can be expressed by the differential inequality (1.2.3). Hence quasiconformal mappings can be treated by the methods of measure and integration.

L^p-integrability for quasiconformal and quasiregular mappings in certain domains in $\mathbf{R}^n$ is an interesting and active branch of quasiconformal and quasiregular mappings. As early as in the 1930s, G.H. Hardy and J.E. Littlewood in [100, 101] studied the L^p-integrability of the real part and the imaginary part of a holomorphic function which is defined in the unit disk in $\mathbf{R}^2$. They also proved that the L^p-norms of the real part and the imaginary part of such a holomorphic function are comparable, see Section 1.5. This is a well-known result and is called Hardy–Littlewood theorem for conjugate harmonic functions in the unit disk. This work opened a new direction of research which we will see later in this chapter.

1.2.2 A-harmonic equations

Historically, the origin of quasiconformal mappings is connected with the developments of the methods of complex functions. Since the power of this concept was first realized, quasiconformal mappings have engaged the attention of many prominent mathematicians and the theory has been greatly expanded to higher dimensions. In higher dimensions one might consider the coordinate function of a quasiconformal mapping. These are rather special solutions of an *A-harmonic equation* [they are coupled by a system of the first-order partial differential equations (the Beltrami system in $\mathbf{R}^n$) like conjugate harmonic functions are coupled by the Cauchy–Riemann system]:

$$d^\star A(x, d\omega) = 0 \tag{1.2.4}$$

for differential forms, where $A : \Omega \times \wedge^l(\mathbf{R}^n) \to \wedge^l(\mathbf{R}^n)$ satisfies the following conditions:

$$|A(x,\xi)| \le a|\xi|^{p-1} \quad \text{and} \quad < A(x,\xi), \xi > \ge |\xi|^p \qquad (1.2.5)$$

for almost every $x \in \Omega$ and all $\xi \in \wedge^l(\mathbf{R}^n)$. Here $a > 0$ is a constant and $1 < p < \infty$ is a fixed exponent associated with (1.2.4). Equation (1.2.4) is often called a *homogeneous A-harmonic equation*. A *solution* ω to (1.2.4) is an element of the Sobolev space $W^{1,p}_{loc}(\Omega, \wedge^{l-1})$ such that

$$\int_\Omega < A(x, d\omega), d\varphi >= 0$$

for all $\varphi \in W^{1,p}(\Omega, \wedge^{l-1})$ with compact support.

Definition 1.2.1. We call u an A-harmonic tensor in Ω if u satisfies the A-harmonic equation (1.2.4) in Ω.

A differential l-form $u \in D'(\Omega, \wedge^l)$ is called a *closed* form if $du = 0$ in Ω. Similarly, a differential $(l+1)$-form $v \in D'(\Omega, \wedge^{l+1})$ is called a *coclosed* form if $d^\star v = 0$. The equation

$$A(x, du) = d^\star v \qquad (1.2.6)$$

is called the *conjugate A-harmonic equation*. For example, $du = d^\star v$ is an analog of a Cauchy–Riemann system in $\mathbf{R}^n$. Clearly, the A-harmonic equation is not affected by adding a closed form to u and coclosed form to v. Therefore, any type of estimates between u and v must be modulo such forms. If u is a solution to (1.2.4) in Ω, then at least locally in a ball B, there exists a form $v \in W^{1,q}(B, \wedge^{l+1})$, $\frac{1}{p} + \frac{1}{q} = 1$, such that (1.2.6) holds. Throughout this chapter, we shall assume that $\frac{1}{p} + \frac{1}{q} = 1$.

Definition 1.2.2. When u and v satisfy (1.2.6) in Ω, and A^{-1} exists in Ω, we call u and v conjugate A-harmonic tensors in Ω.

Equation (1.2.2) is the homogeneous form of the *nonhomogeneous A-harmonic equation*

$$\text{div}\, A(x, \nabla u) = B(x, \nabla u) \qquad (1.2.7)$$

for functions in $\mathbf{R}^n$, where $A: \mathbf{R}^n \times \mathbf{R}^n \to \mathbf{R}^n$ and $B: \mathbf{R}^n \times \mathbf{R}^n \to \mathbf{R}$ are measurable and satisfy

$$|A(x,\xi)| \le a|\xi|^{p-1}, \quad < A(x,\xi), \xi > \ge |\xi|^p, \quad \text{and} \quad |B(x,\xi)| \le b|\xi|^{p-1} \quad (1.2.8)$$

for almost every $x \in \Omega$ and all $\xi \in \mathbf{R}^n$. Here $1 < p < \infty$, and a and b are positive constants. A weak solution to (1.2.7) is a function $u \in W^{1,p}_{loc}(\Omega)$ so that

$$\int_\Omega < A(x, \nabla u), \nabla \varphi > + < B(x, \nabla u), \varphi >= 0 \qquad (1.2.9)$$

for all $\varphi \in C_0^\infty(\Omega)$. Similarly, the *nonhomogeneous A-harmonic equation* for differential forms is written as

$$d^\star A(x, d\omega) = B(x, d\omega) \tag{1.2.10}$$

and a *solution* to (1.2.10) is an element of the Sobolev space $W_{loc}^{1,p}(M, \wedge^{l-1})$ such that

$$\int_M < A(x, d\omega), d\varphi > + < B(x, d\omega), \varphi >= 0$$

for all $\varphi \in W_{loc}^{1,p}(M, \wedge^{l-1})$ with compact support. The study of the nonhomogeneous A-harmonic equation (1.2.10) has just begun; see [72, 102, 103]. The global integrability of the solutions to (1.2.10) was proved recently by C. Nolder in [72].

Choosing A to be special operators, we obtain important examples of A-harmonic equations. For example, let $A(x, \xi) = \xi |\xi|^{p-2}$. Then, A satisfies condition (1.2.8), and we find that (1.2.4) and (1.2.6) reduce to the following *p-harmonic equation*

$$d^\star(du|du|^{p-2}) = 0 \tag{1.2.11}$$

and the *conjugate p-harmonic equation*

$$du|du|^{p-2} = d^\star v, \tag{1.2.12}$$

respectively, where v also satisfies the *q-harmonic equation*

$$d(d^\star v|d^\star v|^{q-2}) = 0 \tag{1.2.13}$$

with $\frac{1}{p} + \frac{1}{q} = 1$.

If we choose $p = 2$ in (1.2.12), then (1.2.12) reduces to the following system:

$$du = d^\star v,$$

which can be considered as an extension of the Cauchy–Riemann system in the plane.

In addition to above harmonic equations for functions and differential forms, another type of the conjugate harmonic equation

$$A(x, g + du) = h + d^\star v \tag{1.2.14}$$

has also received much attention in recent years; see [104, 105], where u, v, g, and h are differential forms.

Definition 1.2.3. If a pair of $(l - 1)$-form u and $(l + 1)$-form v satisfy (1.2.12), then u and v are called conjugate *p*-harmonic differential forms (or conjugate *p*-harmonic tensors).

Example 1.2.4. Let $f(x) = (f^1, f^2, \ldots, f^n)$ be K-quasiregular in $\mathbf{R}^n$, then

$$u = f^l df^1 \wedge df^2 \wedge \cdots \wedge df^{l-1}$$

and

$$v = \star f^{l+1} df^{l+2} \wedge \cdots \wedge df^n \ ,$$

$l = 1, 2, \ldots, n-1$, are conjugate A-harmonic tensors.

Let $f(x) = (f^1, f^2, \ldots, f^n) : \Omega \to \mathbf{R}^n$ be a K-quasiregular mapping, $K \geq 1$. Then, each of the functions

$$u = f^i(x) \ (i = 1, 2, \ldots, n) \ \text{ or } \ u = \log|f(x)|,$$

is a generalized solution of the quasilinear elliptic equation

$$\operatorname{div} A(x, \nabla u) = 0, \quad A = (A_1, A_2, \ldots, A_n), \tag{1.2.15}$$

where

$$A_i(x, \xi) = \frac{\partial}{\partial \xi_i} \left(\sum_{i,j=1}^{n} \theta_{i,j}(x)\xi_i\xi_j \right)^{n/2} \tag{1.2.16}$$

and $\theta_{i,j}$ are some functions, which can be expressed in terms of the differential matrix $Df(x)$ and satisfy

$$C_1(K)|\xi|^2 \leq \sum_{i,j}^{n} \theta_{i,j}\xi_i\xi_j \leq C_2(K)|\xi|^2 \tag{1.2.17}$$

for some constants $C_1(K), C_2(K) > 0$. This is a good example which shows the natural connection between two large sections of analysis: quasiregular mappings theory and the theory of partial differential equations. A significant progress has been made in the study of equation (1.2.15) with condition (1.2.17). From [106], we know that the function $u \in W^{1,n}_{loc}(\Omega)$ is a solution of some equation of the form (1.2.15) with condition (1.2.17) if and only if there exists a differential $(n-1)$-form

$$\theta(x) = \sum_{i=1}^{n} \theta_i(x) dx_1 \wedge \cdots \wedge \hat{dx_i} \wedge \cdots \wedge dx_n \in L^{n/(n-1)}_{loc}(\Omega)$$

with properties (the sign ˆ means that the expression under ˆ is omitted):

a) For every function $g \in W^{1,n}(\Omega)$ with compact support

$$\int_\Omega dg \wedge \theta = 0.$$

b) Almost everywhere on Ω the following inequalities are true:

$$\nu_1 |du(x)|^n \leq \star(du(x) \wedge \theta(x)),$$

where $\star$ denotes the orthogonal complement of a form, and

$$|\theta(x)| \leq \nu_2 |du(x)|^{n-1}$$

with constants $\nu_1, \nu_2 > 0$.

1.3 p-Harmonic equations

In the last section, we have introduced several kinds of harmonic equations, including the p-harmonic equation for differential forms and functions, and the conjugate p-harmonic equations. Using the Hodge star operator, here we will develop a method to find conjugate harmonic tensors in the three-dimensional case.

1.3.1 Two equivalent forms

We recall the *p-harmonic equation*

$$d^\star(du|du|^{p-2}) = 0 \tag{1.3.1}$$

for u and the *q-harmonic equation* for v (the conjugate of u)

$$d(d^\star v|d^\star v|^{q-2}) = 0 \tag{1.3.2}$$

that were introduced in the last section. A solution of the p-harmonic equation is called the *p-harmonic tensor* and a solution of the q-harmonic equation is called the *q-harmonic tensor*. Also, note that if u is a function, equation (1.2.11) becomes the usual p-harmonic equation

$$\operatorname{div}(\nabla u|\nabla u|^{p-2}) = 0. \tag{1.3.3}$$

For the results related to the p-harmonic equation (1.3.3), see [59]. Let

$$\nabla u = (u_{x_1}, u_{x_2}, \ldots, u_{x_n}),$$
$$w = |\nabla u|^{p-2} = (u_{x_1}^2 + u_{x_2}^2 + \cdots + u_{x_n}^2)^{(p-2)/2},$$

then

$$\operatorname{div}(\nabla u|\nabla u|^{p-2}) = \operatorname{div}(\nabla u w)$$
$$= (w u_{x_1})'_{x_1} + (w u_{x_2})'_{x_2} + \cdots + (w u_{x_n})'_{x_n}. \tag{1.3.4}$$

By a simple calculation, we obtain

$$(wu_{x_k})'_{x_k} = (p-2)|\nabla u|^{p-4} \sum_{i=1}^{n} u_{x_k} u_{x_i} u_{x_k x_i} + |\nabla u|^{p-2} u_{x_k x_k} \qquad (1.3.5)$$

for any k with $1 \le k \le n$. Thus, equation (1.3.3) is equivalent to

$$(p-2) \sum_{k=1}^{n} \sum_{i=1}^{n} u_{x_k} u_{x_i} u_{x_k x_i} + |\nabla u|^2 \Delta u = 0. \qquad (1.3.6)$$

Therefore, we have obtained two equivalent forms, (1.3.3) and (1.3.6), either of them is called the *p-harmonic equation for functions*. The solutions of (1.3.3) or (1.3.6) are called *p-harmonic functions*. The work related to the p-harmonic equation can be found in [107–118].

1.3.2 Three-dimensional cases

We first develop a method to find conjugate harmonic tensors in $\mathbf{R}^3$ for $p = 2$, and then consider the general case. If $p = 2$, then equation (1.2.12) reduces to the following simple form:

$$du = d^\star v. \qquad (1.3.7)$$

Here $u = u(x_1, x_2, x_3)$ is any 0-form (function) and v is a 2-form defined by

$$v = v_1 dx_1 \wedge dx_2 + v_2 dx_1 \wedge dx_3 + v_3 dx_2 \wedge dx_3, \qquad (1.3.8)$$

where $v_1 = v_1(x_1, x_2, x_3)$, $v_2 = v_2(x_1, x_2, x_3)$ and $v_3 = v_3(x_1, x_2, x_3)$ are differentiable functions in $\mathbf{R}^3$. Since $n = 3$ and $l = 1$, $d^\star = \star d \star$. Thus, we have

$$
\begin{aligned}
d^\star v &= \star d \star v = \star d(\star v) \\
&= \star d(v_1 dx_3 - v_2 dx_2 + v_3 dx_1) \\
&= \star \left(\left(\tfrac{\partial v_1}{\partial x_1} dx_1 + \tfrac{\partial v_1}{\partial x_2} dx_2 + \tfrac{\partial v_1}{\partial x_3} dx_3 \right) \wedge dx_3 \right. \\
&\quad - \left(\tfrac{\partial v_2}{\partial x_1} dx_1 + \tfrac{\partial v_2}{\partial x_2} dx_2 + \tfrac{\partial v_2}{\partial x_3} dx_3 \right) \wedge dx_2 \\
&\quad + \left. \left(\tfrac{\partial v_3}{\partial x_1} dx_1 + \tfrac{\partial v_3}{\partial x_2} dx_2 + \tfrac{\partial v_3}{\partial x_3} dx_3 \right) \wedge dx_1 \right).
\end{aligned}
\qquad (1.3.9)
$$

We know that $dx_i \wedge dx_j = -dx_j \wedge dx_i$ for $i \ne j$ and $dx_i \wedge dx_i = 0$. Hence, we obtain

$$d^\star v = \left(\frac{\partial v_1}{\partial x_2} + \frac{\partial v_2}{\partial x_3} \right) dx_1 + \left(\frac{\partial v_3}{\partial x_3} - \frac{\partial v_1}{\partial x_1} \right) dx_2 - \left(\frac{\partial v_2}{\partial x_1} + \frac{\partial v_3}{\partial x_2} \right) dx_3.$$

Thus, the equation

$$du = d^\star v$$

is equivalent to the system

$$\frac{\partial u}{\partial x_1} = \frac{\partial v_1}{\partial x_2} + \frac{\partial v_2}{\partial x_3}$$

$$\frac{\partial u}{\partial x_2} = \frac{\partial v_3}{\partial x_3} - \frac{\partial v_1}{\partial x_1} \qquad (1.3.10)$$

$$\frac{\partial u}{\partial x_3} = -\frac{\partial v_2}{\partial x_1} - \frac{\partial v_3}{\partial x_2}.$$

When $q = 2$, the conjugate q-harmonic equation (1.2.13) reduces to

$$d(d^\star v) = 0. \qquad (1.3.11)$$

From (1.3.9), we obtain

$$d(d^\star v) = \left(-\frac{\partial^2 v_1}{\partial x_2^2} - \frac{\partial^2 v_2}{\partial x_2 \partial x_3} - \frac{\partial^2 v_1}{\partial x_1^2} + \frac{\partial^2 v_3}{\partial x_1 \partial x_3}\right) dx_1 \wedge dx_2$$

$$- \left(\frac{\partial^2 v_2}{\partial x_1^2} + \frac{\partial^2 v_3}{\partial x_1 \partial x_2} + \frac{\partial^2 v_1}{\partial x_2 \partial x_3} + \frac{\partial^2 v_2}{\partial x_3^2}\right) dx_1 \wedge dx_3$$

$$+ \left(-\frac{\partial^2 v_2}{\partial x_1 \partial x_2} - \frac{\partial^2 v_3}{\partial x_2^2} + \frac{\partial^2 v_1}{\partial x_1 \partial x_3} - \frac{\partial^2 v_3}{\partial x_3^2}\right) dx_2 \wedge dx_3.$$

So the equation $d(d^\star v) = 0$ is equivalent to the following system:

$$-\frac{\partial^2 v_1}{\partial x_2^2} - \frac{\partial^2 v_2}{\partial x_2 \partial x_3} - \frac{\partial^2 v_1}{\partial x_1^2} + \frac{\partial^2 v_3}{\partial x_1 \partial x_3} = 0$$

$$\frac{\partial^2 v_2}{\partial x_1^2} + \frac{\partial^2 v_3}{\partial x_1 \partial x_2} + \frac{\partial^2 v_1}{\partial x_2 \partial x_3} + \frac{\partial^2 v_2}{\partial x_3^2} = 0 \qquad (1.3.12)$$

$$-\frac{\partial^2 v_2}{\partial x_1 \partial x_2} - \frac{\partial^2 v_3}{\partial x_2^2} + \frac{\partial^2 v_1}{\partial x_1 \partial x_3} - \frac{\partial^2 v_3}{\partial x_3^2} = 0.$$

Thus, in order to find harmonic tensors, we only need to solve system (1.3.10) or (1.3.12). This gives us a method to find conjugate harmonic tensors for the case $p = q = 2$.

Similar to the case of $p = 2$, the equation

$$du|du|^{p-2} = d^\star v$$

is equivalent to (we use the notation $\nabla u = du$)

$$\frac{\partial u}{\partial x_1}|\nabla u|^{p-2} = \frac{\partial v_1}{\partial x_2} + \frac{\partial v_2}{\partial x_3}$$

$$\frac{\partial u}{\partial x_2}|\nabla u|^{p-2} = \frac{\partial v_3}{\partial x_3} - \frac{\partial v_1}{\partial x_1} \tag{1.3.13}$$

$$\frac{\partial u}{\partial x_3}|\nabla u|^{p-2} = -\frac{\partial v_2}{\partial x_1} - \frac{\partial v_3}{\partial x_2}$$

with $p \neq 2$. Note that

$$\nabla u = du = \frac{\partial u}{\partial x_1} dx_1 + \frac{\partial u}{\partial x_2} dx_2 + \frac{\partial u}{\partial x_3} dx_3.$$

Thus, we only need to find a 2-form v defined by (1.3.8) and a p-harmonic function u satisfying (1.3.13). Such pairs of u and v are conjugate p-harmonic tensors. Clearly, system (1.3.13) has infinitely many solutions.

1.3.3 The equivalent system

Now we will develop a method to find p-harmonic tensors in $\mathbf{R}^n$. Similar to the case of $n = 3$, we try to get the equivalent system for the equation

$$du|du|^{p-2} = d^\star v$$

in $\mathbf{R}^n$. Let

$$v = \sum_{i<j} v_{ij} dx_i \wedge dx_j \tag{1.3.14}$$

be a 2-form in $\mathbf{R}^n$, where $v_{ij} = v_{ji}$, $1 \leq i, j \leq n$, then

$$\star v = \sum_{i<j} (-1)^{\frac{2(2+1)}{2}+i+j} v_{ij} dx_{I-\{i,j\}},$$

where $I = \{1, 2, \ldots, n\}$, that is

$$\star v = \sum_{i<j} (-1)^{i+j+1} v_{ij} dx_{I-\{i,j\}}.$$

Thus,

$$d(\star v) = \sum_{i<j} (-1)^{i+j+1} \left((-1)^{i-1} \frac{\partial v_{ij}}{\partial x_i} dx_{I-\{j\}} + (-1)^{j-2} \frac{\partial v_{ij}}{\partial x_j} dx_{I-\{i\}} \right).$$

Therefore, we have

$$d^\star v = \star d(\star v)$$

$$= \sum_{i<j} (-1)^{i+j+1} (-1)^{\frac{n(n-1)}{2} + \frac{n(n+1)}{2}}$$

$$\times \left((-1)^{-j+i-1} \frac{\partial v_{ij}}{\partial x_i} dx_j + (-1)^{-i+j} \frac{\partial v_{ij}}{\partial x_j} dx_i \right)$$

$$= \sum_{i<j} \left((-1)^n \frac{\partial v_{ij}}{\partial x_i} dx_j + (-1)^{n+1} \frac{\partial v_{ij}}{\partial x_j} dx_i \right)$$

$$= (-1)^n \sum_{i<j} \left(\frac{\partial v_{ij}}{\partial x_i} dx_j - \frac{\partial v_{ij}}{\partial x_j} dx_i \right)$$

$$= (-1)^n \left(\sum_{i<j} \frac{\partial v_{ij}}{\partial x_i} dx_j - \sum_{i<j} \frac{\partial v_{ij}}{\partial x_j} dx_i \right)$$

$$= (-1)^n \left(\sum_{j=2}^{n} \sum_{i=1}^{j-1} \frac{\partial v_{ij}}{\partial x_i} dx_j - \sum_{j=2}^{n} \sum_{i=1}^{j-1} \frac{\partial v_{ij}}{\partial x_j} dx_i \right).$$

Since $v_{ij} = v_{ji}$, the second sum can be written as

$$- \sum_{j=2}^{n} \sum_{i=1}^{j-1} \frac{\partial v_{ij}}{\partial x_j} dx_i$$

$$= - \sum_{i=1}^{n-1} \sum_{j=i+1}^{n} \frac{\partial v_{ij}}{\partial x_j} dx_i$$

$$= - \sum_{j=1}^{n-1} \sum_{i=j+1}^{n} \frac{\partial v_{ji}}{\partial x_i} dx_j$$

$$= \sum_{i=2}^{n} \left(-\frac{\partial v_{1i}}{\partial x_i} dx_1 \right) + \sum_{j=2}^{n-1} \left(\sum_{i=j+1}^{n} \left(-\frac{\partial v_{ji}}{\partial x_i} \right) \right) dx_j.$$

Similarly, we have

$$\sum_{j=2}^{n} \left(\sum_{i=1}^{j-1} \frac{\partial v_{ij}}{\partial x_i} \right) dx_j = \sum_{i=1}^{n-1} \frac{\partial v_{in}}{\partial x_i} dx_n + \sum_{j=2}^{n-1} \left(\sum_{i=1}^{j-1} \frac{\partial v_{ij}}{\partial x_i} \right) dx_j.$$

Hence, we obtain

$$d^\star v$$

$$= (-1)^n \left(\sum_{i=1}^{n-1} \frac{\partial v_{in}}{\partial x_i} dx_n + \sum_{j=2}^{n-1} \sum_{i=1}^{j-1} \frac{\partial v_{ij}}{\partial x_i} dx_j - \sum_{i=2}^{n} \frac{\partial v_{1i}}{\partial x_i} dx_1 - \sum_{j=2}^{n-1} \sum_{i=j+1}^{n} \frac{\partial v_{ji}}{\partial x_i} dx_j \right)$$

$$= (-1)^n \left(- \sum_{i=2}^{n} \frac{\partial v_{1i}}{\partial x_i} dx_1 + \sum_{j=2}^{n-1} \left(\sum_{i=1}^{j-1} \frac{\partial v_{ij}}{\partial x_i} - \sum_{i=j+1}^{n} \frac{\partial v_{ji}}{\partial x_i} \right) dx_j + \sum_{i=1}^{n-1} \frac{\partial v_{in}}{\partial x_i} dx_n \right).$$

By (1.2.12), we have the following equation:

$$du |du|^{p-2} \tag{1.3.15}$$

$$= (-1)^n \left(- \sum_{i=2}^{n} \frac{\partial v_{1i}}{\partial x_i} dx_1 + \sum_{j=2}^{n-1} \left(\sum_{i=1}^{j-1} \frac{\partial v_{ij}}{\partial x_i} - \sum_{i=j+1}^{n} \frac{\partial v_{ji}}{\partial x_i} \right) dx_j + \sum_{i=1}^{n-1} \frac{\partial v_{in}}{\partial x_i} dx_n \right).$$

If $u = u(x_1, x_2, \ldots, x_n)$ is a function, then (1.3.15) can be written as the following system:

$$\frac{\partial u}{\partial x_1} |\nabla u|^{p-2} = (-1)^n \left(-\sum_{i=2}^{n} \frac{\partial v_{1i}}{\partial x_i} \right),$$

$$\frac{\partial u}{\partial x_j} |\nabla u|^{p-2} = (-1)^n \left(\sum_{i=1}^{j-1} \frac{\partial v_{ij}}{\partial x_i} - \sum_{i=j+1}^{n} \frac{\partial v_{ji}}{\partial x_i} \right), \quad j = 2, 3, \ldots, n-1,$$

$$\frac{\partial u}{\partial x_n} |\nabla u|^{p-2} = (-1)^n \sum_{i=1}^{n-1} \frac{\partial v_{in}}{\partial x_i}.$$

$$(1.3.16)$$

Thus, we see that (1.2.12) is equivalent to system (1.3.16). If we choose $p = 2$ and $n = 2$ in the above system, then (1.3.16) reduces to the well-known Cauchy–Riemann equations. Thus, system (1.3.16) can be considered as a natural extension of the Cauchy–Riemann equations. In order to find *p*-harmonic tensors, we only need to find u and v satisfying (1.3.16). We conclude this section with the following example of a pair of the conjugate *p*-harmonic tensors, which have nice symmetric properties.

1.3.4 An example

In this section, we introduce an example of a pair of the conjugate harmonic tensors in $\mathbf{R}^3$.

Example 1.3.1. Let

$$u = \frac{3}{\rho} = \frac{3}{\sqrt{x_1^2 + x_2^2 + x_3^2}}$$

be a harmonic function in $\mathbf{R}^3$ and v be a 2-form in $\mathbf{R}^3$ defined by

$$v = v_3 dx_1 \wedge dx_2 + v_2 dx_1 \wedge dx_3 + v_1 dx_2 \wedge dx_3.$$

Note that we have switched v_1 with v_3 in (1.3.8). Then, (1.3.10) becomes

$$\frac{\partial u}{\partial x_1} = \frac{\partial v_3}{\partial x_2} + \frac{\partial v_2}{\partial x_3}$$

$$\frac{\partial u}{\partial x_2} = \frac{\partial v_1}{\partial x_3} - \frac{\partial v_3}{\partial x_1} \qquad (1.3.17)$$

$$\frac{\partial u}{\partial x_3} = -\frac{\partial v_2}{\partial x_1} - \frac{\partial v_1}{\partial x_2}$$

and (1.3.13) reduces to

$$\frac{\partial u}{\partial x_1} |\nabla u|^{p-2} = \frac{\partial v_3}{\partial x_2} + \frac{\partial v_2}{\partial x_3}$$

$$\frac{\partial u}{\partial x_2}|\nabla u|^{p-2} = \frac{\partial v_1}{\partial x_3} - \frac{\partial v_3}{\partial x_1} \tag{1.3.18}$$

$$\frac{\partial u}{\partial x_3}|\nabla u|^{p-2} = -\frac{\partial v_2}{\partial x_1} - \frac{\partial v_1}{\partial x_2}.$$

Here v_1, v_2, and v_3 are defined as follows:

$$v_1 = \frac{x_2 x_3}{\sqrt{\sum x_i^2}} \frac{x_2^4 - x_3^4}{\prod_{i<j}(x_i^2 + x_j^2)} = \frac{x_2 x_3}{\sqrt{x_1^2 + x_2^2 + x_3^2}} \frac{x_2^2 - x_3^2}{(x_1^2 + x_2^2)(x_1^2 + x_3^2)},$$

$$v_2 = \frac{x_1 x_3}{\sqrt{\sum x_i^2}} \frac{x_1^4 - x_3^4}{\prod_{i<j}(x_i^2 + x_j^2)} = \frac{x_1 x_3}{\sqrt{x_1^2 + x_2^2 + x_3^2}} \frac{x_1^2 - x_3^2}{(x_1^2 + x_2^2)(x_2^2 + x_3^2)},$$

$$v_3 = \frac{x_1 x_2}{\sqrt{\sum x_i^2}} \frac{x_1^4 - x_2^4}{\prod_{i<j}(x_i^2 + x_j^2)} = \frac{x_1 x_2}{\sqrt{x_1^2 + x_2^2 + x_3^2}} \frac{x_1^2 - x_2^2}{(x_1^2 + x_3^2)(x_2^2 + x_3^2)}.$$

Then, u and v form a pair of conjugate harmonic tensors. Now, we check that u, v_1, v_2, and v_3 defined above satisfy (1.3.17). For $u = 3/\rho$, system (1.3.17) reduces to

$$\frac{-3x_1}{(x_1^2 + x_2^2 + x_3^2)^{3/2}} = \frac{\partial v_3}{\partial x_2} + \frac{\partial v_2}{\partial x_3},$$

$$\frac{-3x_2}{(x_1^2 + x_2^2 + x_3^2)^{3/2}} = \frac{\partial v_1}{\partial x_3} - \frac{\partial v_3}{\partial x_1}, \tag{1.3.19}$$

$$\frac{-3x_3}{(x_1^2 + x_2^2 + x_3^2)^{3/2}} = -\frac{\partial v_2}{\partial x_1} - \frac{\partial v_1}{\partial x_2}.$$

By a long computation and simplification, we have

$$\frac{\partial v_1}{\partial x_2} = \frac{x_3(x_2^2 - x_3^2)}{(x_1^2 + x_2^2)(x_1^2 + x_2^2 + x_3^2)^{3/2}} + \frac{2x_3 x_2^2}{(x_1^2 + x_2^2)^2(x_1^2 + x_2^2 + x_3^2)^{1/2}},$$

$$\frac{\partial v_1}{\partial x_3} = \frac{x_2(x_2^2 - x_3^2)}{(x_1^2 + x_3^2)(x_1^2 + x_2^2 + x_3^2)^{3/2}} + \frac{-2x_2 x_3^2}{(x_1^2 + x_3^2)^2(x_1^2 + x_2^2 + x_3^2)^{1/2}},$$

$$\frac{\partial v_2}{\partial x_1} = \frac{x_3(x_1^2 - x_3^2)}{(x_1^2 + x_2^2)(x_1^2 + x_2^2 + x_3^2)^{3/2}} + \frac{2x_3 x_1^2}{(x_1^2 + x_2^2)^2(x_1^2 + x_2^2 + x_3^2)^{1/2}},$$

$$\frac{\partial v_2}{\partial x_3} = \frac{x_1(x_1^2 - x_3^2)}{(x_2^2 + x_3^2)(x_1^2 + x_2^2 + x_3^2)^{3/2}} + \frac{-2x_1 x_3^2}{(x_2^2 + x_3^2)^2(x_1^2 + x_2^2 + x_3^2)^{1/2}},$$

$$\frac{\partial v_3}{\partial x_1} = \frac{x_2(x_1^2 - x_2^2)}{(x_1^2 + x_3^2)(x_1^2 + x_2^2 + x_3^2)^{3/2}} + \frac{2x_2 x_1^2}{(x_1^2 + x_3^2)^2(x_1^2 + x_2^2 + x_3^2)^{1/2}},$$

$$\frac{\partial v_3}{\partial x_2} = \frac{x_1(x_1^2 - x_2^2)}{(x_2^2 + x_3^2)(x_1^2 + x_2^2 + x_3^2)^{3/2}} + \frac{-2x_1 x_2^2}{(x_2^2 + x_3^2)^2(x_1^2 + x_2^2 + x_3^2)^{1/2}}.$$

Thus, it follows that

$$\frac{\partial v_3}{\partial x_2} + \frac{\partial v_2}{\partial x_3} = \frac{x_1(2x_1^2-(x_2^2+x_3^2))}{(x_2^2+x_3^2)(x_1^2+x_2^2+x_3^2)^{3/2}} - \frac{2x_1}{(x_2^2+x_3^2)(x_1^2+x_2^2+x_3^2)^{1/2}}$$

$$= \frac{-3x_1(x_2^2+x_3^2)}{(x_2^2+x_3^2)(x_1^2+x_2^2+x_3^2)^{3/2}}$$

$$= \frac{-3x_1}{(x_1^2+x_2^2+x_3^2)^{3/2}}.$$

Hence, the first equation of system (1.3.19) holds. By the same method, we can check that second and third equations of (1.3.19) hold too. Again by a long computation and simplification, it follows that

$$\frac{\partial v_1}{\partial x_1} + \frac{\partial v_2}{\partial x_2} + \frac{\partial v_3}{\partial x_3} = 0,$$

so that v is a closed 2-form.

1.4 Some weight classes

In this section, we present various classes of the weight functions. These weight functions will be used in later chapters to establish weighted norm inequalities. We also discuss some two-weight classes.

1.4.1 $A_r(\Omega)$-weights

We study some basic elements of $A_r(\Omega)$-weights which were introduced by Muckenhoupt (1972) who used this weight class to explore the properties of the Hardy–Littlewood maximal operator. We also discuss some other weight classes. Recall that a function $w(x)$ is called a weight if $w > 0$ a.e. and $w \in L^1_{loc}(\mathbf{R}^n)$.

Definition 1.4.1. w is called a doubling weight and write $w \in D(\Omega)$ if there exists a constant C such that

$$\mu(2B) \leq C\mu(B)$$

for all balls B with $2B \subset \Omega$. Here the measure μ is defined by $d\mu = w(x)dx$. If this condition holds only for all balls B with $4B \subset \Omega$, then w is weak doubling and we write $w \in WD(\Omega)$.

Definition 1.4.2. Let $\sigma > 1$. It is said that w satisfies a weak reverse Hölder inequality and write $w \in WRH_\beta(\Omega)$ when there exist constants C, β, γ with $0 < \gamma < \beta$ such that

$$\left(\frac{1}{|B|}\int_B w^\beta dx\right)^{1/\beta} \le C \left(\frac{1}{|B|}\int_{\sigma B} w^\gamma dx\right)^{1/\gamma} \qquad (1.4.1)$$

for all balls B with $\sigma B \subset \Omega$. We say that w satisfies a reverse Hölder inequality when (1.4.1) holds with $\sigma = 1$, and write $w \in RH_\beta(\Omega)$.

Remark. From [119], we know that if (1.4.1) holds for some γ with $0 < \gamma < \beta$ and any ball B with $\sigma B \subset \Omega$, then it also holds for all γ with $0 < \gamma < \infty$, that is, if

$$\left(\frac{1}{|B|}\int_B w^\beta dx\right)^{1/\beta} \le C \left(\frac{1}{|B|}\int_{\sigma B} w^{\gamma_0} dx\right)^{1/\gamma_0}$$

holds for some γ_0 with $0 < \gamma_0 < \beta$, then

$$\left(\frac{1}{|B|}\int_B w^\beta dx\right)^{1/\beta} \le C \left(\frac{1}{|B|}\int_{\sigma B} w^\gamma dx\right)^{1/\gamma} \qquad (1.4.1)'$$

holds for all γ with $0 < \gamma < \infty$. We should also notice that the space $WRH_\beta(\Omega)$ is independent of $\sigma > 1$, see [120]. We next define Muckenhoupt weights.

Definition 1.4.3. A weight w satisfies the $A_r(\Omega)$-condition in a subset $\Omega \subset \mathbf{R}^n$, where $r > 1$, and write $w \in A_r(\Omega)$ when

$$\sup_B \left(\frac{1}{|B|}\int_B w dx\right) \left(\frac{1}{|B|}\int_B w^{1/(1-r)} dx\right)^{r-1} < \infty,$$

where the supremum is over all balls $B \subset \Omega$.

Notice it follows from Hölder's inequality that if $w \in A_r(\Omega)$ and $0 < \alpha \le 1$, then $w^\alpha \in A_r(\Omega)$. Indeed

$$\sup_B \left(\frac{1}{|B|}\int_B w^\alpha dx\right) \left(\frac{1}{|B|}\int_B w^{\alpha/(1-r)} dx\right)^{r-1}$$

$$\le \sup_B \left(\frac{1}{|B|}\int_B w dx\right)^\alpha \left(\frac{1}{|B|}\int_B w^{1/(1-r)} dx\right)^{\alpha(r-1)} < \infty. \qquad (1.4.2)$$

If w satisfies the A_r-condition for all balls B with $2B \subset E$, we write $w \in A_r^{loc}(E)$. Also we write $A_\infty(E) = \cup_{r>1} A_r(E)$ and $A_\infty^{loc}(E) = \cup_{r>1} A_r^{loc}(E)$.

It is well known that $w \in A_\infty(\Omega)$ if and only if $w \in RH_\beta(\Omega)$ for some $\beta > 1$. This is also true for $A_\infty^{loc}(\Omega)$ and $WRH_\beta(\Omega)$ for some $\beta > 1$. Moreover $A_\infty^{loc}(\Omega) \subset WD(\Omega)$.

1.4.2 $A_r(\lambda, E)$-weights

We have discussed the $A_r(E)$-weights and their properties in section 1.4.1. Now, we introduce the $A_r(\lambda, E)$-weights. These weights have properties similar to those of $A_r(E)$. The following $A_r(\lambda, E)$-weight class was introduced in [121].

Definition 1.4.4. Let w be a locally integrable nonnegative function in $E \subset \mathbf{R}^n$ and assume that $0 < w < \infty$ almost everywhere. We say that w belongs to the $A_r(\lambda, E)$ class, $1 < r < \infty$ and $0 < \lambda < \infty$, or that w is an $A_r(\lambda, E)$-weight, write $w \in A_r(\lambda, E)$ or $w \in A_r(\lambda)$ when it will not cause any confusion, if

$$\sup_B \left(\frac{1}{|B|} \int_B w^\lambda dx\right) \left(\frac{1}{|B|} \int_B \left(\frac{1}{w}\right)^{1/(r-1)} dx\right)^{r-1} < \infty$$

for all balls $B \subset E \subset \mathbf{R}^n$.

It is clear that $A_r(1)$ is the usual A_r-class, see [122, 123, 59] for more properties of A_r-weights. Similar to proofs in [123], we prove some properties of the $A_r(\lambda)$-weights. The following theorem says that $A_r(\lambda)$ is an increasing class with respect to r.

Theorem 1.4.5. If $1 < r < s < \infty$, then $A_r(\lambda) \subset A_s(\lambda)$.

Proof. Let $w \in A_r(\lambda)$. Since $1 < r < s < \infty$, by Hölder's inequality

$$\left(\int_B \left(\tfrac{1}{w}\right)^{1/(s-1)} dx\right)^{s-1} \leq \left(\int_B \left(\tfrac{1}{w}\right)^{1/(r-1)} dx\right)^{r-1} \left(\int_B 1^{1/(s-r)} dx\right)^{s-r}$$

$$= |B|^{s-r} \left(\int_B \left(\tfrac{1}{w}\right)^{1/(r-1)} dx\right)^{r-1}$$

$$= \frac{|B|^{s-1}}{|B|^{r-1}} \left(\int_B \left(\tfrac{1}{w}\right)^{1/(r-1)} dx\right)^{r-1},$$

so that

$$\left(\frac{1}{|B|} \int_B \left(\frac{1}{w}\right)^{1/(s-1)} dx\right)^{s-1} \leq \left(\frac{1}{|B|} \int_B \left(\frac{1}{w}\right)^{1/(r-1)} dx\right)^{r-1}.$$

Thus, we find

$$\sup_B \left(\frac{1}{|B|} \int_B w^\lambda dx\right) \left(\frac{1}{|B|} \int_B \left(\frac{1}{w}\right)^{1/(s-1)} dx\right)^{(s-1)}$$

$$\leq \sup_{B} \left(\frac{1}{|B|} \int_B w^\lambda dx \right) \left(\frac{1}{|B|} \int_B \left(\frac{1}{w} \right)^{1/(r-1)} dx \right)^{(r-1)} < \infty$$

for all balls $B \subset \mathbf{R}^n$ since $w \in A_r(\lambda)$. Therefore, $w \in A_s(\lambda)$, and hence $A_r(\lambda) \subset A_s(\lambda)$. ∎

The following result shows that $A_r(\lambda)$-weights have the property similar to the strong doubling property of A_r-weights.

Theorem 1.4.6. *If $w \in A_r(\lambda)$, $\lambda \geq 1$, and the measure μ is defined by $d\mu = w(x)dx$, then*

$$\frac{|E|^r}{|B|^{\lambda+r-1}} \leq C_{r,\lambda,w} \frac{\mu(E)}{\mu(B)^\lambda}, \tag{1.4.3}$$

where B is a ball in $\mathbf{R}^n$ and E is a measurable subset of B.

Proof. By Hölder's inequality, we have

$$\begin{aligned}
|E| = \int_E dx &= \int_E w^{1/r} w^{-1/r} dx \\
&\leq \left(\int_E w dx \right)^{1/r} \left(\int_E w^{1/(1-r)} dx \right)^{(r-1)/r} \\
&= (\mu(E))^{1/r} \left(\int_E w^{1/(1-r)} dx \right)^{(r-1)/r}.
\end{aligned}$$

This implies

$$|E|^r \leq \mu(E) \left(\int_E w^{1/(1-r)} dx \right)^{r-1}. \tag{1.4.4}$$

Note that $\lambda \geq 1$, by Hölder's inequality again, we have

$$\frac{1}{|B|} \int_B w dx \leq \left(\frac{1}{|B|} \int_B w^\lambda dx \right)^{1/\lambda},$$

so that

$$1 = \frac{1}{\mu(B)} \int_B w dx \leq \frac{|B|}{\mu(B)} \left(\frac{1}{|B|} \int_B w^\lambda dx \right)^{1/\lambda}.$$

Hence, we obtain

$$\mu(B)^\lambda \leq |B|^{\lambda-1} \int_B w^\lambda dx. \tag{1.4.5}$$

Since $w \in A_r(\lambda)$, there exists a constant $C_{r,\lambda,w}$ such that

$$\left(\frac{1}{|B|} \int_B w^\lambda dx \right) \left(\frac{1}{|B|} \int_B \left(\frac{1}{w} \right)^{1/(r-1)} dx \right)^{r-1} \leq C_{r,\lambda,w}. \tag{1.4.6}$$

Combining (1.4.4), (1.4.5), and (1.4.6), we deduce that

$$|E|^r \mu(B)^\lambda \leq \mu(E)|B|^{\lambda-1} \int_B w^\lambda dx \left(\int_E w^{1/(1-r)} dx \right)^{r-1}$$

$$\leq \mu(E)|B|^{\lambda+r-1} \left(\frac{1}{|B|} \int_B w^\lambda dx \right) \left(\frac{1}{|B|} \int_B \left(\frac{1}{w} \right)^{1/(r-1)} dx \right)^{r-1}$$

$$\leq C_{r,\lambda,w} \mu(E)|B|^{\lambda+r-1}.$$

Hence

$$\frac{|E|^r}{|B|^{\lambda+r-1}} \leq C_{r,\lambda,w} \frac{\mu(E)}{\mu(B)^\lambda}. \qquad \blacksquare$$

If we put $\lambda = 1$ in Theorem 1.4.6, then (1.4.3) becomes

$$\frac{|E|^r}{|B|^r} \leq C_{r,w} \frac{\mu(E)}{\mu(B)},$$

which is called the strong doubling property of A_r-weights; see [59]. It is well known that an A_r-weight w satisfies the following reverse Hölder inequality.

Lemma 1.4.7. *If $w \in A_r$, $r > 1$, then there exist constants $\beta > 1$ and C, independent of w, such that*

$$\| w \|_{\beta,Q} \leq C|Q|^{(1-\beta)/\beta} \| w \|_{1,Q}$$

for all cubes $Q \subset \mathbf{R}^n$.

1.4.3 $A_r^\lambda(E)$-weights

The following class of A_r^λ-weights was introduced in [124] and several new weighted integral inequalities for differential forms were proved.

Definition 1.4.8. We say that the weight $w(x) > 0$ satisfies the $A_r^\lambda(E)$-condition, $r > 1$ and $\lambda > 0$, and write $w \in A_r^\lambda(E)$, if

$$\sup_B \left(\frac{1}{|B|} \int_B w dx \right) \left(\frac{1}{|B|} \int_B w^{1/(1-r)} dx \right)^{\lambda(r-1)} < \infty$$

for any ball $B \subset E$. Here E is a subset of $\mathbf{R}^n$.

The $A_r^\lambda(E)$-weights have similar properties as we have discussed for $A_r(\lambda, E)$-weights. We leave it to readers to establish the properties of the $A_r^\lambda(E)$-weights.

Example 1.4.9. Let $\Omega \subset \mathbf{R}^2$ be a bounded domain. For any $x \in \Omega$, we define

$$w(x) = \begin{cases} \frac{1}{|x|^\alpha}, & x \neq 0, \\[2mm] 1, & x = 0, \end{cases}$$

where α is a positive constant with $\alpha < 2$. Then, $w(x)$ is an $A_r^\lambda(\Omega)$-weight for any constant $\lambda > 1$.

To show this let $B \subset \Omega$ be a disc with center x_0 and radius r_0. Then, $r_0 \leq diam(\Omega) < \infty$ since Ω is a bounded domain. We may assume that $x_0 = 0$. Using the polar coordinate substitution, we have

$$\int_B w(x)dx = \int_0^{2\pi} d\theta \int_0^{r_0} \rho^{-\alpha}\rho d\rho$$
$$= 2\pi \int_0^{r_0} \rho^{1-\alpha}d\rho$$
$$= C_1 r_0^{2-\alpha}.$$

Thus,

$$\frac{1}{|B|}\int_B wdx = \frac{C_2}{r_0^\alpha}. \tag{1.4.7}$$

Using the polar coordinate substitution again, it follows that

$$\int_B \left(\frac{1}{w(x)}\right)^{1/(r-1)} dx = C_3 r_0^{2+\alpha/(r-1)}.$$

Hence, we find

$$\left(\frac{1}{|B|}\int_B \left(\frac{1}{w(x)}\right)^{1/(r-1)} dx\right)^{\lambda(r-1)} \leq C_4 r_0^{\alpha\lambda}. \tag{1.4.8}$$

Combining (1.4.7) and (1.4.8), we obtain

$$\sup_B \left(\frac{1}{|B|}\int_B wdx\right)\left(\frac{1}{|B|}\int_B w^{1/(1-r)}dx\right)^{\lambda(r-1)} < \sup_B \left(C_5 r_0^{\alpha(\lambda-1)}\right) < \infty.$$

Therefore, $w(x)$ is an $A_r^\lambda(\Omega)$-weight for any $\lambda > 1$.

Similarly, we have the following example of the $A_r^\lambda(\Omega)$-weights with $\lambda > 0$.

Example 1.4.10. Let $\Omega \subset \mathbf{R}^2$ be a bounded domain. For any $x \in \Omega$, we define $w(x) = |x|^\beta$, where β is a constant with $0 < \beta < 2(r-1)$ and r is the constant in the definition of the $A_r^\lambda(\Omega)$-weights. Then, $w(x)$ is an $A_r^\lambda(\Omega)$-weight for any constant λ with $\lambda > 0$.

1.4.4 Some classes of two-weights

The following class of $A_{r,\lambda}(E)$-weights (or the two-weight) appeared in [125]. It is easy to see that the class of $A_{r,\lambda}(E)$-weights is an extension of the usual class of A_r-weights [59], and also class of $A_r(\lambda)$-weights [121]. See [126–133] for applications of the two-weight.

Definition 1.4.11. A pair of weights (w_1, w_2) satisfy the $A_{r,\lambda}(E)$-condition in a set $E \subset \mathbf{R}^n$, write $(w_1, w_2) \in A_{r,\lambda}(E)$ for some $\lambda \geq 1$ and $1 < r < \infty$ with $1/r + 1/r' = 1$, if

$$\sup_{B \subset E} \left(\frac{1}{|B|} \int_B w_1^\lambda dx \right)^{1/\lambda r} \left(\frac{1}{|B|} \int_B \left(\frac{1}{w_2} \right)^{\lambda r'/r} dx \right)^{1/\lambda r'} < \infty.$$

Definition 1.4.12. A pair of weights (w_1, w_2) satisfy the $A_r^\lambda(E)$-condition in a set $E \subset \mathbf{R}^n$, and write $(w_1, w_2) \in A_r^\lambda(E)$ for some $r > 1$ and $\lambda > 0$, if

$$\sup_B \left(\frac{1}{|B|} \int_B w_1 dx \right) \left(\frac{1}{|B|} \int_B \left(\frac{1}{w_2} \right)^{1/(r-1)} dx \right)^{\lambda(r-1)} < \infty$$

for any ball $B \subset E$.

Definition 1.4.13. A pair of weights (w_1, w_2) satisfy the $A_r(\lambda, E)$-condition in a set $E \subset \mathbf{R}^n$, and write $(w_1, w_2) \in A_r(\lambda, E)$ for some $r > 1$ and $\lambda > 0$, if

$$\sup_B \left(\frac{1}{|B|} \int_B w_1^\lambda dx \right) \left(\frac{1}{|B|} \int_B \left(\frac{1}{w_2} \right)^{1/(r-1)} dx \right)^{r-1} < \infty$$

for any ball $B \subset E$.

Example 1.4.14. Let $\Omega \subset \mathbf{R}^2$ be a bounded domain. For any $x \in \Omega$, we assume that $d(x, \partial\Omega)$ is the distance from x to Ω and define

$$w_1(x) = \frac{1}{d^\alpha(x, \partial\Omega)}$$

and

$$w_2(x) = \begin{cases} \frac{1}{|x|^{\alpha/\lambda} d(x,\partial\Omega)}, & x \neq 0, \\ 1, & x = 0, \end{cases}$$

where α is a constant with $0 < \alpha < 1$ and λ is the parameter appearing in the definition of A_r^λ-weights. Then, (w_1, w_2) is a pair of the $A_r^\lambda(\Omega)$-weights.

To show it, let $B \subset \Omega$ be a ball with center x_0 and radius r_0. We may assume that $x_0 = 0$. Otherwise, we can move the center to the origin by a simple transformation. Then, for any $x \in B$, we have $d(x, \partial\Omega) \geq r_0 - |x|$. Hence,

$$w_1(x) = \frac{1}{d^{\alpha}(x, \partial\Omega)} \leq \frac{1}{(r_0 - |x|)^{\alpha}}$$

and

$$\frac{1}{w_2(x)} = |x|^{\alpha/\lambda} d(x, \partial\Omega) \leq C_1 |x|^{\alpha/\lambda}, \quad x \neq 0$$

since Ω is bounded. Using the polar coordinate substitution, we have

$$\begin{aligned}
\int_B w_1(x) dx &\leq \int_0^{2\pi} d\theta \int_0^{r_0} (r_0 - \rho)^{-\alpha} \rho \, d\rho \\
&= 2\pi \int_0^{r_0} (r_0 - \rho)^{-\alpha} \rho \, d\rho.
\end{aligned} \tag{1.4.9}$$

In (1.4.9) let $r_0 - \rho = t$, to find

$$\begin{aligned}
\frac{1}{|B|} \int_B w_1(x) dx &= \frac{1}{\pi r_0^2} \int_B w_1(x) dx \\
&\leq \frac{2}{r_0^2} \int_0^{r_0} t^{-\alpha}(r_0 - t) dt \\
&= \frac{2}{(1-\alpha)(2-\alpha)} \frac{1}{r_0^{\alpha}} \\
&= \frac{C_2}{r_0^{\alpha}}.
\end{aligned} \tag{1.4.10}$$

By the same method, we obtain

$$\begin{aligned}
\frac{1}{|B|} \int_B \left(\frac{1}{w_2}\right)^{1/(r-1)} dx & \\
\leq \frac{1}{|B|} \int_B \left(C_1 |x|^{\alpha/\lambda}\right)^{1/(r-1)} dx & \\
\leq \frac{C_3 \lambda(r-1)}{\alpha + 2\lambda(r-1)} r_0^{\alpha/(\lambda(r-1))} & \\
= C_4 r_0^{\alpha/(\lambda(r-1))}. &
\end{aligned} \tag{1.4.11}$$

Thus, we have

$$\left(\frac{1}{|B|} \int_B \left(\frac{1}{w_2}\right)^{1/(r-1)} dx\right)^{\lambda(r-1)} \leq C_5 r_0^{\alpha}. \tag{1.4.12}$$

Combining (1.4.10) and (1.4.12), we find that (w_1, w_2) satisfy the conditions of $A_r^{\lambda}(\Omega)$-weights, and hence (w_1, w_2) is a pair of the $A_r^{\lambda}(\Omega)$-weights.

The following result which is proved in [119] can be used to extend the local results to the global cases.

Theorem 1.4.15. *If $\mathcal{V}$ is a collection of cubes in $\mathbf{R}^n$ and C_Q are non-negative numbers associated with the cubes $Q \in \mathcal{V}$ and $w \in A_r$, $d\mu(x) = w(x)dx$, then for $1 \leq p < \infty$ and $N \geq 1$, it follows that*

$$\left(\int_{\mathbf{R}^n} \left(\sum_{Q \in \mathcal{V}} C_Q \cdot \chi_{NQ} \right)^p d\mu(x) \right)^{1/p} \leq B_p \left(\int_{\mathbf{R}^n} \left(\sum_{Q \in \mathcal{V}} C_Q \cdot \chi_Q \right)^p d\mu(x) \right)^{1/p},$$

where B_p is independent of the collection $\mathcal{V}$ and the numbers C_Q.

1.5 Inequalities in John domains

In 1927, M. Riesz proved the following result in [134] for conjugate harmonic functions. Let $f(z) = u(z) + iv(z), v(0) = 0$, be analytic in the unit disk D. Then, for every $p > 1$,

$$\int_0^{2\pi} |u(z)|^p d\varphi \leq C \int_0^{2\pi} |v(z)|^p d\varphi \quad (z = re^{i\varphi}, r < 1),$$

where $C = C(p)$ depends on p only.

In 1932, Hardy and Littlewood (see [100]) proved the following inequality for conjugate harmonic functions in the unit disk D.

Theorem 1.5.1. *For each $p > 0$, there is a constant C such that*

$$\int_D |u - u(0)|^p dxdy \leq C \int_D |v - v(0)|^p dxdy$$

for all analytic functions $f = u + iv$ in the unit disk D.

1.5.1 Local inequalities

This opened a new research direction in analysis. During the last 80 years, their work has attracted several mathematicians and various versions of the Hardy–Littlewood inequalities or theorems have been established. Considering the length of this monograph, we cannot include all of them here; however, we encourage the reader to see [149, 155–158, 177, 103, 178] for these inequalities and their applications. In this section, we first introduce the Hardy–Littlewood inequalities locally, then consider the weighted case in Section 1.5.2. Finally, we extend the result to the global case in δ-John domains in Section 1.5.3.

A proper subdomain $\Omega \subset \mathbf{R}^n$ is called a δ-John domain, $\delta > 0$, if there exists a point $x_0 \in \Omega$ which can be joined with any other point $x \in \Omega$ by a continuous curve $\gamma \subset \Omega$ so that

$$d(\xi, \partial\Omega) \geq \delta|x - \xi|$$

for each $\xi \in \gamma$. Here $d(\xi, \partial\Omega)$ is the Euclidean distance between ξ and $\partial\Omega$.

In [119] Theorem 1.5.1 is generalized to the following global Hardy–Littlewood inequality for K-quasiregular mappings. Let $f = (f^1, f^2, \ldots, f^n)$ be a K-quasiregular mapping in a John domain Ω. If w satisfies the A_r-condition and $0 < p < \infty$, then

$$\|f^i\|^{\#}_{p,\Omega,w} \leq C_p \|f^j\|^{\#}_{p,\Omega,w}$$

for $i, j = 1, 2, \ldots, n$, where C_p is a constant, independent of f. Here

$$\|f\|^{\#}_{p,E,w} = \inf_{a \in \mathbf{R}} \|f - a\|_{p,E,w}.$$

In fact, the above inequality was proved in a class of domains, which is more general than the class of John domains.

The following result was obtained in [99]. Let $D \subset \mathbf{R}^n$ be a bounded, convex domain. To each $y \in D$ there corresponds a linear operator $K_y : C^\infty(D, \wedge^l) \to C^\infty(D, \wedge^{l-1})$ defined by

$$(K_y\omega)(x; \xi_1, \ldots, \xi_{l-1}) = \int_0^1 t^{l-1} \omega(tx + y - ty; x - y, \xi_1, \ldots, \xi_{l-1}) dt$$

and a decomposition

$$\omega = d(K_y\omega) + K_y(d\omega).$$

A homotopy operator $T : C^\infty(D, \wedge^l) \to C^\infty(D, \wedge^{l-1})$ is defined by averaging K_y over all points y in D, i.e.,

$$T\omega = \int_D \varphi(y) K_y \omega dy , \qquad (1.5.1)$$

where $\varphi \in C_0^\infty(D)$ is normalized by $\int_D \varphi(y)dy = 1$. Then, there is also a decomposition

$$\omega = d(T\omega) + T(d\omega). \qquad (1.5.2)$$

The l-form $\omega_D \in D'(D, \wedge^l)$ is defined by

$$\omega_D = |D|^{-1} \int_D \omega(y)dy, \quad l = 0, \quad \text{and} \quad \omega_D = d(T\omega), \quad l = 1, 2, \ldots, n$$

for all $\omega \in L^p(D, \wedge^l)$, $1 \le p < \infty$. By substituting $z = tx + y - ty$ and $t = v/(1 + v)$, we have

$$T\omega(x, \xi) = \int_D \omega(z, \zeta(z, x - z), \xi) dz, \qquad (1.5.3)$$

where the vector function $\zeta : D \times \mathbf{R}^n \to \mathbf{R}^n$ is given by

$$\zeta(z, h) = h \int_0^\infty v^{l-1} (1 + v)^{n-l} \varphi(z - vh) dv.$$

The following imbedding inequality appears in [99]: For each $s > 1$, the integral in (1.5.3) defines a bounded operator

$$T : L^s(D, \wedge^l) \to W^{1,s}(D, \wedge^{l-1}), \quad l = 1, 2, \ldots, n$$

with the norm estimated by

$$\|Tu\|_{W^{1,s}(D)} \le C|D| \|u\|_{s,D}. \qquad (1.5.4)$$

The following L^s-norm estimates for differential forms were also proved in [99]:

(1) Let $u \in L^s_{loc}(D, \wedge^l)$, $l = 1, 2, \ldots, n$, $1 < s < \infty$, be a differential form in a bounded, convex domain $D \subset \mathbf{R}^n$. Then,

$$\|\nabla(Tu)\|_{s,D} \le C|D| \|u\|_{s,D}, \qquad (1.5.5)$$

$$\|Tu\|_{s,D} \le C|D| diam(D) \|u\|_{s,D}. \qquad (1.5.6)$$

(2) Let $u \in D'(D, \wedge^l)$ be such that $du \in L^s(D, \wedge^{l+1})$. Then $u - u_D$ is in $W^{1,s}(D, \wedge^l)$ and

$$\|u - u_D\|_{W^{1,s}(D)} \le C|D| \|du\|_{s,D}. \qquad (1.5.7)$$

Here B is any ball or cube and C is a constant independent of u.

In [71] appears the following local Hardy–Littlewood inequality for solutions to the conjugate A-harmonic equation.

Theorem 1.5.2. *Let u and v be conjugate A-harmonic tensors in $\Omega \subset \mathbf{R}^n$, $\sigma > 1$, and $0 < s, t < \infty$. Then, there exists a constant C, independent of u and v, such that*

$$\|u - u_Q\|_{s,Q} \le C|Q|^\beta \|v - c_1\|_{t,\sigma Q}^{q/p}$$

and

$$\|v - v_Q\|_{t,Q} \le C|Q|^{-\beta p/q} \|u - c_2\|_{s,\sigma Q}^{p/q}$$

for all cubes Q with $\sigma Q \subset \Omega$. Here c_1 is any form in $W_{loc}^{1,q}(\Omega, \wedge)$ with $d^{\star}c_1 = 0$, c_2 is any form in $W_{loc}^{1,p}(\Omega, \wedge)$ with $dc_2 = 0$, and $\beta = 1/s + 1/n - (1/t + 1/n)q/p$.

Theorem 1.5.3 (Covering lemma). *Each Ω has a modified Whitney cover of cubes $\mathcal{V} = \{Q_i\}$ such that*

$$\cup_i Q_i = \Omega,$$

$$\sum_{Q \in \mathcal{V}} \chi_{\sqrt{\frac{5}{4}}Q} \leq N \chi_{\Omega}$$

for all $x \in \mathbf{R}^n$ and some $N > 1$, and if $Q_i \cap Q_j \neq \emptyset$, then there exists a cube R (this cube need not be a member of $\mathcal{V}$) in $Q_i \cap Q_j$ such that $Q_i \cup Q_j \subset NR$. Moreover, if Ω is δ-John, then there is a distinguished cube $Q_0 \in \mathcal{V}$ which can be connected with every cube $Q \in \mathcal{V}$ by a chain of cubes $Q_0, Q_1, \ldots, Q_k = Q$ from $\mathcal{V}$ and such that $Q \subset \rho Q_i$, $i = 0, 1, 2, \ldots, k$, for some $\rho = \rho(n, \delta)$.

In [71] the covering lemma is used to combine the local results into the following global Hardy–Littlewood inequality in John domains.

Theorem 1.5.4. *Let $u \in D^{'}(\Omega, \wedge^0)$ and $v \in D^{'}(\Omega, \wedge^2)$ be conjugate A-harmonic tensors and $0 < s, t < \infty$. If Ω is a δ-John domain, $q \leq p$, $v - c \in L^t(\Omega, \wedge^2)$ and*

$$s = \Phi(t) = \frac{npt}{nq + t(q - p)}, \qquad 0 < t < \infty, \qquad (1.5.8)$$

then $u \in L^s(\Omega, \wedge^0)$ and moreover, there exists a constant C, independent of u and v, such that

$$\| u - u_{Q_0} \|_{s,\Omega} \leq C \| v - c \|_{t,\Omega}^{q/p}.$$

Here c is any form in $W_{loc}^{1,q}(\Omega, \wedge)$ with $d^{\star}c = 0$ and Q_0 is the cube appearing in Theorem 1.5.3.

The covering lemma has been further used in [179–182] [71, 74, 196, 197] to extend the local results into global results.

1.5.2 Weighted inequalities

In [179], Theorems 1.5.2 and 1.5.4 have been generalized to the following weighted versions (Theorems 1.5.5 and 1.5.6).

Theorem 1.5.5. *Let u and v be conjugate A-harmonic tensors in a domain $\Omega \subset \mathbf{R}^n$ and $w \in A_r(\Omega)$. Let $s = \Phi(t)$ be defined by (1.5.8). Then, there exists a constant C, independent of u and v, such that*

$$\left(\int_Q |u - u_Q|^s w dx\right)^{1/s} \leq C \left(\int_{\sigma Q} |v - c|^t w^{pt/qs} dx\right)^{q/pt}$$

for all cubes Q with $\sigma Q \subset \Omega \subset \mathbf{R}^n$ and $\sigma > 1$. Here c is any form in $W_{loc}^{1,q}(\Omega, \wedge)$ with $d^\star c = 0$.

Proof. By Lemma 1.4.7, there exist constants $\alpha > 1$ and C_1, independent of w, such that

$$\| w \|_{\alpha, \sigma Q} \leq C_1 |Q|^{(1-\alpha)/\alpha} \| w \|_{1, \sigma Q}. \tag{1.5.9}$$

Since $1/\alpha s + (\alpha - 1)/\alpha s = 1/s$, by the Hölder inequality, we find that

$$\| u - u_Q \|_{s, Q, w} \leq \| w \|_{\alpha, Q}^{1/s} \cdot \| u - u_Q \|_{\alpha s/(\alpha - 1), Q}. \tag{1.5.10}$$

By Theorem 1.5.2, there is a constant C_2, independent of u and v, such that for any $t' > 0$, we have

$$\| u - u_Q \|_{\alpha s/(\alpha - 1), Q} \leq C_2 |Q|^{\beta'} \cdot \| v - c \|_{t', \sigma Q}^{q/p}, \tag{1.5.11}$$

where $\beta' = (\alpha - 1)/\alpha s + 1/n - (1/t' + 1/n)q/p$. Combining (1.5.10) and (1.5.11), we obtain

$$\| u - u_Q \|_{s, Q, w} \leq C_2 |Q|^{\beta'} \cdot \| w \|_{\alpha, Q}^{1/s} \cdot \| v - c \|_{t', \sigma Q}^{q/p}. \tag{1.5.12}$$

Now, choose $t' = t/k$, where k is to be determined later, and note that $|v - c| = w^{-p/qs}|v - c|w^{p/qs}$. By the Hölder inequality, we get

$$\| v - c \|_{t', \sigma Q} \leq \|(1/w)^{pt/qs}\|_{1/(k-1), \sigma Q}^{1/t} \cdot \left(\int_{\sigma Q} |v - c|^t w^{pt/qs} dx\right)^{1/t}. \tag{1.5.13}$$

From (1.5.9), (1.5.12), and (1.5.13), we find that

$$\| u - u_Q \|_{s, Q, w} \leq C_3 |Q|^{\beta' + (1-\alpha)/\alpha s} \| w \|_{1, \sigma Q}^{1/s} \|(\tfrac{1}{w})^{pt/qs}\|_{1/(k-1), \sigma Q}^{q/pt}$$

$$\times \left(\int_{\sigma Q} |v - c|^t w^{pt/qs} dx\right)^{q/pt}. \tag{1.5.14}$$

We choose $k = 1 + pt(r - 1)/qs$, so that $(k - 1)qs/pt = r - 1$, and since $w \in A_r$, it follows that

$$\| w \|_{1, \sigma Q}^{1/s} \|(\tfrac{1}{w})^{pt/qs}\|_{1/(k-1), \sigma Q}^{q/pt}$$

$$= |\sigma Q|^{1/s + (k-1)q/pt} \left(\tfrac{1}{|\sigma Q|} \int_{\sigma Q} w dx \left(\tfrac{1}{|\sigma Q|} \int_{\sigma Q} \left(\tfrac{1}{w}\right)^{1/(r-1)} dx\right)^{r-1}\right)^{1/s}$$

$$\leq C_4 |Q|^{1/s + (k-1)q/pt}. \tag{1.5.15}$$

By (1.5.14) and (1.5.15), we obtain

$$\| u - u_Q \|_{s,Q,w} \le C_5 |Q|^\gamma \left(\int_{\sigma Q} |v - c|^t w^{pt/qs} dx \right)^{q/pt}, \qquad (1.5.16)$$

where $\gamma = \beta' + (1-\alpha)/\alpha s + 1/s + q(k-1)/pt = -(nq+t(q-p))/npt + 1/s = 0$ by (1.5.8). So (1.5.16) is the same as

$$\| u - u_Q \|_{s,Q,w} \le C \left(\int_{\sigma Q} |v - c|^t w^{pt/qs} dx \right)^{q/pt},$$

that is

$$\left(\int_Q |u - u_Q|^s w dx \right)^{1/s} \le C \left(\int_{\sigma Q} |v - c|^t w^{pt/qs} dx \right)^{q/pt}. \qquad \blacksquare$$

1.5.3 Global inequalities

We discuss the global case in John domains. Using the local result and the properties of John domains, we obtain the following global inequality.

Theorem 1.5.6. *Let $u \in D'(\Omega, \wedge^0)$ and $v \in D'(\Omega, \wedge^2)$ be conjugate A-harmonic tensors. Let $q \le p$, $v - c \in L^t(\Omega, \wedge^2)$, $w \in A_r(\Omega)$, and $s = \Phi(t)$ be defined in (1.5.8). Then, there exists a constant C, independent of u and v, such that*

$$\left(\int_\Omega |u - u_{Q_0}|^s w dx \right)^{1/s} \le C \left(\int_\Omega |v - c|^t w^{pt/qs} dx \right)^{q/pt}$$

for any δ-John domain $\Omega \subset \mathbf{R}^n$. Here c is any form in $W_{loc}^{1,q}(\Omega, \wedge)$ with $d^* c = 0$ and $Q_0 \subset \Omega$ is the cube appearing in Theorem 1.5.3.

Proof. Since $w \in A_r$, we can write $d\mu(x) = w(x) dx$. From Theorem 1.5.5, we have

$$\int_Q |u - u_Q|^s d\mu(x) \le C \left(\int_{\sigma Q} |v - c|^t w^{pt/qs} dx \right)^{qs/pt}. \qquad (1.5.17)$$

We use the notations and the covering $\mathcal{V}$ defined in Theorem 1.5.3, and the properties of the measure $d\mu(x) = w(x) dx$: if $w \in A_r$, then

$$\mu(NQ) \le M N^{nr} \mu(Q) \qquad (1.5.18)$$

for each cube Q with $NQ \subset \mathbf{R}^n$ (see [123]) and

$$\max(\mu(Q_i), \mu(Q_{i+1})) \le M N^{nr} \mu(Q_i \cap Q_{i+1}) \qquad (1.5.19)$$

for the sequence of cubes $Q_i, Q_{i+1}, i = 0, 1, \ldots, k - 1$, described in Theorem 1.5.3. We use the elementary inequality $|a + b|^s \leq 2^s(|a|^s + |b|^s)$ for all $s > 0$. In particular, we have

$$\int_\Omega |u - u_{Q_0}|^s w dx$$
$$= \int_\Omega |u - u_{Q_0}|^s d\mu(x)$$
$$\leq 2^s \sum_{Q \in \mathcal{V}} \int_Q |u - u_Q|^s d\mu(x) + 2^s \sum_{Q \in \mathcal{V}} \int_Q |u_{Q_0} - u_Q|^s d\mu(x).$$
$$(1.5.20)$$

The first sum can be estimated by (1.5.17) and Theorem 1.5.3 as

$$\sum_{Q \in \mathcal{V}} \int_Q |u - u_Q|^s d\mu(x)$$
$$\leq C_1 \sum_{Q \in \mathcal{V}} \left(\int_{\sigma Q} |v - c|^t w^{pt/qs} dx \right)^{qs/pt}$$
$$(1.5.21)$$
$$\leq C_1 N \left(\int_\Omega |v - c|^t w^{pt/qs} dx \right)^{qs/pt}.$$

Now we estimate the second sum in (1.5.20). Fix a cube $Q \in \mathcal{V}$ and let $Q_0, Q_1, \ldots, Q_k = Q$ be the chain in Theorem 1.5.3. Then,

$$|u_{Q_0} - u_Q| \leq \sum_{i=0}^{k-1} |u_{Q_i} - u_{Q_{i+1}}|. \qquad (1.5.22)$$

From (1.5.17) and (1.5.19), we get

$$|u_{Q_i} - u_{Q_{i+1}}|^s = \frac{1}{\mu(Q_i \cap Q_{i+1})} \int_{Q_i \cap Q_{i+1}} |u_{Q_i} - u_{Q_{i+1}}|^s d\mu(x)$$
$$\leq \frac{MN^{nr}}{\max(\mu(Q_i), \mu(Q_{i+1}))} \int_{Q_i \cap Q_{i+1}} |u_{Q_i} - u_{Q_{i+1}}|^s d\mu(x)$$
$$\leq C_2 \sum_{j=i}^{i+1} \frac{1}{\mu(Q_j)} \int_{Q_j} |u - u_{Q_j}|^s d\mu(x)$$
$$\leq C_3 \sum_{j=i}^{i+1} \frac{1}{\mu(Q_j)} \left(\int_{\sigma Q_j} |v - c|^t w^{pt/qs} dx \right)^{qs/pt}.$$

Since $Q \subset NQ_j$ for $j = i, i + 1, 0 \leq i \leq k - 1$, we find

$$|u_{Q_i} - u_{Q_{i+1}}|^s \chi_Q(x) \leq C_3 \sum_{j=i}^{i+1} \frac{\chi_{NQ_j}(x)}{\mu(Q_j)} \left(\int_{\sigma Q_j} |v - c|^t w^{pt/qs} dx \right)^{qs/pt}.$$

Thus, by (1.5.22) we obtain (note $|a + b|^{1/s} \leq 2^{1/s}(|a|^{1/s} + |b|^{1/s})$)

$$|u_{Q_0} - u_Q| \chi_Q(x) \leq C_4 \sum_{R \in \mathcal{V}} \left(\frac{1}{\mu(R)} \left(\int_{\sigma R} |v - c|^t w^{pt/qs} dx \right)^{qs/pt} \right)^{1/s} \cdot \chi_{NR}(x)$$

for every $x \in \mathbf{R}^n$. Hence,

$$\sum_{Q \in \mathcal{V}} \int_Q |u_{Q_0} - u_Q|^s d\mu(x)$$

$$\leq C_5 \int_{\mathbf{R}^n} \left| \sum_{R \in \mathcal{V}} \left(\frac{1}{\mu(R)} \left(\int_{\sigma R} |v - c|^t w^{pt/qs} dx \right)^{qs/pt} \right)^{1/s} \chi_{NR}(x) \right|^s d\mu(x).$$

(1.5.23)

If $0 \leq s \leq 1$, we use the inequality $|\sum t_\alpha|^s \leq \sum |t_\alpha|^s$, (1.5.18), and Theorem 1.5.3 to get

$$\sum_{Q \in \mathcal{V}} \int_Q |u_{Q_0} - u_Q|^s d\mu(x)$$

$$\leq C_6 \sum_{R \in \mathcal{V}} \frac{\mu(NR)}{\mu(R)} \left(\int_{\sigma R} |v - c|^t w^{pt/qs} dx \right)^{\frac{qs}{pt}}$$

$$\leq C_7 \sum_{R \in \mathcal{V}} \left(\int_{\sigma R} |v - c|^t w^{pt/qs} dx \right)^{\frac{qs}{pt}}.$$

Note that $qs/pt \geq 1$ and $\sum t_\alpha^p \leq (\sum t_\alpha)^p$ for $p \geq 1$ and $t_\alpha > 0$. Thus,

$$\sum_{Q \in \mathcal{V}} \int_Q |u_{Q_0} - u_Q|^s d\mu(x)$$

$$\leq C_7 \sum_{R \in \mathcal{V}} \left(\int_\Omega |v - c|^t w^{pt/qs} \chi_{\sigma R}(x) dx \right)^{qs/pt}$$

$$\leq C_7 \left(\int_\Omega |v - c|^t w^{pt/qs} \sum_{R \in \mathcal{V}} \chi_{\sigma R}(x) dx \right)^{qs/pt} \qquad (1.5.24)$$

$$\leq C_7 \left(\int_\Omega |v - c|^t w^{pt/qs} N \chi_\Omega(x) dx \right)^{qs/pt}$$

$$\leq C_8 \left(\int_\Omega |v - c|^t w^{pt/qs} dx \right)^{qs/pt}.$$

Combination of (1.5.20), (1.5.21), and (1.5.24) proves the theorem for the case $0 < s \leq 1$. Now for the case $1 < s < \infty$, by (1.5.23) and Theorem 1.4.16, we have

$$\sum_{Q \in \mathcal{V}} \int_Q |u_{Q_0} - u_Q|^s d\mu(x)$$

$$\leq C_9 \int_{\mathbf{R}^n} \left| \sum_{R \in \mathcal{V}} \left(\frac{1}{\mu(R)} \left(\int_{\sigma R} |v - c|^t w^{pt/qs} dx \right)^{qs/pt} \right)^{1/s} \chi_R(x) \right|^s d\mu(x).$$

Note that

$$\sum_{R \in \mathcal{V}} \chi_R(x) \leq \sum_{R \in \mathcal{V}} \chi_{\sigma R}(x) \leq N \chi_\Omega(x)$$

and thus with the elementary inequality $|\sum_{i=1}^N t_i|^s \leq N^{s-1} \sum_{i=1}^N |t_i|^s$ and Theorem 1.5.3, we obtain

$$\sum_{Q \in \mathcal{V}} \int_Q |u_{Q_0} - u_Q|^s d\mu(x)$$

$$\leq C_{10} \int_{R^n} \left(\sum_{R \in \mathcal{V}} \frac{1}{\mu(R)} \left(\int_{\sigma R} |v - c|^t w^{pt/qs} dx \right)^{qs/pt} \chi_R(x) \right) d\mu(x)$$

$$= C_{10} \sum_{R \in \mathcal{V}} \left(\int_{\sigma R} |v - c|^t w^{pt/qs} dx \right)^{qs/pt}$$

$$\le C_{11} \left(\int_{\Omega} |v - c|^t w^{pt/qs} dx \right)^{qs/pt} . \tag{1.5.25}$$

Finally, combination of (1.5.20), (1.5.21), and (1.5.25) proves the theorem for the case $1 < s < \infty$. ∎

1.6 Inequalities in averaging domains

In the previous section, we have discussed weighted global Hardy–Littlewood inequalities in John domains. One may ask the question, whether there are more general kind of domains such that the Hardy–Littlewood inequality also holds in these domains? We will answer this question in this section.

1.6.1 Averaging domains

In 1989, Susan G. Staples [183] introduced the following L^s-averaging domains.

Definition 1.6.1. A proper subdomain $\Omega \subset \mathbf{R}^n$ is called an L^s-averaging domain, $s \ge 1$, if there exists a constant C such that

$$\left(\frac{1}{|\Omega|} \int_{\Omega} |u - u_{\Omega}|^s dx \right)^{1/s} \le C \sup_{B \subset \Omega} \left(\frac{1}{|B|} \int_{B} |u - u_B|^s dx \right)^{1/s}$$

for all $u \in L^s_{loc}(\Omega)$. Here $|\Omega|$ is the n-dimensional Lebesgue measure of Ω.

In [183] these domains are characterized in terms of the quasihyperbolic metric. Particularly, Staples proved that any John domain is L^s-averaging domain. Recently, S. Ding and C. Nolder [184] have extended the L^s-averaging domains to the following $L^s(\mu)$-averaging domains.

Definition 1.6.2. We call a proper subdomain $\Omega \subset \mathbf{R}^n$ an $L^s(\mu)$-averaging domain, $s \ge 1$, if there exists a constant C such that

$$\left(\frac{1}{\mu(B_0)} \int_{\Omega} |u - u_{B_0,\mu}|^s d\mu \right)^{1/s} \le C \sup_{4B \subset \Omega} \left(\frac{1}{\mu(B)} \int_{B} |u - u_{B,\mu}|^s d\mu \right)^{1/s}$$

for some ball $B_0 \subset \Omega$ and all $u \in L^s_{loc}(\Omega; \mu)$. Here the supremum is over all balls B with $4B \subset \Omega$.

The factor 4 here is for convenience; in fact, these domains are independent of this expansion factor, see [183]. For further properties of $L^s(\mu)$-averaging domains, see [185].

The $L^s(\mu)$-averaging domain can also be characterized by quasihyperbolic metric, which was introduced by F.W. Gehring in 1970s.

Definition 1.6.3. The quasihyperbolic distance between x and y in Ω is given by

$$k(x,y) = k(x,y;\Omega) = \inf_\gamma \int_\gamma \frac{1}{d(z,\partial\Omega)}\, ds,$$

where γ is any rectifiable curve in Ω joining x to y. Here $d(z,\partial\Omega)$ is the Euclidean distance between z and $\partial\Omega$.

F.W. Gehring and B. Osgood [11] have proved that for any two points x and y in Ω there is a quasihyperbolic geodesic arc joining them. For a function u we denote the μ-average over B by

$$u_{B,\mu} = \frac{1}{\mu(B)} \int_B u\, d\mu.$$

1.6.2 $L^s(\mu)$-averaging domains

The results presented in this section show that different versions of Hardy–Littlewood inequality hold both locally and globally in $L^s(\mu)$-averaging domains. In [184], S. Ding and C. Nolder proved the following version of the local Hardy–Littlewood inequality.

Theorem 1.6.4. *Let u and v be conjugate A-harmonic tensors in a domain $\Omega \subset \mathbf{R}^n$ and $w \in A_r^{loc}(\Omega)$. Let $s = \Phi(t)$ be defined by (1.5.8). Then, there is a constant C, independent of u and v, such that*

$$\left(\frac{1}{\mu(B)}\int_B |u - u_B|^s w\, dx\right)^{1/s} \le C|B|^\gamma \left(\int_{\sigma B} |v - c|^{q\kappa} dx\right)^{1/(p\kappa)}$$

for all balls B with $2\sigma B \subset \Omega$. Here $\gamma = 1/(\beta s) + 1/n - (1/t + 1/n)q/p$, σ and c are as in Theorem 1.5.5, $\kappa = \beta ts/(\beta qs - pt)$, where β is the exponent in the weak reverse Hölder inequality for A_r-weights and $\beta qs - pt > 0$.

Proof. By Theorem 1.5.5 and Hölder's inequality, we have

$$\mu(B)^{-1/s} \left(\int_B |u - u_{B,\mu}|^s w\, dx\right)^{1/s}$$

$$\le C_1 \mu(B)^{-1/s} \left(\int_{\sigma B} w^\beta dx\right)^{1/\beta s}$$

$$\times |B|^\gamma \left(\int_{\sigma B} |v - c|^{\beta qts/(\beta qs - pt)} dx\right)^{(\beta qs - pt)/\beta pts}.$$

Since $w \in A_r^{loc}(\Omega)$, it satisfies the weak reverse Hölder inequality (1.4.1) and the weak doubling condition, and so

$$\mu(B)^{-1/s}\left(\int_{\sigma B} w^\beta\right)^{1/\beta s} \leq C_2 \mu(B)^{-1/s}|B|^{(1-\beta)/(\beta s)}\mu(\sigma B)^{1/s}$$

$$\leq C_3|B|^{\frac{1-\beta}{\beta s}}\mu(B)^{-1/s}\mu(B)^{1/s} = C_3|B|^{\frac{1-\beta}{\beta s}}$$

and from this Theorem 1.6.4 follows. $\qquad\blacksquare$

In [184], S. Ding and C. Nolder also obtained the following result which plays an important role in the proof of showing that any John domain is an $L^s(\mu)$-averaging domain.

Theorem 1.6.5. Let $w \in WD(\Omega)$. Suppose that s and q are positive constants and that

$$\mu(\{x \in B : |u(x) - u_{B,\mu}| > t\}) \leq se^{-qt}\mu(B) \qquad (1.6.1)$$

for each $t > 0$ and each ball B with $4B \subset \Omega$. Then, there exists a constant $C = C(s,q,n)$ such that

$$|u_{B(x),\mu} - u_{B(y),\mu}| \leq C(k(x,y)+1) \qquad (1.6.2)$$

for all x and y in Ω. Here $B(x)$ is the ball $B(x, d(x,\partial\Omega)/4)$.

Proof. For each $z \in \Omega$, notice that $4B(z) \subset \Omega$. Fix x and $y \in \Omega$ and choose a quasihyperbolic geodesic arc γ joining x and y in Ω. We define an ordered sequence of points $\{z_j\}$ on γ by induction as follows. First let $z_1 = x$. Next suppose that $z_1, \ldots, z_j$ have been defined, and let $\beta_j = \gamma(z_j, y)$ denote that part of γ from z_j to y and γ_j the component of $\beta_j \cap B(z_j)$ which contains z_j. Define z_{j+1} as the other endpoint of γ_j. We simplify notation as follows: $B_j = B(z_j)$, $d_j = d(z_j, \partial\Omega)$, $r_j = d_j/4 = $ radius of B_j, and $y = z_{m+1}$. From the definition of $\{z_j\}$, we have

$$|z_{j+1} - z_j| = r_j, \quad j = 1, \ldots, m-1, \text{ and } |z_{m+1} - z_m| \leq r_m. \qquad (1.6.3)$$

For $j = 1, \ldots, m-1$, pick $z_j' \in \partial\Omega$ so that $d(z_j, \partial\Omega) = |z_j - z_j'|$. If $z \in \gamma_j \subset B_j$, then

$$d(z, \partial\Omega) \leq |z - z_j'| \leq |z - z_j| + |z_j - z_j'| \leq 4d_j/3.$$

Hence,

$$\int_{\gamma_j} \frac{1}{d(z,\partial\Omega)}ds \geq \frac{3}{4d_j}\int_{\gamma_j}ds \geq \frac{3|z_{j+1} - z_j|}{4d_j} = \frac{3}{16}.$$

Summing the above inequality over j gives

$$\frac{3(m-1)}{16} \leq \sum_{1}^{m-1} \int_{\gamma_j} \frac{1}{d(z,\partial\Omega)} ds \leq \int_{\gamma} \frac{1}{d(z,\partial\Omega)} ds = k(x,y), \qquad (1.6.4)$$

where $m < \infty$.

Consider now the relative size of neighboring balls. Fix j and choose $z, z' \in \partial\Omega$ so that $d_j = |z - z_j|$ and $d_{j+1} = |z' - z_{j+1}|$. Then, $d_j \leq d_{j+1} + r_j$ and $d_{j+1} \leq d_j + r_j$, with the first inequality yielding

$$r_{j+1} = d_{j+1}/4 \geq (d_j - r_j)/4 \geq 3r_j/4,$$

and the second leading to $r_j \geq 3r_{j+1}/4$ if $r_{j+1} \geq r_j$. Hence,

$$3r_j/4 \leq r_{j+1} \leq 4r_j/3. \qquad (1.6.5)$$

Next we shall show that

$$\mu(B_j \cap B_{j+1}) > c_1(\mu(B_j) + \mu(B_{j+1})) \qquad (1.6.6)$$

for a constant c_1. Fix j and let $v = (z_j + z_{j+1})/2$, $s = (\max(r_j, r_{j+1}))/8$, and B be the ball of radius s centered at v. In the case $r_j \geq r_{j+1}$ we have $s = r_j/8 \leq r_{j+1}/6$; similarly, if $r_{j+1} \geq r_j$ we have $s \leq r_j/6$. Thus, for $z \in B$ we obtain

$$|z - z_j| \leq |z - v| + |v - z_j| \leq s + \frac{1}{2}|z_j - z_{j+1}| \leq r_j,$$

and a similar argument shows $|z - z_{j+1}| \leq r_{j+1}$. Hence we conclude that $B \subset B_j \cap B_{j+1}$. Since μ is doubling, we have

$$\mu(B_j \cap B_{j+1}) \geq \mu(B) \geq \mu\left(\frac{1}{8}B_j\right) \geq N_1(w)\mu(B_j)$$

and

$$\mu(B_j \cap B_{j+1}) \geq \mu(B) \geq \mu\left(\frac{1}{8}B_{j+1}\right) \geq N_2(w)\mu(B_{j+1}).$$

Let $c_1 = \frac{1}{2}\min\{N_1(w), N_2(w), 4s\}$. Then, it follows that

$$\mu(B_j \cap B_{j+1}) \geq \frac{1}{2}(N_1(w)\mu(B_j) + N_2(w)\mu(B_{j+1}))$$
$$\geq c_1(\mu(B_j) + \mu(B_{j+1})).$$

For $j = 1, 2, \ldots, m + 1$, let

$$E_j = \{x \in B_j : |u - u_{B_j,\mu}| > t\},$$

where $t = (\log(2s/c_1))/q$. From (1.6.1), we obtain

$$\mu(E_j) \leq c_1\mu(B_j)/2. \qquad (1.6.7)$$

Combination of (1.6.6) and (1.6.7) yields

$$\mu((B_j \cap B_{j+1})\backslash(E_j \cup E_{j+1})) > 0.$$

Therefore, there exists $x \in (B_j \cap B_{j+1})\backslash(E_j \cup E_{j+1})$, and hence

$$|u_{B_j,\mu} - u_{B_{j+1},\mu}| \le |u - u_{B_j,\mu}| + |u - u_{B_{j+1},\mu}| \le 2t.$$

Summing the above inequality and using (1.6.4), we conclude that

$$|u_{B(x),\mu} - u_{B(y),\mu}| \le \sum_1^m |u_{B_j,\mu} - u_{B_{j+1},\mu}| \le 2mt$$

$$\le 32tk(x,y)/3 + 2t$$

$$\le \tfrac{32t}{3}(k(x,y) + 1)$$

$$= \tfrac{32(\log(2s/c_1))}{3q}(k(x,y) + 1)$$

$$= C(k(x,y) + 1). \qquad \blacksquare$$

Finally, we state the following result from [184] which extends Theorem 1.5.6 into the case of $L^s(\mu)$-averaging domains.

Theorem 1.6.6. *Let $u \in D'(\Omega, \wedge^0)$ and $v \in D'(\Omega, \wedge^2)$ be conjugate A-harmonic tensors in an $L^s(\mu)$-averaging domain $\Omega \subset \mathbf{R}^n$. If $v-c \in L^t(\Omega, \wedge^2)$, $w \in A_r^{loc}(\Omega)$ and*

$$t = \Psi(s) = \frac{nqs\beta}{np + (p - q)s\beta},$$

then there exists a constant C, independent of u and v, such that

$$\left(\frac{1}{\mu(\Omega)}\int_\Omega |u - u_{B_0,\mu}|^s w\, dx\right)^{1/s} \le C\left(\int_\Omega |v - c|^{q\kappa} dx\right)^{1/(p\kappa)},$$

where $\kappa = \beta ts/(\beta qs - pt)$ is again as above and $d^\star c = 0$. Here B_0 is some fixed ball.

1.6.3 Other weighted inequalities

Recently, G. Bao [186] proved the following local $A_r(\lambda, \Omega)$-weighted Hardy–Littlewood integral inequality.

Theorem 1.6.7. *Let u and v be conjugate A-harmonic tensors in a domain $\Omega \subset \mathbf{R}^n$ and $w \in A_r(\lambda, \Omega)$ for some $r > 1$ and $\lambda > 1$. Let $0 < s, t < \infty$. Then, there exists a constant C, independent of u and v, such that*

$$\left(\int_B |u - u_B|^s w^{\lambda/p} dx\right)^{1/s} \le C|B|^\gamma \left(\int_{\sigma B} |v - c|^t w^{t/qs} dx\right)^{q/pt}$$

for all balls B with $\sigma B \subset \Omega \subset \mathbf{R}^n$ and $\sigma > 1$. Here c is any form in $W_{loc}^{1,q}(\Omega, \wedge)$ with $d^\star c = 0$ and $\gamma = 1/s + 1/n - (1/t + 1/n)q/p$.

Note that the above inequality can be written in the following symmetric form:

$$\left(\frac{1}{|B|}\int_B |u - u_B|^s w^{\lambda/p} dx\right)^{1/qs} \le C|B|^{(\frac{1}{nq}-\frac{1}{np})} \left(\frac{1}{|B|}\int_{\sigma B} |v - c|^t w^{\frac{t}{qs}} dx\right)^{\frac{1}{pt}}.$$

In [180], parametric versions of the $A_r(\Omega)$-weighted Hardy–Littlewood inequality were established. Corollary 1.6.8 follows from Theorem 1.5.5 using equation (1.4.2).

Corollary 1.6.8. *Let u and v be conjugate A-harmonic tensors in a domain $\Omega \subset \mathbf{R}^n$ and $w \in A_r(\Omega)$ for some $r > 1$. Let $0 < s,t < \infty$. Then, there exists a constant C, independent of u and v, such that*

$$\left(\int_Q |u - u_Q|^s w^\alpha dx\right)^{1/s} \le C|Q|^\gamma \left(\int_{\sigma Q} |v - c|^t w^{pt\alpha/qs} dx\right)^{q/pt}$$

for all cubes Q with $\sigma Q \subset \Omega \subset \mathbf{R}^n$, $\sigma > 1$, and $0 < \alpha \le 1$. Here c is any form in $W_{loc}^{1,q}(\Omega, \wedge)$ with $d^\star c = 0$ and $\gamma = 1/s + 1/n - (1/t + 1/n)q/p$.

Note that $\alpha \in (0,1]$ is arbitrary in Corollary 1.6.8. Hence, for particular values of α, we get different versions of the Hardy–Littlewood inequality. For example, if we let $\alpha = qs$, $qs \le 1$ in Theorem 1.6.8, we obtain the following symmetric inequality.

Corollary 1.6.9. *Let u and v be conjugate A-harmonic tensors in a domain $\Omega \subset \mathbf{R}^n$ and $w \in A_r(\Omega)$ for some $r > 1$. Let $0 < t < \infty$ and $qs \le 1$. Then, there exists a constant C, independent of u and v, such that*

$$\left(\int_Q |u - u_Q|^s w^{qs} dx\right)^{1/qs} \le C|Q|^\gamma \left(\int_{\sigma Q} |v - c|^t w^{pt} dx\right)^{1/pt}$$

for all cubes Q with $\sigma Q \subset \Omega \subset \mathbf{R}^n$ and $\sigma > 1$. Here c is any form in $W_{loc}^{1,q}(\Omega, \wedge)$ with $d^\star c = 0$ and $\gamma = (1/s + 1/n)/q - (1/t + 1/n)/p$.

Similarly, if we choose $\alpha = 1/pt$, $pt \ge 1$ in Theorem 1.6.8, we obtain the following symmetric inequality.

Corollary 1.6.10. *Let u and v be conjugate A-harmonic tensors in a domain $\Omega \subset \mathbf{R}^n$ and $w \in A_r(\Omega)$ for some $r > 1$. Let $0 < s < \infty$ and $pt \ge 1$. Then, there exists a constant C, independent of u and v, such that*

$$\left(\int_Q |u - u_Q|^s w^{1/pt} dx \right)^{1/qs} \le C|Q|^\gamma \left(\int_{\sigma Q} |v - c|^t w^{1/qs} dx \right)^{1/pt}$$

for all cubes Q with $\sigma Q \subset \Omega \subset \mathbf{R}^n$ and $\sigma > 1$. Here c is any form in $W^{1,q}_{loc}(\Omega, \wedge)$ with $d^*c = 0$ and $\gamma = (1/s + 1/n)/q - (1/t + 1/n)/p$.

1.7 Two-weight cases

In the last section, we have discussed different versions of the Hardy–Littlewood inequality, including the A_r-weighted Hardy–Littlewood inequality. These inequalities may be extended into the two-weight case. In order to prove the two-weight inequality, one only needs to replace $w(x)$ on the left-hand side by $w_1(x)$ and on the right-hand side by $w_2(x)$ in single weighted inequalities and slightly adjust the conditions on the weight if necessary. For example, the following is one version of two-weight Hardy–Littlewood inequality that was obtained recently.

1.7.1 Local inequalities

Theorem 1.7.1. *Let u and v be conjugate A-harmonic tensors in a domain $\Omega \subset \mathbf{R}^n$ and $(w_1, w_2) \in A_{r,\lambda}(\Omega)$. Let $s = \Phi(t)$ be as in (1.5.8). Then, there exists a constant C, independent of u and v, such that*

$$\left(\int_B |u - u_B|^s w_1^\alpha dx \right)^{1/s} \le C \left(\int_{\sigma B} |v - c|^t w_2^{\alpha pt/qs} dx \right)^{q/pt} \tag{1.7.1}$$

for all cubes B with $\sigma B \subset \Omega \subset \mathbf{R}^n$ and $\sigma > 1$. Here c is any form in $W^{1,q}_{loc}(\Omega, \wedge)$ with $d^*c = 0$, $0 < \alpha < t$ and $\alpha < \lambda$, where $\lambda \ge 1$.

Proof. Let $\delta = s\lambda/(\lambda - \alpha)$ so that $\delta > s$, and

$$\left(\int_B |u - u_B|^s w_1^\alpha dx \right)^{1/s}$$

$$= \left(\int_B (|u - u_B| w_1^{\alpha/s})^s dx \right)^{1/s}$$

$$\le \left(\int_B |u - u_B|^\delta dx \right)^{1/\delta} \left(\int_B w_1^{\alpha\delta/(\delta - s)} dx \right)^{(\delta - s)/s\delta} \tag{1.7.2}$$

$$= \| u - u_B \|_{\delta,B} \left(\int_B w_1^{\alpha\delta/(\delta - s)} dx \right)^{(\delta - s)/s\delta}.$$

By Theorem 1.5.2, for $q \le p$ and $0 < \delta, \xi < \infty$, we have

$$\| u - u_B \|_{\delta,B} \le C_1 |B|^\beta \| v - c_1 \|^{q/p}_{\xi,\sigma B}, \tag{1.7.3}$$

where $\beta = 1/\delta + 1/n - (1/\xi + 1/n)q/p$ and

$$0 < \xi = \frac{st\lambda q}{(sq - \alpha p)t + \lambda sq}.$$

Note that $\alpha < t$ leads to $qs > pt$ so that $qs > p\alpha$, and hence condition (1.5.8) gives $\xi < t$. Thus, by the Hölder inequality

$$\|v - c_1\|_{\xi,\sigma B}^{q/p}$$

$$= \left(\int_{\sigma B} (|v - c_1| w_2^{\alpha p/sq} w_2^{-\alpha p/sq})^{\xi} dx \right)^{q/\xi p}$$

$$\leq C_2 \left[\left(\int_{\sigma B} (|v - c_1| w_2^{\frac{\alpha p}{sq}})^t dx \right)^{1/t} \left(\int_{\sigma B} (\frac{1}{w_2})^{(\alpha p/sq)(\xi t/(t-\xi))} dx \right)^{\frac{(t-\xi)}{\xi t}} \right]^{q/p}.$$

$$(1.7.4)$$

Substitution of (1.7.3) and (1.7.4) into (1.7.2) gives

$$\left(\int_B |u - u_B|^s w_1^{\alpha} dx \right)^{1/s}$$

$$\leq C_3 |B|^{\beta} \left(\int_{\sigma B} (|v - c_1| w_2^{\alpha p/sq})^t dx \right)^{q/pt}$$

$$\times \left(\int_{\sigma B} (\frac{1}{w_2})^{(\alpha p/sq)(\xi t/(t-\xi))} dx \right)^{(t-\xi)q/\xi tp} \left(\int_B w_1^{\alpha\delta/(\delta-s)} dx \right)^{(\delta-s)/s\delta}.$$

$$(1.7.5)$$

From the choice of δ and ξ, and $sq/\alpha p > 1$, it follows that $r = sq/\alpha p$, and hence

$$\left(\int_{\sigma B} (\frac{1}{w_2})^{(\alpha p/sq)(\xi t/(t-\xi))} dx \right)^{(t-\xi)q/\xi tp} \left(\int_B w_1^{\alpha\delta/(\delta-s)} dx \right)^{(\delta-s)/s\delta}$$

$$\leq C_4 |\sigma B|^{\beta_1} \left(\frac{1}{|\sigma B|} \int_{\sigma B} (\frac{1}{w_2})^{\alpha p\lambda/(sq-\alpha p)} dx \right)^{\frac{(sq-\alpha p)}{\lambda sp}} \left(\frac{1}{|\sigma B|} \int_{\sigma B} w_1^{\lambda} dx \right)^{\alpha/s\lambda}$$

$$\leq C_5 |B|^{\beta_1} \left[\left(\frac{1}{|\sigma B|} \int_{\sigma B} (\frac{1}{w_2})^{\lambda/(r-1)} dx \right)^{(r-1)/r\lambda} \left(\frac{1}{|\sigma B|} \int_{\sigma B} w_1^{\lambda} dx \right)^{1/\lambda r} \right]^{q/p}$$

$$\leq C_6 |B|^{\beta_1},$$

$$(1.7.6)$$

where $\beta_1 = (1/\xi - 1/t)q/p + (1/s - 1/\delta)$. Substituting (1.7.6) into (1.7.5) and using $\beta + \beta_1 = 1/\delta + 1/n - (1/\xi + 1/n)q/p + (1/\xi - 1/t)q/p + (1/s - 1/\delta) = 0$ in view of condition (1.5.8), we finally obtain

$$\left(\int_B |u - u_B|^s w_1^{\alpha} dx \right)^{1/s} \leq C_7 \left(\int_{\sigma B} |v - c_1|^t w_2^{\alpha pt/sq} dx \right)^{q/pt}. \quad \blacksquare$$

1.7.2 Global inequalities

Theorem 1.7.2. *Let $u \in D'(\Omega, \wedge^0)$ and $v \in D'(\Omega, \wedge^2)$ be conjugate A-harmonic tensors. If Ω is a δ-John domain, $q \leq p$, $v - c \in L^t(\Omega, \wedge^2)$, and $s = \Phi(t)$ as in (1.5.8), $\alpha > 0$, and weight $(w_1, w_2) \in A_{r,\lambda}$. Then,*

$$\|u - u_{Q_0}\|_{s,\Omega,w_1^\alpha} \leq C \|v - c\|_{t,\Omega,w_2^{\alpha p t/qs}}^{q/p}, \qquad (1.7.7)$$

where c is any form in $W_{loc}^{1,q}(\Omega, \wedge)$ with $d^\star c = 0$ and Q_0 is a fixed cube.

There are several other versions of the two-weight Hardy–Littlewood inequality, e.g., see [187, 188]. Some work related to the Hardy–Littlewood inequality can also be found in [189–193, 148, 67–69, 73, 74].

1.8 The best integrable condition

In this section, we show that condition (1.5.8) in Theorem 1.5.4 is the best possible.

1.8.1 An example

We illustrate the following example in the three-dimensional space.

Example 1.8.1. Let

$$f(x) = x|x|^\beta = (x_1|x|^\beta, x_2|x|^\beta, x_3|x|^\beta)$$

be a K-quasiregular mapping in $\mathbf{R}^3$. Here $\beta \neq -1$ is a real number. From Example 1.2.4 with $l = 1$, we know that

$$u = f^1 = x_1|x|^\beta$$

and

$$v = \star f^2 df^3 = \star x_2 |x|^\beta d(x_3|x|^\beta)$$

are conjugate A-harmonic tensors. It is easy to compute that

$$v = \beta x_1 x_2 x_3 |x|^{2\beta-2} dx_2 \wedge dx_3 - \beta x_2^2 x_3 |x|^{2\beta-2} dx_1 \wedge dx_3$$
$$+ |x|^{2\beta-2} x_2 (|x|^2 + \beta x_3^2) dx_1 \wedge dx_2,$$

so that

$$|u| = |x_1||x|^\beta$$

and

$$|v| = |x|^{2\beta-1}|x_2|(|x|^2 + (\beta^2 + 2\beta)x_3^2)^{1/2} .$$

Now applying the following spherical coordinate transformation

$$\begin{cases} x_1 &= r\cos\theta\sin\varphi \\ x_2 &= r\sin\theta\sin\varphi \\ x_3 &= r\cos\varphi, \end{cases}$$

where $0 < r \le 1$, $0 \le \varphi \le \pi$, $0 \le \theta \le 2\pi$, we obtain

$$|u|^s = |\sin\varphi|^s|\cos\theta|^s r^{(\beta+1)s} .$$

Let $\Omega = \mathbf{B}^3 \setminus \{0\}$, where $\mathbf{B}^3$ denotes the unit ball in $\mathbf{R}^3$, so that

$$\|u\|_{s,\Omega} = \left(\int_\Omega |u|^s dx\right)^{1/s}$$

$$= \left(\lim_{\varepsilon\to 0}\int_0^{2\pi}|\cos\theta|^s d\theta \int_0^\pi |\sin\varphi|^s \sin\varphi d\varphi \int_\varepsilon^1 r^{(\beta+1)s+2}dr\right)^{1/s}$$

$$= C(s,\beta)\left(\lim_{\varepsilon\to 0}(1 - \varepsilon^{(\beta+1)s+3})\right)^{1/s},$$

where $C(s,\beta)$ is independent of u and v. Similarly, we have

$$\|v\|_{t,\Omega} = \left(\int_\Omega |v|^t dx\right)^{1/t}$$

$$= B(t,\beta)\left(\lim_{\varepsilon\to 0}(1 - \varepsilon^{(2\beta+1)t+3})\right)^{1/t},$$

where $B(t,\beta)$ is also independent of u and v.

Obviously, one sees that $\|u\|_{s,\Omega} < \infty$ if and only if $s > -\frac{3}{\beta+1}$ and $\|v\|_{t,\Omega} < \infty$ if and only if $t > -\frac{3}{2\beta+1}$ for $\beta > -\frac{1}{2}$. Since $n = 3$, $p = \frac{n}{l} = 3$ and $q = \frac{n}{n-l} = \frac{3}{2}$ we find that $\Phi(-\frac{3}{2\beta+1}) = -\frac{3}{\beta+1}$ in (1.5.8). Thus, condition (1.5.8) in Theorem 1.5.4 is the best possible.

Also, from Theorem 3.1 in [71] and viewing the example in another way by comparing integrability exponents, we have

$$\left(\varepsilon^{(\beta+1)s+3}\right)^{1/s} \sim \left(\varepsilon^{(2\beta+1)t+3}\right)^{q/tp} .$$

Thus,

$$\frac{(\beta+1)s+3}{s} = \frac{1}{2}\frac{(2\beta+1)t+3}{t}$$

and hence

$$s = 3t/(3/2 - t/2) \ .$$

On the other hand, substituting $p = 3$ and $q = 3/2$ into $\Phi(t)$ defined in (1.5.8), we have

$$s = \Phi(t) = \frac{npt}{nq + t(q-p)} = 3t/(3/2 - t/2) \ .$$

This again implies that condition (1.5.8) in Theorem 1.5.4 is the best possible.

1.8.2 Remark

Remark. (1) The above example shows that condition (1.5.8) is the best possible in $\mathbf{R}^3$ when $p = 3$ and $q = 3/2$. There exists an example in $\mathbf{R}^2$ which shows that this condition is also best possible for $q > 2$ and all p, see [71] for details.

(2) The above example can be used to show the exactness of exponents in the following result about Lipschitz conditions of conjugate A-harmonic tensors. The result was proved in [71].

If $0 \le k,\ l \le 1$ satisfy $p(k-1) = q(l-1)$, then there exists a constant C such that

$$\|u\|^p_{locLip_k,\Omega}/C \le \|\star v\|^q_{locLip_l,\Omega} \le C\|u\|^p_{locLip_k,\Omega}$$

for all conjugate A-harmonic tensors u and v in Ω. The detailed discussion of this topic is in Chapter 9.

(3) The above example can be used to show that the similar results in $L^s(\mu)$-averaging domains discussed in Section 1.6 are also sharp.

1.9 Inequalities with Orlicz norms

In this section, we introduce the Orlicz norm and discuss various versions of the Hardy–Littlewood inequality with the Orlicz norm.

A continuous and increasing function $\varphi : [0,\infty] \to [0,\infty]$ with $\varphi(0) = 0$ and $\varphi(\infty) = \infty$ is called an *Orlicz function*. The *Orlicz space* $L^\varphi(\Omega)$ consists of all measurable functions f on Ω such that $\int_\Omega \varphi\left(\frac{|f|}{\lambda}\right) dx < \infty$ for some $\lambda = \lambda(f) > 0$. $L^\varphi(\Omega)$ is equipped with the nonlinear *Luxemburg functional*

$$\|f\|_\varphi = \inf\left\{\lambda > 0 : \int_\Omega \varphi\left(\frac{|f|}{\lambda}\right) dx \leq 1\right\}.$$

A convex Orlicz function φ is often called a *Young function*. If φ is a Young function, then $\|\cdot\|_\varphi$ defines a norm in $L^\varphi(\Omega)$, which is called the *Luxemburg norm* or *Orlicz norm*. For $\varphi(t) = t^p \log^\alpha(e+t)$, $0 < p < \infty$ and $\alpha \geq 0$ (note that φ is convex for $1 \leq p < \infty$ and any real α with $\alpha \geq 1 - p$), we have

$$\|f\|_{L^p \log^\alpha L} = \|f\|_{L^p \log^\alpha L(\Omega)} = \inf\left\{k : \int_\Omega |f|^p \log^\alpha\left(e + \frac{|f|}{k}\right) dx \leq k^p\right\}.$$

Let $0 < p < \infty$ and $\alpha \geq 0$ be real numbers and E be any subset of $\mathbf{R}^n$. We define the functional of a measurable function f over E by

$$[f]_{L^p(\log L)^\alpha(E)} = \left(\int_E |f|^p \log^\alpha\left(e + \frac{|f|}{\|f\|_p}\right) dx\right)^{\frac{1}{p}},$$

where $\|f\|_p = \left(\int_E |f(x)|^p dx\right)^{1/p}$.

1.9.1 Norm comparison theorem

It has been proved that the norm $\|f\|_{L^p \log^\alpha L}$ is equivalent to the norm $[f]_{L^p(\log L)^\alpha(\Omega)}$ if $1 \leq p < \infty$ and $\alpha \geq 0$. Following [194], we prove that the norm $\|f\|_{L^p \log^\alpha L}$ is also equivalent to $[f]_{L^p(\log L)^\alpha(\Omega)}$ for $0 < p < 1$ and $\alpha \geq 0$.

Theorem 1.9.1. *For each $f \in L^p(\log L)^\alpha(\Omega)$, $0 < p < \infty$ and $\alpha \geq 0$, we have*

$$\|f\|_p \leq \|f\|_{L^p \log^\alpha L(\Omega)} \leq [f]_{L^p(\log L)^\alpha(\Omega)} \leq C\|f\|_{L^p \log^\alpha L(\Omega)},$$

where $C = 2^{\alpha/p}\left(1 + \left(\frac{\alpha}{ep}\right)^\alpha\right)^{1/p}$ is a constant independent of f.

Proof. Let $K = \|f\|_{L^p \log^\alpha L(\Omega)}$. Then, by the definition of the Luxemburg norm, we have

$$K = \left(\int_\Omega |f|^p \log^\alpha\left(e + \frac{|f|}{K}\right) dx\right)^{1/p}.$$

It is clear that $K \geq \|f\|_p$ and

$$K \leq \left(\int_\Omega |f|^p \log^\alpha\left(e + \frac{|f|}{\|f\|_p}\right) dx\right)^{1/p} = [f]_{L^p(\log L)^\alpha(\Omega)},$$

that is

$$\|f\|_{L^p \log^\alpha L(\Omega)} \le [f]_{L^p(\log L)^\alpha(\Omega)}.$$

On the other hand, using $K \ge \|f\|_p$ and the elementary inequality $|a+b|^s \le 2^s(|a|^s + |b|^s)$, $s \ge 0$, we obtain

$$\int_\Omega |f|^p \log^\alpha \left(e + \frac{|f|}{\|f\|_p}\right) dx$$

$$= \int_\Omega |f|^p \log^\alpha \left(e + \frac{|f|}{K} \cdot \frac{K}{\|f\|_p}\right) dx$$

$$\le \int_\Omega |f|^p \left(\log\left(e + \frac{|f|}{K}\right) + \log\left(\frac{K}{\|f\|_p}\right)\right)^\alpha dx$$

$$\le 2^\alpha \int_\Omega |f|^p \log^\alpha \left(e + \frac{|f|}{K}\right) dx + 2^\alpha \int_\Omega |f|^p \log^\alpha \left(\frac{K}{\|f\|_p}\right) dx$$

$$= 2^\alpha K^p + 2^\alpha \|f\|_p^p \log^\alpha \left(\frac{K}{\|f\|_p}\right).$$

Now since the function $h(t) = t^p \log^\alpha\left(\frac{K}{t}\right)$, $0 < t \le K$, has its maximum value $\left(\frac{\alpha}{ep}\right)^\alpha K^p$ at $t = K/e^{\alpha/p}$, it follows that

$$\|f\|_p^p \log^\alpha \left(\frac{K}{\|f\|_p}\right) \le \left(\frac{\alpha}{ep}\right)^\alpha K^p.$$

Combination of the last two inequalities gives

$$\int_\Omega |f|^p \log^\alpha \left(e + \frac{|f|}{\|f\|_p}\right) dx \le 2^\alpha \left(1 + \left(\frac{\alpha}{ep}\right)^\alpha\right) K^p,$$

which is equivalent to

$$[f]_{L^p(\log L)^\alpha(\Omega)} \le C \|f\|_{L^p \log^\alpha L(\Omega)},$$

where $C = 2^{\alpha/p}\left(1 + \left(\frac{\alpha}{ep}\right)^\alpha\right)^{1/p}$. ∎

1.9.2 $L^p(\log L)^\alpha$-norm inequality

Here, we shall prove the following Hardy–Littlewood inequality with $L^p(\log L)^\alpha$-norm.

Theorem 1.9.2. *Let u and v be solutions to the conjugate A-harmonic equation (1.2.6) in $\Omega \subset \mathbf{R}^n$, $\sigma > 1$, and $0 < s, t < \infty$. Then, there exists a constant C, independent of u and v, such that*

$$\|u - u_B\|_{L^s(\log L)^\alpha(B)} \le C|B|^\gamma \|v - c\|^{q/p}_{L^t(\log L)^\beta(\sigma B)} \qquad (1.9.1)$$

for all balls or cubes B with $\sigma B \subset \Omega$ and all α with $0 < \alpha < s$ and $\beta \ge 0$. Here c is any form in $W^{1,q}_{loc}(\Omega, \wedge)$ with $d^\star c = 0$ and $\gamma = 1/s + 1/n - (1/t + 1/n)q/p - \alpha/s^2$.

Proof. First, using the Hölder inequality with $1/s = 1/(s^2/(s-\alpha)) + 1/(s^2/\alpha)$, we find that

$$\left(\int_B |u - u_B|^s \log^\alpha\left(e + \frac{|u-u_B|}{\|u-u_B\|_s}\right)dx\right)^{\frac{1}{s}}$$
$$= \left(\int_B \left(|u - u_B| \log^{\alpha/s}\left(e + \frac{|u-u_B|}{\|u-u_B\|_s}\right)\right)^s dx\right)^{\frac{1}{s}}$$
$$\le \left(\int_B |u - u_B|^{\frac{s^2}{(s-\alpha)}} dx\right)^{\frac{(s-\alpha)}{s^2}} \left(\int_B \log^s\left(e + \frac{|u-u_B|}{\|u-u_B\|_s}\right)dx\right)^{\frac{\alpha}{s^2}} \qquad (1.9.2)$$
$$= \|u - u_B\|_{s^2/(s-\alpha),B}\left(\int_B \log^s\left(e + \frac{|u-u_B|}{\|u-u_B\|_s}\right)dx\right)^{\alpha/s^2}.$$

Choosing parameter m with $0 < m < t$ and using the Hardy–Littlewood inequality (Theorem 1.5.2), we have

$$\|u - u_B\|_{s^2/(s-\alpha),B} \le C_1|B|^{\gamma_1}\|v - c\|^{q/p}_{m,\sigma B}, \qquad (1.9.3)$$

where $\gamma_1 = (s - \alpha)/s^2 + 1/n - (1/m + 1/n)q/p$ and c is any coclosed form. Applying the Hölder inequality with $1/m = 1/t + (t - m)/mt$, we obtain

$$\|v - c\|_{m,\sigma B}$$
$$= \left(\int_{\sigma B}\left(|v - c| \log^{\frac{\beta}{t}}\left(e + \frac{|v-c|}{\|v-c\|_t}\right) \log^{\frac{-\beta}{t}}\left(e + \frac{|v-c|}{\|v-c\|_t}\right)\right)^m dx\right)^{\frac{1}{m}}$$
$$\le \left(\int_{\sigma B} |v - c|^t \log^\beta\left(e + \frac{|v-c|}{\|v-c\|_t}\right)dx\right)^{1/t}$$
$$\times \left(\int_{\sigma B} \log^{-\beta m/(t-m)}\left(e + \frac{|v-c|}{\|v-c\|_t}\right)dx\right)^{(t-m)/mt} \qquad (1.9.4)$$
$$\le \left(\int_{\sigma B} |v - c|^t \log^\beta\left(e + \frac{|v-c|}{\|v-c\|_t}\right)dx\right)^{1/t}\left(\int_{\sigma B} 1 dx\right)^{(t-m)/mt}$$
$$\le C_2|B|^{(t-m)/mt}\left(\int_{\sigma B} |v - c|^t \log^\beta\left(e + \frac{|v-c|}{\|v-c\|_t}\right)dx\right)^{\frac{1}{t}}.$$

Combination of (1.9.2), (1.9.3), and (1.9.4) yields

$$\left(\int_B |u - u_B|^s \log^\alpha\left(e + \frac{|u-u_B|}{\|u-u_B\|_s}\right)dx\right)^{\frac{1}{s}}$$
$$\le C_3|B|^{\gamma_1 + \frac{q(t-m)}{mpt}}\left(\int_{\sigma B} |v - c|^t \log^\beta\left(e + \frac{|v-c|}{\|v-c\|_t}\right)dx\right)^{q/pt}$$

$$\times \left(\int_B \log^s \left(e + \frac{|u-u_B|}{\|u-u_B\|_s} \right) dx \right)^{\alpha/s^2}. \tag{1.9.5}$$

Now since $x > \log(e+x)$ if $x \geq e$, we have

$$\int_B \log^s \left(e + \frac{|u-u_B|}{\|u-u_B\|_s} \right) dx$$
$$= \int_{\{B: \frac{|u-u_B|}{\|u-u_B\|_s} < e\}} \log^s \left(e + \frac{|u-u_B|}{\|u-u_B\|_s} \right) dx$$
$$+ \int_{\{B: \frac{|u-u_B|}{\|u-u_B\|_s} \geq e\}} \log^s \left(e + \frac{|u-u_B|}{\|u-u_B\|_s} \right) dx \tag{1.9.6}$$
$$\leq C_4 + \int_B \left(\frac{|u-u_B|}{\|u-u_B\|_s} \right)^s dx$$
$$= C_4 + \frac{1}{\|u-u_B\|_s^s} \int_B |u-u_B|^s dx$$
$$= C_5.$$

Substituting (1.9.6) into (1.9.5), we find that

$$\left(\int_B |u-u_B|^s \log^\alpha \left(e + \frac{|u-u_B|}{\|u-u_B\|_s} \right) dx \right)^{\frac{1}{s}}$$
$$\leq C_6 |B|^{\gamma_1 + \frac{q(t-m)}{mpt}} \left(\int_{\sigma B} |v-c|^t \log^\beta \left(e + \frac{|v-c|}{\|v-c\|_t} \right) dx \right)^{q/pt}. \tag{1.9.7}$$

A simple calculation gives

$$\gamma_1 + \frac{q(t-m)}{mpt} = \frac{1}{s} + \frac{1}{n} - \left(\frac{1}{n} + \frac{1}{t} \right) \frac{q}{p} - \frac{\alpha}{s^2}. \tag{1.9.8}$$

Substituting (1.9.8) into (1.9.7), we have

$$\left(\int_B |u-u_B|^s \log^\alpha \left(e + \frac{|u-u_B|}{\|u-u_B\|_s} \right) dx \right)^{\frac{1}{s}}$$
$$\leq C_6 |B|^\gamma \left(\int_{\sigma B} |v-c|^t \log^\beta \left(e + \frac{|v-c|}{\|v-c\|_t} \right) dx \right)^{q/pt}, \tag{1.9.9}$$

where $\gamma = 1/s + 1/n - (1/t + 1/n)q/p - \alpha/s^2$. By the equivalence of the norm $\|f\|_{L^p \log^\alpha L}$ and the functional $[f]_{L^p(\log L)^\alpha(\Omega)}$, we find that (1.9.9) is equivalent to (1.9.1).   ∎

Thus, from Theorem 1.9.2, we have the following estimate for the composition of the homotopy operator T defined in (1.5.1) and the differential operator d:

$$\|T(du)\|_{L^s(\log L)^\alpha(B)} \leq C|B|^\gamma \|v-c\|_{L^t(\log L)^\beta(\sigma B)}^{q/p},$$

provided conditions of Theorem 1.9.2 are satisfied.

1.9.3 $A_r(\Omega)$-weighted case

Parallel to Section 1.5.2, weighted inequalities, we prove the following $A_r(\Omega)$-weighted Hardy–Littlewood inequality with $L^s(\log L)^\alpha$-norm.

Theorem 1.9.3. *In addition to the conditions of Theorem 1.9.2, assume that $w(x) \in A_r(\Omega)$ for some $r > 1$. Then,*

$$\|u - u_B\|_{L^s(\log L)^\alpha(B,w)} \leq C|B|^\gamma \|v - c\|_{L^t(\log L)^\beta(\sigma B, w^{rt/qs})}^{q/p} \qquad (1.9.10)$$

for any ball B, where

$$\gamma = \frac{1}{s} + \frac{1}{n} - \left(\frac{1}{n} + \frac{1}{t}\right)\frac{q}{p} - \frac{\alpha(\lambda-1)}{\lambda s^2}$$

and $\lambda > 1$ is the constant appearing in reverse Hölder inequality for w.

Proof. By the reverse Hölder inequality, there exist constants $\lambda > 1$ and C_1, independent of w, such that

$$\| w \|_{\lambda,B} \leq C_1|B|^{(1-\lambda)/\lambda} \| w \|_{1,B} . \qquad (1.9.11)$$

For any constants $k_i > 0$, $i = 1, 2, 3$, there are constants $m > 0$ and $M > 0$ such that

$$m \log\left(e + \frac{x}{k_1}\right) \leq \log\left(e + \frac{x}{k_2}\right) \leq M \log\left(e + \frac{x}{k_3}\right)$$

for all $x > 0$. Therefore, we obtain

$$m\left(\int_B |u|^t \log^\alpha\left(e + \frac{|u|}{k_1}\right)dx\right)^{\frac{1}{t}}$$
$$\leq \left(\int_B |u|^t \log^\alpha\left(e + \frac{|u|}{k_2}\right)dx\right)^{\frac{1}{t}} \qquad (1.9.12)$$
$$\leq M\left(\int_B |u|^t \log^\alpha\left(e + \frac{|u|}{k_3}\right)dx\right)^{\frac{1}{t}}.$$

By properly selecting constants k_i, we will have different inequalities that we need. Choose $k = \lambda s/(\lambda - 1)$. Applying the Hölder inequality with $1/s = 1/k + (k-s)/ks$ and using (1.9.11) and (1.9.12), we find that

$$\left(\int_B |u - u_B|^s \log^\alpha\left(e + \frac{|u-u_B|}{\|u-u_B\|_s}\right)wdx\right)^{\frac{1}{s}}$$
$$\leq \left(\int_B |u - u_B|^k \log^{\alpha k/s}\left(e + \frac{|u-u_B|}{\|u-u_B\|_s}\right)dx\right)^{\frac{1}{k}}\left(\int_B w^{k/(k-s)}dx\right)^{(k-s)/ks}$$

$$\leq \left(\int_B |u - u_B|^k \log^{\alpha k/s} \left(e + \frac{|u-u_B|}{\|u-u_B\|_s} \right) dx \right)^{\frac{1}{k}} \left(\int_B w^\lambda dx \right)^{1/\lambda s}$$

$$\leq C_2 |B|^{(1-\lambda)/\lambda s} \left(\int_B |u - u_B|^k \log^{\alpha k/s} \left(e + \frac{|u-u_B|}{\|u-u_B\|_k} \right) dx \right)^{\frac{1}{k}} \|w\|_{1,B}^{1/s}.$$

$$(1.9.13)$$

Next, choose $m = qst/(qs + pt(r-1))$. Then, $0 < m < t$. From Theorem 1.9.2, we have

$$\left(\int_B |u - u_B|^k \log^{\frac{\alpha k}{s}} \left(e + \frac{|u-u_B|}{\|u-u_B\|_k} \right) dx \right)^{\frac{1}{k}}$$

$$\leq C_3 |B|^{\gamma'} \left(\int_{\sigma B} |v - c|^m \log^{\beta'} \left(e + \frac{|v-c|}{\|v-c\|_m} \right) dx \right)^{q/pm}, \qquad (1.9.14)$$

where

$$\gamma' = 1/k + 1/n - (1/n + 1/m)q/p - \alpha/ks$$

and the parameter β' will be determined later. Using the Hölder inequality again with $1/m = 1/t + (t-m)/mt$ and (1.9.12), we obtain

$$\left(\int_{\sigma B} |v - c|^m \log^{\beta'} \left(e + \frac{|v-c|}{\|v-c\|_m} \right) dx \right)^{1/m}$$

$$= \left(\int_{\sigma B} \left(|v - c| \log^{\beta'/m} \left(e + \frac{|v-c|}{\|v-c\|_m} \right) w^{p/qs} w^{-p/qs} \right)^m dx \right)^{1/m}$$

$$\leq \left(\int_{\sigma B} |v - c|^t \log^{\beta' t/m} \left(e + \frac{|v-c|}{\|v-c\|_m} \right) w^{pt/qs} dx \right)^{1/t}$$

$$\times \left(\int_{\sigma B} \left(\frac{1}{w} \right)^{mpt/qs(t-m)} dx \right)^{(t-m)/mt} \qquad (1.9.15)$$

$$\leq C_3 \left(\int_{\sigma B} |v - c|^t \log^{\beta' t/m} \left(e + \frac{|v-c|}{\|v-c\|_t} \right) w^{pt/qs} dx \right)^{1/t}$$

$$\times \left(\int_{\sigma B} \left(\frac{1}{w} \right)^{1/(r-1)} dx \right)^{p(r-1)/qs}.$$

Since $w \in A_r$, it follows that

$$\|w\|_{1,B}^{1/s} \left(\int_{\sigma B} \left(\frac{1}{w} \right)^{1/(r-1)} dx \right)^{(r-1)/s}$$

$$\leq |\sigma B|^{r/s} \left(\left(\frac{1}{|\sigma B|} \int_{\sigma B} w dx \right) \left(\frac{1}{|\sigma B|} \int_{\sigma B} \left(\frac{1}{w} \right)^{1/(r-1)} dx \right)^{r-1} \right)^{1/s} \qquad (1.9.16)$$

$$\leq C_4 |B|^{r/s}.$$

Combining (1.9.13), (1.9.14), (1.9.15), and (1.9.16), to conclude that

$$\left(\int_B |u - u_B|^s \log^\alpha \left(e + \frac{|u-u_B|}{||u-u_B||_s} \right) w dx \right)^{\frac{1}{s}}$$
$$\leq C_5 |B|^\gamma \left(\int_{\sigma B} |v - c|^t \log^{\beta' t/m} \left(e + \frac{|v-c|}{||v-c||_m} \right) w^{pt/qs} dx \right)^{q/pt}, \tag{1.9.17}$$

where

$$\gamma = \gamma' + \frac{1-\lambda}{\lambda s} + \frac{r}{s} = \frac{1}{s} + \frac{1}{n} - \left(\frac{1}{n} + \frac{1}{t} \right) \frac{q}{p} - \frac{\alpha(\lambda - 1)}{\lambda s^2}.$$

Selecting $\beta' = \frac{\beta m}{t}$ and using (1.9.12), we have

$$\left(\int_{\sigma B} |v - c|^t \log^{\beta' t/m} \left(e + \frac{|v-c|}{||v-c||_m} \right) w^{pt/qs} dx \right)^{q/pt}$$
$$\leq C_6 \left(\int_{\sigma B} |v - c|^t \log^{\beta} \left(e + \frac{|v-c|}{||v-c||_t} \right) w^{pt/qs} dx \right)^{q/pt}. \tag{1.9.18}$$

Substitution of (1.9.18) into (1.9.17) yields

$$\left(\int_B |u - u_B|^s \log^\alpha \left(e + \frac{|u-u_B|}{||u-u_B||_s} \right) w dx \right)^{\frac{1}{s}}$$
$$\leq C_7 |B|^\gamma \left(\int_{\sigma B} |v - c|^t \log^{\beta} \left(e + \frac{|v-c|}{||v-c||_t} \right) w^{pt/qs} dx \right)^{q/pt}, \tag{1.9.19}$$

which is the same as

$$||u - u_B||_{L^s(\log L)^\alpha(B,w)} \leq C|B|^\gamma ||v - c||^{q/p}_{L^t(\log L)^\beta(\sigma B, w^{pt/qs})}$$

by the equivalence of the norm $||f||_{L^p \log^\alpha L}$ and the functional $[f]_{L^p(\log L)^\alpha(\Omega)}$.

∎

Theorem 1.9.4. *In addition to the conditions of Theorem 1.9.2, assume that* $w \in A_r(\Omega)$ *for some* $r > 1$. *Then,*

$$||u - u_B||_{L^s(\log L)^\alpha(B,w^\eta)} \leq C|B|^\gamma ||v - c||^{q/p}_{L^t(\log L)^\beta(\sigma B, w^{\eta pt/qs})} \tag{1.9.20}$$

for any ball B, where $0 < \eta \leq 1$, $\gamma = \frac{1}{s} + \frac{1}{n} - \left(\frac{1}{n} + \frac{1}{t} \right) \frac{q}{p} - \frac{\alpha(\lambda-1)}{\lambda s^2}$, *and* $\lambda > 1$ *is the constant appearing in reverse Hölder inequality for* w.

Proof. From [59], we know that if $w \in A_r(E)$ and $0 < \eta \leq 1$, then $w^\eta \in A_r(E)$. Therefore, from Theorem 1.9.3, we obtain (1.9.20) immediately. ∎

1.9.4 Global $L^s(\log L)^\alpha$-norm inequality

Finally, we prove the following global Hardy–Littlewood inequality with $L^s(\log L)^\alpha$-norm in δ-John domains.

Theorem 1.9.5. *Let $u \in D'(\Omega, \wedge^0)$ and $v \in D'(\Omega, \wedge^2)$ be solutions to the conjugate A-harmonic equation (1.2.6) in a δ-John domain $\Omega \subset \mathbf{R}^n$. Let $q \leq p$, $v - c \in L^t(\log L)^\beta(\Omega, \wedge^2)$, and $s > 0$ and $t > 0$ satisfy $1/s + 1/n - (1/t + 1/n)q/p - \alpha/s^2 = 0$. Then, there exists a constant C, independent of u and v, such that*

$$||u - u_{B_0}||_{L^s(\log L)^\alpha(\Omega)} \leq C||v - c||_{L^t(\log L)^\beta(\Omega)}^{q/p}, \qquad (1.9.21)$$

where $0 < \alpha < s$ and $\beta > 0$ are constants, c is any form in $W_{loc}^{1,q}(\Omega, \wedge)$ with $d^\star c = 0$ and $B_0 \subset \Omega$ is a fixed ball.

Proof. Applying the Hölder inequality with $1/s = 1/(s^2/(s-\alpha)) + 1/(s^2/\alpha)$, we obtain

$$||u - u_{B_0}||_{L^s(\log L)^\alpha(\Omega)}$$

$$= \left(\int_\Omega |u - u_{B_0}|^s \log^\alpha \left(e + \frac{|u - u_{B_0}|}{||u - u_{B_0}||_{s,\Omega}} \right) dx \right)^{\frac{1}{s}}$$

$$\leq \left(\int_\Omega |u - u_{B_0}|^{s^2/(s-\alpha)} dx \right)^{(s-\alpha)/s^2} \qquad (1.9.22)$$

$$\times \left(\int_\Omega \log^s \left(e + \frac{|u - u_{B_0}|}{||u - u_{B_0}||_{s,\Omega}} \right) dx \right)^{\alpha/s^2}$$

$$= ||u - u_{B_0}||_{s^2/(s-\alpha),\Omega} \left(\int_\Omega \log^s \left(e + \frac{|u - u_{B_0}|}{||u - u_{B_0}||_{s,\Omega}} \right) dx \right)^{\alpha/s^2}.$$

Since $(s - \alpha)/s^2 + 1/n - (1/m + 1/n)q/p = 0$, using Theorem 1.5.4, we have

$$||u - u_{B_0}||_{s^2/(s-\alpha),\Omega} \leq C_1||v - c||_{t,\Omega}^{q/p}. \qquad (1.9.23)$$

Combination of (1.9.22) and (1.9.23) yields

$$||u - u_{B_0}||_{L^s(\log L)^\alpha(\Omega)}$$

$$\leq C_1||v - c||_{t,\Omega}^{q/p} \cdot \left(\int_\Omega \log^s \left(e + \frac{|u - u_{B_0}|}{||u - u_{B_0}||_{s,\Omega}} \right) dx \right)^{\alpha/s^2}. \qquad (1.9.24)$$

Similar to (1.9.6), we can prove that

$$\int_\Omega \log^s \left(e + \frac{|u - u_{B_0}|}{\|u - u_{B_0}\|_{s,\Omega}} \right) dx \leq C_2. \tag{1.9.25}$$

Substituting (1.9.25) into (1.9.24) and using $\log^\beta \left(e + \frac{|v-c|}{\|v-c\|_{t,\Omega}} \right) > 1$, we find that

$$\|u - u_{B_0}\|_{L^s(\log L)^\alpha(\Omega)}$$

$$\leq C_3 \|v - c\|_{t,\Omega}^{q/p}$$

$$\leq C_3 \left(\int_\Omega |v - c|^t \log^\beta \left(e + \frac{|v-c|}{\|v-c\|_{t,\Omega}} \right) dx \right)^{q/pt}$$

$$\leq C_3 \|v - c\|_{L^t(\log L)^\beta(\Omega)}^{q/p}. \qquad \blacksquare$$

Notes to Chapter 1. In this chapter, we have focused our attention on Hardy–Littlewood inequalities on differential forms satisfying the conjugate A-harmonic equation. However, many different versions of the Hardy–Littlewood inequality have been obtained during the recent years. For example, C. Nolder proved Hardy–Littlewood inequalities for solutions of elliptic equations in divergence form in [155] and for quasiregular maps on Carnot groups in [153]. A Carnot group is a connected, simply connected nilpotent Lie group G of topological $\dim G = N \geq 2$ equipped with a graded Lie algebra $\mathcal{G} = V_1 \oplus V_2 \oplus \cdots \oplus V_r$ so that $[V_1, V_i] = V_{i+1}$ for $i = 1, 2, \ldots, r - 1$ and $[V_1, V_r] = 0$. This defines an r-step Carnot group. B. Osikiewiczand and A. Tonge developed an interpolation approach to Hardy–Littlewood inequalities for forms of operators on sequence spaces [158]. See [195–200, 187–190, 18, 148, 67–74] for further results about the Hardy–Littlewood inequality.

Chapter 2
Norm comparison theorems

In the previous chapter, we have discussed various versions of the Hardy–Littlewood inequality for a pair of solutions u and v of the conjugate A-harmonic equation. The purpose of this chapter is to present some norm comparison inequalities for differential forms satisfying the conjugate A-harmonic equations, which have been recently established in [104]. Since the proofs display a general method to obtain L^p-estimates, we include most of them in this chapter. Also, we always assume that $1 < p < \infty$ and $p^{-1} + q^{-1} = 1$ throughout this chapter.

2.1 Introduction

In the first chapter, we have introduced the following conjugate A-harmonic equation:

$$A(x, du) = d^\star v. \tag{2.1.1}$$

In this chapter, we study a more general type of the conjugate A-harmonic equation: the nonhomogeneous conjugate A-harmonic equation

$$A(x, g + du) = h + d^\star v \tag{2.1.2}$$

for differential forms, where $g, h \in D'(\Omega, \wedge^l)$ and $A : \Omega \times \wedge^l(\mathbf{R}^n) \to \wedge^l(\mathbf{R}^n)$ satisfies the following conditions:

$$|A(x, \xi)| \le a|\xi|^{p-1} \quad \text{and} \quad < A(x, \xi), \xi > \ge |\xi|^p \tag{2.1.3}$$

for almost every $x \in \Omega$ and all $\xi \in \wedge^l(\mathbf{R}^n)$. Here $a > 0$ is a constant and $1 < p < \infty$ is a fixed exponent associated with (2.1.2). Some results related to equation (2.1.2) have been obtained in [105]. In this chapter, we first present some local norm inequalities in Section 2.2. Then, we obtain some $A_r(\Omega)$-weighted estimates and the global norm inequalities for solutions of

R.P. Agarwal et al., *Inequalities for Differential Forms*,
DOI 10.1007/978-0-387-68417-8_2, © Springer Science+Business Media, LLC 2009

the nonhomogeneous conjugate A-harmonic equation in Sections 2.3 and 2.4, respectively. Finally, as applications of the results discussed in Sections 2.2, 2.3, and 2.4, we prove the global Sobolev–Poincaré-type imbedding inequality and derive some global L^p-estimates for the gradient operator ∇ and the homotopy operator T from the Banach space $L^s(D, \wedge^l)$ to the Sobolev space $W^{1,s}(D, \wedge^{l-1}), l = 1, 2, \ldots, n$. Some of the results presented in this chapter have nice symmetric properties. These results enrich the L^p-theory of differential forms and can be used to estimate the integrals of differential forms and to study the integrability of differential forms, also to explore the properties of the operators ∇ and T.

2.2 The local unweighted estimates

We develop some unweighted norm inequalities for solutions of conjugate A-harmonic equations (2.1.1) and (2.1.2). These inequalities have rich symmetric properties and play important role in this chapter.

2.2.1 Basic L^p-inequalities

In this section, we present some local norm inequalities which will be extended into the weighted cases and the global cases in next two sections. We first discuss the following norm comparison theorem which describes the relationship between the norm of du and the norm of $d^\star v$.

Theorem 2.2.1. *Let u and v be a pair of solutions to the nonhomogeneous conjugate A-harmonic equation (2.1.2) in a domain $\Omega \subset \mathbf{R}^n$. If $g \in L^p(B, \wedge^l)$ and $h \in L^q(B, \wedge^l)$, then $du \in L^p(B, \wedge^l)$ if and only if $d^\star v \in L^q(B, \wedge^l)$. Moreover, there exist constants C_1, C_2, independent of u and v, such that*

$$\|d^\star v\|_{q,B}^q \leq C_1(\|h\|_{q,B}^q + \|g\|_{p,B}^p + \|du\|_{p,B}^p), \qquad (2.2.1)$$

$$\|du\|_{p,B}^p \leq C_2(\|h\|_{q,B}^q + \|g\|_{p,B}^p + \|d^\star v\|_{q,B}^q) \qquad (2.2.2)$$

for all balls B with $B \subset \Omega \subset \mathbf{R}^n$. Here $p^{-1} + q^{-1} = 1$.

Proof. We only need to prove (2.2.1) and (2.2.2). From equation (2.1.2) and condition (2.1.3), we have

$$|h + d^\star v| = |A(x, g + du)| \leq a|g + du|^{p-1}. \qquad (2.2.3)$$

Hence, we obtain

$$|d^\star v| = |d^\star v + h - h| \le |h| + |d^\star v + h| \le |h| + a|g + du|^{p-1}. \qquad (2.2.4)$$

Using the elementary inequality

$$\left| \sum_{i=1}^{N} t_i \right|^s \le N^{s-1} \sum_{i=1}^{N} |t_i|^s, \qquad (2.2.5)$$

we find that

$$
\begin{aligned}
|d^\star v|^q &\le (|h| + a|g + du|^{p-1})^q \\
&\le 2^{q-1}(|h|^q + a^q|g + du|^{(p-1)q}) \\
&= 2^{q-1}(|h|^q + a^q|g + du|^p) \\
&\le 2^{q-1}(|h|^q + 2^{p-1}a^q(|g|^p + |du|^p)) \\
&\le C_1(|h|^q + |g|^p + |du|^p).
\end{aligned}
\qquad (2.2.6)
$$

Integrating the above inequality over B, we obtain

$$\|d^\star v\|_{q,B}^q \le C_1(\|h\|_{q,B}^q + \|g\|_{p,B}^p + \|du\|_{p,B}^p).$$

This completes the proof of inequality (2.2.1). From (2.1.3) and Schwartz inequality, we have

$$
\begin{aligned}
|g + du|^p &\le\, < A(x, g + du), g + du > \\
&=< h + d^\star v, g + du > \\
&\le |h + d^\star v| \cdot |g + du|.
\end{aligned}
$$

Therefore, we obtain

$$|g + du|^{p-1} \le |h + d^\star v| \le |h| + |d^\star v|. \qquad (2.2.7)$$

Hence,

$$|g + du|^p = |g + du|^{q(p-1)} \le (|h| + |d^\star v|)^q.$$

Using the elementary inequality (2.2.5) again, we find that

$$
\begin{aligned}
|du|^p &= |du + g - g|^p \\
&\le 2^{p-1}(|du + g|^p + |g|^p) \\
&\le 2^{p-1}(|g|^p + (|h| + |d^\star v|)^q) \\
&\le 2^{p-1}(|g|^p + 2^{q-1}(|h|^q + |d^\star v|^q)) \\
&\le C_2(|g|^p + |h|^q + |d^\star v|^q).
\end{aligned}
\qquad (2.2.8)
$$

Integrating inequality (2.2.8) over B, we obtain (2.2.2). ∎

Note that (2.2.1) and (2.2.2) can be used to estimate the integrals of differential forms and to study the integrability of differential forms. From the proof of Theorem 2.2.1, the following corollary is immediate.

Corollary 2.2.2. *Let u and v be a pair of solutions to the nonhomogeneous conjugate A-harmonic equation (2.1.2) in a domain $\Omega \subset \mathbf{R}^n$. Then,*

$$\|h + d^\star v\|_{q,B}^q \leq C_1 \|g + du\|_{p,B}^p,$$

$$\|g + du\|_{p,B}^p \leq C_2 \|h + d^\star v\|_{q,B}^q$$

for all balls B with $B \subset \Omega \subset \mathbf{R}^n$.

2.2.2 Special cases

Applying Theorem 2.2.1 with $g = 0$ and $h = 0$, we obtain the following corollary immediately.

Corollary 2.2.3. *Let u and v be a pair of solutions to the conjugate A-harmonic equation (2.1.1) in a domain $\Omega \subset \mathbf{R}^n$. Then, $du \in L^p(B, \wedge^l)$ if and only if $d^\star v \in L^q(B, \wedge^l)$. Moreover, there exist constants C_1, C_2, independent of u and v, such that*

$$C_1 \|du\|_{p,B}^p \leq \|d^\star v\|_{q,B}^q \leq C_2 \|du\|_{p,B}^p \tag{2.2.9}$$

for all balls B with $B \subset \Omega \subset \mathbf{R}^n$.

Theorem 2.2.4. *Let u and v be a pair of solutions to the conjugate A-harmonic equation (2.1.1) in a domain $\Omega \subset \mathbf{R}^n$. Then, there exists a constant C, independent of u and v, such that*

$$\|d^\star v\|_{q,B}^q \leq C \, diam(B)^{-p} \|u - c\|_{p,\sigma B}^p \tag{2.2.10}$$

for all balls B with $\sigma B \subset \Omega \subset \mathbf{R}^n$. Here c is any closed form and σ is a constant with $\sigma > 1$.

Note that (2.2.10) can be written as

$$\|d^\star v\|_{q,B} \leq C \, diam(B)^{-p/q} \|u - c\|_{p,\sigma B}^{p/q}. \tag{2.2.11}$$

Proof. Since u and v are a pair of solutions to the conjugate A-harmonic equation

$$A(x, du) = d^\star v,$$

it follows that u is a solution to the A-harmonic equation

$$d^\star A(x, du) = 0.$$

Hence, the Caccioppoli inequality is applicable to u. Using Theorem 2.4.1 with $\alpha = 1$ in [180] and Theorem 2.2.1 with $g = 0$ and $h = 0$, we obtain

$$\|d^\star v\|_{q,B}^q \leq C\|du\|_{p,B}^p \leq C diam(B)^{-p}\|u-c\|_{p,\sigma B}^p. \qquad \blacksquare$$

Theorem 2.2.5. *Let u and v be a pair of solutions to the conjugate A-harmonic equation (2.1.1) in a domain $\Omega \subset \mathbf{R}^n$. Then, there exists a constant C, independent of u and v, such that*

$$\|u - u_B\|_{p,B}^p \leq C diam(B)^p\|d^\star v\|_{q,\sigma B}^q \qquad (2.2.12)$$

for all balls B with $\sigma B \subset \Omega \subset \mathbf{R}^n$, where σ is a constant with $\sigma > 1$.
Proof. Similar to the case in Theorem 2.2.4, using Poincaré inequality (Theorem 2.12 with $w(x) = 1$ in [201]) to u, and Theorem 2.2.1 with $g = 0$ and $h = 0$, we obtain

$$\|u - u_B\|_{p,B}^p \leq C diam(B)^p\|du\|_{p,\sigma B}^p \leq C diam(B)^p\|d^\star v\|_{q,\sigma B}^q. \qquad \blacksquare$$

Combining Theorems 2.2.4 and 2.2.5, we have the following norm comparison inequality which has nice symmetric properties. It can be used to estimate the norm of $d^\star v$ in terms of u or $u - c$.

Theorem 2.2.6. *Let u and v be a pair of solutions to the conjugate A-harmonic equation (2.1.1) in a domain $\Omega \subset \mathbf{R}^n$. Then, there exist constants C_1, C_2, independent of u and v, such that*

$$C_1 diam(B)^{-p}\|u - u_B\|_{p,B}^p \leq \|d^\star v\|_{q,\sigma_1 B}^q \leq C_2 diam(B)^{-p}\|u - c\|_{p,\sigma_2 B}^p$$

for all balls B with $\sigma_2 B \subset \Omega \subset \mathbf{R}^n$. Here c is any closed form and p, q, σ_1, σ_2 are constants with $\sigma_2 > \sigma_1 > 1$.

2.3 The local weighted estimates

Here we generalize the inequalities established in the previous section to the $A_r(\Omega)$-weighted versions. We begin with L^s-estimates for $d^\star v$.

2.3.1 L^s-estimates for $d^\star v$

We first discuss L^s-estimates for $d^\star v$ in terms of g, h and du that appear in equation (2.1.2).

Theorem 2.3.1. *Let u and v be a pair of solutions to the nonhomogeneous conjugate A-harmonic equation (2.1.2) in a domain $\Omega \subset \mathbf{R}^n$. Assume that*

$w \in A_r(\Omega)$ *for some* $r > 1$. *Then, there exists a constant* C, *independent of* u *and* v, *such that*

$$\|d^\star v\|_{s,B,w^\alpha}$$

$$\leq C|B|^{\alpha r/s}(\|h\|_{t,B,w^{\alpha t/s}} + \||g|^{p/q}\|_{t,B,w^{\alpha t/s}} + \||du|^{p/q}\|_{t,B,w^{\alpha t/s}}) \quad (2.3.1)$$

for all balls B *with* $B \subset \Omega \subset \mathbf{R}^n$. *Here* α *is any positive constant with* $1 > \alpha r$, $s = (1 - \alpha)q$, *and* $t = s/(1 - \alpha r) = qs/(s - \alpha q(r - 1))$.

Note that (2.3.1) can be written in the following symmetric form:

$$|B|^{-1/s}\|d^\star v\|_{s,B,w^\alpha}$$

$$\leq C|B|^{-1/t}(\|h\|_{t,B,w^{\alpha t/s}} + \||g|^{p/q}\|_{t,B,w^{\alpha t/s}} + \||du|^{p/q}\|_{t,B,w^{\alpha t/s}}).$$

$$(2.3.1)$$

Proof. Since $s = (1 - \alpha)q < q$, using the Hölder inequality, we have

$$\|d^\star v\|_{s,B,w^\alpha} = \left(\int_B \left(|d^\star v|w^{\alpha/s}\right)^s dx\right)^{1/s}$$

$$\leq \left(\int_B |d^\star v|^q dx\right)^{1/q} \left(\int_B \left(w^{\alpha/s}\right)^{qs/(q-s)} dx\right)^{(q-s)/qs} \quad (2.3.2)$$

$$\leq \|d^\star v\|_{q,B} \left(\int_B w dx\right)^{\alpha/s}.$$

Applying the elementary inequality $|\sum_{i=1}^N t_i|^\tau \leq N^{\tau-1}\sum_{i=1}^N |t_i|^\tau$ and Theorem 2.2.1, we obtain

$$\|d^\star v\|_{q,B} \leq C_1 \left(\|h\|_{q,B} + \|g\|_{p,B}^{p/q} + \|du\|_{p,B}^{p/q}\right). \quad (2.3.3)$$

Since $t = qs/(s - \alpha q(r - 1)) > q$, using the Hölder inequality again, we find that

$$\|h\|_{q,B} = \left(\int_B \left(|h|w^{\alpha/s}w^{-\alpha/s}\right)^q dx\right)^{1/q}$$

$$\leq \left(\int_B |h|^t w^{\alpha t/s} dx\right)^{1/t} \left(\int_B \left(\frac{1}{w}\right)^{\alpha qt/s(t-q)} dx\right)^{(t-q)/qt} \quad (2.3.4)$$

$$= \|h\|_{t,B,w^{\alpha t/s}} \left(\int_B \left(\frac{1}{w}\right)^{1/(r-1)} dx\right)^{\alpha(r-1)/s}.$$

Now, choose

$$k = \frac{sp + \alpha pt(r - 1)}{s} = p + \frac{\alpha pt(r - 1)}{s},$$

so that $k > p$. Once again using the Hölder inequality, we have

$$\|g\|_{p,B} = \left(\int_B |g|^p w^{\alpha t/ks} w^{-\alpha t/ks} dx\right)^{1/p}$$

$$\leq \left(\int_B |g|^k w^{\alpha t/s} dx\right)^{1/k} \left(\int_B \left(\tfrac{1}{w}\right)^{\alpha pt/s(k-p)} dx\right)^{(k-p)/kp} \qquad (2.3.5)$$

$$\leq \|g\|_{k,B,w^{\alpha t/s}} \left(\int_B \left(\tfrac{1}{w}\right)^{1/(r-1)} dx\right)^{(k-p)/kp}.$$

After a simple computation, we find that

$$\frac{k-p}{kp} = \frac{\alpha(r-1)}{s} \cdot \frac{st}{ps + \alpha pt(r-1)} = \frac{\alpha q(r-1)}{ps},$$

and hence

$$\|g\|_{p,B}^{p/q} \leq \|g\|_{k,B,w^{\alpha t/s}}^{p/q} \cdot \left(\int_B \left(\frac{1}{w}\right)^{1/(r-1)} dx\right)^{\alpha(r-1)/s}. \qquad (2.3.6)$$

Note that

$$\|g\|_{k,B,w^{\alpha t/s}}^{p/q} = \left(\int_B |g|^k w^{\alpha t/s} dx\right)^{p/kq}$$

$$= \left(\int_B |g|^{(sp+\alpha pt(r-1))/s} w^{\alpha t/s} dx\right)^{ps/(pqs+\alpha pqt(r-1))}$$

$$= \left(\int_B |g|^{pt/q} w^{\alpha t/s} dx\right)^{1/t} \qquad (2.3.7)$$

$$= \||g|^{p/q}\|_{t,B,w^{\alpha t/s}}.$$

Combination of (2.3.6) and (2.3.7) yields

$$\|g\|_{p,B}^{p/q} \leq \||g|^{p/q}\|_{t,B,w^{\alpha t/s}} \cdot \left(\int_B \left(\frac{1}{w}\right)^{1/(r-1)} dx\right)^{\alpha(r-1)/s}. \qquad (2.3.8)$$

Using a similar method, we have

$$\|du\|_{p,B}^{p/q} \leq \||du|^{p/q}\|_{t,B,w^{\alpha t/s}} \cdot \left(\int_B \left(\frac{1}{w}\right)^{1/(r-1)} dx\right)^{\alpha(r-1)/s}. \qquad (2.3.9)$$

Combination of (2.3.2) and (2.3.3) gives

$$\|d^\star v\|_{s,B,w^\alpha} \leq C_1 \left(\|h\|_{q,B} + \|g\|_{p,B}^{p/q} + \|du\|_{p,B}^{p/q}\right) \left(\int_B w dx\right)^{\alpha/s}. \qquad (2.3.10)$$

Substituting $(2.3.4), (2.3.8)$, and $(2.3.9)$ into $(2.3.10)$, we find that

$$\|d^\star v\|_{s,B,w^\alpha} \leq C_1 \left(\|h\|_{t,B,w^{\alpha t/s}} + \||g|^{p/q}\|_{t,B,w^{\alpha t/s}} + \||du|^{p/q}\|_{t,B,w^{\alpha t/s}} \right)$$

$$\times \left(\int_B w dx \right)^{\alpha/s} \left(\int_B \left(\tfrac{1}{w} \right)^{1/(r-1)} dx \right)^{\alpha(r-1)/s}.$$

$$(2.3.11)$$

Since $w(x) \in A_r(\Omega)$, it follows that

$$\left(\int_B w dx \right)^{\alpha/s} \left(\int_B \left(\tfrac{1}{w} \right)^{1/(r-1)} dx \right)^{\alpha(r-1)/s}$$

$$= |B|^{\alpha r/s} \left(\tfrac{1}{|B|} \int_B w dx \right)^{\alpha/s} \left(\tfrac{1}{|B|} \int_B \left(\tfrac{1}{w} \right)^{1/(r-1)} dx \right)^{\alpha(r-1)/s} \qquad (2.3.12)$$

$$\leq C_2 |B|^{\alpha r/s}.$$

Finally, using $(2.3.12)$ and $(2.3.11)$, we obtain

$$\|d^\star v\|_{s,B,w^\alpha}$$

$$\leq C_3 |B|^{\alpha r/s} (\|h\|_{t,B,w^{\alpha t/s}} + \||g|^{p/q}\|_{t,B,w^{\alpha t/s}} + \||du|^{p/q}\|_{t,B,w^{\alpha t/s}}). \qquad \blacksquare$$

2.3.2 L^s-estimates for du

Similar to the proof of Theorem 2.3.1, we have the following weighted L^s-estimate for du.

Theorem 2.3.2. *Let u and v be a pair of solutions to the nonhomogeneous conjugate A-harmonic equation (2.1.2) in a domain $\Omega \subset \mathbf{R}^n$. Assume that $w \in A_r(\Omega)$ for some $r > 1$. Then, there exists a constant C, independent of u and v, such that*

$$\|du\|_{s,B,w^\alpha}$$

$$\leq C|B|^{\alpha r/s} (\|g\|_{t,B,w^{\alpha t/s}} + \||h|^{q/p}\|_{t,B,w^{\alpha t/s}} + \||d^\star v|^{q/p}\|_{t,B,w^{\alpha t/s}})$$

$$(2.3.13)$$

for all balls B with $B \subset \Omega \subset \mathbf{R}^n$. Here α is any positive constant with $1 > \alpha r$, $s = (1-\alpha)p$, and $t = s/(1-\alpha r) = ps/(s - \alpha p(r-1))$.

It is easy to see that inequality $(2.3.13)$ is equivalent to

$$|B|^{-1/s}\|du\|_{s,B,w^\alpha}$$

$$\leq C|B|^{-1/t} (\|g\|_{t,B,w^{\alpha t/s}} + \||h|^{q/p}\|_{t,B,w^{\alpha t/s}} + \||d^\star v|^{q/p}\|_{t,B,w^{\alpha t/s}}).$$

$$(2.3.13)'$$

2.3.3 The norm comparison between d^* and d

Theorem 2.3.3. *Let u and v be a pair of solutions to the conjugate A-harmonic equation (2.1.1) in a domain $\Omega \subset \mathbf{R}^n$. Assume that $w \in A_r(\Omega)$ for some $r > 1$. Then, for all balls B with $B \subset \Omega \subset \mathbf{R}^n$ and any positive constant α with $1 > \alpha r$, there exist constants C_1, C_2, independent of u and v, such that*

$$\|d^\star v\|_{s,B,w^\alpha}^q \leq C_1 |B|^{\alpha q r/s} \|du\|_{pt/q,B,w^{\alpha t/s}}^p \tag{2.3.14}$$

for $s = (1-\alpha)q, t = s/(1-\alpha r) = qs/(s - \alpha q(r-1))$; and

$$\|du\|_{s,B,w^\alpha}^p \leq C_2 |B|^{\alpha p r/s} \|d^\star v\|_{qt/p,B,w^{\alpha t/s}}^q \tag{2.3.15}$$

for $s = (1-\alpha)p, t = s/(1-\alpha r) = ps/(s - \alpha p(r-1))$.

Proof. Applying Theorem 2.3.1 with $g = 0$ and $h = 0$, we obtain

$$\|d^\star v\|_{s,B,w^\alpha} \leq C_1 |B|^{\alpha r/s} \||du|^{p/q}\|_{t,B,w^{\alpha t/s}}. \tag{2.3.16}$$

Note that

$$\||du|^{p/q}\|_{t,B,w^{\alpha t/s}} = \|du\|_{pt/q,B,w^{\alpha t/s}}^{p/q}. \tag{2.3.17}$$

Combination of (2.3.16) and (2.3.17) yields

$$\|d^\star v\|_{s,B,w^\alpha}^q \leq C_1 |B|^{\alpha q r/s} \|du\|_{pt/q,B,w^{\alpha t/s}}^p.$$

Similarly, using Theorem 2.3.2 with $g = 0$ and $h = 0$, we have

$$\|du\|_{s,B,w^\alpha}^p \leq C_2 |B|^{\alpha p r/s} \|d^\star v\|_{qt/p,B,w^{\alpha t/s}}^q. \qquad \blacksquare$$

Theorem 2.3.4. *Let u and v be a pair of solutions to the conjugate A-harmonic equation (2.1.1) in a domain $\Omega \subset \mathbf{R}^n$. Assume that $w \in A_r(\Omega)$ for some $r > 1$. Then, for all balls B with $B \subset \Omega \subset \mathbf{R}^n$ and any positive constant α with $1 > \alpha r$, there exist constants C_1, C_2, independent of u and v, such that*

$$\|u - u_B\|_{s,B,w^\alpha} \leq C_1 diam(B) |B|^{\alpha r/s} \||d^\star v|^{q/p}\|_{t,\sigma B,w^{\alpha t/s}} \tag{2.3.18}$$

for $s = (1-\alpha)p, t = s/(1-\alpha r) = ps/(s - \alpha p(r-1))$; and

$$\|d^\star v\|_{s,B,w^\alpha} \leq C_2 diam(B)^{-p/q} |B|^{\alpha r/s} \||u - c|^{p/q}\|_{t,\sigma B,w^{\alpha t/s}} \tag{2.3.19}$$

for $s = (1-\alpha)q, t = s/(1-\alpha r) = qs/(s - \alpha q(r-1))$. Here $\sigma > 1$ is a constant.

Proof. Applying the Hölder inequality with p and $s = (1-\alpha)p$ and Theorem 2.2.5, we find that

$$\|u - u_B\|_{s,B,w^\alpha} = \left(\int_B \left(|u - u_B| w^{\alpha/s} \right)^s dx \right)^{1/s}$$

$$\leq \left(\int_B |u - u_B|^p dx \right)^{1/p} \left(\int_B \left(w^{\alpha/s} \right)^{ps/(p-s)} dx \right)^{(p-s)/ps}$$

$$\leq \|u - u_B\|_{p,B} \left(\int_B w dx \right)^{\alpha/s}$$

$$\leq C_1 diam(B) \|d^\star v\|_{q,\sigma B}^{q/p} \left(\int_B w dx \right)^{\alpha/s}.$$

$$(2.3.20)$$

Let

$$k = \frac{sq + \alpha qt(r-1)}{s} = q + \frac{\alpha qt(r-1)}{s},$$

so that $k > q$. Using the Hölder inequality again, it follows that

$$\|d^\star v\|_{q,\sigma B}$$

$$= \left(\int_{\sigma B} |d^\star v|^q w^{\alpha t/ks} w^{-\alpha t/ks} dx \right)^{1/q}$$

$$\leq \left(\int_{\sigma B} |d^\star v|^k w^{\alpha t/s} dx \right)^{1/k} \left(\int_{\sigma B} \left(\frac{1}{w} \right)^{\alpha qt/s(k-q)} dx \right)^{(k-q)/kq} \qquad (2.3.21)$$

$$\leq \|d^\star v\|_{k,\sigma B,w^{\alpha t/s}} \left(\int_B \left(\frac{1}{w} \right)^{1/(r-1)} dx \right)^{(k-q)/kq}.$$

Note that

$$\frac{k-q}{kq} = \frac{\alpha(r-1)}{s} \cdot \frac{st}{q + \alpha qt(r-1)} = \frac{\alpha p(r-1)}{qs}.$$

Thus,

$$\|d^\star v\|_{q,\sigma B}^{q/p} \leq \|d^\star v\|_{k,\sigma B,w^{\alpha t/s}}^{q/p} \cdot \left(\int_{\sigma B} \left(\frac{1}{w} \right)^{1/(r-1)} dx \right)^{\alpha(r-1)/s}. \qquad (2.3.22)$$

Also, we have

$$\|d^\star v\|_{k,\sigma B,w^{\alpha t/s}}^{q/p} = \| |d^\star v|^{q/p} \|_{t,\sigma B,w^{\alpha t/s}}. \qquad (2.3.23)$$

Combination of (2.3.20), (2.3.22), and (2.3.23) gives

$$\|u - u_B\|_{s,B,w^\alpha}$$

$$\leq C_1 diam(B) \| |d^\star v|^{q/p} \|_{t,\sigma B,w^{\alpha t/s}} \cdot \left(\int_B w dx \right)^{\alpha/s}$$

$$\times \left(\int_{\sigma B} \left(\frac{1}{w} \right)^{1/(r-1)} dx \right)^{\alpha(r-1)/s}. \qquad (2.3.24)$$

Now, using the condition $w(x) \in A_r(\Omega)$, we find that

$$\left(\int_B w dx\right)^{\alpha/s} \left(\int_{\sigma B} \left(\frac{1}{w}\right)^{1/(r-1)} dx\right)^{\alpha(r-1)/s}$$

$$\leq |\sigma B|^{\alpha r/s} \left(\frac{1}{|\sigma B|} \int_{\sigma B} w dx\right)^{\alpha/s} \left(\frac{1}{|\sigma B|} \int_{\sigma B} \left(\frac{1}{w}\right)^{1/(r-1)} dx\right)^{\alpha(r-1)/s}$$

$$\leq C_2 |B|^{\alpha r/s}.$$

(2.3.25)

Finally, substituting (2.3.25) into (2.3.24) we obtain

$$\|u - u_B\|_{s,B,w^\alpha} \leq C_3 diam(B)|B|^{\alpha r/s} \||d^\star v|^{q/p}\|_{t,\sigma B, w^{\alpha t/s}}.$$

This completes the proof of (2.3.18). The proof of (2.3.19) is similar to that of (2.3.18). ∎

Note. In this section, we have only proved the $A_r(\Omega)$-weighted norm comparison theorems. Using a similar technique, we can obtain the local comparison theorems with other kinds of weights discussed in Section 1.4, including two-weight, etc.

2.4 The global estimates

We have discussed the weighted local estimates for $\|d^\star v\|_{s,B,w^\alpha}$ and $\|du\|_{s,B,w^\alpha}$ in the previous section. In this section, we prove some global norm comparison theorems using the local results.

2.4.1 Global estimates for $d^\star v$

Using the local weighted estimate developed above, we prove the following global estimate for $d^\star v$.

Theorem 2.4.1. *Let u and v be a pair of solutions to the nonhomogeneous conjugate A-harmonic equation (2.1.2) in a bounded domain $\Omega \subset \mathbf{R}^n$. Assume that $w \in A_r(\Omega)$ for some $r > 1$. Then, there exists a constant C, independent of u and v, such that*

$$\|d^\star v\|_{s,\Omega,w^\alpha} \leq C(\|h\|_{t,\Omega,w^{\alpha t/s}} + \||g|^{p/q}\|_{t,\Omega,w^{\alpha t/s}} + \||du|^{p/q}\|_{t,\Omega,w^{\alpha t/s}}), \quad (2.4.1)$$

where α is any positive constant with $1 > \alpha r$, $1 \leq s = (1-\alpha)q$, and $t = s/(1-\alpha r) = qs/(s - \alpha q(r-1))$.

Proof. Applying Theorem 2.3.1 and the covering lemma (Theorem 1.5.3), we have

$$\|d^\star v\|_{s,\Omega,w^\alpha}$$

$$= \left(\int_\Omega |d^\star v|^s w^\alpha dx\right)^{1/s}$$

$$\leq \sum_{B\in\mathcal{V}} \left(\int_B |d^\star v|^s w^\alpha dx\right)^{1/s}$$

$$\leq C_1 \sum_{B\in\mathcal{V}} |B|^{\frac{\alpha r}{s}} \left(\|h\|_{t,B,w^{\frac{\alpha t}{s}}} + \||g|^{\frac{p}{q}}\|_{t,B,w^{\frac{\alpha t}{s}}} + \||du|^{\frac{p}{q}}\|_{t,B,w^{\frac{\alpha t}{s}}}\right)$$

$$\leq C_1 \sum_{B\in\mathcal{V}} |\Omega|^{\frac{\alpha r}{s}} \left(\|h\|_{t,\Omega,w^{\frac{\alpha t}{s}}} + \||g|^{\frac{p}{q}}\|_{t,\Omega,w^{\frac{\alpha t}{s}}} + \||du|^{\frac{p}{q}}\|_{t,\Omega,w^{\frac{\alpha t}{s}}}\right)$$

$$\leq C_2 \left(\|h\|_{t,\Omega,w^{\alpha t/s}} + \||g|^{p/q}\|_{t,\Omega,w^{\frac{\alpha t}{s}}} + \||du|^{p/q}\|_{t,\Omega,w^{\frac{\alpha t}{s}}}\right) \cdot N$$

$$\leq C_3 \left(\|h\|_{t,\Omega,w^{\alpha t/s}} + \||g|^{p/q}\|_{t,\Omega,w^{\alpha t/s}} + \||du|^{p/q}\|_{t,\Omega,w^{\alpha t/s}}\right)$$

since Ω is bounded.   ∎

If we put $g = 0$ and $h = 0$ in Theorem 2.4.1, we obtain the following global norm comparison result for $d^\star v$ and du.

Corollary 2.4.2. *Let u and v be a pair of solutions to the conjugate A-harmonic equation (2.1.1) in a bounded domain $\Omega \subset \mathbf{R}^n$. Assume that $w \in A_r(\Omega)$ for some $r > 1$. Then, there exists a constant C, independent of u and v, such that*

$$\|d^\star v\|_{s,\Omega,w^\alpha} \leq C\||du|^{p/q}\|_{t,\Omega,w^{\alpha t/s}} = C\|du\|_{pt/q,\Omega,w^{\alpha t/s}}^{p/q}, \qquad (2.4.2)$$

where α is any positive constant with $1 > \alpha r$, $s = (1-\alpha)q$, and $t = s/(1 - \alpha r) = qs/(s - \alpha q(r-1))$.

2.4.2 Global estimates for du

Using Theorem 2.3.2 and the covering lemma, we obtain the following global L^s-estimate for du.

Theorem 2.4.3. *Let u and v be a pair of solutions to the nonhomogeneous conjugate A-harmonic equation (2.1.2) in a bounded domain $\Omega \subset \mathbf{R}^n$. Assume that $w \in A_r(\Omega)$ for some $r > 1$. Then, there exists a constant C, independent of u and v, such that*

$$\|du\|_{s,\Omega,w^\alpha} \leq C(\|g\|_{t,\Omega,w^{\alpha t/s}} + \||h|^{q/p}\|_{t,\Omega,w^{\alpha t/s}} + \||d^\star v|^{q/p}\|_{t,\Omega,w^{\alpha t/s}}). \quad (2.4.3)$$

Here α is any positive constant with $1 > \alpha r$, $s = (1-\alpha)p$, and $t = s/(1-\alpha r) = ps/(s - \alpha p(r-1))$.

Similarly, if we choose $g = 0$ and $h = 0$ in Theorem 2.4.3, we obtain the following global norm estimate for du in terms of $d^\star v$.

Corollary 2.4.4. *Let u and v be a pair of solutions to the conjugate A-harmonic equation (2.1.1) in a bounded domain $\Omega \subset \mathbf{R}^n$. Assume that $w \in A_r(\Omega)$ for some $r > 1$. Then, there exists a constant C, independent of u and v, such that*

$$\|du\|_{s,\Omega,w^\alpha} \leq C\||d^\star v|^{q/p}\|_{t,\Omega,w^{\alpha t/s}} = \|d^\star v\|_{qt/p,\Omega,w^{\alpha t/s}}^{q/p}. \tag{2.4.4}$$

Here α is any positive constant with $1 > \alpha r$, $s = (1-\alpha)p$, and $t = s/(1-\alpha r) = ps/(s - \alpha p(r-1))$.

2.4.3 Global L^p-estimates

We denote the space of all l-forms that are L^s-integrable in Ω with respect to the measure μ by $L^s(\Omega, \wedge, \mu)$. Now we prove the following theorem that characterizes the integrability of a pair du and $d^\star v$.

Theorem 2.4.5. *Let u and v be a pair of solutions to the nonhomogeneous conjugate A-harmonic equation (2.1.2) in a bounded domain $\Omega \subset \mathbf{R}^n$. If $g \in L^p(B, \wedge^l, \mu)$ and $h \in L^q(B, \wedge^l, \mu)$, then $du \in L^p(B, \wedge^l, \mu)$ if and only if $d^\star v \in L^q(B, \wedge^l, \mu)$, where the measure μ is defined by $d\mu = w(x)^\alpha dx$, $w(x) \in A_r(\Omega)$ for some $r > 1$. Moreover, there exist constants C_1, C_2, independent of u and v, such that*

$$\|d^\star v\|_{q,\Omega,w^\alpha}^q \leq C_1(\|h\|_{q,\Omega,w^\alpha}^q + \|g\|_{p,\Omega,w^\alpha}^p + \|du\|_{p,\Omega,w^\alpha}^p), \tag{2.4.5}$$

$$\|du\|_{p,\Omega,w^\alpha}^p \leq C_2(\|h\|_{q,\Omega,w^\alpha}^q + \|g\|_{p,\Omega,w^\alpha}^p + \|d^\star v\|_{q,\Omega,w^\alpha}^q) \tag{2.4.6}$$

for any real number α with $\alpha > 0$.

Proof. We only need to prove (2.4.5) and (2.4.6). Multiplying (2.2.6) by w^α, we have

$$|d^\star v|^q w^\alpha \leq C_1(|h|^q w^\alpha + |g|^p w^\alpha + |du|^p w^\alpha) \tag{2.4.7}$$

since $w^\alpha > 0$. Integrating (2.4.7) over Ω we find that (2.4.5) holds. Similarly, from (2.2.8), we obtain

$$|du|^p w^\alpha \leq C_2(|g|^p w^\alpha + |h|^q w^\alpha + |d^\star v|^q w^\alpha). \tag{2.4.8}$$

Hence, inequality (2.4.6) follows immediately.   ■

Setting $g = 0$ and $h = 0$ in Theorem 2.4.5, we have the following global norm comparison theorem that provides a powerful tool for the study of the global integrability of differential forms du and $d^\star v$.

Corollary 2.4.6. *Let u and v be a pair of solutions to the conjugate A-harmonic equation (2.1.1) in a bounded domain $\Omega \subset \mathbf{R}^n$. Then, $du \in L^p(\Omega, \wedge^l, \mu)$ if and only if $d^\star v \in L^q(\Omega, \wedge^l, \mu)$, where the measure μ is defined by $d\mu = w(x)^\alpha dx$, $w \in A_r(\Omega)$ for some $r > 1$. Moreover, there exist constants C_1, C_2, independent of u and v, such that*

$$C_1\|du\|_{p,\Omega,w^\alpha}^p \leq \|d^\star v\|_{q,\Omega,w^\alpha}^q \leq C_2\|du\|_{p,\Omega,w^\alpha}^p \qquad (2.4.9)$$

for any real number α with $\alpha > 0$.

Using Theorem 2.3.4 and the covering lemma, we find the following norm comparison theorem for $u - c$ and $d^\star v$.

Theorem 2.4.7. *Let u and v be a pair of solutions to the conjugate A-harmonic equation (2.1.1) in a bounded domain $\Omega \subset \mathbf{R}^n$. Assume that $w(x) \in A_r(\Omega)$ for some $r > 1$. Then, for any positive constant α with $1 > \alpha r$, there exist constants C_1, C_2, independent of u and v, such that*

$$\|u - u_B\|_{s,\Omega,w^\alpha} \leq C_1\||d^\star v|^{q/p}\|_{t,\Omega,w^{\alpha t/s}} \qquad (2.4.10)$$

for $s = (1 - \alpha)p, t = s/(1 - \alpha r) = ps/(s - \alpha p(r - 1))$; and

$$\|d^\star v\|_{s,\Omega,w^\alpha} \leq C_2\||u - c|^{p/q}\|_{t,\Omega,w^{\alpha t/s}} \qquad (2.4.11)$$

for $s = (1 - \alpha)q, t = s/(1 - \alpha r) = qs/(s - \alpha q(r - 1))$.

2.4.4 Global L^s-estimates

We have obtained the L^p-estimate for du and the L^q-estimate for $d^\star v$ with $p^{-1} + q^{-1} = 1$ in Theorem 2.4.5. However, in applications, we often need L^s-estimates for some s with $1 < s < \infty$. Hence, we now state and prove the following L^s-norm comparison theorem.

Theorem 2.4.8. *Let u and v be a pair of solutions to the nonhomogeneous conjugate A-harmonic equation (2.1.2) in a bounded domain $\Omega \subset \mathbf{R}^n$ and $1 < s < \infty$. Assume that $w \in A_r(\Omega)$ for some $r > 1$. Then,*

$$\|d^\star v\|_{s,\Omega,w^\alpha} \leq C_1(\|h\|_{s,\Omega,w^\alpha} + \|g\|_{s(p-1),\Omega,w^\alpha}^{p-1} + \|du\|_{s(p-1),\Omega,w^\alpha}^{p-1}), \quad (2.4.12)$$

$$\|du\|_{s,\Omega,w^\alpha} \leq C_2(\|g\|_{s,\Omega,w^\alpha} + \|h\|_{s(q-1),\Omega,w^\alpha}^{q-1} + \|d^\star v\|_{s(q-1),\Omega,w^\alpha}^{q-1}), \quad (2.4.13)$$

where C_1, C_2 are constants.

Proof. From (2.2.4), we know that

$$|d^\star v|^s w^\alpha \leq (|h| + a|g + du|^{p-1})^s w^\alpha \qquad (2.4.14)$$

for any weight $w(x) > 0$. Integrating (2.4.14) over Ω and using the Minkowski inequality, we find that

$$\|d^\star v\|_{s,\Omega,w^\alpha} \leq \left(\int_\Omega (|h| + a|g + du|^{p-1})^s w^\alpha dx\right)^{1/s}$$

$$\leq \|h\|_{s,\Omega,w^\alpha} + a\|g + du\|_{s(p-1),\Omega,w^\alpha}^{p-1}$$

$$\leq C_1(\|h\|_{s,\Omega,w^\alpha} + \|g\|_{s(p-1),\Omega,w^\alpha}^{p-1} + \|du\|_{s(p-1),\Omega,w^\alpha}^{p-1}).$$

This completes the proof of (2.4.12). From (2.2.7), we find that

$$|g + du| \leq |h + d^\star v|^{1/(p-1)},$$

and hence

$$\begin{aligned}|du|^s w^\alpha &= |du + g - g|^s w^\alpha \\ &\leq (|du + g| + |g|)^s w^\alpha \\ &\leq (|g| + |h + d^\star v|^{1/(p-1)})^s w^\alpha.\end{aligned} \qquad (2.4.15)$$

Integrating (2.4.15) over Ω and using (2.2.5) and Minkowski inequality again, we obtain

$$\|du\|_{s,\Omega,w^\alpha}$$

$$\leq \left(\int_\Omega \left(|g| + (|h| + |d^\star v|)^{1/(p-1)}\right)^s w^\alpha dx\right)^{1/s}$$

$$\leq \|g\|_{s,\Omega,w^\alpha} + \left(\int_\Omega (|h| + |d^\star v|)^{s/(p-1)} w^\alpha dx\right)^{1/s}$$

$$\leq \|g\|_{s,\Omega,w^\alpha} + \left(C_2 \int_\Omega (|h|^{1/(p-1)} + |d^\star v|^{1/(p-1)})^s w^\alpha dx\right)^{1/s}$$

$$\leq \|g\|_{s,\Omega,w^\alpha} + C_3 \left(\left(\int_\Omega |h|^{s/(p-1)} w^\alpha dx\right)^{1/s} + \left(\int_\Omega |d^\star v|^{s/(p-1)} w^\alpha dx\right)^{1/s}\right)$$

$$\leq C_4 \left(\|g\|_{s,\Omega,w^\alpha} + \|h\|_{s/(p-1),\Omega,w^\alpha}^{1/(p-1)} + \|d^\star v\|_{s/(p-1),\Omega,w^\alpha}^{1/(p-1)}\right)$$

$$\leq C_4(\|g\|_{s,\Omega,w^\alpha} + \|h\|_{s(q-1),\Omega,w^\alpha}^{q-1} + \|d^\star v\|_{s(q-1),\Omega,w^\alpha}^{q-1})$$

since $1/(p-1) = q - 1$. ∎

Remark. We have extended only few local norm inequalities into the global ones. Using similar methods, we can generalize other local results into the global ones. We can also obtain the global comparison theorems with other weights, including two-weight. Also, we notice that the global comparison norm inequalities contain a parameter α. Hence, by choosing different value of α, we have different versions of the global norm inequalities. For example, choosing $\alpha = p > 1$ in (2.4.5) and (2.4.6), we obtain

$$\|d^{\star}v\|_{q,\Omega,w^p}^q \leq C_1(\|h\|_{q,\Omega,w^p}^q + \|g\|_{p,\Omega,w^p}^p + \|du\|_{p,\Omega,w^p}^p)$$

and

$$\|du\|_{p,\Omega,w^p}^p \leq C_2(\|h\|_{q,\Omega,w^p}^q + \|g\|_{p,\Omega,w^p}^p + \|d^{\star}v\|_{q,\Omega,w^p}^q),$$

respectively.

2.5 Applications

In this section, we discuss some applications of the global norm comparison theorems. We prove the global Sobolev–Poincaré-type inequality, and obtain some versions of imbedding theorems for the homotopy operator T and the gradient operator ∇.

2.5.1 Imbedding theorems for differential forms

From Corollary 4.1 in [99], for any $u \in D'(B, \wedge^l)$ with $du \in L^p(B, \wedge^{l+1})$, we have

$$\|u - u_B\|_{W^{1,p}(B,\wedge^l)} \leq C|B|\|du\|_{p,B}. \tag{2.5.1}$$

Combining (2.2.9) and (2.5.1), we obtain the following analog of Sobolev–Poincaré-type imbedding theorem for differential forms satisfying the conjugate A-harmonic equation.

Theorem 2.5.1. *Let u and v be a pair of solutions to the conjugate A-harmonic equation (2.1.1) in a domain $\Omega \subset \mathbf{R}^n$. Then,*

$$\|u - u_B\|_{W^{1,p}(B,\wedge^l)}^p \leq C\|d^{\star}v\|_{q,B}^q \tag{2.5.2}$$

for all balls B with $B \subset \Omega$, where C is a constant.

Using (2.5.2) and the covering lemma, we prove the following global Sobolev–Poincaré-type imbedding theorem.

Theorem 2.5.2. *Let $u \in D'(\Omega, \wedge^0)$ and $v \in D'(\Omega, \wedge^2)$ be a pair of solutions to the conjugate A-harmonic equation (2.1.1) in a δ-John domain $\Omega \subset \mathbf{R}^n$. Then,*

$$\|u - u_{B_0}\|_{W^{1,p}(\Omega,\wedge^l)}^p \leq C\|d^{\star}v\|_{q,\Omega}^q,$$

where C is a constant and $B_0 \subset \Omega$ is a fixed ball.

Notes to Chapter 2. (i) We should note that all the results established in this chapter are about l-forms, $l = 0, 1, \ldots, n$, and that the real functions in $\mathbf{R}^n$ are 0-forms.

(ii) It is known that if $f(x) = (f^1, f^2, \ldots, f^n)$ is K-quasiregular in $\mathbf{R}^n$, then

$$u = f^l df^1 \wedge df^2 \wedge \cdots \wedge df^{l-1}, \quad l = 1, 2, \ldots, n-1,$$

and

$$v = \star f^{l+1} df^{l+2} \wedge \cdots \wedge df^n, \quad l = 1, 2, \ldots, n-1,$$

are solutions of the conjugate A-harmonic equation (2.1.1). Thus, our results, such as Theorems 2.5.1 and 2.5.2, can be used to study the K-quasiregular mapping f.

Chapter 3
Poincaré-type inequalities

3.1 Introduction

We begin this chapter with the following weak reverse Hölder inequality, which is due to C. Nolder [71] and will be used repeatedly later.

Lemma 3.1.1. *Let u be a solution of the nonhomogeneous A-harmonic equation (1.2.10) in a domain Ω and $0 < s, t < \infty$. Then, there exists a constant C, independent of u, such that*

$$\|u\|_{s,B} \le C|B|^{(t-s)/st}\|u\|_{t,\sigma B} \tag{3.1.1}$$

for all balls or cubes B with $\sigma B \subset \Omega$ for some $\sigma > 1$.

Thus, for any s, t with $0 < s, t < \infty$, the local L^s-norm and the local L^t-norm are comparable.

3.2 Inequalities for differential forms

We first discuss the Poincaré inequality for some differential forms. These forms are not necessarily be the solutions of any version of the A-harmonic equation.

3.2.1 Basic inequalities

In 1989, Susan G. Staples [183] proved the following Poincaré inequality for Sobolev functions in L^s-averaging domains. This inequality has been well studied and used in the development of the averaging domains; see [202].

R.P. Agarwal et al., *Inequalities for Differential Forms*,
DOI 10.1007/978-0-387-68417-8_3, © Springer Science+Business Media, LLC 2009

Theorem 3.2.1. *If D is an L^p-averaging domain, $p \geq n$, then there exists a constant C, such that*

$$\left(\frac{1}{m(D)} \int_D |u - u_D|^p dm \right)^{1/p} \leq C|D|^{1/n} \left(\frac{1}{m(D)} \int_D |\nabla u|^p dm \right)^{1/p} \quad (3.2.1)$$

for each Sobolev function u defined in D.

In 1993, the following Poincaré–Sobolev inequality was proved in [99], which can be used to generalize the theory of Sobolev functions to that of differential forms.

Theorem 3.2.2. *Let $u \in D'(Q, \wedge^l)$ and $du \in L^p(Q, \wedge^{l+1})$. Then, $u - u_Q$ is in $L^{np/(n-p)}(Q, \wedge^l)$ and*

$$\left(\int_Q |u - u_Q|^{np/(n-p)} dx \right)^{(n-p)/np} \leq C_p(n) \left(\int_Q |du|^p dx \right)^{1/p} \quad (3.2.2)$$

for Q a cube or a ball in $\mathbf{R}^n$, $l = 0, 1, \ldots, n-1$, and $1 < p < n$.

From Corollary 4.1 in [99], we have the following version of Poincaré inequality for differential forms.

Theorem 3.2.3. *Let $u \in D'(Q, \wedge^l)$ and $du \in L^p(Q, \wedge^{l+1})$. Then, $u - u_Q$ is in $W^{1,p}(Q, \wedge^l)$ with $1 < p < \infty$ and*

$$\|u - u_Q\|_{p,Q} \leq C(n,p)|Q|^{1/n}\|du\|_{p,Q} \quad (3.2.3)$$

for Q a cube or a ball in $\mathbf{R}^n$, $l = 0, 1, \ldots, n-1$.

3.2.2 Weighted inequalities

In 1998, S. Ding and P. Shi [121] proved the following different versions of the local Poincaré inequalities for differential forms. We present these results in Theorems 3.2.4, 3.2.5, 3.2.6, 3.2.7, and 3.2.8. Also, using the local results, they have obtained the global Poincaré inequalities in $L^s(\mu)$-averaging domains.

Theorem 3.2.4. *Let $u \in D'(B, \wedge^l)$ and $du \in L^n(B, \wedge^{l+1})$, $l = 0, 1, \ldots, n-1$. If $1 < s < n$ and $w \in A_{n/s}(n/s)$, then there exists a constant C, independent of u and du, such that*

$$\left(\int_B |u - u_B|^s d\mu \right)^{1/s} \leq C|B|^{1/s} \left(\int_B |du|^n d\mu \right)^{1/n} \quad (3.2.4)$$

for all balls $B \subset \mathbf{R}^n$. Here $d\mu = w(x)dx$.

Note that (3.2.4) can be written as

$$\|u - u_B\|_{s,B,w} \le C|B|^{1/s}\|du\|_{n,B,w}$$

or

$$\left(\frac{1}{|B|}\int_B |u - u_B|^s w(x)dx\right)^{1/s} \le C\left(\int_B |du|^n w(x)dx\right)^{1/n}.$$

Theorem 3.2.5. *Let* $u \in D'(B, \wedge^l)$ *and* $du \in L^p(B, \wedge^{l+1})$, $l = 0, 1, \ldots, n-1$, *and* $1 < p < \infty$. *If* $w \in A_r(t/(t-s))$, *where* $r = t(p-s)/p(t-s)$ *and* $1 < s < t < p$, *then there exists a constant* C, *independent of* u *and* du, *such that*

$$\left(\frac{1}{|B|}\int_B |u - u_B|^s wdx\right)^{1/s} \le C|B|^{1/n}\left(\frac{1}{|B|}\int_B |du|^p w^{p(t-s)/st}dx\right)^{1/p}$$

for all balls $B \subset \mathbf{R}^n$.

Theorem 3.2.6. *Let* $u \in D'(\Omega, \wedge^l)$ *and* $du \in L^n(\Omega, \wedge^{l+1})$, $l = 0, 1, \ldots, n-1$. *If* $1 < s < n < \beta s$ *and* $w \in A_r(1) \cap A_{n/s}(n/s)$, *then there exists a constant* C, *independent of* u *and* du, *such that*

$$\left(\frac{1}{\mu(\Omega)}\int_\Omega |u - u_{B_0}|^s wdx\right)^{1/s} \le C|\Omega|^{1/\beta s}\left(\int_\Omega |du|^\kappa w^{\kappa(s-n)/ns}dx\right)^{1/\kappa}$$

for any $L^s(\mu)$*-averaging domain* Ω *and some ball* B_0 *with* $2B_0 \subset \Omega$. *Here* $\kappa = \beta ns/(\beta s - n)$ *and* β *is the exponent in the reverse Hölder inequality (see Lemma 1.4.7).*

Theorem 3.2.7. *Let* $u \in D'(\Omega, \wedge^l)$ *and* $du \in L^n(\Omega, \wedge^{l+1})$, $l = 0, 1, \ldots, n-1$. *If* $1 < s < n$ *and* $w \in A_{n/s}(n/s)$ *with* $w(x) \ge \alpha > 0$, *then there exists a constant* C, *independent of* u *and* du, *such that*

$$\left(\frac{1}{\mu(\Omega)}\int_\Omega |u - u_{B_0}|^s d\mu\right)^{1/s} \le C\left(\int_\Omega |du|^n d\mu\right)^{1/n}$$

for any $L^s(\mu)$*-averaging domain* Ω *and some ball* B_0 *with* $2B_0 \subset \Omega$. *Here* $d\mu = w(x)dx$.

Theorem 3.2.8. *Let* $u \in D'(\Omega, \wedge^l)$ *and* $du \in L^p(\Omega, \wedge^{l+1})$, $l = 0, 1, \ldots, n-1$. *Let* $1 < s < t < p < \beta s$ *and* $1/p = 1/n + 1/\beta s$, *where* β *is the exponent in the reverse Hölder inequality (see Lemma 1.4.7). If* $w \in A_r(1) \cap A_r(t/(t-s))$ *with* $r = t(p-s)/p(t-s)$, *then there exists a constant* C, *independent of* u *and* du, *such that*

$$\left(\frac{1}{\mu(\Omega)}\int_\Omega |u - u_{B_0}|^s wdx\right)^{1/s} \le C\left(\int_\Omega |du|^n w^{-n/t}dx\right)^{1/n}$$

for any $L^s(\mu)$*-averaging domain* Ω *and some ball* B_0 *with* $2B_0 \subset \Omega$.

3.2.3 Inequalities for harmonic forms

The above Poincaré–Sobolev inequalities are about differential forms. We know that the A-harmonic tensors are differential forms that satisfy the A-harmonic equation. Then naturally, one would ask whether the Poincaré–Sobolev inequalities for A-harmonic tensors are sharper than those for differential forms. The answer is "yes". In [201], S. Ding and C. Nolder proved the following symmetric Poincaré–Sobolev inequalities for solutions of the nonhomogeneous A-harmonic equation (1.2.10).

Theorem 3.2.9. *Let $u \in D'(\Omega, \wedge^l)$ be a solution of the nonhomogeneous A-harmonic equation (1.2.10) in a domain $\Omega \subset \mathbf{R}^n$ and $du \in L^s(\Omega, \wedge^{l+1})$, $l = 0, 1, \ldots, n-1$. Assume that $\sigma > 1$, $0 < s < \infty$ and $w \in A_r$ for some $r > 1$. Then,*

$$\|u - u_B\|_{s,B,w} \le C|B|^{1/n}\|du\|_{s,\sigma B,w} \tag{3.2.5}$$

for all balls B with $\sigma B \subset \Omega$. Here C is a constant independent of u and du.

Note that (3.2.5) is equivalent to

$$\left(\frac{1}{\mu(B)}\int_B |u - u_B|^s d\mu\right)^{1/s} \le C|B|^{1/n}\left(\frac{1}{\mu(B)}\int_{\sigma B}|du|^s d\mu\right)^{1/s}. \tag{3.2.5$'$}$$

Theorem 3.2.10. *Let $u \in D'(\Omega, \wedge^l)$ be a solution of the nonhomogeneous A-harmonic equation (1.2.10) in a domain $\Omega \subset \mathbf{R}^n$ and $du \in L^s(\Omega, \wedge^{l+1})$, $l = 0, 1, \ldots, n-1$. Assume that $\sigma > 1$, $0 < \alpha \le 1$, $1 + \alpha(r-1) < s < \infty$, and $w \in A_r$ for some $r > 1$. Then,*

$$\|u - u_B\|_{s,B,w^\alpha} \le C|B|^{1/n}\|du\|_{s,\sigma B,w^\alpha} \tag{3.2.6}$$

for all balls B with $\sigma B \subset \Omega$. Here C is a constant independent of u and du.

Note that (3.2.6) can be written as

$$\left(\frac{1}{|B|}\int_B |u - u_B|^s w^\alpha dx\right)^{1/s} \le C|B|^{1/n}\left(\frac{1}{|B|}\int_{\sigma B}|du|^s w^\alpha dx\right)^{1/s}. \tag{3.2.6$'$}$$

We have presented Theorems 3.2.4, 3.2.5, 3.2.6, 3.2.7, 3.2.8, and 3.2.9 without proof. Now, we present the proof of Theorem 3.2.10 as follows.

Proof. We assume that $0 < \alpha < 1$ and choose $t = s/(1 - \alpha)$ so that $1 < s < t$. Using the Hölder inequality, we have

$$\left(\int_B |u - u_B|^s w^\alpha dx\right)^{1/s} = \left(\int_B (|u - u_B| w^{\alpha/s})^s dx\right)^{1/s}$$

$$\leq \left(\int_B |u - u_B|^t dx\right)^{1/t} \left(\int_B w^{\alpha t/(t-s)} dx\right)^{(t-s)/st}$$

$$= \|u - u_B\|_{t,B} \left(\int_B w dx\right)^{\alpha/s}.$$

$$(3.2.7)$$

Next, choose

$$m = \frac{s}{\alpha(r-1)+1},$$

so that $m < s$. Since u_B is a closed form, by Lemma 3.1.1 and Theorem 3.2.9 with $w = 1$, we find that

$$\|u - u_B\|_{t,B} \leq C_1 |B|^{(m-t)/mt} \|u - u_B\|_{m,\sigma_1 B}$$

$$\leq C_2 |B|^{(m-t)/mt} |B|^{1/n} \|du\|_{m,\sigma B}$$

$$(3.2.8)$$

for all balls B with $\sigma B \subset \Omega$. Here $\sigma > \sigma_1 > 1$. Now since $1/m = 1/s + (s-m)/sm$, by the Hölder inequality again, we obtain

$$\|du\|_{m,\sigma B} = \left(\int_{\sigma B} \left(|du| w^{\alpha/s} w^{-\alpha/s}\right)^m dx\right)^{1/m}$$

$$\leq \left(\int_{\sigma B} |du|^s w^\alpha dx\right)^{1/s} \left(\int_{\sigma B} \left(\tfrac{1}{w}\right)^{\alpha m/(s-m)} dx\right)^{(s-m)/sm}$$

$$(3.2.9)$$

$$= \left(\int_{\sigma B} |du|^s w^\alpha dx\right)^{1/s} \left(\int_{\sigma B} \left(\tfrac{1}{w}\right)^{1/(r-1)} dx\right)^{\alpha(r-1)/s}.$$

From (3.2.7), (3.2.8), and (3.2.9), we have

$$\left(\int_B |u - u_B|^s w^\alpha dx\right)^{1/s}$$

$$\leq C_2 |B|^{(m-t)/tm} |B|^{1/n} \left(\int_B w dx\right)^{\alpha/s} \left(\int_{\sigma B} \left(\tfrac{1}{w}\right)^{1/(r-1)} dx\right)^{\alpha(r-1)/s}$$

$$\times \left(\int_{\sigma B} |du|^s w^\alpha dx\right)^{1/s}.$$

$$(3.2.10)$$

Since $w \in A_r(\Omega)$, it follows that

$$\left(\int_B w dx\right)^{\alpha/s} \left(\int_{\sigma B} \left(\tfrac{1}{w}\right)^{1/(r-1)} dx\right)^{\alpha(r-1)/s}$$

$$\leq \left(\left(\int_{\sigma B} w dx\right) \left(\int_{\sigma B} \left(\tfrac{1}{w}\right)^{1/(r-1)} dx\right)^{(r-1)}\right)^{\alpha/s}$$

$$= \left(|\sigma B|^{(r-1)+1} \left(\tfrac{1}{|\sigma B|} \int_{\sigma B} w dx\right)\right)$$

$$\times \left(\frac{1}{|\sigma B|} \int_{\sigma B} \left(\frac{1}{w} \right)^{1/(r-1)} dx \right)^{(r-1)} \right)^{\alpha/s}$$

$$\le C_3 |\sigma B|^{\alpha(r-1)/s + \alpha/s} \tag{3.2.11}$$

$$\le C_4 |B|^{\alpha r/s}.$$

Substituting (3.2.11) into (3.2.10) and using $(m-t)/mt = -\alpha/s - \alpha(r-1)/s$, we obtain

$$\left(\int_B |u - u_B|^s w^\alpha dx \right)^{1/s} \le C_5 |B|^{1/n} \left(\int_{\sigma B} |du|^s w^\alpha dx \right)^{1/s}$$

which is equivalent to (3.2.6).

For the case $\alpha = 1$, by Lemma 1.4.7, there exist constants $\beta > 1$ and $C_6 > 0$, such that

$$\| w \|_{\beta,B} \le C_6 |B|^{(1-\beta)/\beta} \| w \|_{1,B} \tag{3.2.12}$$

for any cube or ball $B \subset \mathbf{R}^n$. Choose $t = s\beta/(\beta - 1)$, so that $1 < s < t$ and $\beta = t/(t - s)$. Since $1/s = 1/t + (t - s)/st$, by the Hölder inequality and (3.2.12), we find

$$\left(\int_B |u - u_B|^s w dx \right)^{1/s}$$

$$= \left(\int_B \left(|u - u_B| w^{1/s} \right)^s dx \right)^{1/s}$$

$$\le \left(\int_B |u - u_B|^t dx \right)^{1/t} \left(\int_B \left(w^{1/s} \right)^{st/(t-s)} dx \right)^{(t-s)/st} \tag{3.2.13}$$

$$= \| u - u_B \|_{t,B} \cdot \| w \|_{\beta,B}^{1/s}$$

$$\le C_7 \| u - u_B \|_{t,B} \cdot |B|^{(1-\beta)/\beta s} \| w \|_{1,B}^{1/s} .$$

Now, choose $m = s/r$ so that $m < s$. By the weak reverse Hölder inequality and Theorem 3.2.3, we have

$$\| u - u_B \|_{t,B} \le C_8 |B|^{(m-t)/mt} \| u - u_B \|_{m,\sigma B}$$

$$\le C_9 |B|^{(m-t)/mt} |B|^{1/n} \| du \|_{m,\sigma B}. \tag{3.2.14}$$

Now by the Hölder inequality again, we obtain

$$\| du \|_{m,\sigma B} = \left(\int_{\sigma B} \left(|du| w^{1/s} w^{-1/s} \right)^m dx \right)^{1/m}$$

$$\leq \left(\int_{\sigma B} |du|^s w dx\right)^{1/s} \left(\int_{\sigma B} \left(\tfrac{1}{w}\right)^{m/(s-m)} dx\right)^{(s-m)/sm}$$

$$= \left(\int_{\sigma B} |du|^s w dx\right)^{1/s} \left(\int_{\sigma B} \left(\tfrac{1}{w}\right)^{1/(r-1)} dx\right)^{(r-1)/s} . \tag{3.2.15}$$

Combination of (3.2.14) and (3.2.15) yields

$$\|u - u_B\|_{t,B}$$

$$\leq C_{10}|B|^{(m-t)/mt}|B|^{1/n} \left(\int_{\sigma B} |du|^s w dx\right)^{1/s} \left(\int_{\sigma B} \left(\tfrac{1}{w}\right)^{1/(r-1)} dx\right)^{(r-1)/s} . \tag{3.2.16}$$

Inequality (3.2.11) with $\alpha = 1$ is the same as

$$\left(\int_B w dx\right)^{1/s} \left(\int_{\sigma B} \left(\frac{1}{w}\right)^{1/(r-1)} dx\right)^{(r-1)/s} \leq C_{11}|B|^{r/s}. \tag{3.2.17}$$

Substituting (3.2.16) into (3.2.13) and using (3.2.17), we conclude that

$$\|u - u_B\|_{s,B,w}$$

$$\leq C_{11}|B|^{(1-\beta)/\beta s}|B|^{(m-t)/mt}|B|^{1/n}\|du\|_{s,\sigma B,w}\|w\|_{1,B}^{1/s}$$

$$\times \left(\int_{\sigma B} \left(\tfrac{1}{w}\right)^{1/(r-1)} dx\right)^{(r-1)/s} \tag{3.2.18}$$

$$\leq C_{12}|B|^{(1-\beta)/\beta s}|B|^{(m-t)/mt}|B|^{r/s}|B|^{1/n}\|du\|_{s,\sigma B,w}$$

$$\leq C_{12}|B|^{1/n}\|du\|_{s,\sigma B,w}.$$

Thus (3.2.6) holds for $\alpha = 1$ also.   ∎

Next, we shall prove the following global weighted Poincaré–Sobolev inequality in $L^s(\mu)$-averaging domains.

Theorem 3.2.11. *Let $w \in A_r$ with $w \geq \tau > 0$, $r > 1$, where τ is a constant. Assume that $u \in D'(\Omega, \wedge^0)$ is an A-harmonic tensor and $du \in L^s(\Omega, \wedge^1)$, then*

$$\left(\frac{1}{\mu(\Omega)} \int_\Omega |u - u_{B_0}|^s d\mu\right)^{1/s} \leq C \left(\int_\Omega |du|^s d\mu\right)^{1/s} \tag{3.2.19}$$

for any $L^s(\mu)$-averaging domain Ω and some ball B_0 with $2B_0 \subset \Omega$. Here the measure μ is defined by $d\mu = w(x)dx$ and C is a constant independent of u.

Clearly, we can write (3.2.19) as

$$\|u - u_{B_0}\|_{s,\Omega,w} \leq C\mu(\Omega)^{1/s}\|du\|_{s,\Omega,w}.$$

Proof. Since Ω is an $L^s(\mu)$-averaging domain, it follows that $\mu(\Omega) \leq M$. For any ball $B \in \Omega$,

$$\mu(B) = \int_B d\mu = \int_B w(x)dx \geq \int_B \tau dx = \tau|B|,$$

so that

$$\frac{1}{\mu(B)} \leq \frac{C_1}{|B|}. \tag{3.2.20}$$

Similarly, we have

$$\mu(\Omega) = \int_\Omega d\mu = \int_\Omega w(x)dx \geq \int_\Omega \tau dx = \tau|\Omega|$$

which implies

$$|\Omega| \leq C_2\mu(\Omega) \leq C_3. \tag{3.2.21}$$

By (3.2.20) and (3.2.21), Theorem 3.2.10 with $\alpha = 1$, the definition of $L^s(\mu)$-averaging domains, and noticing $s \geq n$, we find that

$$\left(\tfrac{1}{\mu(\Omega)}\int_\Omega |u - u_{B_0}|^s d\mu\right)^{1/s}$$

$$\leq \left(\tfrac{1}{\mu(B_0)}\int_\Omega |u - u_{B_0}|^s d\mu\right)^{1/s}$$

$$\leq C_4 \sup_{2B\subset\Omega}\left(\tfrac{1}{\mu(B)}\int_B |u - u_B|^s d\mu\right)^{1/s}$$

$$\leq C_5 \sup_{2B\subset\Omega}\left(|B|^{1/n}\left(\tfrac{1}{\mu(B)}\int_{\sigma B} |du|^s d\mu\right)^{1/s}\right)$$

$$\leq C_5 \sup_{2B\subset\Omega}\left(|B|^{1/n}\left(\tfrac{C_1}{|B|}\int_{\sigma B} |du|^s d\mu\right)^{1/s}\right)$$

$$\leq C_6 \sup_{2B\subset\Omega}\left(|B|^{1/n-1/s}\left(\int_{\sigma B} |du|^s d\mu\right)^{1/s}\right)$$

$$\leq C_6 \sup_{2B\subset\Omega}\left(|\Omega|^{1/n-1/s}\left(\int_\Omega |du|^s d\mu\right)^{1/s}\right)$$

$$\leq C_7 \left(\int_\Omega |du|^s d\mu\right)^{1/s}. \quad \blacksquare$$

In [203], we have obtained Poincaré inequalities in which the integral on one side is about Lebesgue measure, but on the other side, the integral is about general measure induced by a weight $w(x)$. We state these results in the following:

Theorem 3.2.12. *Let $u \in D'(\Omega, \wedge^l)$ be an A-harmonic tensor in a domain $\Omega \subset \mathbf{R}^n$ and $du \in L^s_{loc}(\Omega, \wedge^{l+1})$, $l = 0, 1, \ldots, n - 1$. Assume that $\sigma > 1$, $1 < s < \infty$, and $w \in A_r$ for some $r > 1$. Then, there exists a constant C, independent of u, such that*

$$\left(\frac{1}{\mu(B)}\int_B |u - u_B|^s d\mu\right)^{1/s} \leq C|B|^{1/n}\left(\frac{1}{|B|}\int_{\sigma B}|du|^s dx\right)^{1/s} \quad (3.2.22)$$

for all balls B with $\sigma B \subset \Omega$. Here the measure μ is defined by $d\mu = w(x)dx$.

Theorem 3.2.13. *Let $u \in D'(\Omega, \wedge^l)$ be an A-harmonic tensor in a domain $\Omega \subset \mathbf{R}^n$ and $du \in L^s_{loc}(\Omega, \wedge^{l+1})$, $l = 0, 1, \ldots, n-1$. Assume that $\sigma > 1$, $1 < s < \infty$, and $w \in A_r$ for some $r > 1$. Then,*

$$\left(\frac{1}{|B|}\int_B |u - u_B|^s dx\right)^{1/s} \leq C|B|^{1/n}\left(\frac{1}{\mu(\sigma B)}\int_{\sigma B}|du|^s d\mu\right)^{1/s} \quad (3.2.23)$$

for all balls B with $\sigma B \subset \Omega$. Here the measure μ is defined by $d\mu = w(x)dx$ and C is a constant independent of u and du.

3.2.4 Global inequalities in averaging domains

Definition 3.2.14. We call a proper subdomain $\Omega \subset \mathbf{R}^n$ an $L^s(\mu, 0)$-averaging domain, $s \geq 1$, if $\mu(\Omega) < \infty$ and there exists a constant C such that

$$\left(\frac{1}{\mu(B_0)}\int_\Omega |u - u_{B_0}|^s d\mu\right)^{1/s} \leq C \sup_{2B \subset \Omega}\left(\frac{1}{\mu(B)}\int_B |u - u_B|^s d\mu\right)^{1/s}$$
$$(3.2.24)$$

for some ball $B_0 \subset \Omega$ and all $u \in L^s_{loc}(\Omega; \wedge^0, \mu)$. Here the measure μ is defined by $d\mu = w(x)dx$, where $w(x)$ is a weight and $w(x) > 0$ a.e., and the supremum is over all balls B with $2B \subset \Omega$.

Theorem 3.2.15. *Let $w \in A_r$ for some $r > 1$, $u \in D'(\Omega, \wedge^0)$, and $du \in L^s(\Omega, \wedge^1)$. If $s \geq n$, then*

$$\left(\frac{1}{\mu(\Omega)}\int_\Omega |u - u_{B_0}|^s d\mu\right)^{1/s} \leq C|\Omega|^{1/n}\left(\frac{1}{|\Omega|}\int_\Omega |du|^s dx\right)^{1/s} \quad (3.2.25)$$

for any $L^s(\mu, 0)$-averaging domain Ω with $\mu(\Omega) < \infty$ and some ball B_0 with $2B_0 \subset \Omega$. Here the measure μ is defined by $d\mu = w(x)dx$ and C is a constant independent of u and du.

Theorem 3.2.16. *Let $w \in A_r$ with $w(x) \geq \alpha > 0$, $r > 1$, $u \in D'(\Omega, \wedge^0)$, and $du \in L^s(\Omega, \wedge^1)$. If $s \geq n$, then*

$$\left(\frac{1}{|\Omega|}\int_\Omega |u - u_{B_0}|^s dx\right)^{1/s} \leq C|\Omega|^{1/n}\left(\frac{1}{|\Omega|}\int_\Omega |du|^s d\mu\right)^{1/s} \quad (3.2.26)$$

for any L^s-averaging domain Ω and some ball B_0 with $2B_0 \subset \Omega$. Here the measure μ is defined by $d\mu = w(x)dx$ and C is a constant independent of u.

3.2.5 A_r^λ-weighted inequalities

We conclude this section with the following A_r^λ-weighted Poincaré inequality for the solutions of the nonhomogeneous A-harmonic equation.

Theorem 3.2.17. *Let $u \in D'(\Omega, \wedge^l)$ be a differential form satisfying the A-harmonic equation (1.2.10) in a domain $\Omega \subset \mathbf{R}^n$ and $du \in L^s(\Omega, \wedge^{l+1})$, $l = 0, 1, \ldots, n-1$. Suppose that $w \in A_r^\lambda(\Omega)$ for some $r > 1$ and $\lambda > 0$. If $0 < \alpha < 1, \sigma > 1$, and $s > \alpha\lambda(r-1)+1$, then there exists a constant C, independent of u, such that*

$$\left(\int_B |u - u_B|^s w^\alpha dx \right)^{1/s} \le C|B|^{1/n} \left(\int_{\sigma B} |du|^s w^{\alpha\lambda} dx \right)^{1/s}$$

for all balls B with $\sigma B \subset \Omega$. Here u_B is a closed form.

Proof. Choose $t = s/(1-\alpha)$, so that $1 < s < t$. Since $1/s = 1/t + (t-s)/st$, by Hölder's inequality, we find that

$$\left(\int_B |u - u_B|^s w^\alpha dx \right)^{1/s}$$

$$= \left(\int_B (|u - u_B| w^{\alpha/s})^s dx \right)^{1/s}$$

$$\le \left(\int_B |u - u_B|^t dx \right)^{1/t} \left(\int_B w^{\alpha t/(t-s)} dx \right)^{(t-s)/st} \qquad (3.2.27)$$

$$= \|u - u_B\|_{t,B} \left(\int_B w dx \right)^{\alpha/s}.$$

Next, choose

$$m = \frac{s}{\alpha\lambda(r-1)+1},$$

so that $m > 1$. Since u_B is a closed form, by Lemma 3.1.1 and Theorem 3.2.3, we have

$$\|u - u_B\|_{t,B}$$

$$\le C_1 |B|^{(m-t)/mt} \|u - u_B\|_{m,\sigma B} \qquad (3.2.28)$$

$$\le C_2 |B|^{(m-t)/mt} |B|^{1/n} \|du\|_{m,\sigma B}$$

for all balls B with $\sigma B \subset \Omega$. Now since $1/m = 1/s + (s-m)/sm$, by Hölder's inequality again, we obtain

$$\|du\|_{m,\sigma B}$$

$$= \left(\int_{\sigma B} \left(|du| w^{\alpha\lambda/s} w^{-\alpha\lambda/s} \right)^m dx \right)^{1/m}$$

3.2 Inequalities for differential forms

$$\leq \left(\int_{\sigma B} |du|^s w^{\alpha\lambda} dx \right)^{1/s} \left(\int_{\sigma B} \left(\tfrac{1}{w} \right)^{\alpha\lambda m/(s-m)} dx \right)^{(s-m)/sm}$$

$$= \left(\int_{\sigma B} |du|^s w^{\alpha\lambda} dx \right)^{1/s} \left(\int_{\sigma B} \left(\tfrac{1}{w} \right)^{1/(r-1)} dx \right)^{\alpha\lambda(r-1)/s}. \tag{3.2.29}$$

From (3.2.27), (3.2.28), and (3.2.29), we have

$$\left(\int_B |u - u_B|^s w^\alpha dx \right)^{1/s}$$

$$\leq C_2 |B|^{(m-t)/tm} |B|^{1/n} \left(\int_B w dx \right)^{\alpha/s} \tag{3.2.30}$$

$$\times \left(\int_{\sigma B} \left(\tfrac{1}{w} \right)^{1/(r-1)} dx \right)^{\alpha\lambda(r-1)/s} \left(\int_{\sigma B} |du|^s w^{\alpha\lambda} dx \right)^{1/s}.$$

Since $w \in A_r^\lambda(\Omega)$, it follows that

$$\left(\int_B w dx \right)^{\alpha/s} \left(\int_{\sigma B} \left(\tfrac{1}{w} \right)^{1/(r-1)} dx \right)^{\alpha\lambda(r-1)/s}$$

$$\leq \left[\left(\int_{\sigma B} w dx \right) \left(\int_{\sigma B} \left(\tfrac{1}{w} \right)^{1/(r-1)} dx \right)^{\lambda(r-1)} \right]^{\alpha/s}$$

$$= \left[|\sigma B|^{\lambda(r-1)+1} \left(\tfrac{1}{|\sigma B|} \int_{\sigma B} w dx \right) \right.$$

$$\left. \times \left(\tfrac{1}{|\sigma B|} \int_{\sigma B} \left(\tfrac{1}{w} \right)^{1/(r-1)} dx \right)^{\lambda(r-1)} \right]^{\alpha/s} \tag{3.2.31}$$

$$\leq C_3 |\sigma B|^{\alpha\lambda(r-1)/s+\alpha/s}$$

$$\leq C_4 |B|^{\alpha\lambda(r-1)/s+\alpha/s}.$$

Substituting (3.2.31) into (3.2.30) and noting $(m-t)/mt = -\alpha/s - \alpha\lambda(r-1)/s$, we obtain

$$\left(\int_B |u - u_B|^s w^\alpha dx \right)^{1/s} \leq C|B|^{1/n} \left(\int_{\sigma B} |du|^s w^{\alpha\lambda} dx \right)^{1/s}. \quad \blacksquare$$

If we choose $\lambda = 1/\alpha$ in Theorem 3.2.17, we get the following version of the A_r^λ-weighted Poincaré inequality.

Corollary 3.2.18. *Let $u \in D'(\Omega, \wedge^l)$ be a differential form satisfying the A-harmonic equation (1.2.10) in a domain $\Omega \subset \mathbf{R}^n$ and $du \in L^s(\Omega, \wedge^{l+1})$, $l = 0, 1, \ldots, n-1$. Suppose that $w \in A_r^{1/\alpha}(\Omega)$ for some $r > 1$ and $0 < \alpha < 1$. If $s > r$, then there exists a constant C, independent of u, such that*

$$\left(\int_B |u - u_B|^s w^\alpha dx\right)^{1/s} \le C|B|^{1/n} \left(\int_{\sigma B} |du|^s w dx\right)^{1/s}$$

for all balls B with $\sigma B \subset \Omega$. Here u_B is a closed form.

Selecting $\alpha = 1/s$ in Theorem 3.2.17, we obtain the following $A_r^\lambda(\Omega)$-weighted Poincaré inequality.

Corollary 3.2.19. *Let $u \in D'(\Omega, \wedge^l)$ be a differential form satisfying the A-harmonic equation (1.2.10) in a domain $\Omega \subset \mathbf{R}^n$ and $du \in L^s(\Omega, \wedge^{l+1})$, $l = 0, 1, \ldots, n - 1$. Suppose that $w \in A_r^\lambda(\Omega)$ for some $r > 1$ and $\lambda > 0$. If $\sigma > 1$ and $s > \lambda(r-1)/s + 1$, then there exists a constant C, independent of u, such that*

$$\left(\int_B |u - u_B|^s w^{1/s} dx\right)^{1/s} \le C|B|^{1/n} \left(\int_{\sigma B} |du|^s w^{\lambda/s} dx\right)^{1/s} \qquad (3.2.32)$$

for all balls B with $\sigma B \subset \Omega$. Here u_B is a closed form.

It is interesting to note that (3.2.32) reduces to the following $A_r(\Omega)$-weighted inequality when $\lambda = 1$.

Corollary 3.2.20. *Let $u \in D'(\Omega, \wedge^l)$ be a differential form satisfying the A-harmonic equation (1.2.10) in a domain $\Omega \subset \mathbf{R}^n$ and $du \in L^s(\Omega, \wedge^{l+1})$, $l = 0, 1, \ldots, n - 1$. Suppose that $w \in A_r(\Omega)$ for some $r > 1$. If $\sigma > 1$ and $s > \lambda(r-1)/s + 1$, then there exists a constant C, independent of u, such that*

$$\left(\int_B |u - u_B|^s w^{1/s} dx\right)^{1/s} \le C|B|^{1/n} \left(\int_{\sigma B} |du|^s w^{1/s} dx\right)^{1/s}$$

for all balls B with $\sigma B \subset \Omega$. Here u_B is a closed form.

3.3 Inequalities for Green's operator

Let $\wedge^l \Omega$ be the lth exterior power of the cotangent bundle and $C^\infty(\wedge^l \Omega)$ be the space of smooth l-forms in a domain Ω. We use $D'(\Omega, \wedge^l)$ to denote the space of all differential l-forms on Ω and $L^p(\wedge^l \Omega)$ to denote the l-forms

$$\omega(x) = \sum_I \omega_I(x) dx_I = \sum \omega_{i_1 i_2 \cdots i_l}(x) dx_{i_1} \wedge dx_{i_2} \wedge \cdots \wedge dx_{i_l}$$

on Ω satisfying $\int_\Omega |\omega_I|^p < \infty$ for all ordered l-tuples I. The measure on the reference manifold is induced by the volume form *1. Integrals are defined as usual using a partition of unity subordinate to atlas. We write

$$\|u\|_{s,\Omega} = \left(\int_{\Omega} |u|^s \right)^{1/s}.$$

The *Laplace–Beltrami operator* Δ is defined by

$$\Delta = dd^\star + d^\star d.$$

We say that $u \in L^1_{loc}(\wedge^l \Omega)$ has a generalized gradient if, for each coordinate system, the pullbacks of the coordinate function of u have generalized gradient in the familiar sense, see [204]. We write

$$\mathcal{W}(\wedge^l \Omega) = \left\{ u \in L^1_{loc}(\wedge^l \Omega) : u \text{ has generalized gradient} \right\}. \qquad (3.3.1)$$

As usual, the *harmonic l-fields* are defined by

$$\mathcal{H}(\wedge^l \Omega) = \left\{ u \in \mathcal{W}(\wedge^l \Omega) : du = d^\star u = 0, u \in L^p \text{ for some } 1 < p < \infty \right\}.$$

The *orthogonal complement* of $\mathcal{H}$ in L^1 is defined by

$$\mathcal{H}^\perp = \left\{ u \in L^1 : <u, h> = 0 \text{ for all } h \in \mathcal{H} \right\}.$$

We define *Green's operator*

$$G : C^\infty(\wedge^l \Omega) \to \mathcal{H}^\perp \cap C^\infty(\wedge^l \Omega)$$

by setting $G(u)$ equal to the unique element of $\mathcal{H}^\perp \cap C^\infty(\wedge^l \Omega)$ satisfying *Poisson's equation*

$$\Delta G(\omega) = \omega - H(\omega),$$

where H is the *harmonic projection* or the harmonic part of the Hodge decomposition

$$\omega = d\alpha + d^*\beta + H.$$

We also have for $1 < s < \infty$

$$\|d\alpha\|_{s,\Omega} + \|d^*\beta\|_{s,\Omega} + \|H\|_{s,\Omega} \le C_p(\Omega)\|\omega\|_{s,\Omega}. \qquad (3.3.2)$$

We know that G is a bounded self-adjoint linear operator. It has been proved in [21] that for $1 < p < \infty$ a solution exists $\omega \in L^p(\wedge^l \Omega)$.

We will need the following lemma about L^s-estimates for Green's operator which appeared in [21]. The proof of this lemma will be given in Chapter 7, where we will study various operators systematically.

Lemma 3.3.1. Let $u \in C^\infty(\wedge^l \Omega)$, $l = 1, 2, \ldots, n$. For $1 < s < \infty$, there exists a constant C, independent of u, such that

$$\|dd^\star G(u)\|_{s,\Omega} + \|d^\star dG(u)\|_{s,\Omega} + \|dG(u)\|_{s,\Omega}$$

$$+\|d^{\star}G(u)\|_{s,\Omega} + \|G(u)\|_{s,\Omega}$$
$$\leq C\|u\|_{s,\Omega}. \tag{3.3.3}$$

3.3.1 Basic estimates for operators

In the following results we estimate the composition of the Laplace–Beltrami operator and Green's operator.

Theorem 3.3.2. *Let $u \in C^{\infty}(\wedge^{l}\Omega)$, $l = 1, 2, \ldots, n$, and $1 < s < \infty$. Then, there exists a constant C, independent of u, such that*

$$\|\Delta(G(u))\|_{s,\Omega} \leq C\|u\|_{s,\Omega}.$$

Proof. From the definition of the Laplace–Beltrami operator Δ, (3.3.3), and Minkowski's inequality, we have

$$\|\Delta(G(u))\|_{s,\Omega} = \|u - H(u)\|_{s,\Omega}$$
$$\leq C\|u\|_{s,\Omega}.$$

Theorem 3.3.3. *Let $u \in C^{\infty}(\wedge^{l}\Omega)$, $l = 1, 2, \ldots, n$. Assume that $1 < s < \infty$. Then, there exists a constant C, independent of u, such that*

$$\|G(\Delta u)\|_{s,\Omega} \leq C\|u\|_{s,\Omega}.$$

Proof. We know that Green's operator commutes with d and $d^{\star}$ (see [23]), that is, for any differential form $u \in C^{\infty}(\wedge^{l}\Omega)$, we have

$$dG(u) = Gd(u), \quad d^{\star}G(u) = Gd^{\star}(u). \tag{3.3.4}$$

Hence
$$\|G(\Delta u)\|_{s,\Omega} = \|\Delta(G(u))\|_{s,\Omega}.$$

Using Minkowski's inequality and combining Theorems 3.3.2 and 3.3.3, we obtain the following corollary immediately.

Corollary 3.3.4. *Let $u \in C^{\infty}(\wedge^{l}\Omega)$, $l = 1, 2, \ldots, n$. For $1 < s < \infty$, there exists a constant C, independent of u, such that*

$$\|(G\Delta + \Delta G)u\|_{s,\Omega} \leq C\|u\|_{s,\Omega}. \tag{3.3.5}$$

Theorem 3.3.5. *Let $u \in C^{\infty}(\Omega, \wedge^{l})$, $l = 1, 2, \ldots, n$. If $1 < s < \infty$, then there exists a constant C, independent of u, such that*

$$\|(G(u))_{D}\|_{s,D} \leq C|D|\|u\|_{s,D} \tag{3.3.6}$$

for any convex and bounded D with $D \subset \Omega$.

Proof. First, we assume that $1 \leq l \leq n$. We recall the definition of the l-form $\omega_D \in D'(D, \wedge^l)$,

$$\omega_D = |D|^{-1} \int_D \omega(y)dy, \quad l = 0, \quad \text{and} \quad \omega_D = d(T\omega), \quad l = 1, 2, \ldots, n$$

for all $\omega \in L^s(D, \wedge^l)$, and the estimate

$$\|Tu\|_{W^{1,s}(D)} \leq C_1|D|\|u\|_{s,D}. \tag{3.3.7}$$

We find that, using (3.3.3),

$$\begin{aligned}
\|(G(u))_D\|_{s,D} &= \|d(T(G(u)))\|_{s,D} \\
&\leq \|\nabla T(G(u))\|_{s,D} \\
&\leq C|D|\|G(u)\|_{s,D} \\
&\leq C|D|\|u\|_{s,D}.
\end{aligned} \tag{3.3.8}$$

Next, if $l = 0$, using (3.3.3) and Hölder's inequality with $1 = 1/s + 1/q$, we obtain

$$\begin{aligned}
\|(G(u))_D\|_{s,D} &= \left(\int_D |(G(u))_D|^s dx \right)^{1/s} \\
&= \left(\int_D \left| \frac{1}{|D|} \int_D G(u(y))dy \right|^s dx \right)^{1/s} \\
&\leq \left(\left(\frac{1}{|D|} \int_D |G(u(y))|dy \right)^s \int_D 1 dx \right)^{1/s} \\
&= \frac{1}{|D|}|D|^{1/s} \int_D |G(u(y))|dy \tag{3.3.9} \\
&\leq |D|^{1/s-1} \left(\int_D |G(u(y))|^s dy \right)^{1/s} \left(\int_D 1^q dy \right)^{1/q} \\
&= \|G(u)\|_{s,D} \\
&\leq C_6\|u\|_{s,D}. \quad \blacksquare
\end{aligned}$$

Corollary 3.3.6. *Let $u \in C^\infty(\Omega, \wedge^l)$, $l = 1, 2, \ldots, n$. Assume that $1 < s < \infty$. Then, for any convex and bounded D with $D \subset \Omega$, there exists a constant C, independent of u, such that*

$$\|G(u) - (G(u))_D\|_{s,D} \leq C\|G(u) - c\|_{s,D} \tag{3.3.10}$$

for any closed form c, and

$$\|G(u) - (G(u))_D\|_{s,D} \leq C\|u\|_{s,D}. \tag{3.3.11}$$

Proof. We know that $c_D = c$ if c is a closed form. Hence, we have

$$
\begin{aligned}
&\|G(u) - (G(u))_D\|_{s,D} \\
&\quad \leq \|(G(u) - c) - ((G(u))_D - c_D)\|_{s,D} \\
&\quad \leq \|G(u) - c\|_{s,D} + \|(G(u) - c)_D\|_{s,D} \\
&\quad \leq \|G(u) - c\|_{s,D} + C_1\|G(u) - c\|_{s,D} \\
&\quad \leq C_2\|G(u) - c\|_{s,D}.
\end{aligned}
\tag{3.3.12}
$$

Thus, (3.3.10) holds. From (3.3.6) and (3.3.3), we find that

$$
\begin{aligned}
&\|G(u) - (G(u))_D\|_{s,D} \\
&\quad \leq \|G(u)\|_{s,D} + \|(G(u))_D\|_{s,D} \\
&\quad \leq C_3\|u\|_{s,D} + C_4\|u\|_{s,D} \\
&\quad \leq C_5\|u\|_{s,D}. \quad \blacksquare
\end{aligned}
\tag{3.3.13}
$$

3.3.2 Weighted inequality for Green's operator

We have made necessary preparation in the previous section to prove the following Poincaré-type inequality for Green's operator.

Theorem 3.3.7. *Let $u \in C^\infty(\Omega, \wedge^l)$, $l = 1, 2, \ldots, n$. Assume that $1 < s < \infty$. Then, there exists a constant C, independent of u, such that*

$$
\|G(u) - (G(u))_Q\|_{W^{1,s}(Q)} \leq C\|du\|_{s,Q}
\tag{3.3.14}
$$

for all cubes Q with $Q \subset \Omega$.

Proof. For any differential form u, we have decomposition

$$
u = d(Tu) + T(du).
\tag{3.3.15}
$$

Noticing that $u_Q = d(Tu)$ and replacing u by $G(u)$ in (3.3.15), we obtain

$$
\begin{aligned}
&\|G(u) - (G(u))_Q\|_{W^{1,s}(Q)} \\
&\quad = \|Td(G(u))\|_{W^{1,s}(Q)} \\
&\quad \leq C\mu(Q)\|d(G(u))\|_{s,Q} \\
&\quad = C\|G(d(u))\|_{s,Q} \\
&\quad \leq C\|du\|_{s,Q}. \quad \blacksquare
\end{aligned}
$$

A weighted result follows from the following general inequality.

Theorem 3.3.8. *Suppose that $f, g \geq 0$ a.e. in $\Omega, \beta > 1, p > 0, f \in WRH_{\frac{p\beta}{\beta-1}}(\Omega), w \in A_r(\Omega) \cap RH_\beta(\Omega),$ and*

$$\left(\frac{1}{|Q|} \int_{\sigma Q} f^{\frac{p}{r}} \right)^{\frac{r}{p}} \leq C_1 \left(\frac{1}{|Q|} \int_{\sigma^2 Q} g^{\frac{p}{r}} \right)^{\frac{r}{p}}$$

for all cubes Q with $\sigma^2 Q \subset \Omega$. Then, there exists a constant C_2, depending only on $p, \beta, n, r, \sigma,$ and C_1 such that

$$\left(\frac{1}{|Q|} \int_Q f^p w \right)^{\frac{1}{p}} \leq C_2 \left(\frac{1}{|Q|} \int_{\sigma^2 Q} g^p w \right)^{\frac{1}{p}}$$

for all cubes Q with $\sigma^2 Q \subset \Omega$.

Proof. Since $f \in WRH_{\frac{p\beta}{\beta-1}}(\Omega)$, there exist constants C_1 and $0 < \gamma_0 < \frac{p\beta}{\beta-1}$ such that

$$\left(\frac{1}{|Q|} \int_Q f^{\frac{p\beta}{\beta-1}} dx \right)^{\frac{\beta-1}{p\beta}} \leq C_1 \left(\frac{1}{|Q|} \int_{\sigma Q} f^{\gamma_0} dx \right)^{1/\gamma_0}.$$

Thus, from $(1.4.1)'$, we obtain

$$\left(\frac{1}{|Q|} \int_Q f^{\frac{p\beta}{\beta-1}} \right)^{\frac{\beta-1}{p\beta}} \leq C_3 \left(\frac{1}{|Q|} \int_{\sigma Q} f^{\frac{p}{r}} \right)^{\frac{r}{p}}. \tag{3.3.16}$$

We have using (3.3.16), Holder's inequality, and the assumptions that

$$\left(\frac{1}{|Q|} \int_Q f^p w \right)^{\frac{1}{p}}$$

$$\leq \left(\frac{1}{|Q|} \int_Q w^\beta \right)^{\frac{1}{\beta p}} \left(\frac{1}{|Q|} \int_Q f^{\frac{p\beta}{\beta-1}} \right)^{\frac{\beta-1}{p\beta}}$$

$$\leq \left(\frac{1}{|Q|} \int_Q w \right)^{\frac{1}{p}} \left(\frac{1}{|Q|} \int_{\sigma Q} f^{\frac{p}{r}} \right)^{\frac{r}{p}}$$

$$\leq C_1 \left(\frac{1}{|Q|} \int_Q w \right)^{\frac{1}{p}} \left(\frac{1}{|Q|} \int_{\sigma^2 Q} g^{\frac{p}{r}} \right)^{\frac{r}{p}}$$

$$\leq \left(\frac{1}{|Q|} \int_{\sigma^2 Q} w \right)^{\frac{1}{p}} \left(\frac{1}{|Q|} \int_{\sigma^2 Q} w^{\frac{1}{1-r}} \right)^{\frac{r-1}{p}} \left(\frac{1}{|Q|} \int_{\sigma^2 Q} g^p w \right)^{\frac{1}{p}}$$

$$\leq C_2 \left(\int_{\sigma^2 Q} g^p w \right)^{\frac{1}{p}}. \quad \blacksquare$$

Corollary 3.3.9. *Suppose that $p > r, w \in A_r(\Omega) \cap RH_\beta$, and $|Tu|, |\nabla Tu| \in WRH_{\frac{p\beta}{\beta-1}}(\Omega)$. Then*

$$\|G(u) - (G(u))_Q\|_{W^{1,r}(Q),w} \leq C\|du\|_{p,Q,w}.$$

Proof. Since $p > r$, (3.3.14) holds with $s = \frac{p}{r}$. The result follows then from Theorem 3.3.8. ∎

3.3.3 Global inequality for Green's operator

In [103], Y. Wang and C. Wu proved the following global-weighted Poincaré-type inequality for Green's operator that is applied to the solutions of the nonhomogeneous A-harmonic equation (1.2.10).

We will use the following result from [119].

Theorem 3.3.10. *Suppose that Ω is a John domain, $s > 0, \sigma > 1, w \in A_r(\Omega)$, and f is a measurable function. If there exists a constant C_1 and constants b_Q such that*

$$\|f - f_Q\|_{s,Q,w} \leq C_1 b_Q$$

for all cubes Q with $\sigma Q \subset \Omega$, then for some cube Q_0 there exists a constant C_2, independent of f, such that

$$\|f - f_{Q_0}\|_{s,\Omega,w} \leq C_2 \Sigma_Q b_Q.$$

Combining this result with Theorem 3.3.8, we obtain the following result.

Theorem 3.3.11. *If $u \in D'(\Omega, \wedge^0), w \in A_r(\Omega)$ and Ω is a John domain, then for some cube Q_0 there exists a constant C, independent of u, such that*

$$\|G(u) - (G(u))_{Q_0}\|_{p,\Omega,w} \leq C\|\nabla u\|_{p,\Omega,w}.$$

3.4 Inequalities with Orlicz norms

In this section, we will develop the local and global Poincaré inequalities with Orlicz norms. We will assume that u is a solution of the nonhomogeneous A-harmonic equation

$$d^\star A(x, d\omega) = B(x, d\omega) \tag{3.4.1}$$

which has been introduced in Section 1.2. We will need the following definition of $L^\varphi(\mu)$-domains.

Definition 3.4.1. Let φ be a Young function on $[0, \infty)$ with $\varphi(0) = 0$. We call a proper subdomain $\Omega \subset \mathbf{R}^n$ an $L^\varphi(\mu)$-domain, if $\mu(\Omega) < \infty$ and there exists a constant C such that

$$\int_\Omega \varphi(\sigma|u - u_\Omega|)\, d\mu \leq C \sup_{B \subset \Omega} \int_B \varphi(\sigma\tau|u - u_B|)\, d\mu$$

for all u such that $\varphi(|u|) \in L^1_{loc}(\Omega; \mu)$, where the measure μ is defined by $d\mu = w(x)dx$, $w(x)$ is a weight, and τ, σ are constants with $0 < \tau \leq 1$, $0 < \sigma \leq 1$, and the supremum is over all balls $B \subset \Omega$.

One of the main results of this section is the following global-weighted Poincaré inequality for a solution of the nonhomogeneous A-harmonic equation in an $L^\varphi(\mu)$-domain Ω.

Theorem 3.4.2. *Assume that $\Omega \subset \mathbf{R}^n$ is an $L^\varphi(\mu)$-domain with $\varphi(t) = t^p \log^\alpha(e + t/k)$, where $k = \|u - u_{B_0}\|_{p,\Omega}$, $1 < p < \infty$, and $B_0 \subset \Omega$ is a fixed ball. Let $u \in D'(\Omega, \wedge^0)$ be a solution of the nonhomogeneous A-harmonic equation in Ω and $du \in L^p(\Omega, \wedge^1)$, and $w \in A_r(\Omega)$ for some $r > 1$. Then, there is a constant C, independent of u, such that*

$$\|u - u_\Omega\|_{L^p(\log L)^\alpha(\Omega, w)} \leq C|\Omega|^{1/n}\|du\|_{L^p(\log L)^\alpha(\Omega, w)}$$

for any constant $\alpha > 0$.

3.4.1 Local inequality

To prove Theorem 3.4.2, we need the following local Poincaré inequalities, Theorems 3.4.3 and 3.4.4, with Orlicz norms.

Theorem 3.4.3. *Let $u \in D'(\Omega, \wedge^l)$ be a solution of the nonhomogeneous A-harmonic equation in a domain $\Omega \subset \mathbf{R}^n$ and $du \in L^p(\Omega, \wedge^{l+1})$, $l = 0, 1, \ldots, n - 1$. Assume that $1 < p < \infty$. Then, there is a constant C, independent of u, such that*

$$\|u - u_B\|_{L^p(\log L)^\alpha(B)} \leq C|B|^{1/n}\|du\|_{L^p(\log L)^\alpha(\rho B)}$$

for all balls B with $\rho B \subset \Omega$ and $diam(B) \geq d_0$. Here $\alpha > 0$ is any constant, $\rho > 1$ and $d_0 > 0$ are some constants.

Proof. Let $B \subset \Omega$ be a ball with $diam(B) \geq d_0 > 0$. Choose $\varepsilon > 0$ small enough and a constant C_1 such that

$$|B|^{-\varepsilon/p^2} \le C_1. \tag{3.4.2}$$

From Lemma 3.1.1, we find that

$$\|u - u_B\|_{p+\varepsilon, B} \le C_2 |B|^{(p-(p+\varepsilon))/p(p+\varepsilon)} \|u - u_B\|_{p, \sigma_1 B} \tag{3.4.3}$$

for some $\sigma_1 > 1$. We assume that

$$\frac{|u - u_B|}{\|u - u_B\|_{p, B}} \ge 1.$$

Then, for above $\varepsilon > 0$, there exists $C_3 > 0$ such that

$$\log^\alpha \left(e + \frac{|u - u_B|}{\|u - u_B\|_{p, B}} \right) \le C_3 \left(\frac{|u - u_B|}{\|u - u_B\|_{p, \sigma_1 B}} \right)^\varepsilon. \tag{3.4.4}$$

Otherwise, setting

$$B_1 = \left\{ x \in B : \frac{|u - u_B|}{\|u - u_B\|_{p, B}} \ge 1 \right\},$$
$$B_2 = \left\{ x \in B : \frac{|u - u_B|}{\|u - u_B\|_{p, B}} < 1 \right\}$$

and using the elementary inequality $|a + b|^s \le 2^s(|a|^s + |b|^s)$, where $s > 0$ is any constant, we find

$$\|u - u_B\|_{L^p(\log L)^\alpha(B)}$$
$$= \left(\int_B |u - u_B|^p \log^\alpha \left(e + \frac{|u - u_B|}{\|u - u_B\|_{p, B}} \right) dx \right)^{1/p}$$
$$= \left(\int_{B_1} |u - u_B|^p \log^\alpha \left(e + \frac{|u - u_B|}{\|u - u_B\|_{p, B}} \right) dx \right.$$
$$\left. + \int_{B_2} |u - u_B|^p \log^\alpha \left(e + \frac{|u - u_B|}{\|u - u_B\|_{p, B}} \right) dx \right)^{\frac{1}{p}} \tag{3.4.5}$$
$$\le 2^{1/p} \left(\int_{B_1} |u - u_B|^p \log^\alpha \left(e + \frac{|u - u_B|}{\|u - u_B\|_{p, B}} \right) dx \right)^{\frac{1}{p}}$$
$$+ 2^{1/p} \left(\int_{B_2} |u - u_B|^p \log^\alpha \left(e + \frac{|u - u_B|}{\|u - u_B\|_{p, B}} \right) dx \right)^{\frac{1}{p}}.$$

Now, we shall estimate the first term of the right side. Since

$$\frac{|u - u_B|}{\|u - u_B\|_{p, B}} \ge 1$$

on B_1, for $\varepsilon > 0$ in (3.4.2), there exists $C_4 > 0$ such that

$$\log^\alpha \left(e + \frac{|u - u_B|}{\|u - u_B\|_{p, B}} \right) \le C_4 \left(\frac{|u - u_B|}{\|u - u_B\|_{p, \sigma_1 B}} \right)^\varepsilon. \tag{3.4.6}$$

Combining (3.4.3), (3.4.4), and (3.4.6), we obtain

$$\left(\int_{B_1} |u - u_B|^p log^\alpha \left(e + \frac{|u - u_B|}{\|u - u_B\|_{p,B}} \right) dx \right)^{1/p}$$

$$\leq C_5 \left(\frac{1}{\|u - u_B\|_{p,\sigma_1 B}^\varepsilon} \int_{B_1} |u - u_B|^{p+\varepsilon} dx \right)^{1/p}$$

$$\leq C_5 \left(\frac{1}{\|u - u_B\|_{p,\sigma_1 B}^\varepsilon} \int_B |u - u_B|^{p+\varepsilon} dx \right)^{1/p}$$

$$= \frac{C_5}{\|u - u_B\|_{p,\sigma_1 B}^{\varepsilon/p}} \left(\left(\int_B |u - u_B|^{p+\varepsilon} dx \right)^{\frac{1}{p+\varepsilon}} \right)^{(p+\varepsilon)/p} \qquad (3.4.7)$$

$$\leq \frac{C_6}{\|u - u_B\|_{p,\sigma_1 B}^{\varepsilon/p}} \left(|B|^{(p-(p+\varepsilon))/p(p+\varepsilon)} \|u - u_B\|_{p,\sigma_1 B} \right)^{(p+\varepsilon)/p}$$

$$\leq C_7 \|u - u_B\|_{p,\sigma_1 B},$$

where $\sigma_1 > 1$ is a constant. For the second term of (3.4.5), since

$$log^\alpha \left(e + \frac{|u - u_B|}{\|u - u_B\|_{p,B}} \right) \leq M_1 log^\alpha (e + 1) \leq M_2, \quad x \in B_2,$$

we obtain the similar estimate

$$\left(\int_{B_2} |u - u_B|^p log^\alpha \left(e + \frac{|u - u_B|}{\|u - u_B\|_{p,B}} \right) dx \right)^{1/p} \leq C_8 \|u - u_B\|_{p,\sigma_2 B},$$
$$(3.4.8)$$

where $\sigma_2 > 1$ is a constant. From (3.4.5), (3.4.7), and (3.4.8), we have

$$\|u - u_B\|_{L^p (\log L)^\alpha (B)} \leq C_9 \|u - u_B\|_{p,\sigma_3 B}, \qquad (3.4.9)$$

where $\sigma_3 = \max\{\sigma_1, \sigma_2\}$. Applying Theorem 3.2.9 with $w = 1$, we obtain

$$\|u - u_B\|_{p,\sigma_3 B} \leq C_{10} |B|^{1/n} \|du\|_{p,\sigma_4 B}$$

for some $\sigma_4 > \sigma_3$. Note that

$$log^\alpha \left(e + \frac{|u - u_B|}{\|u - u_B\|_{p,\sigma_2 B}} \right) \geq 1, \quad \alpha > 0. \qquad (3.4.10)$$

Now the combination of the last three inequalities yields

$$\|u - u_B\|_{L^p (\log L)^\alpha (B)} \leq C_{11} |B|^{1/n} \|du\|_{p,\sigma_4 B}$$

$$\leq C_{12} |B|^{1/n} \|du\|_{L^p (\log L)^\alpha (\sigma_4 B)}. \qquad \blacksquare$$

3.4.2 Weighted inequalities

It is easy to see that for any constant k, there exist constants $m > 0$ and $M > 0$, such that

$$m \log(e + t) \leq \log \left(e + \frac{t}{k} \right) \leq M \log(e + t), \quad t > 0. \tag{3.4.11}$$

From the weak reverse Hölder inequality (Lemma 3.1.1), we know that the norms $\|u\|_{s,B}$ and $\|u\|_{t,B}$ are comparable when $0 < d_1 \leq diam(B) \leq d_2 < \infty$. Hence, we may assume that $0 < m_1 \leq \|u\|_{s,B} \leq M_1 < \infty$ and $0 < m_2 \leq \|u\|_{t,B} \leq M_2 < \infty$ for some constants m_i and M_i, $i = 1, 2$. Thus, we have

$$C_1 \log \left(e + |u| \right) \leq \log \left(e + \frac{|u|}{\|u\|_{s,B}} \right) \leq C_2 \log \left(e + |u| \right) \tag{3.4.12}$$

and

$$C_3 \log(e + |u|) \leq \log \left(e + \frac{|u|}{\|u\|_{t,B}} \right) \leq C_4 \log \left(e + |u| \right) \tag{3.4.13}$$

for any $s > 0$ and $t > 0$, where C_i are constants, $i = 1, 2, 3, 4$. Using (3.4.12) and (3.4.13), we obtain

$$C_5 \left(\int_B |u|^s \log^\alpha \left(e + \frac{|u|}{\|u\|_{t,B}} \right) dx \right)^{1/s}$$

$$\leq \|u\|_{L^s(\log L)^\alpha(B)} \tag{3.4.14}$$

$$\leq C_6 \left(\int_B |u|^s \log^\alpha \left(e + \frac{|u|}{\|u\|_{t,B}} \right) dx \right)^{1/s}$$

and

$$C_7 \|u\|_{L^t(\log L)^\alpha(B)} \leq \left(\int_B |u|^t \log^\alpha \left(e + \frac{|u|}{\|u\|_{s,B}} \right) dx \right)^{\frac{1}{t}} \leq C_8 \|u\|_{L^t(\log L)^\alpha(B)} \tag{3.4.15}$$

for any ball B and any $s > 0, t > 0$, and $\alpha > 0$. Consequently, we find that $\|u\|_{L^s(\log L)^\alpha(B)} < \infty$ if and only if

$$\left(\int_B |u|^s \log^\alpha \left(e + \frac{|u|}{\|u\|_{t,B}} \right) dx \right)^{1/s} < \infty.$$

Theorem 3.4.4. *Let $u \in D'(\Omega, \wedge^l)$ be a solution of the nonhomogeneous A-harmonic equation in a domain $\Omega \subset \mathbf{R}^n$ and $du \in L^p(\Omega, \wedge^{l+1})$, $l = 0, 1, \ldots, n - 1$. Assume that $1 < p < \infty$ and $w \in A_r(\Omega)$ for some $r > 1$.*

Then, there is a constant C, independent of u, such that

$$\|u - u_B\|_{L^p(\log L)^\alpha(B,w)} \leq C|B|^{1/n}\|du\|_{L^p(\log L)^\alpha(\sigma B,w)}$$

for all balls B with $\sigma B \subset \Omega$ and $\operatorname{diam}(B) \geq d_0$. Here $\alpha > 0$ is any constant, $\sigma > 1$ and $d_0 > 0$ are some constants.

Proof. In view of Lemma 1.4.7, there exist constants $k > 1$ and $C_0 > 0$, such that

$$\| w \|_{k,B} \leq C_0|B|^{(1-k)/k} \| w \|_{1,B} . \tag{3.4.16}$$

Choose $s = kp/(k-1)$, so that $p < s$. Using the Hölder inequality with $1/p = 1/s + (s-p)/sp$, (3.4.16), and (3.4.14), we obtain

$$\|u - u_B\|_{L^p(\log L)^\alpha(B,w)}$$

$$= \left(\int_B \left(|u - u_B| \log^{\frac{\alpha}{p}}\left(e + \frac{|u-u_B|}{\|u-u_B\|_{p,B}}\right) w^{1/p}\right)^p dx\right)^{1/p}$$

$$\leq \left(\int_B |u - u_B|^s \log^{\frac{\alpha s}{p}}\left(e + \frac{|u-u_B|}{\|u-u_B\|_{p,B}}\right) dx\right)^{1/s} \left(\int_B w^{s/(s-p)} dx\right)^{(s-p)/ps}$$

$$\leq C_1 \left(\int_B |u - u_B|^s \log^{\frac{\alpha s}{p}}\left(e + \frac{|u-u_B|}{\|u-u_B\|_{s,B}}\right) dx\right)^{1/s} \left(\left(\int_B w^k dx\right)^{1/k}\right)^{1/p}$$

$$\leq C_2|B|^{(1-k)/kp} \| w \|_{1,B}^{1/p} \left(\int_B |u - u_B|^s \log^{\frac{\alpha s}{p}}\left(e + \frac{|u-u_B|}{\|u-u_B\|_{s,B}}\right) dx\right)^{1/s} . \tag{3.4.17}$$

Applying Theorem 3.4.3, we find that

$$\left(\int_B |u - u_B|^s \log^{\frac{\alpha s}{p}}\left(e + \frac{|u-u_B|}{\|u-u_B\|_{s,B}}\right) dx\right)^{1/s}$$

$$\leq C_3|B|^{1/n}\|du\|_{L^s(\log L)^{\alpha s/p}(\sigma_1 B)}, \tag{3.4.18}$$

where $\sigma_1 > 1$ is some constant. Let $t = p/r$. From Lemma 2.6 in [205] with $\beta = \alpha/r$, we have

$$\left(\int_{\sigma_1 B} |du|^s \log^{\frac{\alpha s}{p}}\left(e + \frac{|du|}{\|du\|_{s,B}}\right) dx\right)^{1/s}$$

$$\leq C_4|B|^{(t-s)/st}\|du\|_{L^t(\log L)^\beta(\sigma_2 B)} \tag{3.4.19}$$

for some $\sigma_2 > \sigma_1$. Using the Hölder inequality again with $1/t = 1/p + (p-t)/pt$, we obtain

$$\|du\|_{L^t(\log L)^\beta(\sigma_2 B)}$$

$$= \left(\int_{\sigma_2 B} |du|^t \log^\beta\left(e + \frac{|du|}{\|du\|_{t,\sigma_2 B}}\right) dx\right)^{1/t}$$

$$= \left(\int_{\sigma_2 B} \left(|du| \log^{\beta/t} \left(e + \frac{|du|}{\|du\|_{t,\sigma_2 B}} \right) w^{1/p} w^{-1/p} \right)^t dx \right)^{1/t}$$

$$\leq \left(\int_{\sigma_2 B} |du|^p \log^{\beta p/t} \left(e + \frac{|du|}{\|du\|_{t,\sigma_2 B}} \right) w dx \right)^{1/p} \left(\int_{\sigma_2 B} \left(\frac{1}{w} \right)^{\frac{t}{p-t}} dx \right)^{(p-t)/pt}$$

$$\leq \|du\|_{L^p(\log L)^\alpha(\sigma_2 B, w)} \left(\int_{\sigma_2 B} \left(\frac{1}{w} \right)^{\frac{1}{r-1}} dx \right)^{(r-1)/p}.$$

(3.4.20)

Combining (3.4.17), (3.4.18), (3.4.19), and (3.4.20), we have

$$\|u - u_B\|_{L^p(\log L)^\alpha(B,w)}$$

$$\leq C_5 |B|^{\frac{1}{n} - \frac{r}{p}} \|du\|_{L^p(\log L)^\alpha(\sigma_2 B, w)} \left(\int_B w dx \left(\int_{\sigma_2 B} \left(\frac{1}{w} \right)^{\frac{1}{r-1}} dx \right)^{(r-1)} \right)^{1/p}$$

$$\leq C_5 |B|^{\frac{1}{n} - \frac{r}{p}} \|du\|_{L^p(\log L)^\alpha(\sigma_2 B, w)} \left(\|w\|_{1,\sigma_2 B} \cdot \|1/w\|_{1/(r-1),\sigma_2 B} \right)^{1/p}.$$

(3.4.21)

Now since $w \in A_r(\Omega)$, it follows that

$$\left(\|w\|_{1,\sigma_2 B} \cdot \|1/w\|_{1/(r-1),\sigma_2 B} \right)^{1/p}$$

$$\leq \left(\left(\int_{\sigma_2 B} w dx \right) \left(\int_{\sigma_2 B} (1/w)^{1/(r-1)} dx \right)^{r-1} \right)^{1/p}$$

$$= \left(|\sigma_2 B|^r \left(\frac{1}{|\sigma_2 B|} \int_{\sigma_2 B} w dx \right) \left(\frac{1}{|\sigma_2 B|} \int_{\sigma_2 B} \left(\frac{1}{w} \right)^{1/(r-1)} dx \right)^{r-1} \right)^{1/p}$$

$$\leq C_6 |B|^{r/p}.$$

(3.4.22)

Substituting (3.4.22) into (3.4.21), we find

$$\|u - u_B\|_{L^p(\log L)^\alpha(B,w)} \leq C_7 |B|^{1/n} \|du\|_{L^p(\log L)^\alpha(\sigma_2 B, w)}. \quad \blacksquare \quad (3.4.23)$$

3.4.3 The proof of the global inequality

Now, we are ready to prove our global inequality given in Theorem 3.4.2.

Proof of Theorem 3.4.2. For any constants $k_i > 0$, $i = 1, 2, 3$, there are constants $C_1 > 0$ and $C_2 > 0$ such that

$$C_1 \log \left(e + \frac{t}{k_1} \right) \leq \log \left(e + \frac{t}{k_2} \right) \leq C_2 \log \left(e + \frac{t}{k_3} \right) \quad (3.4.24)$$

for any $t > 0$. Therefore, we have

$$C_1 \left(\int_B |u|^t \log^\alpha \left(e + \tfrac{|u|}{k_1} \right) dx \right)^{\frac{1}{t}}$$

$$\leq \left(\int_B |u|^t \log^\alpha \left(e + \tfrac{|u|}{k_2} \right) dx \right)^{\frac{1}{t}} \qquad (3.4.25)$$

$$\leq C_2 \left(\int_B |u|^t \log^\alpha \left(e + \tfrac{|u|}{k_3} \right) dx \right)^{\frac{1}{t}}.$$

By properly selecting constants k_i in (3.4.25), we obtain inequalities that we will need. Using the definition of $L^\varphi(\mu)$-domains with $\tau = 1, \sigma = 1$, and $\varphi(t) = t^p \log^\alpha (e + t/k)$, where $k = \|u - u_{B_0}\|_{p,\Omega}$, Theorem 3.4.4, and (3.4.25), we obtain

$$\|u - u_\Omega\|^p_{L^p(\log L)^\alpha(\Omega,w)}$$

$$= \int_\Omega |u - u_\Omega|^p \log^\alpha \left(e + \frac{|u-u_\Omega|}{\|u-u_{B_0}\|_{p,\Omega}} \right) w\,dx$$

$$\leq C_1 \sup_{B \subset \Omega} \int_B |u - u_B|^p \log^\alpha \left(e + \frac{|u-u_B|}{\|u-u_{B_0}\|_{p,\Omega}} \right) w\,dx$$

$$\leq C_2 \sup_{B \subset \Omega} \int_B |u - u_B|^p \log^\alpha \left(e + \frac{|u-u_B|}{\|u-u_B\|_{p,B}} \right) w\,dx$$

$$\leq C_3 \sup_{B \subset \Omega} |B|^{p/n} \|du\|^p_{L^p(\log L)^\alpha(\sigma B,w)}$$

$$\leq C_3 \sup_{B \subset \Omega} |\Omega|^{p/n} \|du\|^p_{L^p(\log L)^\alpha(\Omega,w)}$$

$$\leq C_3 |\Omega|^{p/n} \|du\|^p_{L^p(\log L)^\alpha(\Omega,w)}$$

which is equivalent to

$$\|u - u_\Omega\|_{L^p(\log L)^\alpha(\Omega,w)} \leq C|\Omega|^{1/n} \|du\|_{L^p(\log L)^\alpha(\Omega,w)}. \qquad \blacksquare$$

We can prove that the global inequality also holds in δ-John domains. Specifically, we have the following result.

Theorem 3.4.5. *Let $u \in D'(\Omega, \wedge^0)$ be a solution of the nonhomogeneous A-harmonic equation in a δ-John domain $\Omega \subset \mathbf{R}^n$ and $du \in L^p(\Omega, \wedge^1)$. Assume that $1 < p < \infty$ and $w \in A_r(\Omega)$ for some $r > 1$. Then, there is a constant C, independent of u, such that*

$$\|u - u_\Omega\|_{L^p(\log L)^\alpha(\Omega,w)} \leq C|\Omega|^{1/n} \|du\|_{L^p(\log L)^\alpha(\Omega,w)}$$

for any constant $\alpha > 0$.

Note. In this section, we have only included the global inequalities with Orlicz norms in two kinds of domains, $L^\varphi(\mu)$-domains and δ-John domains.

However, the local results can be extended in some other types of domains also, such as uniform domains and L^s-averaging domains.

3.5 Two-weight inequalities

In recent years, several versions of the two-weight Poincaré inequalities have been developed, see [206, 82, 207], for example. Here we state and prove the results of Y. Wang [206].

3.5.1 Statements of two-weight inequalities

Theorem 3.5.1. *Let $u \in D'(B, \wedge^l)$ and $du \in L^t(B, \wedge^{l+1})$, $l = 0, 1, \ldots, n-1$. Then, there exists a constant $\beta > 1$ such that if $w_1 \in A_r^1$ and $(w_1, w_2) \in A_{t/s}^{s/t}$, where $1 < s < n$, $t = s\beta$, and $r > 1$, we have*

$$\left(\frac{1}{|B|} \int_B |u - u_B|^s w_1 dx \right)^{1/s} \le C \left(\int_B |du|^t w_2 dx \right)^{1/t} \qquad (3.5.1)$$

for all balls $B \subset \mathbf{R}^n$. Here C is a constant independent of u and du.

Theorem 3.5.2. *Let $u \in D'(B, \wedge^l)$ and $du \in L^n(B, \wedge^{l+1})$, $l = 0, 1, \ldots, n-1$. If $1 < s < n$ and $(w_1, w_2) \in A_{n/s}^1$, then there exists a constant C, independent of u and du, such that*

$$\left(\frac{1}{|B|} \int_B |u - u_B|^s w_1^{s/n} dx \right)^{1/s} \le C \left(\int_B |du|^n w_2 dx \right)^{1/n} \qquad (3.5.2)$$

for any ball or cube $B \subset \mathbf{R}^n$.

We remark that the exponents t and n on the right-hand sides of (3.5.1) and (3.5.2) can be improved. In fact, the following result is with the sharper right-hand side.

Theorem 3.5.3. *Let $u \in D'(B, \wedge^l)$ and $du \in L^t(B, \wedge^{l+1})$, $l = 0, 1, \ldots, n-1$. Then, there exists a constant $\beta > 1$ such that if $w_1 \in A_r^1$ and $(w_1, w_2) \in A_r^{\alpha s/t}$, where $1 < s < n$, $t = s + \alpha s(r - 1)$, and $r > 1$, it follows that*

$$\left(\frac{1}{|B|^\beta} \int_B |u - u_B|^s w_1 dx \right)^{1/s} \le C|B|^{\alpha(r-1)/t} \left(\int_B |du|^t w_2^\alpha dx \right)^{1/t} \qquad (3.5.3)$$

for all balls $B \subset \mathbf{R}^n$ and any constant $\alpha > 0$. Here C is a constant independent of u and du.

Clearly, in this result if $\alpha \to 0$, then $t \to s$.

Theorem 3.5.4. *Let $u \in D'(\Omega, \wedge^0)$ and $du \in L^t(\Omega, \wedge^1)$. Then, there exists a constant $\beta > 1$ such that if $w_1 \in A_r^1$ and $(w_1, w_2) \in A_{t/s}^{s/t}$, where $1 < s < n$, $t = s\beta$, $r > 1$, and $w_1 > w_2 \geq \eta > 0$, we have*

$$\left(\frac{1}{\mu_1(\Omega)} \int_\Omega |u - u_{B_0}|^s d\mu_1 \right)^{1/s} \leq C \left(\int_\Omega |du|^t d\mu_2 \right)^{1/t} \tag{3.5.4}$$

for any $L^s(\mu_1)$-averaging domain Ω and some ball B_0 with $2B_0 \subset \Omega$. Here the measures μ_1 and μ_2 are defined by $d\mu_1 = w_1(x)dx$, $d\mu_2 = w_2(x)dx$, and C is a constant independent of u and du.

3.5.2 Proofs of the main theorems

Proof of Theorem 3.5.1. Since $w_1 \in A_r^1$, $r > 1$, by Lemma 1.4.7, there exist constants $\beta > 1$ and $C_1 > 0$, such that

$$\| w_1 \|_{\beta, B} \leq C_1 |B|^{(1-\beta)/\beta} \| w_1 \|_{1,B} \tag{3.5.5}$$

for any cube or ball $B \subset \mathbf{R}^n$. Let $s = n/\beta$, so that $\beta = n/s$. Applying Hölder's inequality, Theorem 3.2.2, and (3.5.5), we find that

$$\left(\int_B |u - u_B|^s w_1 dx \right)^{1/s}$$

$$\leq \left(\int_B \left(w_1^{1/s} \right)^n dx \right)^{1/n} \left(\int_B |u - u_B|^{ns/(n-s)} dx \right)^{(n-s)/ns}$$

$$\leq C_2 \| du \|_{s,B} \left(\int_B w_1^{n/s} dx \right)^{1/n} \tag{3.5.6}$$

$$= C_2 \| du \|_{s,B} \cdot \| w_1 \|_{\beta, B}^{1/s}$$

$$\leq C_3 |B|^{(1-\beta)/s\beta} \| w_1 \|_{1,B}^{1/s} \cdot \| du \|_{s,B}.$$

Setting $t = s\beta$ and using Hölder's inequality again, we obtain

$$\| du \|_{s,B} = \left(\int_B \left(|du| w_2^{1/t} w_2^{-1/t} \right)^s dx \right)^{1/s}$$

$$\leq \left(\int_B \left(|du| w_2^{1/t} \right)^t dx \right)^{1/t} \left(\int_B \left(\frac{1}{w_2} \right)^{s/(t-s)} dx \right)^{(t-s)/ts} \tag{3.5.7}$$

$$= \left(\int_B |du|^t w_2 dx \right)^{1/t} \cdot \left\| (1/w_2)^{1/t} \right\|_{ts/(t-s), B}.$$

Combination of (3.5.6) and (3.5.7) gives

$$\left(\int_B |u - u_B|^s w_1 dx\right)^{1/s}$$

$$\leq C_3 |B|^{(1-\beta)/s\beta} \|w_1\|_{1,B}^{1/s} \cdot \left\|(1/w_2)^{1/t}\right\|_{ts/(t-s),B} \cdot \left(\int_B |du|^t w_2 dx\right)^{1/t}.$$

$$(3.5.8)$$

Using the condition $(w_1, w_2) \in A_{t/s}^{s/t}$, we have

$$\|w_1\|_{1,B}^{1/s} \cdot \left\|(1/w_2)^{1/t}\right\|_{ts/(t-s),B}$$

$$= \left(\int_B w_1 dx\right)^{1/s} \left(\int_B \left(\frac{1}{w_2}\right)^{s/(t-s)} dx\right)^{(t-s)/ts}$$

$$= \left(\left(\int_B w_1 dx\right)\left(\int_B \left(\frac{1}{w_2}\right)^{1/(t/s-1)} dx\right)^{(s/t)(t/s-1)}\right)^{1/s}$$

$$= \left(|B|^{1+(s/t)(t/s-1)}\left(\frac{1}{|B|}\int_B w_1 dx\right)\left(\frac{1}{|B|}\int_B \left(\frac{1}{w_2}\right)^{1/(t/s-1)} dx\right)^{\frac{s}{t}(\frac{t}{s}-1)}\right)^{1/s}$$

$$\leq C_4 |B|^{2/s-1/t}.$$

$$(3.5.9)$$

Now substituting (3.5.9) into (3.5.8) and using $t = s\beta$, we find that

$$\left(\int_B |u - u_B|^s w_1 dx\right)^{1/s} \leq C_5 |B|^{1/s} \left(\int_B |du|^t w_2 dx\right)^{1/t},$$

that is

$$\left(\frac{1}{|B|}\int_B |u - u_B|^s w_1 dx\right)^{1/s} \leq C_5 \left(\int_B |du|^t w_2 dx\right)^{1/t}. \qquad \blacksquare$$

The proof of Theorem 3.5.2 is similar to that of Theorem 3.5.1, and therefore not included here.

Proof of Theorem 3.5.3. Let $s = n/\beta$, where β is the constant appearing in Lemma 1.4.7. Applying Hölder's inequality, Theorem 3.2.2, and (3.5.5), we find that

$$\left(\int_B |u - u_B|^s w_1 dx\right)^{1/s}$$

$$\leq \left(\int_B \left(w_1^{1/s}\right)^n dx\right)^{1/n} \left(\int_B |u - u_B|^{ns/(n-s)} dx\right)^{(n-s)/ns}$$

$$\leq C_2 \|du\|_{s,B} \left(\int_B w_1^{n/s} dx\right)^{1/n}$$

$$= C_2 \|du\|_{s,B} \cdot \|w_1\|_{\beta,B}^{1/s}$$

$$\leq C_3 |B|^{(1-\beta)/s\beta} \|w_1\|_{1,B}^{1/s} \cdot \|du\|_{s,B}. \tag{3.5.10}$$

Let $t = s + \alpha s(r-1)$. Applying Hölder's inequality again, we obtain

$$\|du\|_{s,B} = \left(\int_B \left(|du| w_2^{\alpha/t} w_2^{-\alpha/t} \right)^s dx \right)^{1/s}$$

$$\leq \left(\int_B \left(|du| w_2^{\alpha/t} \right)^t dx \right)^{1/t} \left(\int_B \left(\frac{1}{w_2} \right)^{\alpha s/(t-s)} dx \right)^{(t-s)/ts} \tag{3.5.11}$$

$$= \left(\int_B |du|^t w_2^\alpha dx \right)^{1/t} \cdot \left(\int_B \left(\frac{1}{w_2} \right)^{1/(r-1)} dx \right)^{\alpha(r-1)/t}.$$

Combination of (3.5.10) and (3.5.11) gives

$$\left(\int_B |u - u_B|^s w_1 dx \right)^{1/s}$$

$$\leq C_3 |B|^{(1-\beta)/s\beta} \|w_1\|_{1,B}^{1/s} \cdot \left(\int_B \left(\frac{1}{w_2} \right)^{\frac{1}{r-1}} dx \right)^{\frac{\alpha(r-1)}{t}} \cdot \left(\int_B |du|^n w_2^\alpha dx \right)^{1/t}. \tag{3.5.12}$$

Using the condition $(w_1, w_2) \in A_r^{\alpha s/t}$, we have

$$\|w_1\|_{1,B}^{1/s} \cdot \left(\int_B \left(\frac{1}{w_2} \right)^{1/(r-1)} dx \right)^{\alpha(r-1)/t}$$

$$= \left(|B|^{1+\alpha s(r-1)/t} \left(\frac{1}{|B|} \int_B w_1 dx \right) \left(\frac{1}{|B|} \int_B \left(\frac{1}{w_2} \right)^{\frac{1}{r-1}} dx \right)^{(\alpha s/t)(r-1)} \right)^{1/s}$$

$$\leq C_4 |B|^{1/s + \alpha(r-1)/t}. \tag{3.5.13}$$

Substituting (3.5.13) into (3.5.12), we find that

$$\left(\int_B |u - u_B|^s w_1 dx \right)^{1/s} \leq C_5 |B|^{1/\beta s + \alpha(r-1)/t} \left(\int_B |du|^t w_2^\alpha dx \right)^{1/t}$$

which is equivalent to (3.5.3). ∎

Proof of Theorem 3.5.4. Since $d\mu_2 = w_2(x) dx$ and $w_2 \geq \eta > 0$, we have

$$\mu_2(B) = \int_B w_2 dx \geq \int_B \eta dx = \eta |B|,$$

that is

$$\frac{|B|}{\mu_2(B)} \leq C_1, \tag{3.5.14}$$

where $C_1 = 1/\eta$. From Theorem 3.5.1, definition of $L^s(\mu)$-averaging domains, and inequality (3.5.14), we obtain

$$\left(\frac{1}{\mu_1(\Omega)} \int_\Omega |u - u_{B_0}|^s d\mu_1 \right)^{1/s}$$

$$\leq \left(\frac{1}{\mu_1(B_0)} \int_\Omega |u - u_{B_0}|^s d\mu_1 \right)^{1/s}$$

$$\leq C_2 \sup_{2B \subset \Omega} \left(\frac{1}{\mu_1(B)} \int_B |u - u_B|^s d\mu_1 \right)^{1/s}$$

$$= C_2 \sup_{2B \subset \Omega} \left(\left(\frac{|B|}{\mu_1(B)} \right)^{1/s} \left(\frac{1}{|B|} \int_B |u - u_B|^s d\mu_1 \right)^{1/s} \right)$$

$$\leq C_2 \sup_{2B \subset \Omega} \left(\left(\frac{|B|}{\mu_2(B)} \right)^{1/s} C_3 \left(\int_B |du|^t d\mu_2 \right)^{1/t} \right)$$

$$\leq C_4 \sup_{2B \subset \Omega} \left(\int_\Omega |du|^t d\mu_2 \right)^{1/t}$$

$$= C_4 \left(\int_\Omega |du|^t d\mu_2 \right)^{1/t}. \qquad \blacksquare$$

Remark. (1) Theorems 3.5.2 and 3.5.3 can be extended to the global versions. (2) From [184], we know that John domains are $L^s(\mu)$-averaging domains. Thus, the global results and Theorem 3.5.4 also hold if Ω is a John domain.

3.5.3 $A_r^\lambda(\Omega)$-weighted inequalities

Next, we discuss the following version of two-weight Poincaré inequality for differential forms.

Theorem 3.5.5. *Let $u \in D'(\Omega, \wedge^l)$ be a differential form satisfying the A-harmonic equation (1.2.10) in a domain $\Omega \subset \mathbf{R}^n$ and $du \in L^s(\Omega, \wedge^{l+1})$, $l = 0, 1, \ldots, n-1$. Suppose that $(w_1, w_2) \in A_r^\lambda(\Omega)$ for some $r > 1$ and $\lambda > 0$. If $0 < \alpha < 1, \sigma > 1$, and $s > \alpha\lambda(r-1) + 1$, then there exists a constant C, independent of u, such that*

$$\left(\int_B |u - u_B|^s w_1^\alpha dx \right)^{1/s} \leq C|B|^{1/n} \left(\int_{\sigma B} |du|^s w_2^{\alpha\lambda} dx \right)^{1/s} \tag{3.5.15}$$

for all balls B with $\sigma B \subset \Omega$. Here u_B is a closed form.

Proof. Choose $t = s/(1-\alpha)$, so that $1 < s < t$. Since $1/s = 1/t + (t-s)/st$, by Hölder's inequality, we find that

$$
\left(\int_B |u - u_B|^s w_1^\alpha dx\right)^{1/s}
$$

$$
= \left(\int_B (|u - u_B| w_1^{\alpha/s})^s dx\right)^{1/s}
$$

$$
\leq \left(\int_B |u - u_B|^t dx\right)^{1/t} \left(\int_B w_1^{\alpha t/(t-s)} dx\right)^{(t-s)/st} \tag{3.5.16}
$$

$$
= \|u - u_B\|_{t,B} \left(\int_B w_1 dx\right)^{\alpha/s}.
$$

Next, choose

$$
m = \frac{s}{\alpha\lambda(r-1)+1},
$$

so that $m > 1$. Since u_B is a closed form, using Lemma 3.1.1 and Theorem 3.2.3, we have

$$
\|u - u_B\|_{t,B}
$$

$$
\leq C_1 |B|^{(m-t)/mt} \|u - u_B\|_{m,\sigma B} \tag{3.5.17}
$$

$$
\leq C_2 |B|^{(m-t)/mt} |B|^{1/n} \|du\|_{m,\sigma B}
$$

for all balls B with $\sigma B \subset \Omega$. Now since $1/m = 1/s + (s-m)/sm$, by Hölder's inequality again, we obtain

$$
\|du\|_{m,\sigma B}
$$

$$
= \left(\int_{\sigma B} \left(|du| w_2^{\alpha\lambda/s} w_2^{-\alpha\lambda/s}\right)^m dx\right)^{1/m}
$$

$$
\leq \left(\int_{\sigma B} |du|^s w_2^{\alpha\lambda} dx\right)^{1/s} \left(\int_{\sigma B} \left(\frac{1}{w_2}\right)^{\alpha\lambda m/(s-m)} dx\right)^{(s-m)/sm} \tag{3.5.18}
$$

$$
= \left(\int_{\sigma B} |du|^s w_2^{\alpha\lambda} dx\right)^{1/s} \left(\int_{\sigma B} \left(\frac{1}{w_2}\right)^{1/(r-1)} dx\right)^{\alpha\lambda(r-1)/s}.
$$

From (3.5.16), (3.5.17), and (3.5.18), we have

$$
\left(\int_B |u - u_B|^s w_1^\alpha dx\right)^{1/s}
$$

$$
\leq C_2 |B|^{(m-t)/tm} |B|^{1/n} \left(\int_B w_1 dx\right)^{\alpha/s} \tag{3.5.19}
$$

$$
\times \left(\int_{\sigma B} \left(\frac{1}{w_2}\right)^{1/(r-1)} dx\right)^{\alpha\lambda(r-1)/s} \left(\int_{\sigma B} |du|^s w_2^{\alpha\lambda} dx\right)^{1/s}.
$$

Using the condition $(w_1, w_2) \in A_r^\lambda(\Omega)$, we find

$$\left(\int_B w_1 dx\right)^{\alpha/s} \left(\int_{\sigma B} \left(\tfrac{1}{w_2}\right)^{1/(r-1)} dx\right)^{\alpha\lambda(r-1)/s}$$

$$\leq \left[\left(\int_B w_1 dx\right) \left(\int_{\sigma B} \left(\tfrac{1}{w_2}\right)^{1/(r-1)} dx\right)^{\lambda(r-1)}\right]^{\alpha/s}$$

$$= \left[|\sigma B|^{\lambda(r-1)+1} \left(\tfrac{1}{|\sigma B|} \int_B w_1 dx\right)\right. \tag{3.5.20}$$

$$\left. \times \left(\tfrac{1}{|\sigma B|} \int_{\sigma B} \left(\tfrac{1}{w_2}\right)^{1/(r-1)} dx\right)^{\lambda(r-1)}\right]^{\alpha/s}$$

$$\leq C_3 |\sigma B|^{\alpha\lambda(r-1)/s + \alpha/s}$$

$$\leq C_4 |B|^{\alpha\lambda(r-1)/s + \alpha/s}.$$

Now substituting (3.5.20) in (3.5.19) and using

$$\frac{m-t}{mt} = \frac{-\alpha}{s} - \frac{\alpha\lambda(r-1)}{s},$$

we obtain

$$\left(\int_B |u - u_B|^s w_1^\alpha dx\right)^{1/s} \leq C|B|^{1/n} \left(\int_{\sigma B} |du|^s w_2^{\alpha\lambda} dx\right)^{1/s}. \qquad \blacksquare$$

If we choose $\lambda = 1/\alpha$ in Theorem 3.5.5, we obtain the following version of the A_r^λ-weighted Poincaré inequality.

Corollary 3.5.6. *Let $u \in D'(\Omega, \wedge^l)$ be a differential form satisfying the A-harmonic equation (1.2.10) in a domain $\Omega \subset \mathbf{R}^n$ and $du \in L^s(\Omega, \wedge^{l+1})$, $l = 0, 1, \ldots, n-1$. Suppose that $(w_1, w_2) \in A_r^{1/\alpha}(\Omega)$ for some $r > 1$ and $0 < \alpha < 1$. If $s > r$, then there exists a constant C, independent of u, such that*

$$\left(\int_B |u - u_B|^s w_1^\alpha dx\right)^{1/s} \leq C|B|^{1/n} \left(\int_{\sigma B} |du|^s w_2 dx\right)^{1/s}$$

for all balls B with $\sigma B \subset \Omega$. Here u_B is a closed form.

Selecting $\alpha = 1/s$ in Theorem 3.5.5, we have the following two-weighted Poincaré inequality.

Corollary 3.5.7. *Let* $u \in D'(\Omega, \wedge^l)$ *be a differential form satisfying the A-harmonic equation* (1.2.10) *in a domain* $\Omega \subset \mathbf{R}^n$ *and* $du \in L^s(\Omega, \wedge^{l+1})$, $l = 0, 1, \ldots, n-1$. *Suppose that* $(w_1, w_2) \in A_r^\lambda(\Omega)$ *for some* $r > 1$ *and* $\lambda > 0$. *If* $\sigma > 1$ *and* $s > \lambda(r-1)/s + 1$, *then there exists a constant* C, *independent of* u, *such that*

$$\left(\int_B |u - u_B|^s w_1^{1/s} dx \right)^{1/s} \leq C|B|^{1/n} \left(\int_{\sigma B} |du|^s w_2^{\lambda/s} dx \right)^{1/s} \qquad (3.5.21)$$

for all balls B *with* $\sigma B \subset \Omega$. *Here* u_B *is a closed form.*

When $\lambda = 1$ in Corollary 3.5.7, we obtain the following symmetric two-weighted inequality.

Corollary 3.5.8. *Let* $u \in D'(\Omega, \wedge^l)$ *be a differential form satisfying the A-harmonic equation* (1.2.10) *in a domain* $\Omega \subset \mathbf{R}^n$ *and* $du \in L^s(\Omega, \wedge^{l+1})$, $l = 0, 1, \ldots, n-1$. *Suppose that* $(w_1, w_2) \in A_r(\Omega)$ *for some* $r > 1$. *If* $\sigma > 1$ *and* $s > (r-1)/s + 1$, *then there exists a constant* C, *independent of* u, *such that*

$$\left(\int_B |u - u_B|^s w_1^{1/s} dx \right)^{1/s} \leq C|B|^{1/n} \left(\int_{\sigma B} |du|^s w_2^{1/s} dx \right)^{1/s}$$

for all balls B *with* $\sigma B \subset \Omega$. *Here* u_B *is a closed form.*

3.6 Inequalities for Jacobians

The Jacobian (determinant) has played a critical role in multidimensional analysis and related fields from the very beginning, mainly pertaining to the change of variables in multiple integrals. The recent developments in nonlinear elasticity, weakly differentiable mappings, continuum mechanics, nonlinear PDEs, calculus of variations, and so on have rather clearly shown that the Jacobian is a powerful tool in many branches of science and engineering. For example, Jacobian and the related differential matrix satisfy the Beltrami equation which builds a bridge between mappings of finite distortion and PDEs; in quasiconformal analysis and nonlinear elasticity theory, the so-called stored energy integral of the div-curl field the integrand is a Jacobian. The recent progress in these theories relies basically on estimates of the Jacobians in terms of the cofactor matrix. The present knowledge of Jacobians tells us something more, such as the regularity and topological behavior of the related mappings. Since one of the major applications of Jacobians is to evaluate multiple integrals, the integrability of Jacobians has become a rather important topic.

3.6.1 Some notations

In this section, we establish some Poincaré-type inequalities for Jacobians in some domains, such as $L^s(\mu)$-averaging domain. For this, we recall that earlier in this chapter we have developed some versions of Poincaré-type inequalities for differential forms which do not satisfy any version of the A-harmonic equation. Then, as applications, we estimate subdeterminants of the Jacobian $J(x, f)$ of a mapping $f : \Omega \rightarrow \mathbf{R}^n$, $f = (f^1, \ldots, f^n)$. It is well known that Jacobian $J(x, f)$ is an n-form, specifically,

$$J(x, f)dx = df^1 \wedge \cdots \wedge df^n,$$

where $dx = dx_1 \wedge dx_2 \wedge \cdots \wedge dx_n$. For example, let $f = (f^1, f^2)$ be a differential mapping in $\mathbf{R}^2$. Then,

$$J(x, f)dx \wedge dy = \begin{vmatrix} f_x^1 & f_y^1 \\ f_x^2 & f_y^2 \end{vmatrix} dx \wedge dy$$

$$= \left(f_x^1 f_y^2 - f_y^1 f_x^2 \right) dx \wedge dy$$

and

$$df^1 \wedge df^2 = \left(f_x^1 dx + f_y^1 dy \right) \wedge \left(f_x^2 dx + f_y^2 dy \right)$$

$$= f_y^1 f_x^2 dy \wedge dx + f_x^1 f_y^2 dx \wedge dy$$

$$= \left(f_x^1 f_y^2 - f_y^1 f_x^2 \right) dx \wedge dy,$$

where we have used the property

$$dx_i \wedge dx_j = \begin{cases} 0, & i = j \\ -dx_j \wedge dx_i, & i \neq j. \end{cases}$$

Clearly,

$$J(x, f)dx \wedge dy = df^1 \wedge df^2.$$

Recall the notations we have introduced in Section 1.2, where we assumed that $f : \Omega \rightarrow \mathbf{R}^n$, $f = (f^1, \ldots, f^n)$ is a mapping of Sobolev class $W_{loc}^{1,p}(\Omega, \mathbf{R}^n)$, $1 \leq p < \infty$, whose distributional differential $Df = [\partial f^i / \partial x_j] : \Omega \rightarrow GL(n)$ is a locally integrable function on Ω with values in the space $GL(n)$ of all $n \times n$-matrices. A homeomorphism $f : \Omega \rightarrow \mathbf{R}^n$ of Sobolev class $W_{loc}^{1,n}(\Omega, \mathbf{R}^n)$ is said to be K-quasiconformal, $1 \leq K < \infty$, if its differential matrix $Df(x)$ and the Jacobian determinant

$$J(x, f) = \det Df(x)$$

$$= \begin{vmatrix} f^1_{x_1} & f^1_{x_2} & f^1_{x_3} & \cdots & f^1_{x_n} \\ f^2_{x_1} & f^2_{x_2} & f^2_{x_3} & \cdots & f^2_{x_n} \\ \vdots & \vdots & \vdots & \ddots & \vdots \\ f^n_{x_1} & f^n_{x_2} & f^n_{x_3} & \cdots & f^n_{x_n} \end{vmatrix}$$

satisfy

$$|Df(x)|^n \le KJ(x, f),$$

where $|Df(x)| = \max\{|Df(x)h| : |h| = 1\}$ denotes the norm of the Jacobi matrix $Df(x)$. Let u be the subdeterminant of Jacobian $J(x, f)$ which is obtained by deleting the k rows and k columns, $k = 0, 1, \ldots, n-1$, say,

$$J\left(x_{j_1}, x_{j_2}, \cdots, x_{j_{n-k}}; f^{i_1}, f^{i_2}, \cdots, f^{i_{n-k}}\right)$$

$$= \begin{vmatrix} f^{i_1}_{x_{j_1}} & f^{i_1}_{x_{j_2}} & f^{i_1}_{x_{j_3}} & \cdots & f^{i_1}_{x_{j_{n-k}}} \\ f^{i_2}_{x_{j_1}} & f^{i_2}_{x_{j_2}} & f^{i_2}_{x_{j_3}} & \cdots & f^{i_2}_{x_{j_{n-k}}} \\ \vdots & \vdots & \vdots & \ddots & \vdots \\ f^{i_{n-k}}_{x_{j_1}} & f^{i_{n-k}}_{x_{j_2}} & f^{i_{n-k}}_{x_{j_3}} & \cdots & f^{i_{n-k}}_{x_{j_{n-k}}} \end{vmatrix},$$

which is an $(n-k) \times (n-k)$ subdeterminant of $J(x, f)$, $\{i_1, i_2, \ldots, i_{n-k}\} \subset \{1, 2, \ldots, n\}$, and $\{j_1, j_2, \ldots, j_{n-k}\} \subset \{1, 2, \ldots, n\}$. Also, it is easy to see that $J(x_{j_1}, x_{j_2}, \ldots, x_{j_{n-k}}; f^{i_1}, f^{i_2}, \ldots, f^{i_{n-k}})dx_{j_1} \wedge dx_{j_2} \wedge \cdots \wedge dx_{j_{n-k}}$ is an $(n-k)$-form. Thus, all estimates for differential forms are applicable to the $(n-k)$-form $J\left(x_{j_1}, x_{j_2}, \ldots, x_{j_{n-k}}; f^{i_1}, f^{i_2}, \ldots, f^{i_{n-k}}\right) dx_{j_1} \wedge dx_{j_2} \wedge \cdots \wedge dx_{j_{n-k}}$. For example, if we choose

$$u = J\left(x_{j_1}, x_{j_2}, \ldots, x_{j_{n-k}}; f^{i_1}, f^{i_2}, \ldots, f^{i_{n-k}}\right) dx_{j_1} \wedge dx_{j_2} \wedge \cdots \wedge dx_{j_{n-k}}$$

and use the same method which was developed in the proof of Theorem 3.5.4, we can prove the following Theorems 3.6.1 and 3.6.2.

3.6.2 Two-weight estimates

Theorem 3.6.1. *Let*

$$u = J\left(x_{j_1}, x_{j_2}, \ldots, x_{j_{n-k}}; f^{i_1}, f^{i_2}, \ldots f^{i_{n-k}}\right) dx_{j_1} \wedge dx_{j_2} \wedge \cdots \wedge dx_{j_{n-k}}$$

and

$$du = \sum_{m=1}^{n} \frac{\partial J(x_{j_1}, x_{j_2}, \ldots, x_{j_{n-k}}; f^{i_1}, f^{i_2}, \ldots f^{i_{n-k}})}{\partial x_m} dx_m \wedge dx_{j_1} \wedge \cdots \wedge dx_{j_{n-k}},$$

where $k = 0, 1, \ldots, n-1$. Assume that $s > 1$ and $p > \max\{s, n\}$. Then, there exists a constant $\beta > 1$ such that if $w_1 \in A_r^1$ and $(w_1, w_2) \in A_{p/\gamma}^{s/p}$, where $r > 1$, $s\beta/(\beta - 1) \leq \gamma < p$, and $w_1 > w_2 \geq \eta > 0$, we have

$$\left(\frac{1}{\mu_1(\Omega)} \int_\Omega |u - u_{B_0}|^s w_1 dx \right)^{1/s} \leq C \mu_2(\Omega)^{1/n} \left(\frac{1}{\mu_2(\Omega)} \int_\Omega |du|^p w_2 dx \right)^{1/p}$$

for any $L^s(\mu_1)$-averaging domain Ω and some ball B_0 with $2B_0 \subset \Omega$. Here the measures μ_1 and μ_2 are defined by $d\mu_1 = w_1(x)dx$, $d\mu_2 = w_2(x)dx$, and C is a constant independent of u and du.

Theorem 3.6.2. *Let u and du be as Theorem 3.6.1. Then, there exists a constant $\beta > 1$ such that if $w_1 \in A_r^1$ and $(w_1, w_2) \in A_{n/s}^{s/n}$, where $s = n/\beta$, $r > 1$, and $w_1 > w_2 \geq \eta > 0$, we have*

$$\left(\frac{1}{\mu_1(\Omega)} \int_\Omega |u - u_{B_0}|^s w_1 dx \right)^{1/s} \leq C \mu_2(\Omega)^{1/n} \left(\frac{1}{\mu_2(\Omega)} \int_\Omega |du|^n w_2 dx \right)^{1/n}$$

for any $L^s(\mu_1)$-averaging domain Ω and some ball B_0 with $2B_0 \subset \Omega$. Here the measures μ_1 and μ_2 are defined by $d\mu_1 = w_1(x)dx$, $d\mu_2 = w_2(x)dx$, and C is a constant independent of u and du.

Now recall that

$$u = df^{i_1} \wedge df^{i_2} \wedge \cdots \wedge df^{i_m}$$

is an m-form, where $\{i_1, i_2, \ldots, i_m\} \subset \{1, 2, \ldots, n\}$, $m = 1, 2, \ldots, n$. Thus, any estimate for m-forms can be used to estimate the products of the components of a K-quasiconformal mapping. In fact, similar to Theorems 3.6.1 and 3.6.2, we can prove the following Theorems 3.6.3 and 3.6.4.

Theorem 3.6.3. *Let $u = df^{i_1} \wedge df^{i_2} \wedge \cdots \wedge df^{i_m}$ and $du \in L^p(B, \wedge^{m+1})$, $m = 1, 2, \ldots, n-1$. Assume that $1 < s < p < \infty$. Then, there exists a constant $\beta > 1$ such that if $w_1 \in A_r^1$ and $(w_1, w_2) \in A_{p/k}^{s/p}$ for some $r > 1$ and k with $s\beta/(\beta - 1) \leq k < p$, we have*

$$\left(\frac{1}{|B|} \int_B |u - u_B|^s w_1 dx \right)^{1/s} \leq C |B|^{1/n} \left(\frac{1}{|B|} \int_B |du|^p w_2 dx \right)^{1/p}$$

for all balls $B \subset \mathbf{R}^n$. Here C is a constant independent of u and du.

Theorem 3.6.4. *Let* $u = df^{i_1} \wedge df^{i_2} \wedge \cdots \wedge df^{i_m}$ *and* $du \in L^p(B, \wedge^{m+1})$,
$m = 1, 2, \ldots, n-1$. *Assume that* $s > 1$ *and* $p > \max\{s, n\}$. *Then, there exists
a constant* $\beta > 1$ *such that if* $w_1 \in A_r^1$ *and* $(w_1, w_2) \in A_{p/k}^{s/p}$, *where* $r > 1$,
$s\beta/(\beta - 1) \le k < p$, *and* $w_1 > w_2 \ge \eta > 0$, *we have*

$$\left(\frac{1}{\mu_1(\Omega)} \int_\Omega |u - u_{B_0}|^s w_1 dx \right)^{1/s} \le C\mu_2(\Omega)^{1/n} \left(\frac{1}{\mu_2(\Omega)} \int_\Omega |du|^p w_2 dx \right)^{1/p}$$

for any $L^s(\mu_1)$*-averaging domain* Ω *and some ball* B_0 *with* $2B_0 \subset \Omega$. *Here
the measures* μ_1 *and* μ_2 *are defined by* $d\mu_1 = w_1(x)dx$, $d\mu_2 = w_2(x)dx$, *and
C is a constant independent of u and du.*

Remark. (1) Since $L^s(\mu)$-averaging domains reduce to L^s-averaging domains
if $w = 1$ (so $d\mu = w(x)dx = dx$), Theorems 3.6.1, 3.6.2, and 3.6.4 also hold
if $\Omega \subset \mathbf{R}^n$ is an L^s-averaging domain.
(2) From [184], we know that John domains are $L^s(\mu)$-averaging domains.
Thus, our global results, Theorems 3.6.1, 3.6.2, and 3.6.4 also hold if Ω is a
John domain.

3.7 Inequalities for the projection operator

In this section, we will prove Poincaré inequalities for the projection operator
applied to A-harmonic tensors on Ω. We continue to use the notation and
terms that were introduced in Section 3.3.

3.7.1 Statement of the main theorem

As usual, for $u \in L^1(\Omega, \wedge^l)$, the projection operator H is defined by setting
$H(u)$ to be the unique element of $\mathcal{H}$ such that

$$< u - h(u), h >= 0 \tag{3.7.1}$$

for all $h \in \mathcal{H}$. See [21] for the definition of the projection operator H and
its properties. From [23, Chapter 6] we recall that the projection operator, Green's operator, and the Laplace–Beltrami operator satisfy Poisson's
equation

$$H(u) = u - \Delta G(u). \tag{3.7.2}$$

We first present the following local Poincaré inequality for the projection
operator applied to A-harmonic tensors on Ω.

Theorem 3.7.1. *Let $u \in D'(\Omega, \wedge^l)$ be an A-harmonic tensor on Ω and $du \in L^s(\Omega, \wedge^{l+1})$, $l = 0, 1, \ldots, n - 1$, and H be the projection operator. Assume that $\rho > 1$, $0 < \alpha \leq 1$, $1 + \alpha(r - 1) < s < \infty$, and $w \in A_r$ for some $r > 1$. Then,*

$$\|H(u) - (H(u))_B\|_{s,B,w^\alpha} \leq C|B|^{1/n}\|du\|_{s,\rho B,w^\alpha} \qquad (3.7.3)$$

for all balls B with $\rho B \subset \Omega$. Here C is a constant independent of u.

Note that (3.7.3) can be written in the following symmetric form:

$$\left(\frac{1}{|B|}\int_B |H(u) - (H(u))_B|^s w^\alpha dx\right)^{1/s} \leq C|B|^{1/n}\left(\frac{1}{|B|}\int_{\rho B}|du|^s w^\alpha dx\right)^{1/s}.$$

$$(3.7.3)'$$

3.7.2 Inequality for Δ and G

In order to prove Theorem 3.7.1, we need the following Poincaré inequality for the composition of Δ and G.

Lemma 3.7.2. *Let $u \in D'(\Omega, \wedge^l)$, $l = 0, 1, \ldots, n - 1$, be an A-harmonic tensor on Ω. Assume that $\rho > 1$, $1 < s < \infty$, and $w \in A_r(\Omega)$ for some $r > 1$. Then, there exists a constant C, independent of u, such that*

$$\|\Delta G(u) - (\Delta G(u))_B\|_{s,B,w^\alpha} \leq C\,diam(B)\|du\|_{s,\rho B,w^\alpha} \qquad (3.7.4)$$

for any ball B with $\rho B \subset \Omega$ and any real number α with $0 < \alpha \leq 1$.

Proof. Since G commutes with d (see [23, P225]) and $d(du) = 0$ for any differential form u, we find that

$$\begin{aligned}
d\Delta G(u) &= d(dd^\star + d^\star d)G(u) \\
&= dd^\star dG(u) \\
&= (dd^\star d + d^\star dd)G(u) \\
&= (dd^\star + d^\star d)dG(u) \\
&= \Delta dG(u) \\
&= \Delta G(du).
\end{aligned} \qquad (3.7.5)$$

From the properties of the homotopy operator T (see Section 1.5), we have

$$u = d(Tu) + T(du) \qquad (3.7.6)$$

and

$$\|Tu\|_{s,B} \le C_1 diam(B)\|u\|_{s,B}. \tag{3.7.7}$$

From [21], we know that for any smooth l-form ω,

$$\|dd^\star G(\omega)\|_{s,B} + \|d^\star dG(\omega)\|_{s,B} \le C_2\|\omega\|_{s,B}, \tag{3.7.8}$$

where $1 < s < \infty$ and C_2 is a constant. Choosing $\omega = du$, we find that

$$\|dd^\star G(du)\|_{s,B} + \|d^\star dG(du)\|_{s,B} \le C_2\|du\|_{s,B}. \tag{3.7.9}$$

By the definition of the Laplace–Beltrami operator Δ, Minkowski's inequality, and (3.7.9), we obtain

$$
\begin{aligned}
\|\Delta G(du)\|_{s,B} &= \|(d^\star d + dd^\star)G(du)\|_{s,B} \\
&\le \|d^\star dG(du)\|_{s,B} + \|dd^\star G(du)\|_{s,B} \\
&\le C_2\|du\|_{s,B}.
\end{aligned}
\tag{3.7.10}
$$

Thus, it follows that

$$
\begin{aligned}
\|\Delta G(u) - (\Delta G(u))_B\|_{s,B} \\
= \|Td(\Delta G(u))\|_{s,B} \\
= \|T(\Delta G(du))\|_{s,B} \\
\le C_1 diam(B)\|\Delta G(du)\|_{s,B} \\
\le C_3 diam(B)\|du\|_{s,B}.
\end{aligned}
\tag{3.7.11}
$$

We shall first show that (3.7.4) holds for $0 < \alpha < 1$. Let $t = s/(1-\alpha)$. Using Hölder's inequality and (3.7.11), we have

$$
\begin{aligned}
\|\Delta G(u) - (\Delta G(u))_B\|_{s,B,w^\alpha} \\
= \left(\int_B \left(|\Delta G(u) - (\Delta G(u))_B| w^{\alpha/s} \right)^s dx \right)^{1/s} \\
\le \|\Delta G(u) - (\Delta G(u))_B\|_{t,B} \left(\int_B w^{t\alpha/(t-s)} dx \right)^{(t-s)/st} \\
= \|\Delta G(u) - (\Delta G(u))_B\|_{t,B} \left(\int_B w dx \right)^{\alpha/s} \\
\le C_3 diam(B)\|du\|_{t,B} \left(\int_B w dx \right)^{\alpha/s}.
\end{aligned}
\tag{3.7.12}
$$

Choose $m = s/(1 + \alpha(r-1))$, so that $m < s$. Applying the weak reverse Hölder inequality, we obtain

$$\|du\|_{t,B} \leq C_4|B|^{(m-t)/mt}\|du\|_{m,\rho B}. \tag{3.7.13}$$

Substituting (3.7.13) into (3.7.12), we have

$$\|\Delta G(u) - (\Delta G(u))_B\|_{s,B,w^\alpha}$$
$$\leq C_5|B|^{1/n+(m-t)/mt}\|du\|_{m,\rho B}\left(\int_B w dx\right)^{\alpha/s}. \tag{3.7.14}$$

Using Hölder's inequality again with $1/m = 1/s + (s-m)/sm$, we find that

$$\|du\|_{m,\rho B} = \left(\int_{\rho B}|du|^m dx\right)^{1/m}$$
$$= \left(\int_{\rho B}\left(|du|w^{\alpha/s}w^{-\alpha/s}\right)^m dx\right)^{1/m} \tag{3.7.15}$$
$$\leq \|du\|_{s,\rho B,w^\alpha}\left(\int_{\rho B}\left(\tfrac{1}{w}\right)^{1/(r-1)}dx\right)^{\alpha(r-1)/s}$$

for all balls B with $\rho B \subset \Omega$. Substitution of (3.7.15) into (3.7.14) gives

$$\|\Delta G(u) - (\Delta G(u))_B\|_{s,B,w^\alpha}$$
$$\leq C_6|B|^{\frac{1}{n}+\frac{m-t}{mt}}\|du\|_{s,\rho B,w^\alpha}\left(\int_B w dx\right)^{\alpha/s}\left(\int_{\rho B}\left(\tfrac{1}{w}\right)^{1/(r-1)}dx\right)^{\alpha(r-1)/s}. \tag{3.7.16}$$

Since $w \in A_r(\Omega)$, it follows that

$$\|w\|_{1,B}^{\alpha/s} \cdot \|1/w\|_{1/(r-1),\rho B}^{\alpha/s}$$
$$\leq \left(\left(\int_{\rho B} w dx\right)\left(\int_{\rho B}\left(\tfrac{1}{w}\right)^{1/(r-1)}dx\right)^{r-1}\right)^{\alpha/s}$$
$$= \left(|\rho B|^r\left(\tfrac{1}{|\rho B|}\int_{\rho B} w dx\right)\left(\tfrac{1}{|\rho B|}\int_{\rho B}\left(\tfrac{1}{w}\right)^{1/(r-1)}dx\right)^{r-1}\right)^{\alpha/s} \tag{3.7.17}$$
$$\leq C_7|B|^{\alpha r/s}.$$

Now combination of (3.7.17) and (3.7.16) leads to

$$\|\Delta G(u) - (\Delta G(u))_B\|_{s,B,w^\alpha} \leq C_8|B|^{1/n}\|du\|_{s,\rho B,w^\alpha} \tag{3.7.18}$$

for all balls B with $\rho B \subset \Omega$. Thus, we have proved (3.7.4) for $0 < \alpha < 1$. Next, we shall show that (3.7.4) is also true for $\alpha = 1$, that is, we need to prove that

$$\|\Delta G(u) - (\Delta G(u))_B\|_{s,B,w} \leq C|B|^{1/n}\|du\|_{s,\rho B,w}. \tag{3.7.19}$$

By Lemma 1.4.7, there exist constants $\beta > 1$ and $C_9 > 0$, such that

$$\| w \|_{\beta,B} \leq C_9 |B|^{(1-\beta)/\beta} \| w \|_{1,B} \tag{3.7.20}$$

for any cube or ball $B \subset \mathbf{R}^n$. Set $t = s\beta/(\beta - 1)$, so that $1 < s < t$ and $\beta = t/(t - s)$. Note that $1/s = 1/t + (t - s)/st$. Using Hölder's inequality, (3.7.11), and (3.7.20), we find

$$\left(\int_B |\Delta G(u) - (\Delta G(u))_B|^s w dx \right)^{1/s}$$
$$= \left(\int_B \left(|\Delta G(u) - (\Delta G(u))_B| w^{1/s} \right)^s dx \right)^{1/s}$$
$$\leq \left(\int_B |\Delta G(u) - (\Delta G(u))_B|^t dx \right)^{1/t} \left(\int_B \left(w^{1/s} \right)^{st/(t-s)} dx \right)^{(t-s)/st}$$
$$\leq C_{10} \|\Delta G(u) - (\Delta G(u))_B\|_{t,B} \cdot \|w\|_{\beta,B}^{1/s}$$
$$\leq C_{10} |B|^{1/n} \|du\|_{t,B} \cdot \|w\|_{\beta,B}^{1/s}$$
$$\leq C_{11} |B|^{1/n + (1-\beta)/\beta s} \|w\|_{1,B}^{1/s} \cdot \|du\|_{t,B}$$
$$\leq C_{11} |B|^{1/n - 1/t} \|w\|_{1,B}^{1/s} \cdot \|du\|_{t,B}. \tag{3.7.21}$$

Next, let $m = s/r$. From Lemma 3.1.1, we obtain

$$\|du\|_{t,B} \leq C_{12} |B|^{(m-t)/mt} \|du\|_{m,\rho B}. \tag{3.7.22}$$

Using Lemma 1.1.4, we find that

$$\|du\|_{m,\rho B} = \left(\int_{\rho B} \left(|du| w^{1/s} w^{-1/s} \right)^m dx \right)^{1/m}$$
$$\leq \left(\int_{\rho B} |du|^s w dx \right)^{1/s} \left(\int_{\rho B} \left(\tfrac{1}{w} \right)^{1/(r-1)} dx \right)^{(r-1)/s} \tag{3.7.23}$$

for all balls B with $\rho B \subset \Omega$. Now since $w \in A_r(\Omega)$, similar to (3.7.17), we have

$$\|w\|_{1,B}^{1/s} \cdot \|1/w\|_{1/(r-1),\rho B}^{1/s} \leq C_{13} |B|^{r/s}. \tag{3.7.24}$$

Finally, combining (3.7.21), (3.7.22), (3.7.23), and (3.7.24), it follows that

$$\|\Delta G(u) - (\Delta G(u))_B\|_{s,B,w}$$
$$\leq C_{14} |B|^{1/n - 1/t} \|w\|_{1,B}^{1/s} |B|^{(m-t)/mt} \|du\|_{m,\rho B}$$
$$\leq C_{14} |B|^{1/n - 1/m} \|w\|_{1,B}^{1/s} \cdot \|1/w\|_{1/(r-1),\rho B}^{1/s} \|du\|_{s,\rho B,w}$$
$$\leq C_{15} |B|^{1/n} \|du\|_{s,\rho B,w}$$

for all balls B with $\rho B \subset \Omega$. Hence, (3.7.19) follows. ∎

3.7.3 Proof of the main theorem

Now, we are ready to prove Theorem 3.7.1, the local Poincaré inequality for the projection operator applied to A-harmonic tensors on Ω.

Proof of Theorem 3.7.1. Applying Theorem 3.2.3, Lemma 3.7.2, and using (3.7.2), we obtain

$$\|H(u) - (H(u))_B\|_{s,B,w^\alpha}$$

$$= \|u - \Delta G(u) - (u - \Delta G(u))_B\|_{s,B,w^\alpha}$$

$$= \|u - \Delta G(u) - (u_B - (\Delta G(u))_B)\|_{s,B,w^\alpha}$$

$$\leq \|u - u_B\|_{s,B,w^\alpha} + \|\Delta G(u) - (\Delta G(u))_B\|_{s,B,w^\alpha}$$

$$\leq C_1|B|^{1/n}\|du\|_{s,\rho_2 B,w^\alpha} + C_2|B|^{1/n}\|du\|_{s,\rho_1 B,w^\alpha}$$

$$\leq C_3|B|^{1/n}\|du\|_{s,\rho B,w^\alpha},$$

where $\rho = \max\{\rho_1, \rho_2\}$. ∎

Using Theorems 3.7.1 and 1.5.3, and employing the method similar to the proof of Theorem 1.5.6, we can prove the following global Poincaré inequality for the projection operator.

Theorem 3.7.3. *Let $u \in D'(\Omega, \wedge^0)$ be a solution of the A-harmonic equation (1.2.4) in a δ-John domain Ω, $du \in L^s(\Omega, \wedge^1)$, and H be the projection operator. Assume that $0 < \alpha \leq 1$, $1 + \alpha(r-1) < s < \infty$, and $w \in A_r(\Omega)$ for some $r > 1$. Then,*

$$\|H(u) - (H(u))_{B_0}\|_{s,\Omega,w^\alpha} \leq C diam(\Omega)\|du\|_{s,\Omega,w^\alpha}, \qquad (3.7.25)$$

where $B_0 \subset \Omega$ is a fixed ball and C is a constant independent of u.

3.8 Other Poincaré-type inequalities

In Sections 3.2, 3.3, 3.4, 3.5, 3.6, and 3.7, we have discussed several versions of the Poincaré inequality for differential forms and the related operators applied to differential forms. There are many other versions of the Poincaré inequality which have been widely investigated and used in many fields of mathematics, such as potential theory, partial differential equations. For example, Sobolev and Poincaré inequalities and quasiconformal mappings were discussed in [208]; weighted Poincaré inequalities and Minkowski contents were investigated in [209]; and Poincaré-type inequalities on stratified sets and their applications were studied in [210]. S.M. Buckley and P. Koskela proved the following Sobolev–Poincaré inequality in [211]: if Ω is a John domain and ∇u satisfies a certain mild condition, then

$$\left(\int_\Omega |u - u_{B_0}|^q \right)^{1/q} \leq C \left(\int_\Omega |\nabla u|^p \right)^{1/p}$$

for any function $u \in W_{loc}^{1,1}(\Omega)$, any p with $0 < p < 1$, and appropriate $q > 0$.

Now let (M, g) be a smooth compact Riemannian n-dimensional manifold, $n \geq 3$, and $H_1^2(M)$ be the Sobolev space defined as the completion of $C^\infty(M)$ with respect to

$$\|u\|_{H_1^2}^2 = \int_M |\nabla u|^2 dv_g + \int_M u^2 dv_g.$$

In 2002, E. Hebey [212] investigated the following sharp Sobolev–Poincaré inequality on compact Riemannian manifold M,

$$\left(\int_M |u|^{2n/(n-2)} dv_g \right)^{(n-2)/n} \leq K_n^2 \int_M |\nabla u|^2 dv_g + B \left(\int_M |u| dv_g \right)^2, \quad (3.8.1)$$

where $u \in H_1^2(M)$ and $B > 0$ is independent of u. We say that (3.8.1) is valid if there exists a B such that (3.8.1) holds for all $u \in H_1^2(M)$. E. Hebey provided the validity criterion of inequality (3.8.1).

Recently, in [213–215], P. Koskela and his colleagues investigated the Poincaré inequality on metric spaces, Dirichlet forms, and Steiner symmetrization, respectively. In [216], O. Martio studied John domains, bi-Lipschitz balls, and Poincaré inequality. R. Nibbi discussed Poincaré inequalities and applications to electromagnetism in [151]. M. Pearson investigated Poincaré inequality and entire functions in [160]. Recently, C. Wang studied Poincaré inequality and removable singularities for harmonic maps in [217]. In [218–221, 159], several versions of the Poincaré inequalities for convex domains and planar domains were developed. In [222, 198, 223, 175, 176], Poincaré inequalities on unbounded domains, Lipschitz domains, punctured domains, and trees were treated, respectively. In [224–227], Poincaré inequalities for a class of A_p-weights, with a doubling weights, and with reverse doubling weights were established, respectively. In [228–234], Poincaré-type inequalities for p-Laplacian, pointwise estimates for a class of degenerate elliptic equations, and the Poincaré-type inequalities for Crushin-type operator, etc., were studied. In [235, 214], Poincaré inequalities associated with Dirichlet forms were explored. For several other different versions of Poincaré inequalities, see [236–254, 34, 212, 150, 152, 162, 166, 172]. For some work related to Poincaré inequalities, see [255–263, 238].

Chapter 4
Caccioppoli inequalities

In Chapter 3, we have discussed various versions of the Poincaré-type inequalities in which we have estimated the norm of $u - u_B$ in terms of the corresponding norm of du. In this chapter, we develop a series of estimates which provide upper bounds for the norms of ∇u (if u is a function) or du (if u is a form) in terms of the corresponding norm $u - c$, where c is any closed form. These kinds of estimates are called the Caccioppoli-type estimates or the Caccioppoli inequalities. In Section 4.2, we study the local $A_r(\Omega)$-weighted Caccioppoli inequalities. The local Caccioppoli inequalities with two-weights are discussed in Section 4.3. The global versions of Caccioppoli inequalities on Riemannian manifolds and bounded domains are developed in Sections 4.4 and 4.5, respectively. In Section 4.6, we present Caccioppoli inequalities with Orlicz norms. Finally, in Section 4.7, we address few versions of Caccioppoli inequalities related to the codifferential operator $d^\star$.

4.1 Preliminary results

We assume that Ω is bounded and ψ is any function defined in Ω with values in the extended reals $[-\infty, \infty]$, and ϑ is a function in the Sobolev space $H^{1,p}(\Omega; \mu)$. Let

$$\chi_{\psi,\vartheta} = \chi_{\psi,\vartheta}(\Omega) = \{v \in H^{1,p}(\Omega; \mu) : v \geq \psi \text{ a.e. in } \Omega, v - \vartheta \in H_0^{1,p}(\Omega; \mu)\},$$

where $H_0^{1,p}(\Omega; \mu)$ is the closure of $C_0^\infty(\Omega)$ in $H^{1,p}(\Omega; \mu)$. For $\psi = \vartheta$, we write $\chi_{\psi,\psi}(\Omega) = \chi_\psi(\Omega)$. The problem is to find a u in $\chi_{\psi,\vartheta}$ such that

$$\int_\Omega A(x, \nabla u) \cdot \nabla(v - u)dx \geq 0 \qquad (4.1.1)$$

whenever $v \in \chi_{\psi,\vartheta}$. The function ψ is called an obstacle.

R.P. Agarwal et al., *Inequalities for Differential Forms*,
DOI 10.1007/978-0-387-68417-8_4, © Springer Science+Business Media, LLC 2009

A function $u \in \chi_{\psi,\vartheta}(\Omega)$ which satisfies (4.1.1) for all $v \in \chi_{\psi,\vartheta}(\Omega)$ is called a solution to the obstacle problem with obstacle ψ and boundary values ϑ, or a solution to the obstacle problem in $\chi_{\psi,\vartheta}(\Omega)$. See [264–266, 135] for results on obstacle problems. A function $u \in H^{1,p}_{loc}(\Omega; \mu)$ is called a supersolution of the equation

$$- \operatorname{div} A(x, \nabla u) = 0 \qquad (4.1.2)$$

in Ω if

$$- \operatorname{div} A(x, \nabla u) \geq 0$$

weakly in Ω, that is

$$\int_{\Omega} A(x, \nabla u) \cdot \nabla \varphi dx \geq 0$$

whenever $\varphi \in C_0^{\infty}(\Omega)$ is nonnegative. Here A is an operator satisfying certain conditions. A function v is a subsolution of (4.1.2) if $-v$ is a supersolution of (4.1.2). We begin this chapter with the following Caccioppoli inequality for a solution u to the obstacle problem in $\chi_{\psi,\vartheta}(\Omega)$ with nonpositive obstacle ψ. The proof of Theorem 4.1.1 can be found in [59].

Theorem 4.1.1. *Let $\eta \in C_0^{\infty}(\Omega)$ is nonnegative and $q \geq 0$.*
i) If u is a solution to the obstacle problem in $\chi_{\psi,\vartheta}(\Omega)$ with nonpositive obstacle ψ, then

$$\int_{\Omega} |u^+|^q |\nabla u^+|^p \eta^p d\mu \geq C \int_{\Omega} |u^+|^{p+q} |\nabla \eta|^p d\mu. \qquad (4.1.3)$$

ii) If u is a supersolution of (4.1.2) in Ω, then

$$\int_{\Omega} |u^-|^q |\nabla u^-|^p \eta^p d\mu \geq C \int_{\Omega} |u^-|^{p+q} |\nabla \eta|^p d\mu. \qquad (4.1.4)$$

Here $u^+ = \max(u, 0)$, $u^- = \min(u, 0)$, and C is a constant.

4.2 Local and global weighted cases

In this section, we develop the local $A_r(\Omega)$-weighted Caccioppoli inequalities which will be used to establish the global weighted estimates in later sections. We begin with an elementary Caccioppoli inequality without weights, and then use it to obtain some weighted inequalities.

4.2.1 $A_r(\Omega)$-weighted inequality

C. Nolder [71] obtained the following local Caccioppoli inequality for differential forms satisfying the homogeneous A-harmonic equation

$$d^\star A(x, d\omega) = 0 \tag{4.2.1}$$

in a ball or cube $B \subset \mathbf{R}^n$ (see Section 1.2 for the properties of operator A).

Theorem 4.2.1. *Let u be a differential form satisfying the homogeneous A-harmonic equation in Ω and let $\sigma > 1$. Then, there exists a constant C, independent of u and du, such that*

$$\|du\|_{s,B} \leq \frac{C}{diam(B)} \|u - c\|_{s,\sigma B}$$

for all balls or cubes B with $\sigma B \subset \Omega$ and all closed forms c. Here $1 < s < \infty$ is the exponent associated with the A-harmonic equation (4.2.1).

The above local Caccioppoli inequality was extended to the $A_r(\Omega)$-weighted version in [267].

Theorem 4.2.2. *Let $u \in D'(\Omega, \wedge^l)$, $l = 0, 1, \ldots, n-1$, be an A-harmonic tensor in a domain $\Omega \subset \mathbf{R}^n$ and $\rho > 1$. Assume that $1 < s < \infty$ is the exponent associated with the A-harmonic equation and $w \in A_r$ for some $r > 1$. Then, there exists a constant C, independent of u and du, such that*

$$\|du\|_{s,B,w} \leq \frac{C}{diam(B)} \|u - c\|_{s,\rho B,w} \tag{4.2.2}$$

for all balls B with $\rho B \subset \Omega$ and all closed forms c.

Note that (4.2.2) can be written as

$$\left(\int_B |du|^s w dx \right)^{1/s} \leq \frac{C}{diam(B)} \left(\int_{\rho B} |u - c|^s w dx \right)^{1/s} \tag{4.2.3}$$

or

$$\left(\int_B |du|^s d\mu \right)^{1/s} \leq \frac{C}{diam(B)} \left(\int_{\rho B} |u - c|^s d\mu \right)^{1/s}, \tag{4.2.4}$$

where the measure μ is defined by $d\mu = w(x)dx$ and $w \in A_r$.

Proof. Since $w \in A_r$ for some $r > 1$, by Lemma 1.4.7, there exist constants $\beta > 1$ and $C_1 > 0$, such that

$$\| w \|_{\beta,B} \leq C_1 |B|^{(1-\beta)/\beta} \| w \|_{1,B} \tag{4.2.5}$$

for any cube or ball $B \subset \mathbf{R}^n$. Choose $t = s\beta/(\beta - 1)$, so that $1 < s < t$ and $\beta = t/(t - s)$. Since $1/s = 1/t + (t - s)/st$, by Hölder's inequality, Theorem 4.2.1, and (4.2.5), it follows that

$$
\begin{aligned}
\|du\|_{s,B,w} &= \left(\int_B \left(|du| w^{1/s} \right)^s dx \right)^{1/s} \\
&\leq \left(\int_B |du|^t dx \right)^{1/t} \left(\int_B \left(w^{1/s} \right)^{st/(t-s)} dx \right)^{(t-s)/st} \\
&\leq C_2 \|du\|_{t,B} \cdot \|w\|_{\beta,B}^{1/s} \\
&\leq C_3 diam(B)^{-1} \|u - c\|_{t,\sigma B} \cdot \|w\|_{\beta,B}^{1/s} \\
&\leq C_4 diam(B)^{-1} |B|^{(1-\beta)/\beta s} \|w\|_{1,B}^{1/s} \cdot \|u - c\|_{t,\sigma B} \\
&= C_4 diam(B)^{-1} |B|^{-1/t} \cdot \|w\|_{1,B}^{1/s} \cdot \|u - c\|_{t,\sigma B}
\end{aligned}
\tag{4.2.6}
$$

for all balls B with $\sigma B \subset \Omega$ and all closed forms c. Since c is a closed form and u is an A-harmonic tensor, $u - c$ is also an A-harmonic tensor. Taking $m = s/r$, we find that $m < s < t$. Applying Lemma 3.1.1, we find that

$$
\begin{aligned}
\|u - c\|_{t,\sigma B} &\leq C_5 |B|^{(m-t)/mt} \|u - c\|_{m,\sigma^2 B} \\
&\leq C_5 |B|^{(m-t)/mt} \|u - c\|_{m,\rho B},
\end{aligned}
\tag{4.2.7}
$$

where $\rho = \sigma^2$. Substituting (4.2.7) into (4.2.6), we find

$$
\|du\|_{s,B,w} \leq C_6 diam(B)^{-1} |B|^{-1/m} \cdot \|w\|_{1,B}^{1/s} \cdot \|u - c\|_{m,\rho B}.
\tag{4.2.8}
$$

Now since $1/m = 1/s + (s-m)/sm$, by Hölder's inequality again, we obtain

$$
\begin{aligned}
\|u - c\|_{m,\rho B} \\
= \left(\int_{\rho B} |u - c|^m dx \right)^{1/m} \\
= \left(\int_{\rho B} \left(|u - c| w^{1/s} w^{-1/s} \right)^m dx \right)^{1/m} \\
\leq \left(\int_{\rho B} |u - c|^s w dx \right)^{1/s} \left(\int_{\rho B} \left(\frac{1}{w} \right)^{m/(s-m)} dx \right)^{(s-m)/sm} \\
\leq \|u - c\|_{s,\rho B,w} \cdot \|1/w\|_{m/(s-m),\rho B}^{1/s}
\end{aligned}
\tag{4.2.9}
$$

for all balls B with $\rho B \subset \Omega$ and all closed forms c. Combining (4.2.8) and (4.2.9), we have

$$
\begin{aligned}
\|du\|_{s,B,w} \\
\leq C_6 diam(B)^{-1} |B|^{-1/m} \cdot \|w\|_{1,B}^{1/s} \cdot \|1/w\|_{m/(s-m),\rho B}^{1/s} \cdot \|u - c\|_{s,\rho B,w}.
\end{aligned}
\tag{4.2.10}
$$

Since $w \in A_r$, it follows that

$$\|w\|_{1,B}^{1/s} \cdot \|1/w\|_{m/(s-m),\rho B}^{1/s}$$

$$= \left(\int_B w dx\right)^{1/s} \left(\int_{\rho B} \left(\tfrac{1}{w}\right)^{m/(s-m)} dx\right)^{(s-m)/sm}$$

$$\leq \left(\left(\int_{\rho B} w dx\right) \left(\int_{\rho B} \left(\tfrac{1}{w}\right)^{1/(s/m-1)} dx\right)^{s/m-1}\right)^{1/s} \qquad (4.2.11)$$

$$= \left(|\rho B|^{s/m} \left(\tfrac{1}{|\rho B|} \int_{\rho B} w dx\right) \left(\tfrac{1}{|\rho B|} \int_{\rho B} \left(\tfrac{1}{w}\right)^{1/(r-1)} dx\right)^{r-1}\right)^{1/s}$$

$$\leq C_7 |B|^{1/m}.$$

Substituting (4.2.11) into (4.2.10), we find that

$$\|du\|_{s,B,w} \leq C diam(B)^{-1} \|u - c\|_{s,\rho B,w}$$

for all balls B with $\rho B \subset \Omega$ and all closed forms c. ∎

4.2.2 $A_r(\lambda, \Omega)$-weighted inequality

The following $A_r(\lambda, \Omega)$-weighted Caccioppoli-type estimate for differential forms satisfying the homogeneous A-harmonic equation (4.2.1) was proved by G. Bao in [18] (see Section 1.4 for the definition of $A_r(\lambda, \Omega)$-weights).

Theorem 4.2.3. *Let $u \in D'(\Omega, \wedge^l)$, $l = 0, 1, \ldots, n-1$, be an A-harmonic tensor in a domain $\Omega \subset \mathbf{R}^n$ and $\rho > 1$. Assume that $1 < s < \infty$ is a fixed exponent associated with the A-harmonic equation and $w \in A_r(\lambda, \Omega)$ for some $r > 1$ and $\lambda > 1$. Then, there exists a constant C, independent of u, such that*

$$\left(\int_B |du|^s w^{1/\lambda} dx\right)^{1/s} \leq \frac{C}{diam(B)} \left(\int_{\rho B} |u - c|^s w^{1/\lambda^2} dx\right)^{1/s} \qquad (4.2.12)$$

for all balls B with $\rho B \subset \Omega$ and all closed forms c.

Proof. Choose $t = s\lambda^2/(\lambda^2 - 1)$, so that $1 < s < t$. Since $1/s = 1/t + (t - s)/st$, by Hölder's inequality and Theorem 4.2.1, we have

$$\left(\int_B |du|^s w^{1/\lambda} dx\right)^{1/s}$$

$$= \left(\int_B \left(|du| w^{1/\lambda s}\right)^s dx\right)^{1/s}$$

$$\leq \left(\int_B |du|^t dx\right)^{1/t} \left(\int_B \left(w^{1/\lambda s}\right)^{st/(t-s)} dx\right)^{(t-s)/st} \qquad (4.2.13)$$

$$\leq \|du\|_{t,B} \cdot \left(\int_B w^\lambda dx\right)^{1/\lambda^2 s}$$

$$= C_1 diam(B)^{-1} \|u - c\|_{t,\sigma B} \left(\int_B w^\lambda dx\right)^{1/\lambda^2 s}$$

for all balls B with $\sigma B \subset \Omega$ and all closed forms c. Since c is a closed form and u is an A-harmonic tensor, $u - c$ is also an A-harmonic tensor. Taking

$$m = \frac{\lambda^2 s}{\lambda^2 + r - 1},$$

we find $m < s < t$. Applying Lemma 3.1.1, we have

$$
\begin{aligned}
\|u - c\|_{t,\sigma B} &\leq C_2 |B|^{(m-t)/mt} \|u - c\|_{m,\sigma^2 B} \\
&= C_2 |B|^{(m-t)/mt} \|u - c\|_{m,\rho B},
\end{aligned}
\tag{4.2.14}
$$

where $\rho = \sigma^2$. Substituting (4.2.14) into (4.2.13), we have

$$
\begin{aligned}
&\left(\int_B |du|^s w^{1/\lambda} dx \right)^{1/s} \\
&\qquad \leq C_3 diam(B)^{-1} |B|^{(m-t)/mt} \|u - c\|_{m,\rho B} \left(\int_B w^\lambda dx \right)^{1/\lambda^2 s}.
\end{aligned}
\tag{4.2.15}
$$

Now since $1/m = 1/s + (s - m)/sm$, by Hölder's inequality again, we obtain

$$
\begin{aligned}
&\|u - c\|_{m,\rho B} \\
&\quad = \left(\int_{\rho B} |u - c|^m dx \right)^{1/m} \\
&\quad = \left(\int_{\rho B} \left(|u - c| w^{1/\lambda^2 s} w^{-1/\lambda^2 s} \right)^m dx \right)^{1/m} \\
&\quad \leq \left(\int_{\rho B} |u - c|^s w^{1/\lambda^2} dx \right)^{1/s} \left(\int_{\rho B} \left(\frac{1}{w} \right)^{1/(r-1)} dx \right)^{(r-1)/\lambda^2 s}
\end{aligned}
\tag{4.2.16}
$$

for all balls B with $\rho B \subset \Omega$ and all closed forms c. Combining (4.2.15) and (4.2.16), we obtain

$$
\begin{aligned}
&\left(\int_B |du|^s w^{1/\lambda} dx \right)^{1/s} \\
&\quad \leq \frac{C_3}{diam(B)} |B|^{\frac{m-t}{mt}} \|w\|_{\lambda,B}^{1/\lambda s} \left\| \frac{1}{w} \right\|_{1/(r-1),\rho B}^{1/\lambda^2 s} \left(\int_{\rho B} |u - c|^s w^{1/\lambda^2} dx \right)^{\frac{1}{s}}.
\end{aligned}
\tag{4.2.17}
$$

Since $w \in A_r(\lambda)$, it follows that

$$
\begin{aligned}
&\|w\|_{\lambda,B}^{1/\lambda s} \cdot \|1/w\|_{1/(r-1),\rho B}^{1/\lambda^2 s} \\
&\quad \leq \left(\left(\int_{\rho B} w^\lambda dx \right) \left(\int_{\rho B} (1/w)^{1/(r-1)} dx \right)^{r-1} \right)^{1/\lambda^2 s} \\
&\quad = \left(|\rho B|^r \left(\frac{1}{|\rho B|} \int_{\rho B} w^\lambda dx \right) \left(\frac{1}{|\rho B|} \int_{\rho B} \left(\frac{1}{w} \right)^{1/(r-1)} dx \right)^{r-1} \right)^{1/\lambda^2 s} \\
&\quad \leq C_4 |B|^{r/\lambda^2 s}.
\end{aligned}
\tag{4.2.18}
$$

Substituting (4.2.18) into (4.2.17), we find that

$$\left(\int_B |du|^s w^{1/\lambda} dx\right)^{1/s} \leq \frac{C}{diam(B)} \left(\int_{\rho B} |u - c|^s w^{1/\lambda^2} dx\right)^{1/s}$$

for all balls B with $\rho B \subset \Omega$ and all closed forms c. ∎

4.2.3 $A_r^\lambda(\Omega)$-weighted inequality

We have presented the $A_r(\lambda, \Omega)$-weighted Caccioppoli-type estimate in the previous section. A similar result holds for $A_r^\lambda(\Omega)$-weights.

Theorem 4.2.4. *Let $u \in D'(\Omega, \wedge^l)$ be a differential form satisfying the A-harmonic equation (4.2.1) in a domain $\Omega \subset \mathbf{R}^n$, $l = 0, 1, \ldots, n - 1$ and $\rho > 1$. Assume that $w \in A_r^\lambda(\Omega)$ for some $r > 1$ and $\lambda > 1$. If $1 < s < \infty$ is a fixed exponent associated with the A-harmonic equation, then there exists a constant C, independent of u, such that*

$$\left(\int_B |du|^s w^{1/\lambda} dx\right)^{1/s} \leq C diam(B)^{-1} \left(\int_{\rho B} |u - c|^s w dx\right)^{1/s}$$

for all balls B with $\rho B \subset \Omega$ and all closed forms c.

Proof. Choose $t = \lambda s/(\lambda - 1)$. Applying Hölder's inequality and Theorem 4.2.1, we have

$$\left(\int_B |du|^s w^{1/\lambda} dx\right)^{1/s} = \left(\int_B (|du| w^{1/\lambda s})^s dx\right)^{1/s}$$

$$\leq \left(\int_B |du|^t dx\right)^{1/t} \left(\int_B w^{t/\lambda(t-s)} dx\right)^{(t-s)/st}$$

$$= \|du\|_{t,B} \left(\int_B w dx\right)^{1/\lambda s} \tag{4.2.19}$$

$$\leq C_1 diam(B)^{-1} \|u - c\|_{t,\sigma B} \left(\int_B w dx\right)^{1/\lambda s}.$$

Next, set $m = s/r$, so that $m < s < t$. Note that if u is a solution of equation (4.2.1) and c is a closed form, then $u - c$ is also a solution of equation (4.2.1). Using Lemma 3.1.1, we obtain

$$\|u - c\|_{t,\sigma B} \leq C_2 |B|^{(m-t)/mt} \|u - c\|_{m,\rho B}, \tag{4.2.20}$$

where $\rho = \sigma^2$. Substituting (4.2.20) into (4.2.19) we find that

$$\left(\int_B |du|^s w^{1/\lambda} dx\right)^{1/s}$$

$$\leq C_3 diam(B)^{-1} |B|^{(m-t)/mt} \|u - c\|_{m,\rho B} \left(\int_B w dx\right)^{1/\lambda s}. \tag{4.2.21}$$

Now using Hölder's inequality with $1/m = 1/s + (s-m)/ms$, we have

$$\|u - c\|_{m,\rho B}$$

$$= \left(\int_{\rho B} \left(|u-c| w^{1/s} w^{-1/s} \right)^m dx \right)^{1/m}$$

$$\leq \left(\int_{\rho B} |u-c|^s w dx \right)^{1/s} \left(\int_{\rho B} \left(\tfrac{1}{w} \right)^{m/(s-m)} dx \right)^{(s-m)/ms} \qquad (4.2.22)$$

$$\leq \left(\int_{\rho B} |u-c|^s w dx \right)^{1/s} \left(\int_{\rho B} \left(\tfrac{1}{w} \right)^{1/(r-1)} dx \right)^{(r-1)/s}$$

for all balls B with $\rho B \subset \Omega$ and all closed forms c. Combining (4.2.21) and (4.2.22), we obtain

$$\left(\int_B |du|^s w^{1/\lambda} dx \right)^{1/s}$$

$$\leq C_3 diam(B)^{-1} |B|^{(m-t)/mt} \left(\int_B w dx \right)^{1/\lambda s} \left(\int_{\rho B} \left(\tfrac{1}{w} \right)^{1/(r-1)} dx \right)^{(r-1)/s}$$

$$\times \left(\int_{\rho B} |u-c|^s w dx \right)^{1/s}.$$

$$(4.2.23)$$

Next, using the condition $w \in A_r^\lambda(\Omega)$, we find

$$\left(\int_B w dx \right)^{1/\lambda s} \left(\int_{\rho B} (1/w)^{1/(r-1)} dx \right)^{(r-1)/s}$$

$$\leq \left(\left(\int_{\rho B} w dx \right) \left(\int_{\rho B} (1/w)^{1/(r-1)} dx \right)^{\lambda(r-1)} \right)^{1/\lambda s}$$

$$\leq |\rho B|^{\frac{r-1}{s} + \frac{1}{\lambda s}} \left(\left(\tfrac{1}{|\rho B|} \int_B w dx \right) \left(\tfrac{1}{|\rho B|} \int_{\rho B} (1/w)^{1/(r-1)} dx \right)^{\lambda(r-1)} \right)^{1/\lambda s}$$

$$\leq C_4 |\rho B|^{(r-1)/s + 1/\lambda s}$$

$$= C_5 |B|^{(r-1)/s + 1/\lambda s}.$$

$$(4.2.24)$$

Finally, substituting (4.2.24) into (4.2.23), we obtain

$$\left(\int_B |du|^s w^{1/\lambda} dx \right)^{1/s} \leq C diam(B)^{-1} \left(\int_{\rho B} |u-c|^s w dx \right)^{1/s}. \qquad \blacksquare$$

4.2.4 Parametric version

The following more general $A_r(\Omega)$-weighted result is proved in [180] recently.

Theorem 4.2.5. *Let $u \in D'(\Omega, \wedge^l)$, $l = 0, 1, \ldots, n-1$, be an A-harmonic tensor in a domain $\Omega \subset \mathbf{R}^n$ and $\rho > 1$. Assume that $1 < s < \infty$ is the* *exponent associated with the A-harmonic equation and $w \in A_r(\Omega)$ for some $r > 1$. Then, there exists a constant C, independent of u, such that*

$$\left(\int_B |du|^s w^\alpha dx \right)^{1/s} \leq \frac{C}{diam(B)} \left(\int_{\rho B} |u - c|^s w^\alpha dx \right)^{1/s} \qquad (4.2.25)$$

for all balls B with $\rho B \subset \Omega$ and all closed forms c. Here α is any constant with $0 < \alpha \leq 1$.

Proof. Theorem 4.2.5 follows from Theorem 4.2.2 and equation (1.4.2).

4.2.5 Inequalities with two parameters

The following parametric version of the $A_r(\lambda, \Omega)$-weighted Caccioppoli-type inequality was obtained recently. This is a very general result since it contains two parameters, α and λ. In fact, for some particular values of α and λ our result reduces to several known results.

Theorem 4.2.6. *Let $u \in D'(\Omega, \wedge^l)$, $l = 0, 1, \ldots, n-1$, be an A-harmonic tensor in a domain $\Omega \subset \mathbf{R}^n$, $\rho > 1$, and $0 < \alpha < 1$. Assume that $1 < s < \infty$ is a fixed exponent associated with the A-harmonic equation and $w \in A_r(\lambda, \Omega)$ for some $r > 1$ and $\lambda > 0$. Then, there exists a constant C, independent of u, such that*

$$\left(\int_B |du|^s w^{\alpha\lambda} dx \right)^{1/s} \leq \frac{C}{diam(B)} \left(\int_{\rho B} |u - c|^s w^\alpha dx \right)^{1/s} \qquad (4.2.26)$$

for all balls B with $\rho B \subset \Omega$ and all closed forms c.

4.3 Local and global two-weight cases

We have presented some Caccioppoli inequalities with one weight in the previous section The purpose of this section is to study two-weighted local Caccioppoli inequalities for solutions of the nonhomogeneous A-harmonic equation. For this, we recall that throughout this chapter M is a compact Riemannian manifold on $\mathbf{R}^n$. As usual, we begin our discussion with unweighted inequalities.

4.3.1 An unweighted inequality

We first prove the following elementary version of the local Caccioppoli inequality.

Theorem 4.3.1. *Let $u \in D'(\Omega, \wedge^l)$ be a solution to the nonhomogeneous A-harmonic equation (1.2.10) in a bounded domain Ω, and $\sigma > 1$ be a constant. Then, there exists a constant C, independent of u, such that*

$$\|du\|_{p,B} \leq C \, diam(B)^{-1}\|u - c\|_{p,\sigma B} \qquad (4.3.1)$$

for all balls or cubes B with $\sigma B \subset \Omega$ and all closed forms c. Here $1 < p < \infty$.

Proof. Let $\sigma > 1$ be a real number and $\eta \in C_0^\infty(\sigma B), \eta \equiv 1$ in B, and

$$|\nabla \eta| \leq \frac{C_1}{diam(B)}|B|^{-1/n},$$

where C_1 is a constant. Choosing the test form $\varphi = -u\eta^p$ for (1.2.10), we have

$$d\varphi = -\eta^p du - pu\eta^{p-1}d\eta.$$

From the definition of solutions to equation (1.2.10) in Section 1.2, we obtain

$$\int_B A(x, du) \cdot (-\eta^p du) + \int_B A(x, du) \cdot (-pu\eta^{p-1}d\eta) + \int_B B(x, du) \cdot (-u\eta^p) = 0. \qquad (4.3.2)$$

Thus, it follows that

$$\int_{\sigma B} A(x, du) \cdot (\eta^p du)$$
$$= -\int_{\sigma B} A(x, du) \cdot (pu\eta^{p-1}d\eta) - \int_{\sigma B} B(x, du) \cdot (u\eta^p). \qquad (4.3.3)$$

Applying (1.2.8), we have

$$\int_{\sigma B} |A(x, du) \cdot (\eta^p du)|$$
$$= \int_{\sigma B} |\eta^p A(x, du) \cdot du| = \int_{\sigma B} |A(x, du) \cdot du| \geq \int_{\sigma B} |du|^p. \qquad (4.3.4)$$

Note the fact that $|\nabla \eta| = |d\eta|$ and $diam(B) \leq diam(\Omega) < \infty$. Using (4.3.4), (4.3.3), (1.2.8), and the Hölder inequality, we obtain

$$\int_{\sigma B} |\eta|^p|du|^p$$
$$\leq \int_{\sigma B} |A(x, du) \cdot (\eta^p du)|$$
$$\leq \int_{\sigma B} |A(x, du)| \cdot (p|u||\eta|^{p-1}|d\eta|) + \int_{\sigma B} |B(x, du)| \cdot (|u||\eta|^p)$$
$$\leq \int_{\sigma B} a|du|^{p-1} \cdot (p|u||\eta|^{p-1}|\nabla\eta|) + \int_{\sigma B} b|du|^{p-1} \cdot (|u||\eta|^p)$$
$$\leq \frac{C_2}{diam(B)} \int_{\sigma B} |du|^{p-1}|\eta|^{p-1}|u| + C_3 \int_{\sigma B} |du|^{p-1}|u||\eta|^p$$

$$\leq \frac{C_2}{diam(B)}\|u\|_{p,\sigma B}\left(\int_{\sigma B}((|du|\eta)^{p-1})^{\frac{p}{p-1}}\right)^{\frac{p-1}{p}}$$

$$+C_3\|\eta u\|_{p,\sigma B}\left(\int_{\sigma B}((|du|\eta)^{p-1})^{\frac{p}{p-1}}\right)^{\frac{p-1}{p}}$$

$$\leq \left(\frac{C_2}{diam(B)}\|u\|_{p,\sigma B}+C_3\|\eta u\|_{p,\sigma B}\right)\left(\int_{\sigma B}(\eta|du|)^p\right)^{\frac{p-1}{p}}$$

$$\leq \left(\frac{C_2}{diam(B)}\|u\|_{p,\sigma B}+C_3\|u\|_{p,\sigma B}\right)\left(\int_{\sigma B}(\eta|du|)^p\right)^{\frac{p-1}{p}} \qquad (4.3.5)$$

$$\leq \frac{C_2+C_3 diam(B)}{diam(B)}\|u\|_{p,\sigma B}\left(\int_{\sigma B}(\eta du)^p\right)^{\frac{p-1}{p}}$$

$$\leq \frac{C_4}{diam(B)}\|u\|_{p,\sigma B}\left(\int_{\sigma B}|\eta|^p|du|^p\right)^{\frac{p-1}{p}}$$

which is equivalent to

$$\|\eta du\|_{p,\sigma B}\leq \frac{C_4}{diam(B)}\|u\|_{p,\sigma B}.$$

Thus, we have

$$\|du\|_{p,B}=\|\eta du\|_{p,B}$$
$$\leq \|\eta du\|_{p,\sigma B}$$
$$\leq \frac{C_4}{diam(B)}\|u\|_{p,\sigma B}.$$

It is clear that if u is a solution of (1.2.10) and c is any closed form, then $u-c$ is also a solution of (1.2.10). Thus, it follows that

$$\|du\|_{p,B}\leq Cdiam(B)^{-1}\|u-c\|_{p,\sigma B}. \qquad \blacksquare$$

4.3.2 Two-weight inequalities

Using the elementary version of the local Caccioppoli inequality obtained above, we prove the following two-weight Caccioppoli inequality.

Theorem 4.3.2. *Let $u \in D'(\Omega, \wedge^l)$, $l = 0, 1, \ldots, n-1$, be a solution of the nonhomogeneous A-harmonic equation in a domain $\Omega \subset \mathbf{R}^n$ and $\rho > 1$. Assume that $1 < s < \infty$ is the exponent associated with the A-harmonic equation and $(w_1, w_2) \in A_{r,\lambda}(\Omega)$ for some $\lambda \geq 1$ and $1 < r < \infty$. Then, there exists a constant C, independent of u, such that*

$$\left(\int_B |du|^s w_1^\alpha dx\right)^{1/s} \leq \frac{C}{diam(B)} \left(\int_{\rho B} |u-c|^s w_2^\alpha dx\right)^{1/s} \tag{4.3.6}$$

or

$$\|du\|_{s,B,w_1^\alpha} \leq C diam(B)^{-1} \|u-c\|_{s,\rho B,w_2^\alpha} \tag{4.3.6}'$$

for all balls B with $\rho B \subset \Omega$ and all closed forms c. Here α is any constant with $0 < \alpha < \lambda$.

Proof. Choose $t = \lambda s/(\lambda - \alpha)$. Since $1/s = 1/t + (t-s)/st$, from the Hölder inequality and Theorem 4.3.1, we obtain

$$\left(\int_B |du|^s w_1^\alpha dx\right)^{1/s}$$
$$= \left(\int_B \left(|du| w_1^{\alpha/s}\right)^s dx\right)^{1/s}$$
$$\leq \left(\int_B |du|^t dx\right)^{1/t} \left(\int_B \left(w_1^{\alpha/s}\right)^{st/(t-s)} dx\right)^{(t-s)/st} \tag{4.3.7}$$
$$\leq \|du\|_{t,B} \cdot \left(\int_B w_1^\lambda dx\right)^{\alpha/\lambda s}$$
$$\leq C_1 diam(B)^{-1} \|u-c\|_{t,\sigma B} \left(\int_B w_1^\lambda dx\right)^{\alpha/\lambda s}$$

for all balls B with $\sigma B \subset \Omega$ and all closed forms c. Let $m = \lambda s/(\lambda + \alpha(r-1))$, so that $m < s < t$. Using Lemma 3.1.1 and the fact that $u-c$ is also a solution of (1.2.10), we have

$$\|u-c\|_{t,\sigma B} \leq C_2 |B|^{(m-t)/mt} \|u-c\|_{m,\rho B}, \tag{4.3.8}$$

where $\rho > \sigma > 1$. Substituting (4.3.8) into (4.3.7), we obtain

$$\left(\int_B |du|^s w_1^\alpha dx\right)^{1/s}$$
$$\leq C_3 diam(B)^{-1} |B|^{(m-t)/mt} \|u-c\|_{m,\rho B} \left(\int_B w_1^\lambda dx\right)^{\alpha/\lambda s}. \tag{4.3.9}$$

The Hölder inequality with $1/m = 1/s + (s-m)/sm$ again yields

$$\|u-c\|_{m,\rho B}$$
$$= \left(\int_{\rho B} |u-c|^m dx\right)^{1/m}$$
$$= \left(\int_{\rho B} \left(|u-c| w_2^{\alpha/s} w_2^{-\alpha/s}\right)^m dx\right)^{1/m} \tag{4.3.10}$$
$$\leq \left(\int_{\rho B} |u-c|^s w_2^\alpha dx\right)^{1/s} \left(\int_{\rho B} \left(\frac{1}{w_2}\right)^{\lambda/(r-1)} dx\right)^{\alpha(r-1)/\lambda s}$$

for all balls B with $\rho B \subset \Omega$ and all closed forms c. Substituting (4.3.10) into (4.3.9), we find

$$\left(\int_B |du|^s w_1^\alpha dx\right)^{1/s}$$
$$\leq C_3 diam(B)^{-1}|B|^{(m-t)/mt}\|w_1\|_{\lambda,B}^{\alpha/s}\|1/w_2\|_{\lambda/(r-1),\rho B}^{\alpha/s} \qquad (4.3.11)$$
$$\times \left(\int_{\rho B} |u-c|^s w_2^\alpha dx\right)^{1/s}.$$

Using the condition $(w_1, w_2) \in A_{r,\lambda}(M)$, we obtain

$$\|w_1\|_{\lambda,B}^{\alpha/s}\|1/w_2\|_{\lambda/(r-1),\rho B}^{\alpha/s}$$
$$\leq \left(\left(\int_{\rho B} w_1^\lambda dx\right)\left(\int_{\rho B}(1/w_2)^{\lambda/(r-1)}dx\right)^{r-1}\right)^{\alpha/\lambda s}$$
$$= \left(|\rho B|^r \left(\frac{1}{|\rho B|}\int_{\rho B} w_1^\lambda dx\right)\left(\frac{1}{|\rho B|}\int_{\rho B}\left(\frac{1}{w_2^\lambda}\right)^{\frac{1}{r-1}}dx\right)^{r-1}\right)^{\alpha/\lambda s} \qquad (4.3.12)$$
$$\leq C_4|B|^{\alpha r/\lambda s}.$$

Combining (4.3.11) and (4.3.12), we have

$$\left(\int_B |du|^s w_1^\alpha dx\right)^{1/s} \leq \frac{C_5}{diam(B)}\left(\int_{\rho B} |u-c|^s w_2^\alpha dx\right)^{1/s} \qquad (4.3.13)$$

for all balls B with $\rho B \subset \Omega$ and all closed forms c. ∎

Note that Theorem 4.3.2 contains two weights, $w_1(x)$ and $w_2(x)$, and two parameters, λ and α. These features make this result more flexible and useful. In fact, the existing versions of the Caccioppoli-type inequality in [186, 267, 71] are special cases of Theorem 4.3.2 with suitable choices of weights and parameters.

Theorem 4.3.3. *Let $u \in D'(\Omega, \wedge^l)$, $l = 0, 1, \ldots, n-1$, be a solution of the nonhomogeneous A-harmonic equation in a domain $\Omega \subset \mathbf{R}^n$, $\rho > 1$, and $0 < \alpha < 1$. Assume that $1 < s < \infty$ is a fixed exponent associated with the A-harmonic equation and $(w_1, w_2) \in A_r(\lambda, \Omega)$ for some $r > 1$ and $\lambda > 0$. Then, there exists a constant C, independent of u, such that*

$$\left(\int_B |du|^s w_1^{\alpha\lambda} dx\right)^{1/s} \leq \frac{C}{diam(B)}\left(\int_{\rho B} |u-c|^s w_2^\alpha dx\right)^{1/s} \qquad (4.3.14)$$

for all balls B with $\rho B \subset \Omega$ and all closed forms c.

Proof. Choose $t = s/(1-\alpha)$, so that $1 < s < t$. Since $1/s = 1/t + (t-s)/st$, by the Hölder inequality and Theorem 4.2.1, it follows that

$$\left(\int_B |du|^s w_1^{\alpha\lambda} dx \right)^{1/s}$$

$$= \left(\int_B \left(|du| w_1^{\alpha\lambda/s} \right)^s dx \right)^{1/s}$$

$$\leq \left(\int_B |du|^t dx \right)^{1/t} \left(\int_B \left(w_1^{\alpha\lambda/s} \right)^{st/(t-s)} dx \right)^{(t-s)/st} \qquad (4.3.15)$$

$$\leq \|du\|_{t,B} \cdot \left(\int_B w_1^{\lambda} dx \right)^{\alpha/s}$$

$$\leq C_1 diam(B)^{-1} \|u - c\|_{t,\sigma B} \left(\int_B w_1^{\lambda} dx \right)^{\alpha/s}$$

for all balls B with $\sigma B \subset \Omega$ and all closed forms c. Now since c is a closed form and u is an A-harmonic tensor, $u - c$ is also an A-harmonic tensor. Taking $m = s/(1 + \alpha(r-1))$, so that $m < s < t$. Thus, an application of Lemma 3.1.1 yields

$$\|u - c\|_{t,\sigma B} \leq C_2 |B|^{(m-t)/mt} \|u - c\|_{m,\sigma^2 B}$$
$$= C_2 |B|^{(m-t)/mt} \|u - c\|_{m,\rho B}, \qquad (4.3.16)$$

where $\rho = \sigma^2$. Substituting (4.3.16) into (4.3.15), we have

$$\left(\int_B |du|^s w_1^{\alpha\lambda} dx \right)^{1/s}$$
$$\leq C_3 diam(B)^{-1} |B|^{(m-t)/mt} \|u - c\|_{m,\rho B} \left(\int_B w_1^{\lambda} dx \right)^{\alpha/s}. \qquad (4.3.17)$$

Now since $1/m = 1/s + (s-m)/sm$, by the Hölder inequality again, we obtain

$$\|u - c\|_{m,\rho B}$$

$$= \left(\int_{\rho B} |u - c|^m dx \right)^{1/m}$$

$$= \left(\int_{\rho B} \left(|u - c| w_2^{\alpha/s} w_2^{-\alpha/s} \right)^m dx \right)^{1/m} \qquad (4.3.18)$$

$$\leq \left(\int_{\rho B} |u - c|^s w_2^{\alpha} dx \right)^{1/s} \left(\int_{\rho B} \left(\frac{1}{w_2} \right)^{1/(r-1)} dx \right)^{\alpha(r-1)/s}$$

for all balls B with $\rho B \subset \Omega$ and all closed forms c. Combining (4.3.17) and (4.3.18), it follows that

$$\left(\int_B |du|^s w_1^{\alpha\lambda} dx \right)^{1/s}$$
$$\leq C_3 diam(B)^{-1} |B|^{(m-t)/mt} \|w_1\|_{\lambda,B}^{\alpha\lambda/s} \|1/w_2\|_{1/(r-1),\rho B}^{\alpha/s} \qquad (4.3.19)$$
$$\times \left(\int_{\rho B} |u - c|^s w_2^{\alpha} dx \right)^{1/s}.$$

Since $(w_1, w_2) \in A_r(\lambda, \Omega)$, we have

$$\|w_1\|_{\lambda,B}^{\alpha\lambda/s} \cdot \|1/w_2\|_{1/(r-1),\rho B}^{\alpha/s}$$

$$\leq \left(\left(\int_{\rho B} w_1^\lambda dx \right) \left(\int_{\rho B} (1/w_2)^{1/(r-1)} dx \right)^{r-1} \right)^{\alpha/s}$$

$$= \left(|\rho B|^r \left(\frac{1}{|\rho B|} \int_{\rho B} w_1^\lambda dx \right) \left(\frac{1}{|\rho B|} \int_{\rho B} \left(\frac{1}{w_2} \right)^{1/(r-1)} dx \right)^{r-1} \right)^{\alpha/s} \quad (4.3.20)$$

$$\leq C_4 |B|^{\alpha r/s}.$$

Substituting (4.3.20) into (4.3.19), we find

$$\left(\int_B |du|^s w_1^{\alpha\lambda} dx \right)^{1/s} \leq \frac{C}{diam(B)} \left(\int_{\rho B} |u - c|^s w_2^\alpha dx \right)^{1/s} \quad (4.3.21)$$

for all balls B with $\rho B \subset \Omega$ and all closed forms c. ∎

Note that if $\lambda = 1$, then $A_r(\lambda, \Omega) = A_r(1, \Omega)$ becomes the usual $A_r(\Omega)$ weight.

4.4 Inequalities with Orlicz norms

The purpose of this section is to develop various versions of the Caccioppoli inequalities with Orlicz norms $\|\cdot\|_{L^p(\log L)^\alpha(E)}$. We begin with unweighted $\|\cdot\|_{L^p(\log L)^\alpha(E)}$ norm inequalities for du. Then, we discuss the $A_r(M)$-weighted cases. All the results of this section can be extended rather easily to the two-weight cases.

4.4.1 Basic $\|\cdot\|_{L^p(\log L)^\alpha(E)}$ estimates

The purpose of this section is to prove the following Caccioppoli-type estimate with Orlicz norms for solutions to the nonhomogeneous A-harmonic equation on a Riemannian manifold.

Theorem 4.4.1. Let $u \in L^p(\log L)^\alpha(\Omega, \wedge^l)$, $l = 0, 1, \ldots, n-1$, be a solution to the nonhomogeneous A-harmonic equation in a domain $\Omega \subset \mathbf{R}^n$. Then, there exists a constant C, independent of u, such that

$$\|du\|_{L^p(\log L)^\alpha(B)} \leq C|B|^{-1/n} \|u - c\|_{L^p(\log L)^\alpha(\sigma B)} \quad (4.4.1)$$

for some constant $\sigma > 1$ and all balls B with $\sigma B \subset \Omega$ and $diam(B) \geq d_0 > 0$. Here d_0, $1 < p < \infty$, and $\alpha > 0$ are constants.

Proof. Let $B \subset M$ be a ball with $diam(B) \geq d_0 > 0$. Let $\varepsilon > 0$ be small enough and the constant C_1 large enough, so that

$$|B|^{-\varepsilon/p^2} \leq C_1. \tag{4.4.2}$$

Applying Lemma 2.9 in [201], we have

$$\|du\|_{p+\varepsilon,B} \leq C_2 |B|^{(p-(p+\varepsilon))/p(p+\varepsilon)} \|du\|_{p,\sigma B} \tag{4.4.3}$$

for some $\sigma > 1$. We may assume that $|du|/\|du\|_{p,B} \geq 1$ on B. Otherwise, setting

$$B_1 = \{x \in B : |du|/\|du\|_{p,B} \geq 1\},$$
$$B_2 = \{x \in B : |du|/\|du\|_{p,B} < 1\}$$

and using the elementary inequality $|a + b|^s \leq 2^s(|a|^s + |b|^s)$, where $s > 0$ is an arbitrary constant, we obtain

$$\|du\|_{L^p(\log L)^\alpha(B)}$$

$$= \left(\int_B |du|^p \log^\alpha \left(e + \frac{|du|}{\|du\|_{p,B}} \right) dx \right)^{1/p}$$

$$= \left(\int_{B_1} |du|^p \log^\alpha \left(e + \frac{|du|}{\|du\|_{p,B}} \right) dx \right.$$

$$+ \left. \int_{B_2} |du|^p \log^\alpha \left(e + \frac{|du|}{\|du\|_{p,B}} \right) dx \right)^{1/p} \tag{4.4.4}$$

$$\leq 2^{1/p} \left(\left(\int_{B_1} |du|^p \log^\alpha \left(e + \frac{|du|}{\|du\|_{p,B}} \right) dx \right)^{\frac{1}{p}} \right.$$

$$+ \left. \left(\int_{B_2} |du|^p \log^\alpha \left(e + \frac{|du|}{\|du\|_{p,B}} \right) dx \right)^{\frac{1}{p}} \right).$$

Now, we shall estimate the first term on the right-hand side of (4.4.4). Since $|du|/\|du\|_{p,B} \geq 1$ on B_1, for $\varepsilon > 0$ in (4.4.2), there exists $C_3 > 0$ such that

$$\log^\alpha \left(e + \frac{|du|}{\|du\|_{p,B}} \right) \leq C_3 \left(\frac{|du|}{\|du\|_{p,\sigma_1 B}} \right)^\varepsilon. \tag{4.4.5}$$

Combining (4.4.2), (4.4.3), and (4.4.5), we obtain

$$\left(\int_{B_1} |du|^p \log^\alpha \left(e + \frac{|du|}{\|du\|_{p,B}} \right) dx \right)^{1/p}$$

$$\leq C_4 \left(\frac{1}{\|du\|_{p,\sigma_1 B}^\varepsilon} \int_{B_1} |du|^{p+\varepsilon} dx \right)^{1/p}$$

$$\leq C_4 \left(\frac{1}{\|du\|^{\varepsilon}_{p,\sigma_1 B}} \int_B |du|^{p+\varepsilon} dx \right)^{1/p}$$

$$= \frac{C_4}{\|du\|^{\varepsilon/p}_{p,\sigma_1 B}} \left(\left(\int_B |du|^{p+\varepsilon} dx \right)^{\frac{1}{p+\varepsilon}} \right)^{(p+\varepsilon)/p}$$

$$\leq \frac{C_5}{\|du\|^{\varepsilon/p}_{p,\sigma_1 B}} \left(|B|^{(p-(p+\varepsilon))/p(p+\varepsilon)} \|du\|_{p,\sigma_1 B} \right)^{(p+\varepsilon)/p}$$

$$\leq C_6 \|du\|_{p,\sigma_1 B},$$

$$\text{(4.4.6)}$$

where $\sigma_1 > 1$ is a constant. Next, since

$$\log^{\alpha} \left(e + \frac{|du|}{\|du\|_{p,B}} \right) \leq M_1 \log^{\alpha}(e+1) \leq M_2, \quad x \in B_2,$$

we can estimate the second term, to find

$$\left(\int_{B_2} |du|^p \log^{\alpha} \left(e + \frac{|du|}{\|du\|_{p,B}} \right) dx \right)^{1/p} \leq C_7 \|du\|_{p,\sigma_2 B}, \qquad \text{(4.4.6)}'$$

where $\sigma_2 > 1$ is a constant. From (4.4.4), (4.4.6), and (4.4.6)$'$, we have

$$\|du\|_{L^p(\log L)^{\alpha}(B)} \leq C_8 \|du\|_{p,\sigma_3 B}, \qquad \text{(4.4.7)}$$

where $\sigma_3 = \max\{\sigma_1, \sigma_2\}$. Note that $diam(B) = C_9 |B|^{1/n}$, and hence from Theorem 4.3.2, we obtain

$$\|du\|_{p,\sigma_3 B} \leq C_{10} |B|^{-1/n} \|u - c\|_{p,\sigma_4 B}$$

for some $\sigma_4 > \sigma_3$ and all closed forms c. It is easy to see that

$$\log^{\alpha} \left(e + \frac{|u-c|}{\|u-c\|_{p,\sigma_2 B}} \right) \geq 1, \quad \alpha > 0. \qquad \text{(4.4.8)}$$

Now combining last three inequalities, we find

$$\|du\|_{L^p(\log L)^{\alpha}(B)} \leq C_{11} |B|^{-1/n} \|u - c\|_{p,\sigma_4 B}$$

$$\leq C_{11} |B|^{-1/n} \|u - c\|_{L^p(\log L)^{\alpha}(\sigma_4 B)}. \qquad \blacksquare$$

$$\text{(4.4.9)}$$

If we revise (4.4.5) and (4.4.8) in the proof of Theorem 4.4.1, we obtain the following version of the Cacciopoli-type estimate.

Corollary 4.4.2. *Let $u \in L^p(\log L)^{\alpha}(\Omega, \wedge^l)$, $l = 0, 1, \ldots, n-1$, be a solution to the nonhomogeneous A-harmonic equation in a domain $\Omega \subset \mathbf{R}^n$. Then, there exists a constant C, independent of u, such that*

$$\left(\int_B |du|^p \log^\alpha \left(e + \frac{|du|}{\|du\|_{p,M}} \right) dx \right)^{1/p}$$

$$\leq \frac{C}{diam(B)} \left(\int_{\sigma B} |u - c|^p \log^\alpha \left(e + \frac{|u-c|}{\|u-c\|_{p,M}} \right) dx \right)^{1/p}$$

for some constant $\sigma > 1$ and all balls B with $\sigma B \subset \Omega$ and $diam(B) \geq d_0 > 0$. Here d_0, $1 < p < \infty$, and $\alpha > 0$ are constants.

4.4.2 Weak reverse Hölder inequalities

We now prove the weak reverse Hölder inequalities with $\| \cdot \|_{L^s(\log L)^\alpha(E)}$ norms. These results will be used later to establish some weighted inequalities.

Lemma 4.4.3. *Let u be an A-harmonic tensor in a domain $\Omega \subset \mathbf{R}^n$, $\sigma > 1$ and $0 < s,t < \infty$. Then, there exists a constant C, independent of u, such that*

$$\|u\|_{L^s(\log L)^\alpha(B)} \leq C|B|^{(t-s)/st}\|u\|_{L^t(\log L)^\beta(\sigma B)}$$

for all constants $\alpha > 0$ and $\beta > 0$, and all balls B with $\sigma B \subset \Omega$ and $diam(B) \geq d_0 > 0$, where d_0 is a fixed constant.

Proof. For any ball $B \subset M$ with $diam(B) \geq d_0 > 0$, we may choose $\varepsilon > 0$ small enough and the constant C_1 large enough so that

$$|B|^{-\varepsilon/st} \leq C_1. \tag{4.4.10}$$

From Lemma 3.1.1, we have

$$\|u\|_{s+\varepsilon,B} \leq C_2|B|^{(t-(s+\varepsilon))/t(s+\varepsilon)}\|u\|_{t,\sigma B} \tag{4.4.11}$$

for some $\sigma > 1$. Similar to (4.4.5) in the proof of Theorem 4.4.1, we can assume that $\frac{|u|}{\|u\|_{t,B}} \geq 1$ on B. For above $\varepsilon > 0$, there exists $C_3 > 0$ such that

$$\log^\alpha \left(e + \frac{|u|}{\|u\|_{s,B}} \right) \leq C_3 \left(\frac{|u|}{\|u\|_{t,\sigma B}} \right)^\varepsilon. \tag{4.4.12}$$

From (4.4.11) and (4.4.12), it follows that

$$\|u\|_{L^s(\log L)^\alpha(B)}$$

$$= \left(\int_B |u|^s \log^\alpha \left(e + \frac{|u|}{\|u\|_{s,B}} \right) dx \right)^{1/s}$$

$$\leq C_4 \left(\frac{1}{\|u\|_{t,\sigma B}^\varepsilon} \int_B |u|^{s+\varepsilon} dx \right)^{1/s}$$

$$\leq \frac{C_4}{\|u\|_{t,\sigma B}^{\varepsilon/s}} \left(\left(\int_B |u|^{s+\varepsilon} dx \right)^{\frac{1}{s+\varepsilon}} \right)^{(s+\varepsilon)/s}$$

$$\leq \frac{C_5}{\|u\|_{t,\sigma B}^{\varepsilon/s}} \left(|B|^{(t-(s+\varepsilon))/t(s+\varepsilon)} \|u\|_{t,\sigma B} \right)^{(s+\varepsilon)/s} \tag{4.4.13}$$

$$\leq C_6 |B|^{\frac{t-s-\varepsilon}{st}} \|u\|_{t,\sigma B}.$$

Now from (4.4.10) and (4.4.13) and using

$$\log^\beta \left(e + \frac{|u|}{\|u\|_{t,\sigma B}} \right) \geq 1, \quad \beta > 0,$$

we obtain

$$\|u\|_{L^s (\log L)^\alpha (B)}$$

$$\leq C_6 |B|^{\frac{t-s-\varepsilon}{st}} \|u\|_{t,\sigma B}$$

$$\leq C_7 |B|^{\frac{t-s-\varepsilon}{st}} \left(\int_{\sigma B} |u|^t \log^\beta \left(e + \frac{|u|}{\|u\|_{t,\sigma B}} \right) dx \right)^{1/t} \tag{4.4.14}$$

$$\leq C_7 |B|^{\frac{t-s-\varepsilon}{st}} \|u\|_{L^t (\log L)^\beta (\sigma B)}$$

$$\leq C_8 |B|^{\frac{t-s}{st}} \|u\|_{L^t (\log L)^\beta (\sigma B)}. \quad \blacksquare$$

Using a similar method developed in the proof of Lemma 4.4.3 and from Lemma 2.9 in [268], we can prove the following version of the weak reverse Hölder inequality with Orlicz norms. Note that this result cannot be obtained by replacing u by du in Lemma 4.4.3 since du may not be a solution of (1.2.10).

Lemma 4.4.4. *Let u be an A-harmonic function in a domain $\Omega \subset \mathbf{R}^n$, $\sigma > 1$, and $0 < s, t < \infty$. Then, there exists a constant C, independent of u, such that*

$$\|du\|_{L^s (\log L)^\alpha (B)} \leq C |B|^{(t-s)/st} \|du\|_{L^t (\log L)^\beta (\sigma B)}$$

for all balls B with $\sigma B \subset \Omega$ and $diam(B) \geq d_0 > 0$. Here d_0 is a fixed constant, $\alpha > 0$ and $\beta > 0$ are arbitrary constants.

4.4.3 $A_r(M)$-weighted cases

Theorem 4.4.5. *Let $u \in L^p (\log L)^\alpha (\Omega, \wedge^l)$, $l = 0, 1, \ldots, n-1$, be a solution to the nonhomogeneous A-harmonic equation in a domain $\Omega \subset \mathbf{R}^n$ and $w \in A_r(\Omega)$ for some $r > 1$. Also assume that $diam\Omega \leq \infty$. Then, there exists a constant C, independent of u, such that*

$$\|du\|_{L^p(\log L)^\alpha(B,w)} \leq C|B|^{-1/n}\|u-c\|_{L^p(\log L)^\alpha(\sigma B,w)} \tag{4.4.15}$$

for any closed form c, some constant $\sigma > 1$, and all balls B with $\sigma B \subset \Omega$ and $diam(B) \geq d_0 > 0$. Here d_0, $1 < p < \infty$, and $\alpha > 0$ are constants.

Proof. Let B be a ball with $\sigma B \subset \Omega$ and $diam(B) \geq d_0 > 0$. Also $d_0 \leq diam(B) \leq diam(\Omega) < \infty$. Thus, $0 < v_1 \leq |B| \leq v_2 < \infty$ for some constants v_1 and v_2. By Lemma 3.1.1, we have

$$m_1\|u\|_{s,\rho_1 B} \leq \|u\|_{t,B} \leq m_2\|u\|_{s,\rho_2 B} \tag{4.4.16}$$

for any solution u of (1.2.10) and any constants $s, t > 0$, where $0 < \rho_1 < 1$, $\rho_2 > 1$, $0 < m_1 < 1$, and $m_2 > 1$ are some constants. By Lemma 1.4.7, there exist constants $k > 1$ and $C_0 > 0$, such that

$$\| w \|_{k,B} \leq C_0 |B|^{(1-k)/k} \| w \|_{1,B} . \tag{4.4.17}$$

Choose $s = pk/(k-1)$, so that $1 < p < s$ and $k = s/(s-p)$. We know that $u \in L^p(\log L)^\alpha(\Omega, \wedge^l)$ implies $u \in L^p(\Omega, \wedge^l)$. Thus, for any closed form c, it follows that $u - c \in L^p(\Omega, \wedge^l)$ since M is compact. By Caccioppoli inequality with L^p-norms, we know that $du \in L^p(M, \wedge^l)$ which implies $\|du\|_{p,\Omega} = N < \infty$. If $\|du\|_{p,B} = 0$, then $du = 0$ a.e. on B and Theorem 4.4.3 follows. Thus, we assume that $0 < m_1 \leq \|du\|_{s,B} < M_1$ and $0 < m_2 \leq \|du\|_{p,B} < M_2$ by (4.4.16). Since $1/p = 1/s + (s-p)/ps$, by the Hölder inequality, (4.4.17), and (3.4.14), we find

$\|du\|_{L^p(\log L)^\alpha(B,w)}$

$$= \left(\int_B |du|^p \log^\alpha \left(e + \frac{|du|}{\|du\|_{p,B}} \right) w dx \right)^{1/p}$$

$$= \left(\int_B \left(|du| \log^{\frac{\alpha}{p}} \left(e + \frac{|du|}{\|du\|_{p,B}} \right) w^{1/p} \right)^p dx \right)^{1/p}$$

$$\leq \left(\int_B |du|^s \log^{\frac{\alpha s}{p}} \left(e + \frac{|du|}{\|du\|_{p,B}} \right) dx \right)^{\frac{1}{s}} \left(\int_B w^{\frac{s}{s-p}} dx \right)^{\frac{s-p}{sp}} \tag{4.4.18}$$

$$\leq C_1 \left(\int_B |du|^s \log^{\frac{\alpha s}{p}} \left(e + \frac{|du|}{\|du\|_{s,B}} \right) dx \right)^{\frac{1}{s}} \left(\left(\int_B w^k dx \right)^{1/k} \right)^{\frac{1}{p}}$$

$$\leq C_2 |B|^{\frac{1-k}{kp}} \| w \|_{1,B}^{\frac{1}{p}} \left(\int_B |du|^s log^{\frac{\alpha s}{p}} \left(e + \frac{|du|}{\|du\|_{s,B}} \right) dx \right)^{\frac{1}{s}} .$$

Now an application of Theorem 4.4.1 yields

$$\left(\int_B |du|^s \log^{\frac{\alpha s}{p}} \left(e + \frac{|du|}{\|du\|_{s,B}} \right) dx \right)^{1/s}$$

$$\leq C_3 |B|^{-1/n} \|u-c\|_{L^s(\log L)^{\alpha s/p}(\sigma_1 B)}. \tag{4.4.19}$$

Here c is any closed form. Next, choose $t = \frac{p}{r}$. Using (4.4.19) and Lemma 4.4.3 with $\beta = \frac{\alpha}{r}$, we obtain

$$\left(\int_B |du|^s \log^{\frac{\alpha s}{r}} \left(e + \frac{|du|}{\|du\|_{p,B}} \right) dx \right)^{1/s}$$

$$\leq C_4 |B|^{-1/n} |B|^{(t-s)/st} \|u - c\|_{L^t(\log L)^\beta(\sigma_2 B)}$$

(4.4.20)

for some $\sigma_2 > \sigma_1$. Using the Hölder inequality again with $1/t = 1/p + (p-t)/pt$, we have

$$\|u - c\|_{L^t(\log L)^\beta(\sigma_2 B)}$$

$$= \left(\int_{\sigma_2 B} |u - c|^t \log^\beta \left(e + \frac{|u-c|}{\|u-c\|_{t,\sigma_2 B}} \right) dx \right)^{1/t}$$

$$= \left(\int_{\sigma_2 B} \left(|u - c| \log^{\frac{\beta}{t}} \left(e + \frac{|u-c|}{\|u-c\|_{t,\sigma_2 B}} \right) w^{1/p} w^{-1/p} \right)^t dx \right)^{1/t}$$

$$\leq \left(\int_{\sigma_2 B} |u - c|^p \log^{\beta p/t} \left(e + \frac{|u-c|}{\|u-c\|_{t,\sigma_2 B}} \right) w dx \right)^{\frac{1}{p}}$$

(4.4.21)

$$\times \left(\int_{\sigma_2 B} \left(\frac{1}{w} \right)^{\frac{t}{p-t}} dx \right)^{\frac{p-t}{pt}}$$

$$\leq \|u - c\|_{L^p(\log L)^\alpha(\sigma_2 B, w)} \left(\int_{\sigma_2 B} \left(\frac{1}{w} \right)^{\frac{1}{r-1}} dx \right)^{(r-1)/p}.$$

Combining (4.4.18), (4.4.20), and (4.4.21), we find

$$\|du\|_{L^p(\log L)^\alpha(B, w)}$$

$$\leq C_5 |B|^{-r/p - 1/n} \|u - c\|_{L^p(\log L)^\alpha(\sigma_2 B, w)}$$

$$\times \left(\int_B w dx \left(\int_{\sigma_2 B} \left(\frac{1}{w} \right)^{\frac{1}{r-1}} dx \right)^{(r-1)} \right)^{1/p}$$

(4.4.22)

$$\leq C_5 |B|^{-r/p - 1/n} \|u - c\|_{L^p(\log L)^\alpha(\sigma_2 B, w)}$$

$$\times \left(\|w\|_{1,\sigma_2 B} \cdot \|1/w\|_{1/(r-1), \sigma_2 B} \right)^{1/p}.$$

Now since $w \in A_r(\Omega)$, it follows that

$$\left(\|w\|_{1,\sigma_2 B} \cdot \|1/w\|_{1/(r-1), \sigma_2 B} \right)^{1/p}$$

$$= \left(\left(\int_{\sigma_2 B} w dx \right) \left(\int_{\sigma_2 B} (1/w)^{1/(r-1)} dx \right)^{r-1} \right)^{1/p}$$

$$= \left(|\sigma_2 B|^r \left(\frac{1}{|\sigma_2 B|} \int_{\sigma_2 B} w \, dx \right) \left(\frac{1}{|\sigma_2 B|} \int_{\sigma_2 B} \left(\frac{1}{w} \right)^{\frac{1}{r-1}} dx \right)^{r-1} \right)^{\frac{1}{p}}$$

$$\leq C_6 |B|^{r/p}. \tag{4.4.23}$$

Finally, substituting (4.4.23) into (4.4.22), we obtain

$$\|du\|_{L^p(\log L)^\alpha(B,w)} \leq C_7 |B|^{-1/n} \|u - c\|_{L^p(\log L)^\alpha(\sigma_2 B, w)}. \qquad \blacksquare$$

4.5 Inequalities with the codifferential operator

In this section, we study the properties of the codifferential operator $d^\star$ that is applied to a differential $(l+1)$-form $v \in D'(\Omega, \wedge^{l+1})$ satisfying the conjugate A-harmonic equation

$$A(x, du) = d^\star v, \tag{4.5.1}$$

where $A : \Omega \times \wedge^l(\mathbf{R}^n) \to \wedge^l(\mathbf{R}^n)$ satisfies the following conditions:

$$|A(x, \xi)| \leq a|\xi|^{p-1} \quad \text{and} \quad < A(x, \xi), \xi > \geq |\xi|^p \tag{4.5.2}$$

for almost every $x \in \Omega$ and all $\xi \in \wedge^l(\mathbf{R}^n)$. Here $a > 0$ is a constant and $1 < p < \infty$ is a fixed exponent associated with (4.5.1).

4.5.1 L^q-estimate for $d^\star v$

We have explored some properties of a pair of solutions (u, v) to the above equation in Chapter 1. In this section, we focus on the integral properties of the codifferential operator $d^\star$.

Theorem 4.5.1. Let $u \in D'(\Omega, \wedge^{l-1})$ and $v \in D'(\Omega, \wedge^{l+1})$ be a pair of A-harmonic tensors in a domain $\Omega \subset \mathbf{R}^n$ and $\sigma > 1$ be a constant. Then, there exists a constant C, independent of v, such that

$$\|d^\star v\|_{q,B} \leq C|B|^{-1/n} \|v - c_1\|_{q,\sigma B} \tag{4.5.3}$$

for all balls or cubes B with $\sigma B \subset \Omega$ and all co-closed forms c_1. Here $1 < q < \infty$.

Proof. Choosing $s = p, t = q$, and $Q = B$ in the local Hardy–Littlewood inequality (Theorem 1.5.2), we have

$$\|u - u_B\|_{p,B}^p \leq C_1 |B|^{\frac{p-q}{n}} \|v - c_1\|_{q,\sigma_1 B}^q. \tag{4.5.4}$$

From (4.5.1) and (4.5.2), we obtain

$$
\begin{aligned}
\int_B |d^\star v|^q dx &= \int_B |A(x, du)|^q dx \\
&\le C_2 \int_B |du|^{q(p-1)} dx \\
&= C_2 \int_B |du|^p dx,
\end{aligned}
\tag{4.5.5}
$$

and hence

$$
\|d^\star v\|_{q,B}^q \le C_2 \|du\|_{p,B}^p.
$$

From Theorem 4.3.1, it follows that

$$
\|d^\star v\|_{q,B}^q \le C_3 diam(B)^{-p} \|u - c\|_{p,\sigma_2 B}^p
$$

for any closed form c. Since u_B is a closed form for any ball, we may choose $c = u_B$. Now since $diam(B) = C_4 |B|^{1/n}$, we have

$$
\|d^\star v\|_{q,B}^q \le C_5 |B|^{-p/n} \|u - u_B\|_{p,\sigma_2 B}^p.
\tag{4.5.6}
$$

Finally, a combination of (4.5.4) and (4.5.6) yields

$$
\|d^\star v\|_{q,B}^q \le C_6 |B|^{-q/n} \|v - c_1\|_{q,\sigma_3 B}^q
$$

which is the same as

$$
\|d^\star v\|_{q,B} \le C |B|^{-1/n} \|v - c_1\|_{q,\sigma_3 B}. \quad \blacksquare
$$

4.5.2 Two-weight estimate for $d^\star v$

Theorem 4.5.2. *Let $u \in D'(\Omega, \wedge^{l-1})$ and $v \in D'(\Omega, \wedge^{l+1})$ be a pair of A-harmonic tensors in a domain $\Omega \subset \mathbf{R}^n$ and $\sigma > 1$ be a constant. Assume that $(w_1, w_2) \in A_{r,\lambda}(\Omega)$ for some $r > 1$ and $\lambda > 0$. Then, there exists a constant C, independent of v, such that*

$$
\|d^\star v\|_{q,B,w_1^\alpha} \le C |B|^{-1/n} \|v - c_1\|_{q,\sigma B, w_2^\alpha}
\tag{4.5.7}
$$

for all balls or cubes B with $\sigma B \subset \Omega$ and all co-closed forms c_1. Here $1 < q < \infty$ and $0 < \alpha < \lambda$ are constants.

Proof. Choose $s = \lambda q / (\lambda - \alpha)$. Since $1/q = 1/s + (s - q)/sq$, using the Hölder inequality, we obtain

$$
\left(\int_B |d^\star v|^q w_1^\alpha dx \right)^{1/q}
$$

$$
= \left(\int_B \left(|d^\star v| w_1^{\alpha/q} \right)^q dx \right)^{1/q}
$$

$$\leq \left(\int_B |d^\star v|^s dx\right)^{1/s} \left(\int_B \left(w_1^{\alpha/q}\right)^{sq/(s-q)} dx\right)^{(s-q)/sq}$$

$$= \|d^\star v\|_{s,B} \cdot \left(\int_B w_1^\lambda dx\right)^{\alpha/\lambda q} \qquad (4.5.8)$$

for all balls B with $B \subset M$. Now let $t = \lambda q/(\lambda + \alpha(r-1))$, so that $t < q < s$. Using the weak reverse Hölder inequality for $d^\star v$, it follows that

$$\|d^\star v\|_{s,B} \leq C_1 |B|^{(t-s)/st} \|d^\star v\|_{t,\sigma_1 B}. \qquad (4.5.9)$$

Substituting (4.5.9) into (4.5.8) and using Theorem 4.5.1, we have

$$\left(\int_B |d^\star v|^q w_1^\alpha dx\right)^{1/q}$$

$$\leq C_2 |B|^{(t-s)/st} \|d^\star v\|_{t,\sigma_1 B} \left(\int_B w_1^\lambda dx\right)^{\alpha/\lambda q} \qquad (4.5.10)$$

$$\leq C_3 |B|^{(t-s)/st} |B|^{-1/n} \|v - c_1\|_{t,\sigma_2 B} \left(\int_B w_1^\lambda dx\right)^{\alpha/\lambda q}.$$

Using the Hölder inequality with $1/t = 1/q + (q-t)/qt$ again, we obtain

$$\|v - c_1\|_{t,\sigma_2 B}$$

$$= \left(\int_{\sigma_2 B} |v - c_1|^t dx\right)^{1/t}$$

$$= \left(\int_{\sigma_2 B} \left(|v - c_1| w_2^{\alpha/q} w_2^{-\alpha/q}\right)^t dx\right)^{1/t} \qquad (4.5.11)$$

$$\leq \left(\int_{\sigma_2 B} |v - c_1|^q w_2^\alpha dx\right)^{1/q} \left(\int_{\sigma_2 B} \left(\frac{1}{w_2}\right)^{\lambda/(r-1)} dx\right)^{\alpha(r-1)/\lambda q}$$

for all balls B with $\sigma_2 B \subset \Omega$ and all co-closed forms c_1. Substituting (4.5.11) into (4.5.10), we find

$$\left(\int_B |d^\star v|^q w_1^\alpha dx\right)^{1/q}$$

$$\leq C_4 |B|^{-1/n} |B|^{(t-s)/st} \|w_1\|_{\lambda,B}^{\alpha/q} \|1/w_2\|_{\lambda/(r-1),\sigma_2 B}^{\alpha/q} \qquad (4.5.12)$$

$$\times \left(\int_{\sigma_2 B} |v - c_1|^q w_2^\alpha dx\right)^{1/q}.$$

Now using the condition $(w_1, w_2) \in A_{r,\lambda}(\Omega)$, we have

$$\|w_1\|_{\lambda,B}^{\alpha/q} \|1/w_2\|_{\lambda/(r-1),\sigma_2 B}^{\alpha/q}$$

$$\leq \left(\left(\int_{\sigma_2 B} w_1^\lambda dx\right) \left(\int_{\sigma_2 B} (1/w_2)^{\lambda/(r-1)} dx\right)^{r-1}\right)^{\alpha/\lambda q}$$

$$= \left(|\sigma_2 B|^r \left(\tfrac{1}{|\sigma_2 B|} \int_{\sigma_2 B} w_1^\lambda dx \right) \left(\tfrac{1}{|\sigma_2 B|} \int_{\sigma_2 B} \left(\tfrac{1}{w_2^\lambda} \right)^{1/(r-1)} dx \right)^{r-1} \right)^{\alpha/\lambda q}$$

$$\leq C_5 |B|^{\alpha r/\lambda q}.$$

(4.5.13)

Finally, combining (4.5.12) and (4.5.13) and noting that

$$\frac{t-s}{st} + \frac{\alpha r}{\lambda q} = 0,$$

we obtain

$$\left(\int_B |d^\star v|^q w_1^\alpha dx \right)^{1/q} \leq C_6 |B|^{-1/n} \left(\int_{\sigma_2 B} |v - c_1|^q w_2^\alpha dx \right)^{1/q} \quad (4.5.14)$$

for all balls B with $\sigma_2 B \subset \Omega$ and all co-closed forms c_1.   ∎

Notes to Chapter 4. Many mathematicians have made significant contributions to this topic. In particular, I. Perić and D. Žubrinić established a Caccioppoli-type estimate for quasilinear elliptic operator in [161]; M. Giaquinta and J. Souček obtained a new Caccioppoli inequality in [269]; G.A. Serëgin provided a local estimate of the Caccioppoli-type inequality for extremal variational problems of Hencky plasticity in [173], and R.F. Gariepy gave a new proof of the partial regularity minimizers based on the Caccioppoli inequality [270]. For more versions of Caccioppoli-type inequalities, see [205, 271–273, 136, 170].

Chapter 5
Imbedding theorems

5.1 Introduction

In recent years various versions of imbedding theorems for differential forms have been established. The imbedding theorems for functions can be found in almost every book on partial differential equations, see Sections 7.7 and 7.8 in [274], for example. For different versions of imbedding theorems, see [20, 275–280, 32, 268, 81]. Many results for Sobolev functions have been extended to differential forms in $\mathbf{R}^n$. The imbedding theorems play crucial role in generalizing the theory of Sobolev functions into the theory of differential forms. The objective of this chapter is to discuss several other versions of imbedding theorems for differential forms. We also explore some imbedding theorems related to operators, such as the homotopy operator T and Green's operator G. We first study the imbedding theorems for quasiconformal mappings in Section 5.2. Then, we establish an imbedding theorem for differential forms satisfying the nonhomogeneous A-harmonic equation in Section 5.3. In Section 5.4, we present the $A_r(\Omega)$-weighted imbedding theorems related to the gradient operator and the homotopy operator. In Section 5.5, we explore some $A_r(\lambda, \Omega)$-weighted imbedding theorems and $A_r^\lambda(\Omega)$-weighted imbedding theorems. In Section 5.6, we develop some L^s-estimates and imbedding theorems for the compositions of operators. Finally, in Section 5.7, we study the two-weight imbedding inequalities.

5.2 Quasiconformal mappings

The following three theorems can be found in [59]. In Theorem 5.2.1, the measure μ is defined by A_r-weights and in Theorem 5.2.3, the measure μ is defined by the Jacobian $J(x, f)$ of a K-quasiconformal mapping $f : \mathbf{R}^n \to \mathbf{R}^n$. Theorem 5.2.2 implies that the Jacobian $J(x, f)$ of a K-quasiconformal mapping f belongs to A_r for some $r > 1$.

R.P. Agarwal et al., *Inequalities for Differential Forms*,
DOI 10.1007/978-0-387-68417-8_5, © Springer Science+Business Media, LLC 2009

Theorem 5.2.1. *Let $w \in A_r$, $r_0 < q < r < nq$, and $\kappa = n/(n - r/q) > 1$. Then, there exists a constant C such that*

$$\left(\frac{1}{\mu(B)} \int_B |\varphi|^{\kappa r} d\mu \right)^{1/\kappa r} \leq C \, diam(B) \left(\frac{1}{\mu(B)} \int_B |\nabla \varphi|^r d\mu \right)^{1/r},$$

where B is a ball, $d\mu = w(x)dx$, and $\varphi \in C_0^\infty(B)$.

Theorem 5.2.2. *The Jacobian $J(x, f)$ of a K-quasiconformal mapping $f : \mathbf{R}^n \to \mathbf{R}^n$ is an A_∞-weight with constants depending only on n and K.*

Theorem 5.2.3. *Let $1 < r < n$, $f : \mathbf{R}^n \to \mathbf{R}^n$ be a K-quasiconformal mapping and $w(x) = J(x, f)^{1-r/n}$. Then, there are constants $\kappa = \kappa(n, K) > 1$ and $C = C(n, K, r) > 0$ such that*

$$\left(\frac{1}{\mu(B)} \int_B |\varphi|^{\kappa r} d\mu \right)^{1/\kappa r} \leq C \, diam(B) \left(\frac{1}{\mu(B)} \int_B |\nabla \varphi|^r d\mu \right)^{1/r},$$

where B is a ball, $d\mu = w(x)dx$, and $\varphi \in C_0^\infty(B)$.

5.3 Solutions to the nonhomogeneous equation

We have presented the imbedding results for quasiconformal mappings in Section 5.2. In the remaining part of this chapter, we focus our attention on the weighted imbedding theorems for differential forms satisfying the nonhomogeneous A-harmonic equation

$$d^\star A(x, d\omega) = B(x, d\omega) \tag{5.3.1}$$

which we have introduced in Section 1.2. We will also develop the imbedding inequalities for some operators.

Theorem 5.3.1. *Let $u \in D'(\Omega, \wedge^l)$ be a solution of the nonhomogeneous A-harmonic equation (5.3.1) in a bounded domain $\Omega \subset \mathbf{R}^n$ and $du \in L^s(\Omega, \wedge^{l+1})$, $l = 0, 1, \ldots, n-1$. Assume that $1 < s < \infty$ and $w \in A_r(\Omega)$ for some $r > 1$. Then,*

$$\|u - u_B\|_{W^{1,s}(B),w} \leq C \|du\|_{s,\sigma B,w}$$

for all balls B with $\sigma B \subset \Omega$, $\sigma > 1$. Here C is some constant independent of u and du.

Proof. Replacing u by du in (1.5.5) and (1.5.6), we obtain

$$\|\nabla T(du)\|_{s,B} \leq C_1 |B| \|du\|_{s,B} \tag{5.3.2}$$

and
$$\|T(du)\|_{s,B} \le C_2|B|diam(B)\|du\|_{s,B}. \tag{5.3.3}$$

Using the similar skill developed in the proofs of inequalities (5.4.5) and (5.4.20) and the weak reverse Hölder inequality for du (Theorem 6.2.12), we can extend inequalities (5.3.2) and (5.3.3) into the weighted versions

$$\|\nabla T(du)\|_{s,B,w} \le C_3|B|\|du\|_{s,\sigma_1 B,w} \tag{5.3.4}$$

and
$$\|T(du)\|_{s,B,w} \le C_4|B|diam(B)\|du\|_{s,\sigma_2 B,w}, \tag{5.3.5}$$

where $\sigma_1, \sigma_2 > 1$. From (3.3.15), (5.3.4), and (5.3.5), we obtain

$$\|u - u_B\|_{W^{1,s}(B),w}$$

$$= \|T(du)\|_{s,B,w}$$

$$= (diam(B))^{-1}\|T(du)\|_{s,B,w} + \|\nabla(T(du))\|_{s,B,w}$$

$$\le (diam(B))^{-1}C_4|B|diam(B)\|du\|_{s,\sigma_1 B,w} + C_3|B|\|du\|_{s,\sigma_2 B,w}$$

$$\le C_5|B|\|du\|_{s,\sigma B,w},$$

$$\le C_6\|du\|_{s,\sigma B,w},$$

that is
$$\|u - u_B\|_{W^{1,s}(B),w} \le C_6\|du\|_{s,\sigma B,w},$$

where $\sigma = \max\{\sigma_1, \sigma_2\}$.   ∎

It is well known that if $w \in A_r(E)$ and α is any constant with $0 < \alpha \le 1$, then $w^\alpha \in A_r(E)$. Thus, inequality (5.3.2) still holds if we replace $w(x)$ by $w^\alpha(x)$ in (5.3.2). Hence, for any solution u of the nonhomogeneous A-harmonic equation (5.3.1), we have

$$\|u - u_B\|_{W^{1,s}(B),w^\alpha} \le C\|du\|_{s,\sigma B,w^\alpha} \tag{5.3.6}$$

for any constant α with $0 < \alpha \le 1$.

5.4 Imbedding inequalities for operators

We have established the $A_r(\Omega)$-weighted imbedding inequality for solutions of the nonhomogeneous A-harmonic equation in the previous section. The main purpose of this section is to develop the $A_r(\Omega)$-weighted imbedding inequality for the homotopy operator. For this, first some estimates related to the gradient operator and the composition of the gradient operator and the homotopy operator will be obtained.

5.4.1 The gradient and homotopy operators

Let u be a differential form in the Sobolev space $W^{1,p}(E, \wedge^l)$ of l-forms defined over a bounded and convex E, $0 < p < \infty$ and $w(x)$ be a weight. The weighted norm of Tu is denoted by $\|Tu\|_{W^{1,p}(E,\wedge^l)}$ and defined as

$$\|Tu\|_{W^{1,p}(E,\wedge^l),\ w^\alpha} = diam(E)^{-1}\|Tu\|_{p,E,w^\alpha} + \|\nabla(Tu)\|_{p,E,w^\alpha}, \quad (5.4.1)$$

where α is a real number. Some elementary estimates for Tu and $\nabla(Tu)$ are known, in particular from [99], we have the following lemma.

Lemma 5.4.1. Let $u \in L^s_{loc}(B, \wedge^l)$, $l = 1, 2, \ldots, n$, $1 < s < \infty$, be a differential form in a ball $B \subset \mathbf{R}^n$. Then,

$$\|\nabla(Tu)\|_{s,B} \le C|B|\|u\|_{s,B}, \quad (5.4.2)$$

$$\|Tu\|_{s,B} \le C|B|diam(B)\|u\|_{s,B}. \quad (5.4.3)$$

Now we prove the following local weighted imbedding theorem for differential forms under the homotopy operator T.

Theorem 5.4.2. Let $u \in L^s_{loc}(\Omega, \wedge^l)$, $l = 1, 2, \ldots, n$, $1 < s < \infty$, be a differential form satisfying the nonhomogeneous A-harmonic equation (5.3.1) in a bounded, convex domain $\Omega \subset \mathbf{R}^n$ and $T : L^s(\Omega, \wedge^l) \to W^{1,s}(\Omega, \wedge^{l-1})$ be the homotopy operator defined in (1.5.1). Assume that $\rho > 1$ and $w \in A_r(\Omega)$ for some $r > 1$. Then, there exists a constant C, independent of u, such that

$$\|Tu\|_{W^{1,s}(B),\ w^\alpha} \le C|B|\|u\|_{s,\rho B,w^\alpha} \quad (5.4.4)$$

for all balls B with $\rho B \subset \Omega$ and any real number α with $0 < \alpha \le 1$.

In order to prove Theorem 5.4.2, we need the following $A_r(\Omega)$-weighted imbedding theorem [277].

Theorem 5.4.3. Let $u \in L^s_{loc}(\Omega, \wedge^l)$, $l = 1, 2, \ldots, n$, $1 < s < \infty$, be a differential form satisfying (5.3.1) in a bounded, convex domain $\Omega \subset \mathbf{R}^n$ and $T : C^\infty(\Omega, \wedge^l) \to C^\infty(\Omega, \wedge^{l-1})$ be the homotopy operator defined in (1.5.1). Assume that $\rho > 1$ and $w \in A_r(\Omega)$ for some $r > 1$. Then, there exists a constant C, independent of u, such that

$$\left(\int_B |\nabla(Tu)|^s w^\alpha dx\right)^{1/s} \le C|B|\left(\int_{\rho B} |u|^s w^\alpha dx\right)^{1/s} \quad (5.4.5)$$

for any real number α with $0 < \alpha \le 1$.

Note that (5.4.5) can be written as

$$\|\nabla(Tu)\|_{s,B,w^\alpha} \le C|B|\|u\|_{s,\rho B,w^\alpha}. \qquad (5.4.5)'$$

Proof. First, we shall show that inequality (5.4.5) holds for $0 < \alpha < 1$. Let $t = s/(1-\alpha)$. From Theorem 1.1.4, we have

$$\left(\int_B |\nabla(Tu)|^s w^\alpha dx\right)^{1/s}$$

$$= \left(\int_B \left(|\nabla(Tu)|w^{\alpha/s}\right)^s dx\right)^{1/s}$$

$$\le \|\nabla(Tu)\|_{t,B} \left(\int_B w^{t\alpha/(t-s)}dx\right)^{(t-s)/st} \qquad (5.4.6)$$

$$= \|\nabla(Tu)\|_{t,B} \left(\int_B wdx\right)^{\alpha/s}.$$

Thus, from Lemma 5.4.1, we obtain

$$\|\nabla(Tu)\|_{t,B} \le C_1|B|\|u\|_{t,B}. \qquad (5.4.7)$$

Choose $m = s/(1+\alpha(r-1))$, so that $m < s$. Substituting (5.4.7) into (5.4.6) and using Lemma 3.1.1, we find

$$\left(\int_B |\nabla(Tu)|^s w^\alpha dx\right)^{1/s}$$

$$\le C_1|B|\|u\|_{t,B} \left(\int_B wdx\right)^{\alpha/s} \qquad (5.4.8)$$

$$\le C_2|B||B|^{(m-t)/mt}\|u\|_{m,\rho B} \left(\int_B wdx\right)^{\alpha/s}.$$

From Theorem 1.1.4 with $1/m = 1/s + (s-m)/sm$, we have

$$\|u\|_{m,\rho B}$$

$$= \left(\int_{\rho B} |u|^m dx\right)^{1/m}$$

$$= \left(\int_{\rho B} \left(|u|w^{\alpha/s}w^{-\alpha/s}\right)^m dx\right)^{1/m} \qquad (5.4.9)$$

$$\le \left(\int_{\rho B} |u|^s w^\alpha dx\right)^{1/s} \left(\int_{\rho B} \left(\tfrac{1}{w}\right)^{1/(r-1)} dx\right)^{\alpha(r-1)/s}$$

for all balls B with $\rho B \subset \Omega$. Substituting (5.4.9) into (5.4.8), we obtain

$$\left(\int_B |\nabla(Tu)|^s w^\alpha dx\right)^{1/s}$$

$$\le C_2|B||B|^{(m-t)/mt} \left(\int_{\rho B} |u|^s w^\alpha dx\right)^{1/s}$$

$$\times \left(\int_B w dx\right)^{\alpha/s} \left(\int_{\rho B} \left(\tfrac{1}{w}\right)^{1/(r-1)} dx\right)^{\alpha(r-1)/s}. \tag{5.4.10}$$

Since $w \in A_r(\Omega)$, we find

$$\|w\|_{1,B}^{\alpha/s} \cdot \|1/w\|_{1/(r-1),\rho B}^{\alpha/s}$$

$$\leq \left(\left(\int_{\rho B} w dx\right)\left(\int_{\rho B}(1/w)^{1/(r-1)}dx\right)^{r-1}\right)^{\alpha/s}$$

$$= \left(|\rho B|^r \left(\tfrac{1}{|\rho B|}\int_{\rho B} w dx\right)\left(\tfrac{1}{|\rho B|}\int_{\rho B}\left(\tfrac{1}{w}\right)^{1/(r-1)}dx\right)^{r-1}\right)^{\alpha/s} \tag{5.4.11}$$

$$\leq C_3|B|^{\alpha r/s}.$$

Combining (5.4.11) and (5.4.10), it follows that

$$\left(\int_B |\nabla(Tu)|^s w^\alpha dx\right)^{1/s} \leq C_4|B| \left(\int_{\rho B} |u|^s w^\alpha dx\right)^{1/s} \tag{5.4.12}$$

for all balls B with $\rho B \subset \Omega$. Thus, (5.4.5) holds for $0 < \alpha < 1$.

Next, we shall prove (5.4.5) for $\alpha = 1$, that is, we need to show that

$$\|\nabla(Tu)\|_{s,B,w} \leq C|B|\|u\|_{s,\rho B,w}. \tag{5.4.13}$$

By Lemma 1.4.7, there exist constants $\beta > 1$ and $C_5 > 0$, such that

$$\| w \|_{\beta,B} \leq C_5|B|^{(1-\beta)/\beta} \| w \|_{1,B} \tag{5.4.14}$$

for any cube or ball $B \subset \mathbf{R}^n$. Choose $t = s\beta/(\beta - 1)$, so that $1 < s < t$ and $\beta = t/(t - s)$. Since $1/s = 1/t + (t - s)/st$, by Theorem 1.1.4, Lemma 5.4.1, and (5.4.14), we have

$$\left(\int_B |\nabla(Tu)|^s w dx\right)^{1/s}$$

$$= \left(\int_B \left(|\nabla(Tu)|w^{1/s}\right)^s dx\right)^{1/s}$$

$$\leq \left(\int_B |\nabla(Tu)|^t dx\right)^{1/t} \left(\int_B \left(w^{1/s}\right)^{st/(t-s)} dx\right)^{(t-s)/st}$$

$$\leq C_6\|\nabla(Tu)\|_{t,B} \cdot \|w\|_{\beta,B}^{1/s} \tag{5.4.15}$$

$$\leq C_6|B|\|u\|_{t,B} \cdot \|w\|_{\beta,B}^{1/s}$$

$$\leq C_7|B||B|^{(1-\beta)/\beta s}\|w\|_{1,B}^{1/s} \cdot \|u\|_{t,B}$$

$$\leq C_7|B||B|^{-1/t}\|w\|_{1,B}^{1/s} \cdot \|u\|_{t,B}.$$

In Lemma 3.1.1, let $m = s/r$, to have

$$\|u\|_{t,B} \leq C_8 |B|^{(m-t)/mt} \|u\|_{m,\rho B}.$$ (5.4.16)

Now Theorem 1.1.4 yields

$$\|u\|_{m,\rho B}$$
$$= \left(\int_{\rho B} \left(|u| w^{1/s} w^{-1/s} \right)^m dx \right)^{1/m}$$ (5.4.17)
$$\leq \left(\int_{\rho B} |u|^s w dx \right)^{1/s} \left(\int_{\rho B} \left(\tfrac{1}{w} \right)^{1/(r-1)} dx \right)^{(r-1)/s}$$

for all balls B with $\rho B \subset \Omega$. Next, since $w \in A_r(\Omega)$, it follows that

$$\|w\|_{1,B}^{1/s} \cdot \|1/w\|_{1/(r-1),\rho B}^{1/s}$$
$$\leq \left(\left(\int_{\rho B} w dx \right) \left(\int_{\rho B} (1/w)^{1/(r-1)} dx \right)^{r-1} \right)^{1/s}$$
$$= \left(|\rho B|^r \left(\tfrac{1}{|\rho B|} \int_{\rho B} w dx \right) \left(\tfrac{1}{|\rho B|} \int_{\rho B} \left(\tfrac{1}{w} \right)^{1/(r-1)} dx \right)^{r-1} \right)^{1/s}$$ (5.4.18)
$$\leq C_9 |B|^{r/s}.$$

Combining (5.4.15), (5.14.16), (5.4.17), and (5.4.18), we have

$$\|\nabla (Tu)\|_{s,B,w}$$
$$\leq C_{10} |B| |B|^{-1/t} \|w\|_{1,B}^{1/s} |B|^{(m-t)/mt} \|u\|_{m,\rho B}$$
$$\leq C_{10} |B| |B|^{-1/m} \|w\|_{1,B}^{1/s} \cdot \|1/w\|_{1/(r-1),\rho B}^{1/s} \|u\|_{s,\rho B,w}$$ (5.4.19)
$$\leq C_{11} |B| \|u\|_{s,\rho B,w}$$

for all balls B with $\rho B \subset \Omega$. Hence, (5.4.13) holds. ∎

At this point, we should notice that Theorem 5.4.3 can also be proved in a different way. Indeed, we first prove that inequality (5.4.5) holds for any $A_r(\Omega)$-weight $w(x)$, and then using the property of $A_r(\Omega)$-weight, we can replace $w(x)$ by $w^\alpha(x)$ with $0 < \alpha \leq 1$ as we did in the proof of inequality (5.3.6). However, the method that we have used to prove Theorem 5.4.3 requires some insight into the treatment of weights. Using the method similar to the proof of Theorem 5.4.3, we obtain

$$\|Tu\|_{s,B,w^\alpha} \leq C |B| diam(B) \|u\|_{s,\rho B,w^\alpha},$$ (5.4.20)

where α is any real number with $0 < \alpha \leq 1$ and $\rho > 1$.

Proof of Theorem 5.4.2. From (5.4.1), (5.4.5)′, and (5.4.20), we have

$$
\begin{aligned}
\|Tu\|&_{W^{1,s}(B),\ w^\alpha} \\
&= diam(B)^{-1}\|Tu\|_{s,B,w^\alpha} + \|\nabla(Tu)\|_{s,B,w^\alpha} \\
&\leq diam(B)^{-1}\left(C_1|B|diam(B)\|u\|_{s,\rho B,w^\alpha}\right) + C_2|B|\|u\|_{s,\rho B,w^\alpha} \\
&\leq C_1|B|\|u\|_{s,\rho B,w^\alpha} + C_2|B|\|u\|_{s,\rho B,w^\alpha} \\
&\leq C_3|B|\|u\|_{s,\rho B,w^\alpha},
\end{aligned}
$$

which is equivalent to (5.4.4).   ∎

5.4.2 Some special cases

Note that the parameter α in both Theorems 5.4.2 and 5.4.3 is any real number with $0 < \alpha \leq 1$. Therefore, we can deduce different versions of the weighted imbedding inequality by choosing different values of α. For example, when $t = 1 - \alpha$ in Theorem 5.4.3 and $d\mu = w(x)dx$, then inequality (5.4.5) becomes

$$
\left(\int_B |\nabla(Tu)|^s w^{-t} d\mu\right)^{1/s} \leq C|B| \left(\int_{\rho B} |u|^s w^{-t} d\mu\right)^{1/s}. \tag{5.4.21}
$$

If we choose $\alpha = 1/r$ in Theorem 5.4.3, then (5.4.5) reduces to

$$
\left(\int_B |\nabla(Tu)|^s w^{1/r} dx\right)^{1/s} \leq C|B| \left(\int_{\rho B} |u|^s w^{1/r} dx\right)^{1/s}. \tag{5.4.22}
$$

If we choose $\alpha = 1/s$ in Theorem 5.4.3, then $0 < \alpha < 1$ since $1 < s < \infty$, and (5.4.5) reduces to the following symmetric version:

$$
\left(\int_B |\nabla(Tu)|^s w^{1/s} dx\right)^{1/s} \leq C|B| \left(\int_{\rho B} |u|^s w^{1/s} dx\right)^{1/s}. \tag{5.4.23}
$$

Finally, if we choose $\alpha = 1$ in Theorem 5.4.3, we have the following weighted imbedding inequality:

$$
\|\nabla(Tu)\|_{s,B,w} \leq C|B|\|u\|_{s,\rho B,w}. \tag{5.4.24}
$$

Similarly, when $\alpha = 1$ in Theorem 5.4.2, we have

$$
\|Tu\|_{W^{1,s}(B),\ w} \leq C|B|\|u\|_{s,B,w}. \tag{5.4.25}
$$

5.4.3 Global imbedding theorems

We shall prove the following global $A_r(\Omega)$-weighted imbedding theorem for the operators applied to the differential forms satisfying the nonhomogeneous A-harmonic equation (5.3.1) in a domain Ω.

Theorem 5.4.4.   *Let $u \in L^s(\Omega, \wedge^l)$, $l = 1, 2, \ldots, n$, $1 < s < \infty$, be a differential form satisfying (5.3.1) in a bounded, convex domain $\Omega \subset \mathbf{R}^n$ and $T : L^s(\Omega, \wedge^l) \to W^{1,s}(\Omega, \wedge^{l-1})$ be the homotopy operator defined in (1.5.1). Assume that $w \in A_r(\Omega)$ for some $r > 1$. Then, there exists a constant C, independent of u, such that*

$$\|\nabla(Tu)\|_{s,\Omega,w^\alpha} \le C\|u\|_{s,\Omega,w^\alpha}, \qquad (5.4.26)$$

$$\|Tu\|_{W^{1,s}(\Omega),w^\alpha} \le C\|u\|_{s,\Omega,w^\alpha} \qquad (5.4.27)$$

for any real number α with $0 < \alpha \le 1$.

Proof.   From the facts on page 16 in [204], we know that any open subset Ω of $\mathbf{R}^n$ is the union of a sequence of mutually disjoint Whitney cubes. Also, cubes are convex. Thus, the definition of the homotopy operator T can be extended into any open subset Ω of $\mathbf{R}^n$. Using (5.4.5) and the covering lemma (Theorem 1.5.3), we find that

$$
\begin{aligned}
&\|\nabla(Tu)\|_{s,\Omega,w^\alpha} \\
&= \left(\int_\Omega |\nabla(Tu)|^s w^\alpha dx\right)^{1/s} \\
&\le \sum_{Q \in \mathcal{V}} \left(C_1 |Q| \left(\int_{\rho Q} |u|^s w^\alpha dx\right)^{1/s}\right) \\
&\le C_1 |\Omega| \sum_{Q \in \mathcal{V}} \left(\int_{\rho Q} |u|^s w^\alpha dx\right)^{1/s} \qquad (5.4.28) \\
&\le C_2 \sum_{Q \in \mathcal{V}} \left(\int_\Omega |u|^s w^\alpha dx\right)^{1/s} \\
&\le C_3 \left(\int_\Omega |u|^s w^\alpha dx\right)^{1/s} \\
&= C_3 \|u\|_{s,\Omega,w^\alpha}
\end{aligned}
$$

since Ω is bounded. Thus, (5.4.26) holds. Similarly, using Theorem 1.5.3 and (5.4.20), we have

$$\|Tu\|_{s,\Omega,w^\alpha} \le C_4 diam(\Omega)\|u\|_{s,\Omega,w^\alpha}. \qquad (5.4.29)$$

Now combining (5.4.1), (5.4.28), and (5.4.29), we obtain

$$\|Tu\|_{W^{1,s}(\Omega),\, w^\alpha}$$
$$= diam(\Omega)^{-1}\|Tu\|_{s,\Omega,w^\alpha} + \|\nabla(Tu)\|_{s,\Omega,w^\alpha}$$
$$\leq C_4\|u\|_{s,\Omega,w^\alpha} + C_3\|u\|_{s,\Omega,w^\alpha} \tag{5.4.30}$$
$$\leq C_5\|u\|_{s,\Omega,w^\alpha}$$

and hence (5.4.27) holds. ∎

Remark. Similar to the local case, when α assumes some particular values in (5.4.26) and (5.4.27), we obtain different versions of the global result. For example, if we choose $\alpha = 1$, we find that (5.4.26) and (5.4.27) reduce to

$$\|\nabla(Tu)\|_{s,\Omega,w} \leq C\|u\|_{s,\Omega,w}, \tag{5.4.31}$$

$$\|Tu\|_{W^{1,s}(\Omega),\, w} \leq C\|u\|_{s,\Omega,w}, \tag{5.4.32}$$

respectively.

5.5 Other weighted cases

In the previous section, we have discussed both local and global $A_r(\Omega)$-weighted imbedding theorems for differential forms satisfying the nonhomogeneous A-harmonic equation (5.3.1) in a domain Ω. In this section, we study the $A_r(\lambda, \Omega)$-weighted imbedding theorems for operators applied to the solutions of the nonhomogeneous A-harmonic equation (5.3.1).

5.5.1 L^s-estimates for T

First, we recall the definition of $A_r(\lambda, \Omega)$-weights which was introduced in [121]. A weight $w(x)$ is called an $A_r(\lambda, \Omega)$-weight for some $r > 1$ and $\lambda > 0$ in a domain Ω, write $w \in A_r(\lambda, \Omega)$, if $w(x) > 0$ a.e., and

$$\sup_B \left(\frac{1}{|B|}\int_B w^\lambda dx\right)\left(\frac{1}{|B|}\int_B \left(\frac{1}{w}\right)^{1/(r-1)} dx\right)^{(r-1)} < \infty \tag{5.5.1}$$

for any ball $B \subset \Omega$.

It is worth noticing that there exists some difference between $A_r(\lambda, \Omega)$-weights and $A_r(\lambda)$-weights. We say that a weight $w(x)$ is an $A_r(\lambda)$-weight if inequality (5.5.1) holds for any ball $B \subset \mathbf{R}^n$. The condition $w(x) \in A_r(\lambda, \Omega)$, however, requires that (5.5.1) holds for all balls $B \subset \Omega$. We should also note that $A_r(1, \Omega) = A_r(\Omega)$ is the usual $A_r(\Omega)$-weights which have been widely studied. The following results are obtained in [268] recently.

Theorem 5.5.1. *Let* $u \in L^s_{loc}(\Omega, \wedge^l)$, $l = 1, 2, \ldots, n$, $1 < s < \infty$, *be a differential form satisfying the nonhomogeneous A-harmonic equation* (5.3.1) *in a bounded, convex domain* $\Omega \subset \mathbf{R}^n$ *and* $T : C^\infty(\Omega, \wedge^l) \to C^\infty(\Omega, \wedge^{l-1})$ *be the homotopy operator defined in* (1.5.1). *Assume that* $\rho > 1$ *and* $w \in A_r(\lambda, \Omega)$ *for some* $r > 1$ *and* $\lambda > 0$. *Then, there exists a constant* C, *independent of* u, *such that*

$$\left(\int_B |Tu|^s w^{\alpha \lambda} dx \right)^{1/s} \leq C|B|diam(B) \left(\int_{\rho B} |u|^s w^\alpha dx \right)^{1/s} \tag{5.5.2}$$

for all balls B *with* $\rho B \subset \Omega$ *and any real number* α *with* $0 < \alpha < 1$.

Note that (5.5.2) can be written as

$$\|Tu\|_{s,B,w^{\alpha\lambda}} \leq C|B|diam(B)\|u\|_{s,\rho B,w^\alpha}. \tag{5.5.2}'$$

Proof. Let $t = s/(1 - \alpha)$. Using the Hölder inequality and Lemma 5.4.1, we have

$$
\begin{aligned}
\left(\int_B |Tu|^s w^{\alpha\lambda} dx \right)^{1/s} \\
= \left(\int_B \left(|Tu| w^{\alpha\lambda/s} \right)^s dx \right)^{1/s} \\
\leq \|Tu\|_{t,B} \left(\int_B w^{t\alpha\lambda/(t-s)} dx \right)^{(t-s)/st} \\
\leq C_1 |B| diam(B) \|u\|_{t,B} \left(\int_B w^\lambda dx \right)^{\alpha/s}.
\end{aligned}
\tag{5.5.3}
$$

Choose $m = s/(1 + \alpha(r - 1))$, so that $m < s$. Using Lemma 3.1.1, we obtain

$$\|u\|_{t,B} \leq C_2 |B|^{(m-t)/mt} \|u\|_{m,\rho B}, \tag{5.5.4}$$

where $\rho > 1$. Substituting (5.5.4) into (5.5.3), we find

$$
\begin{aligned}
\left(\int_B |Tu|^s w^{\alpha\lambda} dx \right)^{1/s} \\
\leq C_3 |B| diam(B) |B|^{(m-t)/mt} \|u\|_{m,\rho B} \left(\int_B w^\lambda dx \right)^{\alpha/s}.
\end{aligned}
\tag{5.5.5}
$$

Now since $1/m = 1/s + (s-m)/sm$, by the Hölder inequality again, we obtain

$$
\begin{aligned}
\|u\|_{m,\rho B} \\
= \left(\int_{\rho B} |u|^m dx \right)^{1/m} \\
= \left(\int_{\rho B} \left(|u| w^{\alpha/s} w^{-\alpha/s} \right)^m dx \right)^{1/m}
\end{aligned}
$$

$$\leq \left(\int_{\rho B} |u|^s w^\alpha dx\right)^{1/s} \left(\int_{\rho B} \left(\tfrac{1}{w}\right)^{1/(r-1)} dx\right)^{\alpha(r-1)/s} \tag{5.5.6}$$

for all balls B with $\rho B \subset \Omega$. Combining (5.5.5) and (5.5.6), it follows that

$$\left(\int_B |Tu|^s w^{\alpha\lambda} dx\right)^{1/s}$$

$$\leq C_3 |B| diam(B) |B|^{(m-t)/mt} \|w\|_{\lambda,B}^{\alpha\lambda/s} \|1/w\|_{1/(r-1),\rho B}^{\alpha/s} \tag{5.5.7}$$

$$\times \left(\int_{\rho B} |u|^s w^\alpha dx\right)^{1/s}.$$

Since $w \in A_r(\lambda, \Omega)$, we have

$$\|w\|_{\lambda,B}^{\alpha\lambda/s} \cdot \|1/w\|_{1/(r-1),\rho B}^{\alpha/s}$$

$$\leq \left(\left(\int_{\rho B} w^\lambda dx\right)\left(\int_{\rho B}(1/w)^{1/(r-1)}dx\right)^{r-1}\right)^{\alpha/s}$$

$$= \left(|\rho B|^r \left(\tfrac{1}{|\rho B|}\int_{\rho B} w^\lambda dx\right)\left(\tfrac{1}{|\rho B|}\int_{\rho B}\left(\tfrac{1}{w}\right)^{1/(r-1)}dx\right)^{r-1}\right)^{\alpha/s} \tag{5.5.8}$$

$$\leq C_4 |B|^{\alpha r/s}.$$

Substituting (5.5.8) into (5.5.7), we find that

$$\left(\int_B |Tu|^s w^{\alpha\lambda} dx\right)^{1/s} \leq C|B| diam(B) \left(\int_{\rho B} |u|^s w^\alpha dx\right)^{1/s} \tag{5.5.9}$$

for all balls B with $\rho B \subset \Omega$. $\blacksquare$

5.5.2 L^s-estimates for $\nabla \circ T$

Theorem 5.5.2. *Let* $u \in L^s_{loc}(\Omega, \wedge^l)$, $l = 1, 2, \ldots, n$, $1 < s < \infty$, *be a differential form satisfying the nonhomogeneous A-harmonic equation* (5.3.1) *in a bounded, convex domain* $\Omega \subset \mathbf{R}^n$ *and* $T : C^\infty(\Omega, \wedge^l) \to C^\infty(\Omega, \wedge^{l-1})$ *be the homotopy operator defined in* (1.5.1). *Assume that* $\rho > 1$ *and* $w \in A_r(\lambda, \Omega)$ *for some* $r > 1$ *and* $\lambda > 0$. *Then, there exists a constant* C, *independent of* u, *such that*

$$\left(\int_B |\nabla(Tu)|^s w^{\alpha\lambda} dx\right)^{1/s} \leq C|B| \left(\int_{\rho B} |u|^s w^\alpha dx\right)^{1/s} \tag{5.5.10}$$

for all balls B *with* $\rho B \subset \Omega$ *and any real number* α *with* $0 < \alpha < 1$.

Note that (5.5.10) can be written as

$$\|\nabla(Tu)\|_{s,B,w^{\alpha\lambda}} \le C|B|\|u\|_{s,\rho B,w^{\alpha}}. \tag{5.5.10}'$$

Proof. Let $t = s/(1-\alpha)$. Using Hölder's inequality, we have

$$\left(\int_B |\nabla(Tu)|^s w^{\alpha\lambda} dx\right)^{1/s}$$
$$= \left(\int_B \left(|\nabla(Tu)|w^{\alpha\lambda/s}\right)^s dx\right)^{1/s}$$
$$\le \|\nabla(Tu)\|_{t,B} \left(\int_B w^{t\alpha\lambda/(t-s)} dx\right)^{(t-s)/st} \tag{5.5.11}$$
$$= \|\nabla(Tu)\|_{t,B} \left(\int_B w^\lambda dx\right)^{\alpha/s}.$$

Thus, from Lemma 5.4.1, it follows that

$$\|\nabla(Tu)\|_{t,B} \le C_1|B|\|u\|_{t,B}. \tag{5.5.12}$$

Set $m = s/(1+\alpha(r-1))$, so that $m < s$. Substituting (5.5.12) into (5.5.11) and using Lemma 3.1.1, we have

$$\left(\int_B |\nabla(Tu)|^s w^{\alpha\lambda} dx\right)^{1/s}$$
$$\le C_1|B|\|u\|_{t,B} \left(\int_B w^\lambda dx\right)^{\alpha/s} \tag{5.5.13}$$
$$\le C_2|B||B|^{(m-t)/mt}\|u\|_{m,\rho B} \left(\int_B w^\lambda dx\right)^{\alpha/s}.$$

Using the Hölder inequality again with $1/m = 1/s + (s-m)/sm$, we find

$$\|u\|_{m,\rho B}$$
$$= \left(\int_{\rho B} |u|^m dx\right)^{1/m}$$
$$= \left(\int_{\rho B} \left(|u|w^{\alpha/s}w^{-\alpha/s}\right)^m dx\right)^{1/m} \tag{5.5.14}$$
$$\le \left(\int_{\rho B} |u|^s w^\alpha dx\right)^{1/s} \left(\int_{\rho B} \left(\tfrac{1}{w}\right)^{1/(r-1)} dx\right)^{\alpha(r-1)/s}$$

for all balls B with $\rho B \subset \Omega$. Substituting (5.5.14) into (5.5.13), we obtain

$$\left(\int_B |\nabla(Tu)|^s w^{\alpha\lambda} dx\right)^{1/s}$$
$$\le C_2|B||B|^{(m-t)/mt} \left(\int_{\rho B} |u|^s w^\alpha dx\right)^{1/s} \tag{5.5.15}$$
$$\times \left(\int_B w^\lambda dx\right)^{\alpha/s} \left(\int_{\rho B} \left(\tfrac{1}{w}\right)^{1/(r-1)} dx\right)^{\alpha(r-1)/s}.$$

Now using the condition $w \in A_r(\lambda, \Omega)$, we obtain

$$\|w\|_{\lambda,B}^{\alpha\lambda/s} \cdot \|1/w\|_{1/(r-1),\rho B}^{\alpha/s}$$

$$\leq \left(\left(\int_{\rho B} w^\lambda dx \right) \left(\int_{\rho B} (1/w)^{1/(r-1)} dx \right)^{r-1} \right)^{\alpha/s}$$

$$= \left(|\rho B|^r \left(\frac{1}{|\rho B|} \int_{\rho B} w^\lambda dx \right) \left(\frac{1}{|\rho B|} \int_{\rho B} \left(\frac{1}{w} \right)^{1/(r-1)} dx \right)^{r-1} \right)^{\alpha/s}$$

$$\leq C_3 |B|^{\alpha r/s}. \tag{5.5.16}$$

Combining (5.5.16) and (5.5.15), we find that

$$\left(\int_B |\nabla(Tu)|^s w^{\alpha\lambda} dx \right)^{1/s} \leq C_4 |B| \left(\int_{\rho B} |u|^s w^\alpha dx \right)^{1/s} \tag{5.5.17}$$

for all balls B with $\rho B \subset \Omega$.   ∎

In Theorems 5.5.1 and 5.5.2, the parameters α and λ are any real numbers with $0 < \alpha < 1$ and $\lambda > 0$. Therefore, we can deduce different versions of the weighted imbedding inequalities by choosing particular values of α and λ. In fact, some existing results are the special cases of our theorems. For example, when $\alpha = 1/r$ in Theorem 5.5.2, then (5.5.10) reduces to

$$\left(\int_B |\nabla(Tu)|^s w^{\lambda/r} dx \right)^{1/s} \leq C|B| \left(\int_{\rho B} |u|^s w^{1/r} dx \right)^{1/s}. \tag{5.5.18}$$

If we choose $\alpha = 1/s$ in Theorem 5.5.2, then $0 < \alpha < 1$ since $1 < s < \infty$, and (5.5.10) reduces to

$$\left(\int_B |\nabla(Tu)|^s w^{\lambda/s} dx \right)^{1/s} \leq C|B| \left(\int_{\rho B} |u|^s w^{1/s} dx \right)^{1/s}, \tag{5.5.19}$$

where $w \in A_r(\lambda, \Omega)$. If $w \in A_r(1, \Omega)$, from (5.5.19) we obtain the following symmetric version:

$$\left(\int_B |\nabla(Tu)|^s w^{1/s} dx \right)^{1/s} \leq C|B| \left(\int_{\rho B} |u|^s w^{1/s} dx \right)^{1/s}. \tag{5.5.20}$$

5.5.3 $A_r(\lambda, \Omega)$-weighted imbedding theorems

Using Theorems 5.5.1 and 5.5.2, it is easy to prove the following local $A_r(\lambda, \Omega)$-weighted imbedding inequality for differential forms under the homotopy operator T.

Theorem 5.5.3. *Let* $u \in L^s_{loc}(\Omega, \wedge^l)$, $l = 1, 2, \ldots, n$, $1 < s < \infty$, *be a differential form satisfying* (5.3.1) *in a bounded, convex domain* $\Omega \subset \mathbf{R}^n$ *and* $T : L^s(\Omega, \wedge^l) \to W^{1,s}(\Omega, \wedge^{l-1})$ *be the operator defined in* (1.5.1). *Assume that* $\rho > 1$ *and* $w \in A_r(\lambda, \Omega)$ *for some* $r > 1$ *and* $\lambda > 0$. *Then, there exists a constant* C, *independent of* u, *such that*

$$\|Tu\|_{W^{1,s}(B), \, w^{\alpha\lambda}} \le C|B|\|u\|_{s,\rho B, w^\alpha} \tag{5.5.21}$$

for all balls B *with* $\rho B \subset \Omega$ *and any real number* α *with* $0 < \alpha < 1$.

Proof. Applying $(5.5.2)'$ and $(5.5.10)'$, we obtain

$$\|Tu\|_{W^{1,s}(B,\wedge^l), \, w^{\alpha\lambda}}$$
$$= diam(B)^{-1}\|Tu\|_{s,B,w^{\alpha\lambda}} + \|\nabla(Tu)\|_{s,B,w^{\alpha\lambda}}$$
$$\le diam(B)^{-1}\left[C_1|B|diam(B)\|u\|_{s,\rho B, w^\alpha}\right] + C_2|B|\|u\|_{s,\rho B, w^\alpha}$$
$$\le C_1|B|\|u\|_{s,\rho B, w^\alpha} + C_2|B|\|u\|_{s,\rho B, w^\alpha}$$
$$\le C_3|B|\|u\|_{s,\rho B, w^\alpha},$$

which is equivalent to inequality (5.5.21). ∎

Using Theorem 5.5.3, we can prove the following Sobolev–Poincaré imbedding inequality for differential forms.

Theorem 5.5.4. *Let* $du \in L^s_{loc}(\Omega, \wedge^{l+1})$, $l = 0, 1, 2, \ldots, n-1$, $1 < s < \infty$, *be a differential form satisfying* (5.3.1) *in a bounded, convex domain* $\Omega \subset \mathbf{R}^n$. *Assume that* $\rho > 1$ *and* $w \in A_r(\lambda, \Omega)$ *for some* $r > 1$ *and* $\lambda > 0$. *Then, there exists a constant* C, *independent of* u, *such that*

$$\|u - u_B\|_{W^{1,s}(B), \, w^{\alpha\lambda}} \le C|B|\|du\|_{s,\rho B, w^\alpha} \tag{5.5.22}$$

for all balls B *with* $\rho B \subset \Omega$ *and any real number* α *with* $0 < \alpha < 1$.

Proof. Since $\omega = du \in L^s_{loc}(\Omega, \wedge^{l+1})$ satisfies (5.3.1), using (5.5.21) and $u_B = u - T(du)$, we have

$$\|u - u_B\|_{W^{1,s}(B), \, w^{\alpha\lambda}}$$
$$= \|T(du)\|_{W^{1,s}(B), \, w^{\alpha\lambda}}$$
$$\le C|B|\|du\|_{s,\rho B, w^\alpha}. ∎$$

5.5.4 Some corollaries

Choosing $\lambda = 1, \alpha = 1/s$, and $\alpha = 1/\lambda$ with $\lambda > 1$ in Theorem 5.5.1, we have the following corollaries, respectively.

Corollary 5.5.5. *Let $u \in L^s_{loc}(\Omega, \wedge^l)$, $l = 1, 2, \ldots, n$, $1 < s < \infty$, be a differential form satisfying (5.3.1) in a bounded, convex domain $\Omega \subset \mathbf{R}^n$ and $T : C^\infty(\Omega, \wedge^l) \to C^\infty(\Omega, \wedge^{l-1})$ be the homotopy operator defined in (1.5.1). Assume that $\rho > 1$ and $w \in A_r(\Omega)$ for some $r > 1$. Then, there exists a constant C, independent of u, such that*

$$\left(\int_B |Tu|^s w^\alpha dx \right)^{1/s} \leq C|B|diam(B) \left(\int_{\rho B} |u|^s w^\alpha dx \right)^{1/s}$$

for all balls B with $\rho B \subset \Omega$ and any real number α with $0 < \alpha < 1$.

Corollary 5.5.6. *Let $u \in L^s_{loc}(\Omega, \wedge^l)$, $l = 1, 2, \ldots, n$, $1 < s < \infty$, be a differential form satisfying (5.3.1) in a bounded, convex domain $\Omega \subset \mathbf{R}^n$ and $T : C^\infty(\Omega, \wedge^l) \to C^\infty(\Omega, \wedge^{l-1})$ be the homotopy operator defined in (1.5.1). Assume that $\rho > 1$ and $w \in A_r(\lambda, \Omega)$ for some $r > 1$ and $\lambda > 0$. Then, there exists a constant C, independent of u, such that*

$$\left(\int_B |Tu|^s w^{\lambda/s} dx \right)^{1/s} \leq C|B|diam(B) \left(\int_{\rho B} |u|^s w^{1/s} dx \right)^{1/s}$$

for all balls B with $\rho B \subset \Omega$.

Corollary 5.5.7. *Let $u \in L^s_{loc}(\Omega, \wedge^l)$, $l = 1, 2, \ldots, n$, $1 < s < \infty$, be a differential form satisfying (5.3.1) in a bounded, convex domain $\Omega \subset \mathbf{R}^n$ and $T : C^\infty(\Omega, \wedge^l) \to C^\infty(\Omega, \wedge^{l-1})$ be the homotopy operator defined in (1.5.1). Assume that $\rho > 1$ and $w \in A_r(\lambda, \Omega)$ for some $r > 1$ and $\lambda > 1$. Then, there exists a constant C, independent of u, such that*

$$\left(\int_B |Tu|^s w dx \right)^{1/s} \leq C|B|diam(B) \left(\int_{\rho B} |u|^s w^{1/\lambda} dx \right)^{1/s}$$

for all balls B with $\rho B \subset \Omega$.

Next, choosing $\lambda = 1$ in Corollary 5.5.6, we have the following symmetric imbedding inequality.

Corollary 5.5.8. *Let $u \in L^s_{loc}(\Omega, \wedge^l)$, $l = 1, 2, \ldots, n$, $1 < s < \infty$, be a differential form satisfying (5.3.1) in a bounded, convex domain $\Omega \subset \mathbf{R}^n$ and $T : C^\infty(\Omega, \wedge^l) \to C^\infty(\Omega, \wedge^{l-1})$ be the homotopy operator defined in (1.5.1). Assume that $\rho > 1$ and $w \in A_r(\Omega)$ for some $r > 1$. Then, there exists a constant C, independent of u, such that*

$$\left(\int_B |Tu|^s w^{1/s} dx \right)^{1/s} \leq C|B|diam(B) \left(\int_{\rho B} |u|^s w^{1/s} dx \right)^{1/s}$$

for all balls B with $\rho B \subset \Omega$.

Now assume that $w \in A_r(\Omega)$ for some $r > 1$. Then, from Theorem 5.5.3, we have

$$\|Tu\|_{W^{1,s}(B),\, w^{\alpha}} \le C|B|\|u\|_{s,\rho B, w^{\alpha}} \qquad (5.5.23)$$

for all balls B with $\rho B \subset \Omega$ and any real number α with $0 < \alpha < 1$.

Putting $\alpha = 1/s$, $s > 1$, in (5.5.23), we find

$$\|Tu\|_{W^{1,s}(B),\, w^{1/s}} \le C|B|\|u\|_{s,\rho B, w^{1/s}}. \qquad (5.5.24)$$

Let $\lambda > 1$ and $\alpha = 1/\lambda$ in Theorem 5.5.3. Then, inequality (5.5.21) becomes

$$\|Tu\|_{W^{1,s}(B),\, w} \le C|B|\|u\|_{s,\rho B, w^{1/\lambda}}, \qquad (5.5.25)$$

where B is any ball with $\rho B \subset \Omega$ and $w \in A_r(\lambda, \Omega)$ for some $r > 1$ and $\lambda > 1$.

Setting $\alpha = 1/s$ in Theorem 5.5.3, we have the following corollary.

Corollary 5.5.9. *Let* $u \in L^s_{loc}(\Omega, \wedge^l)$, $l = 1, 2, \ldots, n$, $1 < s < \infty$, *be a differential form satisfying* (5.3.1) *in a bounded, convex domain* $\Omega \subset \mathbf{R}^n$ *and* $T : L^s(\Omega, \wedge^l) \to W^{1,s}(\Omega, \wedge^{l-1})$ *be the operator defined in* (1.5.1). *Assume that* $\rho > 1$ *and* $w \in A_r(\lambda, \Omega)$ *for some* $r > 1$ *and* $\lambda > 0$. *Then, there exists a constant* C, *independent of* u, *such that*

$$\|Tu\|_{W^{1,s}(B),\, w^{\lambda/s}} \le C|B|\|u\|_{s,\rho B, w^{1/s}}$$

for all balls B *with* $\rho B \subset \Omega$.

Similarly, from Theorem 5.5.4, we have the following Sobolev–Poincaré imbedding inequalities for differential forms.

Corollary 5.5.10. *Let* $du \in L^s_{loc}(\Omega, \wedge^{l+1})$, $l = 0, 1, 2, \ldots, n-1$, $1 < s < \infty$, *be a differential form satisfying* (5.3.1) *in a bounded, convex domain* $\Omega \subset \mathbf{R}^n$. *Assume that* $\rho > 1$ *and* $w \in A_r(\Omega)$ *for some* $r > 1$. *Then, there exists a constant* C, *independent of* u, *such that*

$$\|u - u_B\|_{W^{1,s}(B),\, w^{\alpha}} \le C|B|\|du\|_{s,\rho B, w^{\alpha}}$$

for all balls B *with* $\rho B \subset \Omega$ *and any real number* α *with* $0 < \alpha < 1$.

Corollary 5.5.11. *Let* $du \in L^s_{loc}(\Omega, \wedge^{l+1})$, $l = 0, 1, 2, \ldots, n-1$, $1 < s < \infty$, *be a differential form satisfying* (5.3.1) *in a bounded, convex domain* $\Omega \subset \mathbf{R}^n$. *Assume that* $\rho > 1$ *and* $w \in A_r(\lambda, \Omega)$ *for some* $r > 1$ *and* $\lambda > 1$. *Then, there exists a constant* C, *independent of* u, *such that*

$$\|u - u_B\|_{W^{1,s}(B),\, w} \le C|B|\|du\|_{s,\rho B, w^{1/\lambda}}$$

for all balls B *with* $\rho B \subset \Omega$.

Corollary 5.5.12. *Let* $du \in L^s_{loc}(\Omega, \wedge^{l+1})$, $l = 0, 1, 2, \ldots, n-1$, $1 < s < \infty$, *be a differential form satisfying (5.3.1) in a bounded, convex domain* $\Omega \subset \mathbf{R}^n$. *Assume that* $\rho > 1$ *and* $w \in A_r(\lambda, \Omega)$ *for some* $r > 1$ *and* $\lambda > 0$. *Then, there exists a constant* C, *independent of* u, *such that*

$$\|u - u_B\|_{W^{1,s}(B), \, w^{\lambda/s}} \leq C|B| \|du\|_{s, \rho B, w^{1/s}}$$

for all balls B *with* $\rho B \subset \Omega$.

Choosing $\lambda = 1$ in Corollary 5.5.12, we have the following symmetric inequality.

Corollary 5.5.13. *Let* $du \in L^s_{loc}(\Omega, \wedge^{l+1})$, $l = 0, 1, 2, \ldots, n-1$, $1 < s < \infty$, *be a differential form satisfying (5.3.1) in a bounded, convex domain* $\Omega \subset \mathbf{R}^n$. *Assume that* $\rho > 1$ *and* $w \in A_r(\Omega)$ *for some* $r > 1$. *Then, there exists a constant* C, *independent of* u, *such that*

$$\|u - u_B\|_{W^{1,s}(B), \, w^{1/s}} \leq C|B| \|du\|_{s, \rho B, w^{1/s}}$$

for all balls B *with* $\rho B \subset \Omega$.

5.5.5 Global imbedding theorems

From Theorem 1.5.3 (covering lemma) and the local imbedding inequalities, we have the following global $A_r(\lambda, \Omega)$-weighted imbedding inequalities in a bounded domain Ω.

Theorem 5.5.14. *Let* $u \in L^s(\Omega, \wedge^l)$, $l = 1, 2, \ldots, n$, $1 < s < \infty$, *be a differential form satisfying (5.3.1) in a bounded, convex domain* $\Omega \subset \mathbf{R}^n$ *and* $T : C^\infty(\Omega, \wedge^l) \to C^\infty(\Omega, \wedge^{l-1})$ *be the homotopy operator defined in (1.5.1). Assume that* $w \in A_r(\lambda, \Omega)$ *for some* $r > 1$ *and* $\lambda > 0$. *Then, there exists a constant* C, *independent of* u, *such that*

$$\|Tu\|_{s, \Omega, w^{\alpha\lambda}} \leq C\|u\|_{s, \Omega, w^\alpha} \tag{5.5.26}$$

and

$$\|\nabla(Tu)\|_{s, \Omega, w^{\alpha\lambda}} \leq C\|u\|_{s, \Omega, w^\alpha} \tag{5.5.27}$$

for any real number α *with* $0 < \alpha < 1$.

Proof. Applying (5.5.2) and Theorem 1.5.3, we find that

$$\left(\int_\Omega |Tu|^s w^{\alpha\lambda} dx\right)^{1/s}$$

$$\leq \sum_{Q \in \mathcal{V}} \left(C_1 |Q| diam(Q) \left(\int_{\rho Q} |u|^s w^\alpha dx\right)^{1/s}\right)$$

$$\leq C_1 |\Omega| diam(\Omega) \sum_{Q \in \mathcal{V}} \left(\int_{\rho Q} |u|^s w^\alpha dx\right)^{1/s}$$

$$\leq C_2 \sum_{Q \in \mathcal{V}} \left(\int_\Omega |u|^s w^\alpha dx\right)^{1/s}$$

$$\leq C_3 \left(\int_\Omega |u|^s w^\alpha dx\right)^{1/s},$$

which shows that (5.5.26) holds. From (5.5.10) and Theorem 1.5.3, we can prove (5.5.27) similarly. ∎

Theorem 5.5.15. *Let* $u \in L^s(\Omega, \wedge^l)$, $l = 1, 2, \ldots, n$, $1 < s < \infty$, *be a differential form satisfying* (5.3.1) *in a bounded, convex domain* $\Omega \subset \mathbf{R}^n$ *and* $T : L^s(\Omega, \wedge^l) \to W^{1,s}(\Omega, \wedge^{l-1})$ *be the homotopy operator defined in* (1.5.1). *Assume that* $w \in A_r(\lambda, \Omega)$ *for some* $r > 1$ *and* $\lambda > 0$. *Then, there exists a constant* C, *independent of* u, *such that*

$$\|Tu\|_{W^{1,s}(\Omega), w^{\alpha\lambda}} \leq C\|u\|_{s,\Omega,w^\alpha} \tag{5.5.28}$$

for any real number α *with* $0 < \alpha < 1$.

Proof. Using (5.5.21) and Theorem 1.5.3, we find that

$$\|Tu\|_{W^{1,s}(\Omega), w^{\alpha\lambda}}$$

$$\leq \sum_{Q \in \mathcal{V}} \left(\|Tu\|_{W^{1,s}(Q), w^{\alpha\lambda}}\right)$$

$$\leq \sum_{Q \in \mathcal{V}} \left(C_1 |Q| \|u\|_{s,\rho Q, w^\alpha}\right)$$

$$\leq C_1 |\Omega| \sum_{Q \in \mathcal{V}} \left(\int_{\rho Q} |u|^s w^\alpha dx\right)^{1/s}$$

$$\leq C_2 \sum_{Q \in \mathcal{V}} \left(\int_\Omega |u|^s w^\alpha dx\right)^{1/s}$$

$$\leq C_3 \left(\int_\Omega |u|^s w^\alpha dx\right)^{1/s}$$

which shows that inequality (5.5.28) holds. ∎

Next, we shall prove the following global Sobolev–Poincaré imbedding inequality for differential functions.

Theorem 5.5.16. *Let* $du \in L^s_{loc}(\Omega, \wedge^1)$, $1 < s < \infty$, *be a differential function satisfying* (5.3.1) *in a bounded, convex domain* $\Omega \subset \mathbf{R}^n$ *and* $T : L^s(\Omega, \wedge^l) \to W^{1,s}(\Omega, \wedge^{l-1})$, $l = 1, 2 \ldots, n$, *be the homotopy operator defined in* (1.5.1). *Assume that* $w \in A_r(\lambda, \Omega)$ *for some* $r > 1$ *and* $\lambda > 0$. *Then, there exists a constant* C, *independent of* u, *such that*

$$\|u - u_{Q_0}\|_{W^{1,s}(\Omega), \, w^{\alpha\lambda}} \leq C \|du\|_{s, \Omega, w^{\alpha}} \tag{5.5.29}$$

for any real number α *with* $0 < \alpha < 1$. *Here* Q_0 *is some cube in* Ω.

Proof. From the local Sobolev–Poincaré imbedding inequality (5.5.22) and Theorem 1.5.3, we find that

$$\|u - u_{Q_0}\|_{W^{1,s}(\Omega), \, w^{\alpha\lambda}}$$

$$\leq \sum_{Q \in \mathcal{V}} \left(\|u - u_B\|_{W^{1,s}(Q), \, w^{\alpha\lambda}} \right)$$

$$\leq \sum_{Q \in \mathcal{V}} \left(C_1 |Q| \left(\int_{\rho Q} |du|^s w^{\alpha} dx \right)^{1/s} \right)$$

$$\leq C_1 |\Omega| \sum_{Q \in \mathcal{V}} \left(\int_{\rho Q} |du|^s w^{\alpha} dx \right)^{1/s}$$

$$\leq C_2 \sum_{Q \in \mathcal{V}} \left(\int_{\Omega} |du|^s w^{\alpha} dx \right)^{1/s}$$

$$\leq C_3 \left(\int_{\Omega} |du|^s w^{\alpha} dx \right)^{1/s}.$$

Thus, the imbedding inequality (5.5.29) holds. ∎

5.5.6 $A^\lambda_r(\Omega)$-weighted estimates

B. Liu studied the $A^\lambda_r(\Omega)$-weighted estimates in [182] and obtained some versions of $A^\lambda_r(\Omega)$-weighted imbedding inequalities for differential forms satisfying the homogeneous A-harmonic equation. These inequalities are similar to the results presented in Theorems 5.5.1, 5.5.2, and 5.5.3 for the nonhomogeneous A-harmonic equation. Using the known results for the solutions of the nonhomogeneous A-harmonic equation, we can prove Theorems 5.5.17 and 5.5.18 with $A^\lambda_r(\Omega)$-weights.

Theorem 5.5.17. *Let* $u \in L^s_{loc}(\Omega, \wedge^l)$, $l = 1, 2, \ldots, n$, $1 < s < \infty$, *be a differential form satisfying* (5.3.1) *in a bounded, convex domain* $\Omega \subset \mathbf{R}^n$ *and* $T : C^\infty(\Omega, \wedge^l) \to C^\infty(\Omega, \wedge^{l-1})$ *be the homotopy operator defined in* (1.5.1). *Suppose that* $\rho > 1$ *and* $w \in A^\lambda_r(\Omega)$ *for some* $r > 1$ *and* $\lambda > 0$. *Then, there exists a constant* C, *independent of* u, *such that*

$$\left(\int_B |Tu|^s w^\alpha dx\right)^{1/s} \leq C|B|diam(B)\left(\int_{\rho B} |u|^s w^{\alpha\lambda} dx\right)^{1/s}$$

and

$$\left(\int_B |\nabla(Tu)|^s w^\alpha dx\right)^{1/s} \leq C|B|\left(\int_{\rho B} |u|^s w^{\alpha\lambda} dx\right)^{1/s}$$

for all balls B with $\rho B \subset \Omega$ and any real number α with $0 < \alpha < 1$.

Theorem 5.5.18. *Let $u \in L^s_{loc}(\Omega, \wedge^l)$, $l = 1, 2, \ldots, n$, $1 < s < \infty$, be a differential form satisfying (5.3.1) in a bounded, convex domain $\Omega \subset \mathbf{R}^n$ and $T : L^s(\Omega, \wedge^l) \to W^{1,s}(\Omega, \wedge^{l-1})$ be the operator defined in (1.5.1). Assume that $\rho > 1$ and $w \in A_r^\lambda(\Omega)$ for some $r > 1$ and $\lambda > 0$. Then, there exists a constant C, independent of u, such that*

$$\|Tu\|_{W^{1,s}(B),\ w^\alpha} \leq C|B|\|u\|_{s,\rho B, w^{\alpha\lambda}}$$

for all balls B with $\rho B \subset \Omega$ and any real number α with $0 < \alpha < 1$.

As we have discussed in the previous sections, the parameters α and λ make Theorems 5.5.17 and 5.5.18 more flexible and useful.

5.6 Compositions of operators

In this section, we study different versions of the L^s-estimates for the compositions of T, d, G and ∇, T, G acted on differential forms which are solutions of the nonhomogeneous A-harmonic equation in a domain Ω.

5.6.1 $A_r(\Omega)$-weighted estimates for $T \circ d \circ G$

Theorem 5.6.1. *Let $u \in L^s_{loc}(\Omega, \wedge^l)$, $l = 1, 2, \ldots, n$, $1 < s < \infty$, be a smooth differential form satisfying (5.3.1) in a bounded, convex domain Ω and $T : C^\infty(\Omega, \wedge^l) \to C^\infty(\Omega, \wedge^{l-1})$ be the homotopy operator defined in (1.5.1). Assume that $\rho > 1$ and $w(x) \in A_r(\Omega)$ for some $1 < r < \infty$. Then, $T(d(G(u))) \in L^s_{loc}(\Omega, \wedge^l)$. Moreover, there exists a constant C, independent of u, such that*

$$\|T(d(G(u)))\|_{s,B,w^\alpha} \leq C|B|diam(B)\|u\|_{s,\rho B, w^\alpha} \tag{5.6.1}$$

for all balls B with $\rho B \subset \Omega$ and any real number α with $0 < \alpha \leq 1$.

Proof. The L^s-integrability of $T(d(G(u)))$ follows from inequality (5.6.1), and hence it suffices to show that (5.6.1) holds. From Lemma 5.4.1 and (3.3.3), we obtain

$$\|T(d(G(u)))\|_{s,B}$$

$$\leq C_1 |B| diam(B) \|d(G(u))\|_{s,B} \tag{5.6.2}$$

$$\leq C_2 |B| diam(B) \|u\|_{s,B}.$$

We shall first prove (5.6.1) for $0 < \alpha < 1$. Let $t = s/(1 - \alpha)$. Using Theorem 1.1.4, we have

$$\|T(d(G(u)))\|_{s,B,w^\alpha}$$

$$= \left(\int_B \left(|T(d(G(u)))| w^{\alpha/s} \right)^s dx \right)^{1/s}$$

$$\leq \|T(d(G(u)))\|_{t,B} \left(\int_B w^{t\alpha/(t-s)} dx \right)^{(t-s)/st} \tag{5.6.3}$$

$$= \|T(d(G(u)))\|_{t,B} \left(\int_B w dx \right)^{\alpha/s}.$$

Thus, from (5.6.2), it follows that

$$\|T(d(G(u)))\|_{t,B} \leq C_1 |B| diam(B) \|u\|_{t,B}. \tag{5.6.4}$$

Choose $m = s/(1 + \alpha(r-1))$, so that $m < s$. Substituting (5.6.4) into (5.6.3) and using Lemma 3.1.1, we obtain

$$\|T(d(G(u)))\|_{s,B,w^\alpha}$$

$$\leq C_1 |B| diam(B) \|u\|_{t,B} \left(\int_B w dx \right)^{\alpha/s} \tag{5.6.5}$$

$$\leq C_2 |B| diam(B) |B|^{(m-t)/mt} \|u\|_{m,\rho B} \left(\int_B w dx \right)^{\alpha/s}.$$

Now using the Hölder inequality with $1/m = 1/s + (s - m)/sm$, we find that

$$\|u\|_{m,\rho B}$$

$$= \left(\int_{\rho B} |u|^m dx \right)^{1/m}$$

$$= \left(\int_{\rho B} \left(|u| w^{\alpha/s} w^{-\alpha/s} \right)^m dx \right)^{1/m} \tag{5.6.6}$$

$$\leq \left(\int_{\rho B} |u|^s w^\alpha dx \right)^{1/s} \left(\int_{\rho B} \left(\frac{1}{w} \right)^{1/(r-1)} dx \right)^{\alpha(r-1)/s}$$

for all balls B with $\rho B \subset \Omega$. Substituting (5.6.6) into (5.6.5), we obtain

$$\|T(d(G(u)))\|_{s,B,w^\alpha}$$

$$\leq C_2|B|diam(B)|B|^{(m-t)/mt}\left(\int_{\rho B}|u|^s w^\alpha dx\right)^{1/s} \tag{5.6.7}$$

$$\times \left(\int_B w dx\right)^{\alpha/s}\left(\int_{\rho B}\left(\tfrac{1}{w}\right)^{1/(r-1)}dx\right)^{\alpha(r-1)/s}.$$

Next, since $w \in A_r(\Omega)$, it is easy to conclude that

$$\|w\|_{1,B}^{\alpha/s}\cdot\|1/w\|_{1/(r-1),\rho B}^{\alpha/s}$$

$$\leq \left(\left(\int_{\rho B}w dx\right)\left(\int_{\rho B}(1/w)^{1/(r-1)}dx\right)^{r-1}\right)^{\alpha/s}$$

$$= \left(|\rho B|^r\left(\tfrac{1}{|\rho B|}\int_{\rho B}w dx\right)\left(\tfrac{1}{|\rho B|}\int_{\rho B}\left(\tfrac{1}{w}\right)^{1/(r-1)}dx\right)^{r-1}\right)^{\alpha/s} \tag{5.6.8}$$

$$\leq C_3|B|^{\alpha r/s}.$$

Combining (5.6.8) and (5.6.7), we have

$$\|T(d(G(u)))\|_{s,B,w^\alpha}\leq C_4|B|diam(B)\|u\|_{s,\rho B,w^\alpha} \tag{5.6.9}$$

for all balls B with $\rho B \subset \Omega$. Thus, (5.6.1) holds for $0 < \alpha < 1$.

Next, we shall prove (5.6.1) for $\alpha = 1$, that is, we need to show that

$$\|T(d(G(u)))\|_{s,B,w}\leq C|B|diam(B)\|u\|_{s,\rho B,w}. \tag{5.6.10}$$

By Lemma 1.4.7, there exist constants $\beta > 1$ and $C_5 > 0$, such that

$$\|\,w\,\|_{\beta,B}\leq C_5|B|^{(1-\beta)/\beta}\,\|\,w\,\|_{1,B}\,. \tag{5.6.11}$$

Choose $t = s\beta/(\beta - 1)$, so that $1 < s < t$ and $\beta = t/(t-s)$. Since $1/s = 1/t + (t-s)/st$, by the Hölder inequality, (5.6.2), and (5.6.11), it follows that

$$\left(\int_B |T(d(G(u)))|^s w dx\right)^{1/s}$$

$$= \left(\int_B \left(|T(d(G(u)))|w^{1/s}\right)^s dx\right)^{1/s}$$

$$\leq \left(\int_B |T(d(G(u)))|^t dx\right)^{1/t}\left(\int_B \left(w^{1/s}\right)^{st/(t-s)}dx\right)^{(t-s)/st}$$

$$\leq C_6\|T(d(G(u)))\|_{t,B}\cdot\|w\|_{\beta,B}^{1/s} \tag{5.6.12}$$

$$\leq C_6|B|diam(B)\|u\|_{t,B}\cdot\|w\|_{\beta,B}^{1/s}$$

$$\leq C_7|B|diam(B)|B|^{(1-\beta)/\beta s}\|w\|_{1,B}^{1/s}\cdot\|u\|_{t,B}$$

$$\leq C_7|B|diam(B)|B|^{-1/t}\|w\|_{1,B}^{1/s}\cdot\|u\|_{t,B}.$$

Set $m = s/r$ in Lemma 3.1.1, to have

$$\|u\|_{t,B} \le C_8 |B|^{(m-t)/mt} \|u\|_{m,\rho B}. \qquad (5.6.13)$$

Using the Hölder inequality again, we find

$$\|u\|_{m,\rho B}$$
$$= \left(\int_{\rho B} \left(|u| w^{1/s} w^{-1/s} \right)^m dx \right)^{1/m} \qquad (5.6.14)$$
$$\le \left(\int_{\rho B} |u|^s w dx \right)^{1/s} \left(\int_{\rho B} \left(\tfrac{1}{w} \right)^{1/(r-1)} dx \right)^{(r-1)/s}.$$

Now since $w \in A_r(\Omega)$, we have

$$\|w\|_{1,B}^{1/s} \cdot \|1/w\|_{1/(r-1),\rho B}^{1/s}$$
$$\le \left(\left(\int_{\rho B} w dx \right) \left(\int_{\rho B} (1/w)^{1/(r-1)} dx \right)^{r-1} \right)^{1/s}$$
$$= \left(|\rho B|^r \left(\tfrac{1}{|\rho B|} \int_{\rho B} w dx \right) \left(\tfrac{1}{|\rho B|} \int_{\rho B} \left(\tfrac{1}{w} \right)^{1/(r-1)} dx \right)^{r-1} \right)^{1/s} \qquad (5.6.15)$$
$$\le C_9 |B|^{r/s}.$$

Combining (5.6.12), (5.6.13), (5.6.14), and (5.6.15), we obtain

$$\|T(d(G(u)))\|_{s,B,w}$$
$$\le C_{10} |B| diam(B) |B|^{-1/t} \|w\|_{1,B}^{1/s} |B|^{(m-t)/mt} \|u\|_{m,\rho B}$$
$$\le C_{10} |B| diam(B) |B|^{-1/m} \|w\|_{1,B}^{1/s} \cdot \|1/w\|_{1/(r-1),\rho B}^{1/s} \|u\|_{s,\rho B,w}$$
$$\le C_{11} |B| diam(B) \|u\|_{s,\rho B,w}$$

for all balls B with $\rho B \subset \Omega$. Hence, (5.6.10) holds. ∎

Theorem 5.6.2. *Let $u \in L^s_{loc}(\Omega, \wedge^l)$, $l = 1, 2, \ldots, n$, $1 < s < \infty$, be a smooth differential form satisfying (5.3.1) in bounded, convex domain Ω and $T : C^\infty(\Omega, \wedge^l) \to C^\infty(\Omega, \wedge^{l-1})$ be the homotopy operator defined in (1.5.1). Then, there exists a constant C, independent of u, such that*

$$\|T(d(G(u)))\|_{W^{1,s}(B)} \le C|B|\|u\|_{s,B} \qquad (5.6.16)$$

for all balls B with $B \subset \Omega$.

Proof. From (3.3.14), Lemma 5.4.1, and (3.3.3), we obtain

$$\|T(d(G(u)))\|_{W^{1,s}(B)}$$

$$= diam(B)^{-1}\|T(d(G(u)))\|_{s,B} + \|\nabla(T(d(G(u))))\|_{s,B}$$

$$\leq diam(B)^{-1} \cdot C_1|B|diam(B)\|d(G(u))\|_{s,B} + C_2|B|\|d(G(u))\|_{s,B}$$

$$\leq C_3|B|\|d(G(u))\|_{s,B}$$

$$\leq C_4|B|\|u\|_{s,B}. \qquad \blacksquare$$

5.6.2 $A_r(\Omega)$-weighted estimates for $\nabla \circ T \circ G$

Similar to Theorem 5.6.1, we have the following result for $\nabla \circ T \circ G$.

Theorem 5.6.3. Let $u \in L^s_{loc}(\Omega, \wedge^l)$, $l = 1, 2, \ldots, n$, $1 < s < \infty$, be a smooth differential form satisfying (5.3.1) in a bounded, convex domain Ω and $T : C^\infty(\Omega, \wedge^l) \to C^\infty(\Omega, \wedge^{l-1})$ be the homotopy operator defined in (1.5.1). Assume that $\rho > 1$ and $w(x) \in A_r(\Omega)$ for some $1 < r < \infty$. Then, $\nabla(T(G(u))) \in L^s_{loc}(\Omega, \wedge^l)$. Moreover, there exists a constant C, independent of u, such that

$$\|\nabla(T(G(u)))\|_{s,B,w^\alpha} \leq C|B|\|u\|_{s,\rho B,w^\alpha} \qquad (5.6.17)$$

for all balls B with $\rho B \subset \Omega$ and any real number α with $0 < \alpha \leq 1$.

Proof. From (5.6.17), it is clear that $\nabla(T(G(u))) \in L^s_{loc}(\Omega, \wedge^l)$. Hence, we only need to prove (5.6.17). From Lemma 5.4.1 and (3.3.3), we find that

$$\|\nabla(T(G(u)))\|_{s,B} \leq C|B|\|G(u)\|_{s,B} \leq C|B|\|u\|_{s,B}. \qquad (5.6.18)$$

The rest of the proof is similar to that of Theorem 5.6.1, and hence omitted.
$\blacksquare$

Theorem 5.6.4. Let $u \in L^s_{loc}(\Omega, \wedge^l)$, $l = 1, 2, \ldots, n$, $1 < s < \infty$, be a smooth differential form satisfying (5.3.1) in a bounded, convex domain Ω and $T : C^\infty(\Omega, \wedge^l) \to C^\infty(\Omega, \wedge^{l-1})$ be the homotopy operator defined in (1.5.1). Assume that $\rho > 1$ and $w(x) \in A_r(\Omega)$ for some $1 < r < \infty$. Then, there exists a constant C, independent of u, such that

$$\|T(d(G(u)))\|_{W^{1,s}(B),w^\alpha} \leq C|B|\|u\|_{s,\rho B,w^\alpha} \qquad (5.6.19)$$

for all balls B with $\rho B \subset \Omega$ and any real number α with $0 < \alpha \leq 1$.

Proof. Using (3.3.15), and Theorems 5.6.1 and 5.6.3, we obtain

$$\|T(d(G(u)))\|_{W^{1,s}(B),w^\alpha}$$

$$= diam(B)^{-1}\|T(d(G(u)))\|_{s,B,w^\alpha} + \|\nabla(T(d(G(u))))\|_{s,B,w^\alpha}$$

$$\leq diam(B)^{-1}C_1 diam(B)|B|\|u\|_{s,\rho_1 B,w^\alpha} + C_2|B|\|u\|_{s,\rho_2 B,w^\alpha} \qquad (5.6.20)$$

$$\leq C_3|B|\|u\|_{s,\rho B,w^\alpha},$$

where $\rho = max\{\rho_1, \rho_2\}$. ∎

Next, as applications of the local results, we prove the following global $A_r(\Omega)$-weighted estimates for the compositions of operators.

Theorem 5.6.5. *Let $u \in L^s(\Omega, \wedge^l)$, $l = 1, 2, \ldots, n$, $1 < s < \infty$, be a smooth differential form satisfying (5.3.1) in a bounded, convex domain Ω and $T : C^\infty(\Omega, \wedge^l) \to C^\infty(\Omega, \wedge^{l-1})$ be the homotopy operator defined in (1.5.1). Assume that $w \in A_r(\Omega)$ for some $r > 1$. Then, there exists a constant C, independent of u, such that*

$$\|T(d(G(u)))\|_{s,\Omega,w^\alpha} \leq C|\Omega|diam(\Omega)\|u\|_{s,\Omega,w^\alpha} \qquad (5.6.21)$$

for any real number α with $0 < \alpha \leq 1$.

Proof. Let $\mathcal{V} = \{Q_i\}$ be a Whitney cover of Ω. Applying the covering lemma and Theorem 5.6.1, we obtain

$$\|T(d(G(u)))\|_{s,\Omega,w^\alpha}$$

$$\leq \sum_{Q\in\mathcal{V}} \|T(d(G(u)))\|_{s,Q,w^\alpha}$$

$$\leq \sum_{Q\in\mathcal{V}} \left(C_1|Q|diam(Q)\|u\|_{s,\rho Q,w^\alpha}\right)$$

$$\leq \sum_{Q\in\mathcal{V}} \left(C_2|\Omega|diam(\Omega)\|u\|_{s,\Omega,w^\alpha}\right) \qquad (5.6.22)$$

$$\leq \left(C_2|\Omega|diam(\Omega)\|u\|_{s,\Omega,w^\alpha}\right) \cdot N$$

$$\leq C_3|\Omega|diam(\Omega)\|u\|_{s,\Omega,w^\alpha}. \quad ∎$$

Theorem 5.6.6. *Let $u \in L^s(\Omega, \wedge^l)$, $l = 1, 2, \ldots, n$, $1 < s < \infty$, be a smooth differential form satisfying (5.3.1) in a bounded, convex domain Ω and $T : C^\infty(\Omega, \wedge^l) \to C^\infty(\Omega, \wedge^{l-1})$ be the homotopy operator defined in (1.5.1). Assume that $w \in A_r(\Omega)$ for some $r > 1$. Then, there exists a constant C, independent of u, such that*

$$\|\nabla(T(G(u)))\|_{s,\Omega,w^\alpha} \leq C|\Omega|\|u\|_{s,\Omega,w^\alpha} \qquad (5.6.23)$$

for any real number α with $0 < \alpha \leq 1$.

Proof. The proof requires Theorems 1.5.3 and 5.6.3 and is similar to that of Theorem 5.6.5. ∎

5.6.3 Imbedding for $T \circ d \circ G$

At this stage, we are ready to prove the following global $A_r(\Omega)$-weighted Sobolev–Poincaré imbedding theorem for the composition of operators applied to the solutions of the nonhomogeneous A-harmonic equation.

Theorem 5.6.7. *Let $u \in L^s(\Omega, \wedge^l)$, $l = 1, 2, \ldots, n$, $1 < s < \infty$, be a smooth differential form satisfying (5.3.1) in a bounded, convex domain Ω. Assume that $T : C^\infty(\Omega, \wedge^l) \to C^\infty(\Omega, \wedge^{l-1})$ is the homotopy operator defined in (1.5.1) and $w(x) \in A_r(\Omega)$ for some $1 < r < \infty$. Then, $T(d(G(u))) \in W^{1,s}(\Omega, \wedge^l)$. Moreover, there exists a constant C, independent of u, such that*

$$\|T(d(G(u)))\|_{W^{1,s}(\Omega), w^\alpha} \leq C|\Omega| \|u\|_{s, \Omega, w^\alpha} \qquad (5.6.24)$$

for any real number α with $0 < \alpha \leq 1$.

Proof. Applying (3.3.15) and (3.3.3), and Theorems 5.6.5 and 5.6.6, we find that

$$\|T(d(G(u)))\|_{W^{1,s}(\Omega), w^\alpha}$$

$$= diam(\Omega)^{-1} \|T(d(G(u)))\|_{s, \Omega, w^\alpha} + \|\nabla(T(d(G(u))))\|_{s, \Omega, w^\alpha}$$

$$\leq C_1 diam(\Omega)^{-1} \cdot |\Omega| diam(\Omega) \|u\|_{s, \Omega, w^\alpha} + C_2 |\Omega| \|u\|_{s, \Omega, w^\alpha}$$

$$\leq C_1 |\Omega| \|u\|_{s, \Omega, w^\alpha} + C_2 |\Omega| \|u\|_{s, \Omega, w^\alpha}$$

$$\leq C_3 |\Omega| \|u\|_{s, \Omega, w^\alpha}. \qquad \blacksquare$$

For $\alpha = 1$, Theorems 5.6.5, 5.6.6, and 5.6.7 reduce to the following interesting corollaries.

Corollary 5.6.8. *Let $u \in L^s(\Omega, \wedge^l)$, $l = 1, 2, \ldots, n$, $1 < s < \infty$, be a smooth differential form satisfying (5.3.1) in a bounded, convex domain Ω and $T : C^\infty(\Omega, \wedge^l) \to C^\infty(\Omega, \wedge^{l-1})$ be the homotopy operator defined in (1.5.1). Assume that $w \in A_r(\Omega)$ for some $r > 1$. Then, there exists a constant C, independent of u, such that*

$$\|T(d(G(u)))\|_{s, \Omega, w} \leq C|\Omega| diam(\Omega) \|u\|_{s, \Omega, w}. \qquad (5.6.25)$$

Corollary 5.6.9. *Let $u \in L^s(\Omega, \wedge^l)$, $l = 1, 2, \ldots, n$, $1 < s < \infty$, be a smooth differential form satisfying (5.3.1) in a bounded, convex domain Ω and*

$T : C^\infty(\Omega, \wedge^l) \to C^\infty(\Omega, \wedge^{l-1})$ *be the homotopy operator defined in* (1.5.1). *Assume that* $w \in A_r(\Omega)$ *for some* $r > 1$. *Then, there exists a constant* C, *independent of* u, *such that*

$$\|\nabla(T(G(u)))\|_{s,\Omega,w} \leq C|\Omega|\|u\|_{s,\Omega,w}. \qquad (5.6.26)$$

Corollary 5.6.10. *Let* $u \in L^s(\Omega, \wedge^l)$, $l = 1, 2, \ldots, n$, $1 < s < \infty$, *be a smooth differential form satisfying* (5.3.1) *in a bounded, convex domain* Ω. *Assume that* $T : C^\infty(\Omega, \wedge^l) \to C^\infty(\Omega, \wedge^{l-1})$ *is a homotopy operator defined in* (1.5.1) *and* $w(x) \in A_r(\Omega)$ *for some* $1 < r < \infty$. *Then,* $T(d(G(u))) \in W^{1,s}(\Omega, \wedge^l)$. *Moreover, there exists a constant* C, *independent of* u, *such that*

$$\|T(d(G(u)))\|_{W^{1,s}(\Omega),w} \leq C|\Omega|\|u\|_{s,\Omega,w}. \qquad (5.6.27)$$

Next, since $1 < s < \infty$, when $\alpha = 1/s$ in (5.6.21), (5.6.23), and (5.6.24), we obtain, respectively

$$\|T(d(G(u)))\|_{s,\Omega,w^{1/s}} \leq C|\Omega|diam(\Omega)\|u\|_{s,\Omega,w^{1/s}},$$

$$\|\nabla(T(G(u)))\|_{s,\Omega,w^{1/s}} \leq C|\Omega|\|u\|_{s,\Omega,w^{1/s}},$$

$$\|T(d(G(u)))\|_{W^{1,s}(\Omega),w^{1/s}} \leq C|\Omega|\|u\|_{s,\Omega,w^{1/s}}.$$

Similarly, when $\alpha = s/(r+s)$ in (5.6.21), (5.6.23), and (5.6.24), we find that

$$\|T(d(G(u)))\|_{s,\Omega,w^{s/(r+s)}} \leq C|\Omega|diam(\Omega)\|u\|_{s,\Omega,w^{s/(r+s)}},$$

$$\|\nabla(T(G(u)))\|_{s,\Omega,w^{s/(r+s)}} \leq C|\Omega|\|u\|_{s,\Omega,w^{s/(r+s)}},$$

$$\|T(d(G(u)))\|_{W^{1,s}(\Omega),w^{s/(r+s)}} \leq C|\Omega|\|u\|_{s,\Omega,w^{s/(r+s)}}.$$

Finally, from Theorems 5.6.5 and 5.6.6, the following corollary for the compositions of operators immediately follows.

Corollary 5.6.11. *The compositions* $T \circ d \circ G$ *and* $\nabla \circ T \circ G$ *are bounded operators if* Ω *is a bounded, convex domain.*

5.7 Two-weight cases

We have already discussed various versions of the L^s-estimates and imbedding inequalities for the solutions of the A-harmonic equation and the related operators applied to these solutions. In this section, we study the two-weighted imbedding inequalities.

5.7.1 Two-weight imbedding for the operator T

We begin with the following local weighted imbedding theorem for differential forms under the homotopy operator T.

Theorem 5.7.1. *Assume that $u \in L^s_{loc}(\Omega, \wedge^l)$, $l = 1, 2, \ldots, n$, $1 < s < \infty$, is a differential form satisfying the nonhomogeneous A-harmonic equation (5.3.1) in a bounded, convex domain $\Omega \subset \mathbf{R}^n$ and $T : L^s(\Omega, \wedge^l) \rightarrow W^{1,s}(\Omega, \wedge^{l-1})$ is the homotopy operator defined in (1.5.1). If $\rho > 1$ and $(w_1, w_2) \in A_r(\Omega)$ for some $r > 1$, then there exists a constant C, independent of u, such that*

$$\|Tu\|_{W^{1,s}(B),\, w_1^\alpha} \leq C|B|\|u\|_{s,\rho B, w_2^\alpha} \tag{5.7.1}$$

for all balls B with $\rho B \subset \Omega$ and any real number α with $0 < \alpha < 1$. Moreover, if $(w_1, w_2) \in A_r(\Omega)$ and $w_1 \in RH(\Omega)$, we have

$$\|Tu\|_{W^{1,s}(B),\, w_1} \leq C|B|\|u\|_{s,\rho B, w_2} \tag{5.7.2}$$

for all balls B with $\rho B \subset \Omega$.

In order to prove Theorem 5.7.1, first we need to prove the following two-weighted imbedding inequality.

Theorem 5.7.2. *Let $u \in L^s_{loc}(\Omega, \wedge^l)$, $l = 1, 2, \ldots, n$, $1 < s < \infty$, be a differential form satisfying (5.3.1) in a bounded, convex domain $\Omega \subset \mathbf{R}^n$ and $T : C^\infty(\Omega, \wedge^l) \rightarrow C^\infty(\Omega, \wedge^{l-1})$ be the homotopy operator defined in (1.5.1). Assume that $\rho > 1$ and $(w_1, w_2) \in A_r(\Omega)$ for some $r > 1$. Then, there exists a constant C, independent of u, such that*

$$\left(\int_B |\nabla(Tu)|^s w_1^\alpha dx \right)^{1/s} \leq C|B| \left(\int_{\rho B} |u|^s w_2^\alpha dx \right)^{1/s} \tag{5.7.3}$$

for any real number α with $0 < \alpha < 1$. Moreover, if $(w_1, w_2) \in A_r(\Omega)$ and $w_1 \in RH(\Omega)$, we have

$$\left(\int_B |\nabla(Tu)|^s w_1 dx \right)^{1/s} \leq C|B| \left(\int_{\rho B} |u|^s w_2 dx \right)^{1/s} \tag{5.7.4}$$

for all balls B with $\rho B \subset \Omega$.

Note that (5.7.3) can be written as

$$\|\nabla(Tu)\|_{s,B,w_1^\alpha} \leq C|B|\|u\|_{s,\rho B, w_2^\alpha}. \tag{5.7.4'}$$

Proof. First, we shall prove that inequality (5.7.3) holds for $0 < \alpha < 1$. Let $t = s/(1-\alpha)$. From Theorem 1.1.4, we find that

$$
\begin{aligned}
&\left(\int_B |\nabla(Tu)|^s w_1^\alpha dx\right)^{1/s} \\
&= \left(\int_B \left(|\nabla(Tu)|w_1^{\alpha/s}\right)^s dx\right)^{1/s} \\
&\leq \|\nabla(Tu)\|_{t,B} \left(\int_B w_1^{t\alpha/(t-s)} dx\right)^{(t-s)/st} \\
&= \|\nabla(Tu)\|_{t,B} \left(\int_B w_1 dx\right)^{\alpha/s}.
\end{aligned}
\tag{5.7.5}
$$

Thus, from Lemma 5.4.1, it follows that

$$
\|\nabla(Tu)\|_{t,B} \leq C_1 |B| \|u\|_{t,B}.
\tag{5.7.6}
$$

Choose $m = s/(1+\alpha(r-1))$, so that $m < s$. Substituting (5.7.6) into (5.7.5) and using Lemma 3.1.1, we obtain

$$
\begin{aligned}
&\left(\int_B |\nabla(Tu)|^s w_1^\alpha dx\right)^{1/s} \\
&\leq C_1 |B| \|u\|_{t,B} \left(\int_B w_1 dx\right)^{\alpha/s} \\
&\leq C_2 |B| |B|^{(m-t)/mt} \|u\|_{m,\rho B} \left(\int_B w_1 dx\right)^{\alpha/s}.
\end{aligned}
\tag{5.7.7}
$$

Now from Theorem 1.1.4 with $1/m = 1/s + (s-m)/sm$, we have

$$
\begin{aligned}
&\|u\|_{m,\rho B} \\
&= \left(\int_{\rho B} |u|^m dx\right)^{1/m} \\
&= \left(\int_{\rho B} \left(|u|w_2^{\alpha/s} w_2^{-\alpha/s}\right)^m dx\right)^{1/m} \\
&\leq \left(\int_{\rho B} |u|^s w_2^\alpha dx\right)^{1/s} \left(\int_{\rho B} \left(\frac{1}{w_2}\right)^{1/(r-1)} dx\right)^{\alpha(r-1)/s}
\end{aligned}
\tag{5.7.8}
$$

for all balls B with $\rho B \subset \Omega$. Substituting (5.7.8) into (5.7.7), it follows that

$$
\begin{aligned}
&\left(\int_B |\nabla(Tu)|^s w_1^\alpha dx\right)^{1/s} \\
&\leq C_2 |B| |B|^{(m-t)/mt} \left(\int_{\rho B} |u|^s w_2^\alpha dx\right)^{1/s} \\
&\quad \times \left(\int_B w_1 dx\right)^{\alpha/s} \left(\int_{\rho B} \left(\frac{1}{w_2}\right)^{1/(r-1)} dx\right)^{\alpha(r-1)/s}.
\end{aligned}
\tag{5.7.9}
$$

Next, since $(w_1, w_2) \in A_r(\Omega)$, we find that

$$\|w_1\|_{1,B}^{\alpha/s} \cdot \|1/w_2\|_{1/(r-1),\rho B}^{\alpha/s}$$

$$\leq \left(\left(\int_{\rho B} w_1 dx \right) \left(\int_{\rho B} (1/w_2)^{1/(r-1)} dx \right)^{r-1} \right)^{\alpha/s}$$

$$= \left(|\rho B|^r \left(\frac{1}{|\rho B|} \int_{\rho B} w_1 dx \right) \left(\frac{1}{|\rho B|} \int_{\rho B} \left(\frac{1}{w_2} \right)^{1/(r-1)} dx \right)^{r-1} \right)^{\alpha/s} \quad (5.7.10)$$

$$\leq C_3 |B|^{\alpha r/s}.$$

Combination of (5.7.10) and (5.7.9) yields

$$\left(\int_B |\nabla(Tu)|^s w_1^\alpha dx \right)^{1/s} \leq C_4 |B| \left(\int_{\rho B} |u|^s w_2^\alpha dx \right)^{1/s} \quad (5.7.11)$$

for all balls B with $\rho B \subset \Omega$. Thus, (5.7.3) holds if $0 < \alpha < 1$.

Next, we show that (5.7.4) holds if $w_1 \in RH(\Omega)$. Since $w_1 \in RH(\Omega)$, there exist constants $\beta > 1$ and $C_5 > 0$, such that

$$\| w_1 \|_{\beta,B} \leq C_5 |B|^{(1-\beta)/\beta} \| w_1 \|_{1,B} \quad (5.7.12)$$

for any cube or ball $B \subset \mathbf{R}^n$. Choose $t = s\beta/(\beta-1)$, so that $1 < s < t$ and $\beta = t/(t-s)$. Since $1/s = 1/t + (t-s)/st$, by Theorem 1.1.4, Lemma 5.4.1, and (5.7.12), we obtain

$$\left(\int_B |\nabla(Tu)|^s w_1 dx \right)^{1/s}$$

$$= \left(\int_B \left(|\nabla(Tu)| w_1^{1/s} \right)^s dx \right)^{1/s}$$

$$\leq \left(\int_B |\nabla(Tu)|^t dx \right)^{1/t} \left(\int_B \left(w_1^{1/s} \right)^{st/(t-s)} dx \right)^{(t-s)/st}$$

$$\leq C_6 \|\nabla(Tu)\|_{t,B} \cdot \|w_1\|_{\beta,B}^{1/s} \quad (5.7.13)$$

$$\leq C_6 |B| \|u\|_{t,B} \cdot \|w_1\|_{\beta,B}^{1/s}$$

$$\leq C_7 |B| |B|^{(1-\beta)/\beta s} \|w_1\|_{1,B}^{1/s} \cdot \|u\|_{t,B}$$

$$\leq C_7 |B| |B|^{-1/t} \|w_1\|_{1,B}^{1/s} \cdot \|u\|_{t,B}.$$

Now, from Lemma 3.1.1 with $m = s/r$, we have

$$\|u\|_{t,B} \leq C_8 |B|^{(m-t)/mt} \|u\|_{m,\rho B}, \quad (5.7.14)$$

and hence from Theorem 1.1.4, we find that

$$
\begin{aligned}
&\|u\|_{m,\rho B} \\
&= \left(\int_{\rho B} \left(|u| w_2^{1/s} w_2^{-1/s} \right)^m dx \right)^{1/m} \\
&\leq \left(\int_{\rho B} |u|^s w_2 dx \right)^{1/s} \left(\int_{\rho B} \left(\frac{1}{w_2} \right)^{1/(r-1)} dx \right)^{(r-1)/s}
\end{aligned}
\tag{5.7.15}
$$

for all balls B with $\rho B \subset \Omega$. Using the condition $(w_1, w_2) \in A_r(\Omega)$, we have

$$
\begin{aligned}
&\|w_1\|_{1,B}^{1/s} \cdot \|1/w_2\|_{1/(r-1),\rho B}^{1/s} \\
&\leq \left(\left(\int_{\rho B} w_1 dx \right) \left(\int_{\rho B} (1/w_2)^{1/(r-1)} dx \right)^{r-1} \right)^{1/s} \\
&= \left(|\rho B|^r \left(\frac{1}{|\rho B|} \int_{\rho B} w_1 dx \right) \left(\frac{1}{|\rho B|} \int_{\rho B} \left(\frac{1}{w_2} \right)^{1/(r-1)} dx \right)^{r-1} \right)^{1/s} \\
&\leq C_9 |B|^{r/s}.
\end{aligned}
\tag{5.7.16}
$$

Combining $(5.7.13), (5.7.14), (5.7.15), and (5.7.16)$, if follows that

$$
\begin{aligned}
&\|\nabla(Tu)\|_{s,B,w_1} \\
&\leq C_{10} |B| |B|^{-1/t} \|w_1\|_{1,B}^{1/s} |B|^{(m-t)/mt} \|u\|_{m,\rho B} \\
&\leq C_{10} |B| |B|^{-1/m} \|w_1\|_{1,B}^{1/s} \cdot \|1/w_2\|_{1/(r-1),\rho B}^{1/s} \|u\|_{s,\rho B,w_2} \\
&\leq C_{11} |B| \|u\|_{s,\rho B,w_2}
\end{aligned}
\tag{5.7.17}
$$

for all balls B with $\rho B \subset \Omega$. Hence, (5.7.4) holds.   ∎

Using the method similar to the proof of Theorem 5.7.2, we obtain

$$
\|Tu\|_{s,B,w_1^\alpha} \leq C|B| diam(B) \|u\|_{s,\rho B,w_2^\alpha},
\tag{5.7.18}
$$

where $(w_1, w_2) \in A_r(\Omega)$ for some $r > 1$, $w_1 \in RH(\Omega)$, and α is any real number with $0 < \alpha \leq 1$ and $\rho > 1$.

Proof of Theorem 5.7.1. From (5.4.1), (5.7.4)′, and (5.7.18), we have

$$
\begin{aligned}
\|Tu\|_{W^{1,s}(B),\, w_1^\alpha} \\
&= diam(B)^{-1}\|Tu\|_{s,B,w_1^\alpha} + \|\nabla(Tu)\|_{s,B,w_1^\alpha} \\
&\le diam(B)^{-1}\left(C_1|B|diam(B)\|u\|_{s,\rho B,w_2^\alpha}\right) + C_2|B|\|u\|_{s,\rho B,w_2^\alpha} \\
&\le C_1|B|\|u\|_{s,\rho B,w_2^\alpha} + C_2|B|\|u\|_{s,\rho B,w_2^\alpha} \\
&\le C_3|B|\|u\|_{s,\rho B,w_2^\alpha}
\end{aligned}
$$

which is equivalent to (5.7.1).   ∎

Note that the parameter α in both Theorems 5.7.1 and 5.7.2 is any real number with $0 < \alpha < 1$. Therefore, we can have different versions of the weighted imbedding inequality by choosing some particular values of α. For example, when $t = 1 - \alpha$, $d\mu = w_1(x)dx$, and $d\nu = w_2(x)dx$, then inequality (5.7.3) becomes

$$
\left(\int_B |\nabla(Tu)|^s w_1^{-t} d\mu\right)^{1/s} \le C|B|\left(\int_{\rho B} |u|^s w_2^{-t} d\nu\right)^{1/s}. \tag{5.7.19}
$$

If we choose $\alpha = 1/r$, then (5.7.3) reduces to

$$
\left(\int_B |\nabla(Tu)|^s w_1^{1/r} dx\right)^{1/s} \le C|B|\left(\int_{\rho B} |u|^s w_2^{1/r} dx\right)^{1/s}. \tag{5.7.20}
$$

Similarly, if we choose $\alpha = 1/s$, then $0 < \alpha < 1$ since $1 < s < \infty$, and (5.7.3) reduces to the following symmetric version:

$$
\left(\int_B |\nabla(Tu)|^s w_1^{1/s} dx\right)^{1/s} \le C|B|\left(\int_{\rho B} |u|^s w_2^{1/s} dx\right)^{1/s}. \tag{5.7.21}
$$

Finally, if we choose $\alpha = r/(r+s)$ in Theorem 5.7.2, we have the following weighted imbedding inequality:

$$
\|\nabla(Tu)\|_{s,B,w_1^{r/(r+s)}} \le C|B|\|u\|_{s,\rho B,w_2^{r/(r+s)}}. \tag{5.7.22}
$$

Similarly, selecting $\alpha = s/(r+s)$ in Theorem 5.7.1, we have

$$
\|Tu\|_{W^{1,s}(B),\, w_1^{s/(r+s)}} \le C|B|\|u\|_{s,B,w_2^{s/(r+s)}}. \tag{5.7.23}
$$

5.7.2 $A_{r,\lambda}(E)$-weighted imbedding

Theorem 5.7.3.   *Let* $u \in L^s_{loc}(\Omega, \wedge^l)$, $l = 1, 2, \dots, n$, $1 < s < \infty$, *be a differential form satisfying (5.3.1) in a bounded, convex domain* $\Omega \subset \mathbf{R}^n$ *and* $T : C^\infty(\Omega, \wedge^l) \to C^\infty(\Omega, \wedge^{l-1})$ *be the homotopy operator defined in (1.5.1). Assume that* $\rho > 1$ *and* $(w_1, w_2) \in A_{r,\lambda}(\Omega)$ *for some* $\lambda \geq 1$ *and* $1 < r < \infty$. *Then, there exists a constant* C, *independent of* u, *such that*

$$\left(\int_B |\nabla(Tu)|^s w_1^\alpha dx \right)^{1/s} \leq C|B| \left(\int_{\rho B} |u|^s w_2^\alpha dx \right)^{1/s} \tag{5.7.24}$$

for all balls B *with* $\rho B \subset \Omega$ *and any real number* α *with* $0 < \alpha < \lambda$.

Note that (5.7.24) can be written as

$$\|\nabla(Tu)\|_{s,B,w_1^\alpha} \leq C|B| \|u\|_{s,\rho B, w_2^\alpha}. \tag{5.7.25}$$

Proof.   Choose $t = \lambda s/(\lambda - \alpha)$. Since $1/s = 1/t + (t-s)/st$, from the Hölder inequality, we obtain

$$\left(\int_B |\nabla(Tu)|^s w_1^\alpha dx \right)^{1/s}$$

$$= \left(\int_B \left(|\nabla(Tu)| w_1^{\alpha/s} \right)^s dx \right)^{1/s}$$

$$\leq \left(\int_B |\nabla(Tu)|^t dx \right)^{1/t} \left(\int_B \left(w_1^{\alpha/s} \right)^{st/(t-s)} dx \right)^{(t-s)/st} \tag{5.7.26}$$

$$\leq \|\nabla(Tu)\|_{t,B} \cdot \left(\int_B w_1^\lambda dx \right)^{\alpha/\lambda s}$$

for all balls $B \subset \Omega$. Thus, from Lemma 5.4.1, it follows that

$$\|\nabla(Tu)\|_{t,B} \leq C_1 |B| \|u\|_{t,B}. \tag{5.7.27}$$

Choose $m = \lambda s/(\lambda + \alpha(r-1))$, so that $m < s < t$. Substituting (5.7.27) into (5.7.26) and using Lemma 3.1.1, we find that

$$\left(\int_B |\nabla(Tu)|^s w_1^\alpha dx \right)^{1/s}$$

$$\leq C_1 |B| \|u\|_{t,B} \left(\int_B w_1^\lambda dx \right)^{\alpha/\lambda s} \tag{5.7.28}$$

$$\leq C_2 |B| |B|^{(m-t)/mt} \|u\|_{m,\rho B} \left(\int_B w_1^\lambda dx \right)^{\alpha/\lambda s}.$$

Now, using Theorem 1.1.4 with $1/m = 1/s + (s-m)/sm$, we have

$$
\begin{aligned}
\|u\|_{m,\rho B} \\
&= \left(\int_{\rho B} |u|^m dx \right)^{1/m} \\
&= \left(\int_{\rho B} \left(|u| w_2^{\alpha/s} w_2^{-\alpha/s} \right)^m dx \right)^{1/m} \\
&\leq \left(\int_{\rho B} |u|^s w_2^{\alpha} dx \right)^{1/s} \left(\int_{\rho B} \left(\frac{1}{w_2} \right)^{\lambda/(r-1)} dx \right)^{\alpha(r-1)/\lambda s}
\end{aligned}
\tag{5.7.29}
$$

for all balls B with $\rho B \subset \Omega$. Substituting (5.7.29) into (5.7.28), it follows that

$$
\begin{aligned}
\left(\int_B |\nabla(Tu)|^s w_1^{\alpha} dx \right)^{1/s} \\
&\leq C_3 |B| |B|^{(m-t)/mt} \left(\int_{\rho B} |u|^s w_2^{\alpha} dx \right)^{1/s} \\
&\quad \times \left(\int_B w_1^{\lambda} dx \right)^{\alpha/\lambda s} \left(\int_{\rho B} \left(\frac{1}{w_2} \right)^{\lambda/(r-1)} dx \right)^{\alpha(r-1)/\lambda s}.
\end{aligned}
\tag{5.7.30}
$$

Next, using the condition $(w_1, w_2) \in A_{r,\lambda}(\Omega)$, we obtain

$$
\begin{aligned}
\left(\int_B w_1^{\lambda} dx \right)^{\alpha/\lambda s} \left(\int_{\rho B} \left(\frac{1}{w_2} \right)^{\lambda/(r-1)} dx \right)^{\alpha(r-1)/\lambda s} \\
&\leq \left(\left(\int_{\rho B} w_1^{\lambda} dx \right) \left(\int_{\rho B} (1/w_2)^{\lambda/(r-1)} dx \right)^{r-1} \right)^{\alpha/\lambda s} \\
&= \left(|\rho B|^r \left(\frac{1}{|\rho B|} \int_{\rho B} w_1^{\lambda} dx \right) \left(\frac{1}{|\rho B|} \int_{\rho B} \left(\frac{1}{w_2} \right)^{\frac{1}{r-1}} dx \right)^{r-1} \right)^{\alpha/\lambda s} \\
&\leq C_4 |B|^{\alpha r/\lambda s}.
\end{aligned}
\tag{5.7.31}
$$

Combining (5.7.30) and (5.7.31), we have

$$
\left(\int_B |\nabla(Tu)|^s w_1^{\alpha} dx \right)^{1/s} \leq C|B| \left(\int_{\rho B} |u|^s w_2^{\alpha} dx \right)^{1/s}
\tag{5.7.32}
$$

for all balls B with $\rho B \subset \Omega$. ∎

We are now ready to prove the following imbedding inequality with two-weights.

Theorem 5.7.4. *Let $u \in L^s_{loc}(\Omega, \wedge^l)$, $l = 1, 2, \ldots, n$, $1 < s < \infty$, be a differential form satisfying (5.3.1) in a bounded, convex domain $\Omega \subset \mathbf{R}^n$ and $T : C^\infty(\Omega, \wedge^l) \to C^\infty(\Omega, \wedge^{l-1})$ be the homotopy operator defined in (1.5.1). Assume that $\rho > 1$ and $(w_1, w_2) \in A_{r,\lambda}(\Omega)$ for some $\lambda \geq 1$ and $1 < r < \infty$. Then, there exists a constant C, independent of u, such that*

$$\|Tu\|_{W^{1,s}(B), \, w_1^\alpha} \leq C|B| \|u\|_{s,\rho B, w_2^\alpha} \tag{5.7.33}$$

for all balls B with $\rho B \subset \Omega$ and any real number α with $0 < \alpha < \lambda$.

Proof. Using the same method developed in the proof of Theorem 5.7.3, we have

$$\|Tu\|_{s,B,w_1^\alpha} \leq C_1 |B| diam(B) \|u\|_{s,\rho B, w_2^\alpha}, \tag{5.7.34}$$

where α is any real number with $0 < \alpha < \lambda$ and $\rho > 1$. Combining (1.1.1), (5.7.25), and (5.7.34), we obtain

$$\|Tu\|_{W^{1,s}(B), \, w_1^\alpha}$$

$$= diam(B)^{-1} \|Tu\|_{s,B,w_1^\alpha} + \|\nabla(Tu)\|_{s,B,w_1^\alpha}$$

$$\leq diam(B)^{-1} \left(C_2 |B| diam(B) \|u\|_{s,\rho B, w_2^\alpha} \right) + C_3 |B| \|u\|_{s,\rho B, w_2^\alpha}$$

$$\leq C_2 |B| \|u\|_{s,\rho B, w_2^\alpha} + C_3 |B| \|u\|_{s,\rho B, w_2^\alpha}$$

$$\leq C_4 |B| \|u\|_{s,\rho B, w_2^\alpha}$$

which is equivalent to (5.7.33). ∎

Note that Theorem 5.7.3 contains two weights, $w_1(x)$ and $w_2(x)$, and two parameters, λ and α. This makes the imbedding inequalities more flexible and useful. In fact, many existing versions of the imbedding inequality are special cases of Theorem 5.7.3 with suitable choices of weights and parameters. For example, if we consider the case $(w_1, w_2) \in A_{r,\lambda}(\Omega)$ for some $\lambda > 1$ in Theorem 5.7.3, then we can choose $\alpha = 1$ and obtain the following corollary.

Corollary 5.7.5. *Let $u \in L^s_{loc}(\Omega, \wedge^l)$, $l = 1, 2, \ldots, n$, $1 < s < \infty$, be a differential form satisfying (5.3.1) in a bounded, convex domain $\Omega \subset \mathbf{R}^n$ and $T : C^\infty(\Omega, \wedge^l) \to C^\infty(\Omega, \wedge^{l-1})$ be the homotopy operator defined in (1.5.1). Assume that $\rho > 1$ and $(w_1, w_2) \in A_{r,\lambda}(\Omega)$ for some $\lambda > 1$ and $1 < r < \infty$. Then, there exists a constant C, independent of u, such that*

$$\left(\int_B |\nabla(Tu)|^s w_1 dx \right)^{1/s} \leq C|B| \left(\int_{\rho B} |u|^s w_2 dx \right)^{1/s} \tag{5.7.35}$$

for all balls B with $\rho B \subset \Omega$.

If we let $w_1(x) = w_2(x) = w(x)$ in (5.7.35), we obtain

$$\left(\int_B |\nabla(Tu)|^s w dx \right)^{1/s} \leq C|B| \left(\int_{\rho B} |u|^s w dx \right)^{1/s}, \tag{5.7.36}$$

where the weight $w(x)$ satisfies

$$\sup_{B \subset E} \left(\frac{1}{|B|} \int_B (w)^\lambda dx \right)^{1/\lambda r} \left(\frac{1}{|B|} \int_B \left(\frac{1}{w} \right)^{\lambda r'/r} dx \right)^{1/\lambda r'} < \infty,$$

which is a generalization of the usual A_r-weights.

We know that the $A_{r,\lambda}(\Omega)$-weight reduces to the usual $A_r(\Omega)$-weight if $w_1(x) = w_2(x)$ and $\lambda = 1$. Hence, when $w_1(x) = w_2(x) = w(x)$ and $\lambda = 1$ in Theorem 5.7.3, it reduces to the following local $A_r(\Omega)$-weighted imbedding inequality.

Theorem 5.7.6. *Let* $u \in L^s_{loc}(\Omega, \wedge^l)$, $l = 1, 2, \ldots, n$, $1 < s < \infty$, *be a differential form satisfying (5.3.1) in a bounded, convex domain* $\Omega \subset \mathbf{R}^n$ *and* $T : C^\infty(\Omega, \wedge^l) \to C^\infty(\Omega, \wedge^{l-1})$ *be the homotopy operator defined in (1.5.1). Assume that* $\rho > 1$ *and* $w \in A_r(\Omega)$ *for some* r *with* $1 < r < \infty$. *Then, there exists a constant* C, *independent of* u, *such that*

$$\left(\int_B |\nabla(Tu)|^s w^\alpha dx \right)^{1/s} \leq C|B| \left(\int_{\rho B} |u|^s w^\alpha dx \right)^{1/s}$$

for all balls B *with* $\rho B \subset \Omega$ *and any real number* α *with* $0 < \alpha < 1$.

Similarly, if we choose $\alpha = 1/s$ in Theorem 5.7.3, then inequality (5.7.25) reduces to

$$\left(\int_B |\nabla(Tu)|^s w_1^{1/s} dx \right)^{1/s} \leq C|B| \left(\int_{\rho B} |u|^s w_2^{1/s} dx \right)^{1/s}. \tag{5.7.37}$$

Finally, for $\alpha = 1/r$ in Theorem 5.7.3, inequality (5.7.25) reduces to

$$\left(\int_B |\nabla(Tu)|^s w_1^{1/r} dx \right)^{1/s} \leq C|B| \left(\int_{\rho B} |u|^s w_2^{1/r} dx \right)^{1/s}. \tag{5.7.38}$$

5.7.3 $A_r(\lambda, E)$-weighted imbedding

Theorem 5.7.7. *Let* $u \in L^s_{loc}(\Omega, \wedge^l)$, $l = 1, 2, \ldots, n$, $1 < s < \infty$, *be a solution of the nonhomogeneous A-harmonic equation (5.3.1) in a bounded,* *convex domain* $\Omega \subset \mathbf{R}^n$ *and* $T : C^\infty(\Omega, \wedge^l) \to C^\infty(\Omega, \wedge^{l-1})$ *be a homotopy operator defined in (1.5.1). Assume that* $\rho > 1$ *and* $(w_1, w_2) \in A_r(\lambda, \Omega)$ *for some* $\lambda > 0$ *and* $1 < r < \infty$. *Then, there exists a constant* C, *independent of* u, *such that*

$$\left(\int_B |\nabla(Tu)|^s w_1^{\alpha\lambda} dx \right)^{1/s} \leq C|B| \left(\int_{\rho B} |u|^s w_2^\alpha dx \right)^{1/s} \tag{5.7.39}$$

for all balls B *with* $\rho B \subset \Omega$ *and any real number* α *with* $0 < \alpha < 1$.

Note that (5.7.39) can be written as

$$\|\nabla(Tu)\|_{s,B,w_1^{\alpha\lambda}} \leq C|B| \|u\|_{s,\rho B,w_2^\alpha}. \tag{5.7.39}'$$

Proof. Select $t = s/(1-\alpha)$, so that $1 < s < t$. Noticing $1/s = 1/t + (t-s)/st$ and using Hölder's inequality, we obtain

$$\left(\int_B |\nabla(Tu)|^s w_1^{\alpha\lambda} dx \right)^{1/s}$$

$$= \left(\int_B \left(|\nabla(Tu)| w_1^{\alpha\lambda/s} \right)^s dx \right)^{1/s}$$

$$\leq \left(\int_B |\nabla(Tu)|^t dx \right)^{1/t} \left(\int_B \left(w_1^{\alpha\lambda/s} \right)^{st/(t-s)} dx \right)^{(t-s)/st} \tag{5.7.40}$$

$$\leq \|\nabla(Tu)\|_{t,B} \cdot \left(\int_B w_1^\lambda dx \right)^{\alpha/s}$$

for all balls $B \subset \Omega$. Applying Lemma 5.4.1, we have

$$\|\nabla(Tu)\|_{t,B} \leq C_1 |B| \|u\|_{t,B}. \tag{5.7.41}$$

Choose $m = s/(1 + \alpha(r-1))$, so that $m < s < t$. Substituting (5.7.41) into (5.7.40) and using Lemma 3.1.1, we obtain

$$\left(\int_B |\nabla(Tu)|^s w_1^{\alpha\lambda} dx \right)^{1/s}$$

$$\leq C_1 |B| \|u\|_{t,B} \left(\int_B w_1^\lambda dx \right)^{\alpha/s} \tag{5.7.42}$$

$$\leq C_2 |B| |B|^{(m-t)/mt} \|u\|_{m,\rho B} \left(\int_B w_1^\lambda dx \right)^{\alpha/s}.$$

Now using the Hölder inequality with $1/m = 1/s + (s-m)/sm$, we find that

$$
\begin{aligned}
\|u\|_{m,\rho B} \\
= \left(\int_{\rho B} |u|^m dx \right)^{1/m} \\
= \left(\int_{\rho B} \left(|u| w_2^{\alpha/s} w_2^{-\alpha/s} \right)^m dx \right)^{1/m} \\
\leq \left(\int_{\rho B} |u|^s w_2^\alpha dx \right)^{1/s} \left(\int_{\rho B} \left(\frac{1}{w_2} \right)^{1/(r-1)} dx \right)^{\alpha(r-1)/s}
\end{aligned} \tag{5.7.43}
$$

for all balls B with $\rho B \subset \Omega$. Substituting (5.7.43) into (5.7.42), it follows that

$$
\begin{aligned}
\left(\int_B |\nabla(Tu)|^s w_1^{\alpha\lambda} dx \right)^{1/s} \\
\leq C_3 |B| |B|^{(m-t)/mt} \left(\int_{\rho B} |u|^s w_2^\alpha dx \right)^{1/s} \\
\times \left(\int_B w_1^\lambda dx \right)^{\alpha/s} \left(\int_{\rho B} \left(\frac{1}{w_2} \right)^{1/(r-1)} dx \right)^{\alpha(r-1)/s} .
\end{aligned} \tag{5.7.44}
$$

Next, using the condition $(w_1(x), w_2(x)) \in A_r(\lambda, \Omega)$, we obtain

$$
\begin{aligned}
\left(\int_B w_1^\lambda dx \right)^{\alpha/s} \left(\int_{\rho B} \left(\frac{1}{w_2} \right)^{1/(r-1)} dx \right)^{\alpha(r-1)/s} \\
\leq \left(\left(\int_{\rho B} w_1^\lambda dx \right) \left(\int_{\rho B} (1/w_2)^{1/(r-1)} dx \right)^{r-1} \right)^{\alpha/s} \\
= \left(|\rho B|^r \left(\frac{1}{|\rho B|} \int_{\rho B} w_1^\lambda dx \right) \left(\frac{1}{|\rho B|} \int_{\rho B} \left(\frac{1}{w_2} \right)^{1/(r-1)} dx \right)^{r-1} \right)^{\alpha/s} \\
\leq C_4 |B|^{\alpha r/s}.
\end{aligned} \tag{5.7.45}
$$

Combining (5.7.44) and (5.7.45), we find that

$$
\left(\int_B |\nabla(Tu)|^s w_1^{\alpha\lambda} dx \right)^{1/s} \leq C|B| \left(\int_{\rho B} |u|^s w_2^\alpha dx \right)^{1/s} \tag{5.7.46}
$$

for all balls B with $\rho B \subset \Omega$. ∎

Applying Theorem 5.7.7 and the same method used in the proof of Theorem 5.7.4, we can prove the following imbedding inequality with two-weights in the weight class $A_r(\lambda, \Omega)$.

Theorem 5.7.8. *Let* $u \in L^s_{loc}(\Omega, \wedge^l)$, $l = 1, 2, \ldots, n$, $1 < s < \infty$, *be a solution of the nonhomogeneous A-harmonic equation* (5.3.1) *in a bounded, convex domain* $\Omega \subset \mathbf{R}^n$ *and* $T : C^\infty(\Omega, \wedge^l) \to C^\infty(\Omega, \wedge^{l-1})$ *be the homotopy operator defined in* (1.5.1). *Assume that* $\rho > 1$ *and* $(w_1, w_2) \in A_r(\lambda, \Omega)$ *for some* $\lambda > 0$ *and* $1 < r < \infty$. *Then, there exists a constant* C, *independent of* u, *such that*

$$\|Tu\|_{W^{1,s}(B), \, w_1^{\alpha\lambda}} \le C|B|\|u\|_{s, \rho B, w_2^\alpha} \tag{5.7.47}$$

for all balls B *with* $\rho B \subset \Omega$ *and any real number* α *with* $0 < \alpha < 1$.

5.7.4 Global imbedding theorem

It should be noticed that, using the covering lemma, many of the above local imbedding inequalities can be extended to the global cases on the bounded, convex domains. For example, Theorems 5.7.7 and 5.7.8 can be stated as follows.

Theorem 5.7.9. *Let* $u \in L^s_{loc}(\Omega, \wedge^l)$, $l = 1, 2, \ldots, n$, $1 < s < \infty$, *be a solution of the nonhomogeneous A-harmonic equation* (5.3.1) *on the bounded, convex domain* Ω *in* $\mathbf{R}^n$ *and* $T : C^\infty(\Omega, \wedge^l) \to C^\infty(\Omega, \wedge^{l-1})$ *be the homotopy operator defined in* (1.5.1). *Assume that* $\rho > 1$ *and* $(w_1, w_2) \in A_r(\lambda, \Omega)$ *for some* $\lambda > 0$ *and* $1 < r < \infty$. *Then, there exists a constant* C, *independent of* u, *such that*

$$\left(\int_\Omega |\nabla(Tu)|^s w_1^{\alpha\lambda} dx \right)^{1/s} \le C|\Omega| \left(\int_\Omega |u|^s w_2^\alpha dx \right)^{1/s} \tag{5.7.48}$$

for any real number α *with* $0 < \alpha < 1$.

Theorem 5.7.10. *Let* $u \in L^s_{loc}(\Omega, \wedge^l)$, $l = 1, 2, \ldots, n$, $1 < s < \infty$, *be a solution of the nonhomogeneous A-harmonic equation* (5.3.1) *on the bounded, convex domain* Ω *in* $\mathbf{R}^n$ *and* $T : C^\infty(\Omega, \wedge^l) \to C^\infty(\Omega, \wedge^{l-1})$ *be the homotopy operator defined in* (1.5.1). *Assume that* $\rho > 1$ *and* $(w_1, w_2) \in A_r(\lambda, \Omega)$ *for some* $\lambda > 0$ *and* $1 < r < \infty$. *Then, there exists a constant* C, *independent of* u, *such that*

$$\|Tu\|_{W^{1,s}(\Omega), \, w_1^{\alpha\lambda}} \le C|\Omega|\|u\|_{s, \Omega, w_2^\alpha} \tag{5.7.49}$$

for any real number α *with* $0 < \alpha < 1$.

Using Theorem 3.3.8 and imitating the proof of Theorem 3.2.10, we obtain the following global Sobolev imbedding inequality for Green's operator that is applied to the solutions of the nonhomogeneous A-harmonic equation in the John domain, see [103].

Theorem 5.7.11. *Let $u \in D'(\Omega, \wedge^0)$ be a solution of the nonhomogeneous A-harmonic equation (1.2.10), G be Green's operator, and $w \in A_r(\Omega)$ for some $1 < r < \infty$. Assume that s is a fixed exponent associated with the A-harmonic equation (1.2.10), $r < s < \infty$. Then, there exists a constant C, independent of u, such that*

$$\|G(u) - (G(u))_{Q_0}\|_{W^{1,s}(\Omega),w} \leq C\|du\|_{s,\Omega,w} \qquad (5.7.50)$$

for any δ-John domain $\Omega \subset \mathbf{R}^n$. Here $Q_0 \subset \Omega$ is a fixed cube.

Chapter 6
Reverse Hölder inequalities

In this chapter, we will present various versions of the reverse Hölder inequality which serve as powerful tools in mathematical analysis. The original study of the reverse Hölder inequality can be traced back in Muckenhoupt's work in [145]. During recent years, different versions of the reverse Hölder inequality have been established for different classes of functions, such as eigenfunctions of linear second-order elliptic operators [281], functions with discrete-time variable [282], and continuous exponential martingales [119].

6.1 Preliminaries

We begin this chapter with some preliminaries which form the bases for the study of the reverse Hölder inequalities. These results will be extended to the solutions of the A-harmonic equation in this chapter.

6.1.1 Gehring's lemma

We first state Gehring's lemma, which plays a crucial role in the study of the reverse Hölder inequalities for $A_r(\Omega)$-weights. This lemma has also been used to improve the degree of integrability of solutions of certain partial differential equations.

Theorem 6.1.1. *Let $1 < p < \infty$, and f be a nonnegative measurable function on a domain $\Omega \subset \mathbf{R}^n$ satisfying*

$$\frac{1}{|B|} \int_B f^p dx \leq C_p \left(\frac{1}{|B|} \int_B f dx \right)^p \qquad (6.1.1)$$

R.P. Agarwal et al., *Inequalities for Differential Forms*,
DOI 10.1007/978-0-387-68417-8_6, © Springer Science+Business Media, LLC 2009

for all balls B with $\sigma B \subset \Omega$, where C_p is a constant independent of the balls B and $\sigma > 1$. Then, there exist another exponent $s > p$ and a constant C_s depending only on n, p, and C_p such that

$$\frac{1}{|B|}\int_B f^s dx \leq C_s \left(\frac{1}{|B|}\int_B f dx\right)^s \tag{6.1.2}$$

for every ball $B \subset \Omega$.

Remark. Let s, t, σ, and C_1 be some positive numbers with $0 < t < s < \infty$ and $\sigma > 1$. If

$$\left(\frac{1}{|B|}\int_B f^s dx\right)^{1/s} \leq C_1 \left(\frac{1}{|B|}\int_{\sigma B} f^t dx\right)^{1/t}$$

holds for any ball B with $\sigma B \subset \Omega$, then there exists $\varepsilon > 0$ such that

$$\left(\frac{1}{|B|}\int_B f^{s+\varepsilon} dx\right)^{1/(s+\varepsilon)} \leq C_1 \left(\frac{1}{|B|}\int_{\sigma B} f^r dx\right)^{1/r}$$

for any number $r > 0$ and any ball B with $\sigma B \subset \Omega$.

In 1985, C. Nolder proved the following theorem which can be used to establish the global integrability of measurable functions in a John domain Ω.

Theorem 6.1.2. *Suppose that f and g are measurable functions in a John domain Ω, $q > 0$, $\sigma > 1$, and $w \in A_r(\Omega)$. If there exists a constant A such that*

$$\int_B |f - f_{B,\mu}|^q d\mu \leq A \int_{\sigma B} |g|^q d\mu \tag{6.1.3}$$

for all balls B with $\sigma B \subset \Omega$, then there is a constant B, independent of f and g, such that

$$\int_\Omega |f - f_{B_0,\mu}|^q d\mu \leq B \int_\Omega |g|^q d\mu \tag{6.1.4}$$

for some ball $B_0 \subset \Omega$.

6.1.2 Inequalities for supersolutions

In this chapter, we shall focus our attention to different versions of the weak Hölder inequality for the solutions of the A-harmonic equation. For this, first we shall state the weak Hölder inequality for the positive supersolutions. Recall that a function u in the weighted Sobolev space $W^{1,p}_{loc}(\Omega, \mu)$ is called a (weak) solution of the A-harmonic equation

$$-\operatorname{div} A(x, \nabla u) = 0 \qquad (6.1.5)$$

in Ω if

$$\int_\Omega A(x, \nabla u) \cdot \nabla \varphi(x) dx = 0 \qquad (6.1.6)$$

whenever $\varphi \in C_0^\infty(\Omega)$. A function u in the weighted Sobolev space $W_{loc}^{1,p}(\Omega, \mu)$ is called a supersolution of the A-harmonic equation (6.1.5) in Ω if

$$-\operatorname{div} A(x, \nabla u) \geq 0$$

weakly in Ω, i.e.,

$$\int_\Omega A(x, \nabla u) \cdot \nabla \varphi(x) dx \geq 0 \qquad (6.1.7)$$

whenever $\varphi \in C_0^\infty(\Omega)$ is nonnegative. A function v is a subsolution of (6.1.5) if $-v$ is a supersolution of (6.1.5). We shall need the following result which has been established in [59].

Theorem 6.1.3. *Suppose that u is a positive supersolution in $2B$, where B is a ball. If*

$$0 < q \leq s < \kappa(p-1),$$

then

$$\left(\frac{1}{\mu(B)} \int_B u^s d\mu \right)^{\frac{1}{s}} \leq C \left(\frac{1}{\mu(2B)} \int_{2B} u^q d\mu \right)^{\frac{1}{q}},$$

where C is a constant independent of u and

$$\kappa = \begin{cases} \dfrac{n}{n-p} & \text{if } \quad p < n \\[2mm] 2 & \text{if } \quad p \geq n. \end{cases}$$

Now we state the following reverse Hölder inequality for $A_r(\Omega)$-weights.

Theorem 6.1.4. *If $w \in A_r(\Omega)$, then there exist constants $\beta > 1$ and C, such that*

$$\left(\frac{1}{|B|} \int_B w^\beta dx \right)^{\frac{1}{\beta}} \leq C \left(\frac{1}{|B|} \int_B w dx \right) \qquad (6.1.8)$$

for all balls $B \subset \Omega$.

6.2 The first weighted case

In [71], C. Nolder proved the weak reverse Hölder inequality which has played an important role in the establishment of the integral estimates for

A-harmonic tensors. Nolder's result has been introduced earlier in this monograph; however, for the convenience of readers, we restate the weak reverse Hölder inequality as follows.

Theorem 6.2.1. *Let u be an A-harmonic tensor in Ω, $\sigma > 1$, and $0 < s$, $t < \infty$. Then, there exists a constant C, independent of u, such that*

$$\|u\|_{s,B} \leq C|B|^{(t-s)/st}\|u\|_{t,\sigma B}$$

for all balls or cubes B with $\sigma B \subset \Omega$.

6.2.1 $A_r(\Omega)$-weighted inequalities

In 1999, S. Ding [267] extended the above weak reverse Hölder inequality to the following $A_r(\Omega)$-weighted version.

Theorem 6.2.2. *Let $u \in D'(\Omega, \wedge^l)$, $l = 0, 1, \ldots, n$, be an A-harmonic tensor in a domain $\Omega \subset \mathbf{R}^n$, $\sigma > 1$. Assume that $0 < s, t < \infty$, and $w \in A_r(\Omega)$ for some $r > 1$. Then, there exists a constant C, independent of u, such that*

$$\left(\int_B |u|^s w dx\right)^{1/s} \leq C|B|^{(t-s)/st} \left(\int_{\sigma B} |u|^t w^{t/s} dx\right)^{1/t} \tag{6.2.1}$$

for all balls B with $\sigma B \subset \Omega$.

Proof. By Theorem 6.1.4, there exist constants $\beta > 1$ and $C_1 > 0$, such that

$$\| w \|_{\beta,B} \leq C_1 |B|^{(1-\beta)/\beta} \| w \|_{1,B} \tag{6.2.2}$$

for any cube or ball $B \subset \mathbf{R}^n$. Select $k = s\beta/(\beta - 1)$, so that $s < k$ and $\beta = k/(k - s)$. From (6.2.2) and the Hölder inequality, we obtain

$$\left(\int_B |u|^s w dx\right)^{1/s}$$

$$= \left(\int_B (|u|w^{1/s})^s dx\right)^{1/s}$$

$$\leq \left(\int_B |u|^k dx\right)^{1/k} \left(\int_B \left(w^{1/s}\right)^{ks/(k-s)} dx\right)^{(k-s)/ks}$$

$$\leq \|u\|_{k,B} \cdot \|w\|_{\beta,B}^{1/s} \tag{6.2.3}$$

$$\leq C_2 |B|^{(1-\beta)/\beta s}\|w\|_{1,B}^{1/s}\|u\|_{k,B}$$

$$= C_2 |B|^{-1/k}\|w\|_{1,B}^{1/s}\|u\|_{k,B}$$

for all balls B with $B \subset \Omega$. Choosing $m = st/(s + t(r - 1))$, then from Theorem 6.2.1, we have

$$\|u\|_{k,B} \leq C_3 |B|^{(m-k)/km} \|u\|_{m,\sigma B}. \tag{6.2.4}$$

Combination of (6.2.3) and (6.2.4) gives

$$\|u\|_{s,B,w} \leq C_4 |B|^{-1/m} \|w\|_{1,B}^{1/s} \|u\|_{m,\sigma B}. \tag{6.2.5}$$

Now since $m < t$, using the Hölder inequality, it follows that

$$\|u\|_{m,\sigma B}$$

$$= \left(\int_{\sigma B} \left(|u| w^{1/s} w^{-1/s} \right)^m dx \right)^{1/m}$$

$$\leq \left(\int_{\sigma B} |u|^t w^{t/s} dx \right)^{1/t} \left(\int_{\sigma B} \left(\frac{1}{w} \right)^{mt/(s(t-m))} dx \right)^{(t-m)/mt} \tag{6.2.6}$$

$$\leq \|1/w\|_{mt/(s(t-m)),\sigma B}^{1/s} \left(\int_{\sigma B} |u|^t w^{t/s} dx \right)^{1/t}.$$

From the choice of m, we have $r - 1 = s(t - m)/mt$. Since $w \in A_r$, we find that

$$\|w\|_{1,B}^{1/s} \cdot \|1/w\|_{mt/(s(t-m)),\sigma B}^{1/s}$$

$$= \left(\left(\int_B w dx \right) \left(\int_{\sigma B} \left(\frac{1}{w} \right)^{mt/(s(t-m))} dx \right)^{s(t-m)/mt} \right)^{1/s}$$

$$\leq \left(|\sigma B|^{1+s(t-m)/mt} \left(\frac{1}{|\sigma B|} \int_B w dx \right) \left(\frac{1}{|\sigma B|} \int_{\sigma B} \left(\frac{1}{w} \right)^{1/(r-1)} dx \right)^{r-1} \right)^{1/s}$$

$$\leq C_5 |B|^{1/s + 1/m - 1/t}. \tag{6.2.7}$$

Now the combination of (6.2.5), (6.2.6), and (6.2.7) yields

$$\|u\|_{s,B,w}$$

$$\leq C_4 |B|^{-1/m} \|w\|_{1,B}^{1/s} \|1/w\|_{mt/(s(t-m)),\sigma B}^{1/s} \left(\int_{\sigma B} |u|^t w^{t/s} dx \right)^{1/t} \tag{6.2.8}$$

$$\leq C_6 |B|^{1/s - 1/t} \left(\int_{\sigma B} |u|^t w^{t/s} dx \right)^{1/t}.$$

It is easy to see that (6.2.8) is equivalent to inequality (6.2.1). ∎

In 1999, B. Liu and S. Ding established the following local $A_r(\Omega)$-weighted weak reverse Hölder inequality for A-harmonic tensors on any subdomain of $\Omega \subset \mathbf{R}^n$.

Theorem 6.2.3. *Let u be an A-harmonic tensor in a domain $\Omega \subset \mathbf{R}^n$. Assume that $0 < s$, $t < \infty$, $\sigma > 1$, and $w \in A_r$ for some $r > 1$, then there exists a constant C, independent of u, such that*

$$\left(\frac{1}{\mu(B)} \int_B |u|^s \, wdx\right)^{1/s} \leq C \left(\frac{1}{\mu(\sigma B)} \int_{\sigma B} |u|^t \, wdx\right)^{1/t} \tag{6.2.9}$$

for all balls B with $\sigma B \subset \Omega$.

Proof. Since $w \in A_r(\Omega)$ for some $r > 1$, from Theorem 6.1.4, there exists a constant $\beta > 1$, such that (6.1.8) holds. For any $s > 0$, we can write

$$\frac{1}{s} = \frac{\beta - 1}{\beta s} + \frac{1}{\beta s}.$$

Thus, by the Hölder inequality, there exists a constant C_1 such that

$$\begin{aligned}
&\left(\int_B |u|^s \, wdx\right)^{\frac{1}{s}} \\
&= \left(\int_B (|u| \, w^{\frac{1}{s}})^s dx\right)^{\frac{1}{s}} \\
&\leq \left(\int_B |u|^{\frac{\beta s}{\beta-1}} dx\right)^{\frac{\beta-1}{\beta s}} \left(\int_B w^\beta dx\right)^{\frac{1}{\beta s}} \\
&\leq C_1 |B|^{(1-\beta)/\beta s} \|w\|_{1,B}^{1/s} \|u\|_{\beta s/(\beta-1),B}.
\end{aligned} \tag{6.2.10}$$

Now, we choose $k = s(r-1)/t$ and $m = st/(s+tk)$, so that $m < t$. Thus, from Theorem 6.2.1, we have

$$\|u\|_{\frac{\beta s}{\beta-1},B} \leq C_2 |B|^{(m(\beta-1)-\beta s)/\beta s m} \|u\|_{m,\sigma B}$$

which yields

$$\begin{aligned}
&\left(\int_B |u|^s \, wdx\right)^{\frac{1}{s}} \\
&\leq C_3 |B|^{\frac{1-\beta}{\beta s} + \frac{m(\beta-1)-\beta s}{\beta s m}} \|w\|_{1,B}^{\frac{1}{s}} \|u\|_{m,\sigma B} \\
&= C_3 |B|^{-1/m} (\mu(B))^{1/s} \|u\|_{m,\sigma B}.
\end{aligned} \tag{6.2.11}$$

Therefore,

$$\left(\frac{1}{\mu(B)} \int_B |u|^s \, wdx\right)^{\frac{1}{s}} \leq C_3 |B|^{-1/m} \|u\|_{m,\sigma B}. \tag{6.2.12}$$

On the other hand, noting that $(t-m)/m = r-1$ and using the Hölder inequality again, we obtain

$$\|u\|_{m,\sigma B}$$

$$= \left(\int_{\sigma B} (|u|\, w^{\frac{1}{t}} w^{-\frac{1}{t}})^m dx \right)^{1/m}$$

$$\leq \left(\int_{\sigma B} (|u|\, w^{1/t})^t dx \right)^{1/t} \left(\int_{\sigma B} (w^{-1/t})^{\frac{mt}{t-m}} dx \right)^{(t-m)/mt} \qquad (6.2.13)$$

$$= \left(\int_{\sigma B} |u|^t\, wdx \right)^{\frac{1}{t}} \|w^{-1}\|_{1/(r-1),\sigma B}^{1/t}.$$

Now using the property that $w \in A_r(\Omega)$, it follows that

$$\|w\|_{1,\sigma B}^{\frac{1}{t}} \|1/w\|_{\frac{1}{r-1},\sigma B}^{\frac{1}{t}}$$

$$= \left[\left(\int_{\sigma B} wdx \right) \left(\int_{\sigma B} (\tfrac{1}{w})^{\frac{1}{r-1}} dx \right)^{r-1} \right]^{\frac{1}{t}}$$

$$\leq \left[|\sigma B|^r \left(\tfrac{1}{|\sigma B|} \int_{\sigma B} wdx \right) \left(\tfrac{1}{|\sigma B|} \int_{\sigma B} (\tfrac{1}{w})^{\frac{1}{r-1}} dx \right)^{r-1} \right]^{\frac{1}{t}} \qquad (6.2.14)$$

$$\leq C_4 |\sigma B|^{r/t},$$

which gives

$$\|1/w\|_{\frac{1}{r-1},\sigma B}^{\frac{1}{t}} \leq C_4 |\sigma B|^{\frac{r}{t}} \|w\|_{1,\sigma B}^{-\frac{1}{t}} \leq C_5 |B|^{\frac{r}{t}} \|w\|_{1,\sigma B}^{-\frac{1}{t}}. \qquad (6.2.15)$$

Combining (6.2.12), (6.2.13), (6.2.14), and (6.2.15), and noticing that $r/t - 1/m = 0$, we finally obtain

$$\left(\frac{1}{\mu(B)} \int_B |u|^s\, wdx \right)^{1/s} \leq C_6 \left(\frac{1}{\mu(\sigma B)} \int_{\sigma B} |u|^t\, wdx \right)^{1/t}. \qquad \blacksquare$$

6.2.2 Inequalities in $L^s(\mu)$-averaging domains

As an application of the local result, Theorem 6.2.3, we prove the following global reverse Hölder inequality in $L^s(\mu)$-averaging domains. However, we remark that this result can also be proved by the generalized Hölder inequality.

Theorem 6.2.4. *Let u be an A-harmonic tensor in an $L^s(\mu)$-averaging domain $\Omega \subset \mathbf{R}^n$ and $1 < s \leq t < \infty$. Assume that $w \in A_r$ for some $r > 1$ and the measure μ is defined by $d\mu = w(x)dx$. Then, there exists a constant C, independent of u, such that*

$$\left(\frac{1}{\mu(\Omega)}\int_\Omega |u|^s\, w dx\right)^{1/s} \leq C\left(\frac{1}{\mu(\Omega)}\int_\Omega |u|^t\, w dx\right)^{1/t}. \qquad (6.2.16)$$

Proof. We may assume that

$$\sup_{2B\subset\Omega}\left(\frac{1}{\mu(B)}\int_B |u|^t\, w dx\right)^{1/t} < \infty.$$

Choose a ball $B_0 \subset \Omega$ and a constant C_1 large enough, such that

$$\sup_{2B\subset\Omega}\left(\tfrac{1}{\mu(B)}\int_B |u|^t\, w dx\right)^{1/t}$$
$$\leq \sup_{2B\subset\Omega}\left(\tfrac{1}{\mu(B)}\int_{\sigma B} |u|^t\, w dx\right)^{1/t} \qquad (6.2.17)$$
$$\leq C_1\left(\tfrac{1}{\mu(B_0)}\int_{\sigma B_0} |u|^t\, w dx\right)^{1/t}.$$

Moreover, since

$$\mu(B) = \int_B w(x)dx \leq \int_{\sigma B} w(x)dx = \mu(\sigma B),$$

from Theorem 6.2.3, we obtain

$$\left(\tfrac{1}{\mu(\Omega)}\int_\Omega |u|^s\, w dx\right)^{\frac{1}{s}}$$
$$= \left(\tfrac{1}{\mu(\Omega)}\right)^{\frac{1}{s}}\left[\mu(B_0)\tfrac{1}{\mu(B_0)}\int_\Omega |u|^s\, w dx\right]^{\frac{1}{s}}$$
$$\leq \left(\tfrac{1}{\mu(\Omega)}\right)^{\frac{1}{s}}(\mu(B_0))^{\frac{1}{s}}\sup_{2B\subset\Omega}\left[\tfrac{1}{\mu(B)}\int_B |u|^s\, w dx\right]^{\frac{1}{s}}$$
$$\leq \left(\tfrac{1}{\mu(\Omega)}\right)^{\frac{1}{s}}(\mu(B_0))^{\frac{1}{s}}\sup_{2B\subset\Omega}\left[\tfrac{1}{\mu(\sigma B)}\int_{\sigma B} |u|^t\, w dx\right]^{\frac{1}{t}}$$
$$\leq \left(\tfrac{1}{\mu(\Omega)}\right)^{\frac{1}{s}}(\mu(B_0))^{\frac{1}{s}}\sup_{2B\subset\Omega}\left[\tfrac{C_3}{\mu(B)}\int_{\sigma B} |u|^t\, w dx\right]^{\frac{1}{t}}$$
$$\leq C_4\left(\tfrac{1}{\mu(\Omega)}\right)^{\frac{1}{s}}(\mu(B_0))^{\frac{1}{s}}\left[\tfrac{1}{\mu(B_0)}\int_{\sigma B_0} |u|^t\, w dx\right]^{\frac{1}{t}}$$
$$\leq C_5\left(\tfrac{1}{\mu(\Omega)}\right)^{\frac{1}{s}}(\mu(B_0))^{\frac{1}{s}-\frac{1}{t}}\left[\int_{\sigma B_0} |u|^t\, w dx\right]^{\frac{1}{t}}$$
$$\leq C_6\left(\tfrac{1}{\mu(\Omega)}\right)^{\frac{1}{s}}(\mu(\Omega))^{\frac{1}{s}-\frac{1}{t}}\left[\int_\Omega |u|^t\, w dx\right]^{\frac{1}{t}}$$
$$= C_6\left[\tfrac{1}{\mu(\Omega)}\int_\Omega |u|^t\, w dx\right]^{\frac{1}{t}}$$

which implies that inequality (6.2.16) holds. ∎

6.2.3 Inequalities in John domains

We know that a δ-John domain is an $L^s(\mu)$-averaging domain when w satisfies the $A_r(\Omega)$-condition. Thus, Theorem 6.2.4 holds if $\Omega \subset \mathbf{R}^n$ is a δ-John domain. Therefore, the following result is a special case of Theorem 6.2.4. However, as a further application of Theorem 6.2.3, we provide its proof. Once again we notice that it can also be proved by the generalized Hölder inequality.

Theorem 6.2.5. *Let $\Omega \subset \mathbf{R}^n$ be a δ-John domain, u be any A-harmonic tensor, $0 < s < t < \infty$, and $w \in A_r$ for some $r > 1$. Then, there exists a constant C, independent of u, such that*

$$\left(\frac{1}{\mu(\Omega)} \int_\Omega |u|^s \, wdx \right)^{1/s} \leq C \left(\frac{1}{\mu(\Omega)} \int_\Omega |u|^t \, wdx \right)^{1/t}. \tag{6.2.18}$$

Proof. By Theorem 6.2.3 and the covering lemma, there exists a covering $\mathcal{V}$ of Ω such that

$$\|u\|_{s,\Omega,w}^s = \int_\Omega |u|^s \, wdx$$

$$\leq \sum_{Q\in\mathcal{V}} \int_Q |u|^s \, wdx$$

$$\leq C_1 \sum_{Q\in\mathcal{V}} \mu(Q)\mu(\sigma Q)^{-\frac{s}{t}} \left(\int_{\sigma Q} |u|^t \, wdx \right)^{\frac{s}{t}}$$

$$\leq C_1 \sum_{Q\in\mathcal{V}} \mu(\sigma Q)^{1-\frac{s}{t}} \left(\int_{\sigma Q} |u|^t \, wdx \right)^{\frac{s}{t}}$$

$$\leq C_1 \mu(\Omega)^{(t-s)/t} \sum_{Q\in\mathcal{V}} \left(\int_{\sigma Q} |u|^t \, wdx \right)^{\frac{s}{t}}$$

$$\leq C_2 \mu(\Omega)^{(t-s)/t} \left(\int_\Omega |u|^t \, wdx \right)^{\frac{s}{t}}.$$

Thus,

$$\left(\int_\Omega |u|^s \, wdx \right)^{1/s} \leq C_3 \mu(\Omega)^{(t-s)/st} \left(\int_\Omega |u|^t \, wdx \right)^{1/t}$$

which is equivalent to (6.2.18). ∎

Remark. (1) Our local result, Theorem 6.2.3, holds in any type of domains. (2) We only applied Theorem 6.2.3 to $L^s(\mu)$-averaging domains and δ-John domains, and this has resulted in Theorems 6.2.4 and 6.2.5, respectively. Clearly, Theorem 6.2.3 can also be applied to other kind of domains, such as the domains with Whitney covers.

6.2.4 Parametric inequalities

In 2001, S. Ding proved the following $A_r(\Omega)$-weighted weak reverse Hölder inequality for A-harmonic tensors with a parameter α.

Theorem 6.2.6. *Let $u \in D'(\Omega, \wedge^l)$, $l = 0, 1, \ldots, n$, be an A-harmonic tensor in a domain $\Omega \subset \mathbf{R}^n$, $\sigma > 1$. Assume that $0 < s, t < \infty$, and $w \in A_r(\Omega)$ for some $r > 1$. Then, there exists a constant C, independent of u, such that*

$$\left(\frac{1}{|B|} \int_B |u|^s w^\alpha dx \right)^{1/s} \leq C \left(\frac{1}{|B|} \int_{\sigma B} |u|^t w^{\alpha t/s} dx \right)^{1/t} \qquad (6.2.19)$$

for all balls B with $\sigma B \subset \Omega$ and any real number α with $0 < \alpha \leq 1$.

Proof. First, we suppose that $0 < \alpha < 1$. Let $k = s/(1-\alpha)$. From Hölder's inequality, we find that

$$\left(\int_B |u|^s w^\alpha dx \right)^{1/s}$$

$$= \left(\int_B \left(|u| w^{\alpha/s} \right)^s dx \right)^{1/s}$$

$$\leq \left(\int_B |u|^k dx \right)^{1/k} \left(\int_B \left(w^{\alpha/s} \right)^{ks/(k-s)} dx \right)^{(k-s)/ks} \qquad (6.2.20)$$

$$= \|u\|_{k,B} \left(\int_B w dx \right)^{\alpha/s}$$

for all balls B with $B \subset \Omega$. Let $m = st/(s + \alpha t(r-1))$. By Theorem 6.2.1, we obtain

$$\|u\|_{k,B} \leq C_1 |B|^{(m-k)/km} \|u\|_{m,\sigma B}. \qquad (6.2.21)$$

Now the Hölder inequality with $1/m = 1/t + (t-m)/mt$ yields

$$\|u\|_{m,\sigma B}$$

$$= \left(\int_{\sigma B} \left(|u| w^{\alpha/s} w^{-\alpha/s} \right)^m dx \right)^{1/m}$$

$$\leq \left(\int_{\sigma B} |u|^t w^{\alpha t/s} dx \right)^{1/t} \left(\int_{\sigma B} (1/w)^{\alpha m t/s(t-m)} dx \right)^{(t-m)/mt} \qquad (6.2.22)$$

$$= \left(\int_{\sigma B} |u|^t w^{\alpha t/s} dx \right)^{1/t} \left(\int_{\sigma B} (1/w)^{1/(r-1)} dx \right)^{\alpha(r-1)/s}.$$

Combining (6.2.20), (6.2.21), and (6.2.22), we find that

$$\left(\int_B |u|^s w^\alpha dx\right)^{1/s}$$

$$\leq C_1 |B|^{(m-k)/km} \left(\int_{\sigma B} |u|^t w^{\alpha t/s} dx\right)^{1/t} \qquad (6.2.23)$$

$$\times \left(\int_B w dx\right)^{\alpha/s} \left(\int_{\sigma B} (1/w)^{1/(r-1)} dx\right)^{\alpha(r-1)/s}.$$

Next, since $w \in A_r(\Omega)$, it follows that

$$\left(\int_B w dx\right)^{\alpha/s} \left(\int_{\sigma B} (1/w)^{1/(r-1)} dx\right)^{\alpha(r-1)/s}$$

$$= \left(\left(\int_B w dx\right) \left(\int_{\sigma B} (1/w)^{1/(r-1)} dx\right)^{r-1}\right)^{\alpha/s}$$

$$\leq |\sigma B|^{\alpha r/s} \left(\left(\frac{1}{|\sigma B|}\int_B w dx\right) \left(\frac{1}{|\sigma B|}\int_{\sigma B} (1/w)^{1/(r-1)} dx\right)^{r-1}\right)^{\alpha/s}$$

$$\leq C_2 |\sigma B|^{\alpha r/s}$$

$$= C_3 |B|^{\alpha r/s}.$$

$$(6.2.24)$$

Substituting (6.2.24) into (6.2.23), we obtain

$$\left(\int_B |u|^s w^\alpha dx\right)^{1/s} \leq C_4 |B|^{(t-s)/st} \left(\int_{\sigma B} |u|^t w^{\alpha t/s} dx\right)^{1/t}.$$

Therefore, inequality (6.2.19) holds for $0 < \alpha < 1$.

For the case $\alpha = 1$, by Theorem 6.1.4, there exist constants $\beta > 1$ and $C_5 > 0$, such that

$$\| w \|_{\beta,B} \leq C_5 |B|^{(1-\beta)/\beta} \| w \|_{1,B} \qquad (6.2.25)$$

for any cube or ball $B \subset \mathbf{R}^n$. Choose $k = s\beta/(\beta-1)$, so that $s < k$ and $\beta = k/(k-s)$. By (6.2.25) and Hölder's inequality, we find that

$$\left(\int_B |u|^s w dx\right)^{1/s}$$

$$\leq \left(\int_B |u|^k dx\right)^{1/k} \left(\int_B (w^{1/s})^{sk/(k-s)} dx\right)^{(k-s)/sk}$$

$$= \|u\|_{k,B} \cdot \|w\|_{\beta,B}^{1/s} \qquad (6.2.26)$$

$$\leq C_6 |B|^{(1-\beta)/\beta s} \|w\|_{1,B}^{1/s} \cdot \|u\|_{k,B}$$

$$= C_6 |B|^{-1/k} \|w\|_{1,B}^{1/s} \cdot \|u\|_{k,B}.$$

Choosing $m = st/(s + t(r - 1))$ and repeating the same procedure as for the case $0 < \alpha < 1$, we find that (6.2.19) is also true for $\alpha = 1$. ∎

As a particular case, we choose $\alpha = 1$ in Theorem 6.2.6 to obtain the following version of the reverse Hölder inequality.

Corollary 6.2.7. *Let $u \in D'(\Omega, \wedge^l)$, $l = 0, 1, \ldots, n$, be an A-harmonic tensor in a domain $\Omega \subset \mathbf{R}^n$, $\sigma > 1$. Assume that $0 < s, t < \infty$, and $w \in A_r(\Omega)$ for some $r > 1$. Then, there exists a constant C, independent of u, such that*

$$\left(\frac{1}{|B|} \int_B |u|^s w \, dx \right)^{1/s} \leq C \left(\frac{1}{|B|} \int_{\sigma B} |u|^t w^{t/s} dx \right)^{1/t}$$

for all balls B with $\sigma B \subset \Omega$.

The choice $\alpha = s$, $0 < s \leq 1$, in Theorem 6.2.6 leads to the following symmetric version of the reverse Hölder inequality.

Corollary 6.2.8. *Let $u \in D'(\Omega, \wedge^l)$, $l = 0, 1, \ldots, n$, be an A-harmonic tensor in a domain $\Omega \subset \mathbf{R}^n$, $\sigma > 1$. Assume that $0 < t < \infty$, $0 < s \leq 1$, and $w \in A_r(\Omega)$ for some $r > 1$. Then, there exists a constant C, independent of u, such that*

$$\left(\frac{1}{|B|} \int_B |u|^s w^s \, dx \right)^{1/s} \leq C \left(\frac{1}{|B|} \int_{\sigma B} |u|^t w^t dx \right)^{1/t}$$

for all balls B with $\sigma B \subset \Omega$.

Finally, Theorem 6.2.6 for $\alpha = 1/t$, $t \geq 1$ reduces to the following corollary.

Corollary 6.2.9. *Let $u \in D'(\Omega, \wedge^l)$, $l = 0, 1, \ldots, n$, be an A-harmonic tensor in a domain $\Omega \subset \mathbf{R}^n$, $\sigma > 1$. Assume that $t \geq 1$, $0 < s < \infty$, and $w \in A_r(\Omega)$ for some $r > 1$. Then, there exists a constant C, independent of u, such that*

$$\left(\frac{1}{|B|} \int_B |u|^s w^{1/t} dx \right)^{1/s} \leq C \left(\frac{1}{|B|} \int_{\sigma B} |u|^t w^{1/s} dx \right)^{1/t}$$

for all balls B with $\sigma B \subset \Omega$.

Now we shall prove the following global reverse Hölder inequality with $A_r(\Omega)$-weights for A-harmonic tensors.

Theorem 6.2.10. *Let $u \in D'(\Omega, \wedge^l)$, $l = 0, 1, \ldots, n$, be an A-harmonic tensor in a domain $\Omega \subset \mathbf{R}^n$ with $|\Omega| < \infty$. Assume that $0 < s \leq t < \infty$ and $w \in A_r(\Omega)$ for some $r > 1$. Then,*

$$\left(\frac{1}{|\Omega|}\int_\Omega |u|^s w^\alpha dx\right)^{1/s} \le \left(\frac{1}{|\Omega|}\int_\Omega |u|^t w^{\alpha t/s} dx\right)^{1/t} \tag{6.2.27}$$

for any real number α with $0 < \alpha \le 1$.

Proof. It is clear that (6.2.27) holds if $s = t$. Now we assume that $s < t$. Using the Hölder inequality with $1/s = 1/t + (t - s)/st$, we have

$$
\begin{aligned}
&\left(\int_\Omega |u|^s w^\alpha dx\right)^{1/s} \\
&= \left(\int_\Omega \left(|u|w^{\alpha/s}\right)^s dx\right)^{1/s} \\
&\le \left(\int_\Omega 1 dx\right)^{(t-s)/st} \left(\int_\Omega \left(|u|w^{\alpha/s}\right)^t dx\right)^{1/t} \\
&= |\Omega|^{(t-s)/st} \left(\int_\Omega |u|^t w^{\alpha t/s} dx\right)^{1/t},
\end{aligned}
\tag{6.2.28}
$$

which is equivalent to (6.2.27). ∎

Remark. As in Theorem 6.2.4, Theorem 6.2.10 can be proved as an application of Theorem 6.2.6. Although, Theorem 6.2.10 requires a stronger condition $0 < s \le t < \infty$, the resulting inequality (6.2.27) is sharper than (6.2.19). We also remark that in Theorem 6.2.10 particular values of α lead to global results which are analogous to those we have listed for the local case.

We have discussed the weak reverse Hölder inequality for differential forms u satisfying the A-harmonic equation, which is useful when we estimate the L^s-norm $\|u\|_{s,B}$. But, when we study the solutions of the A-harmonic equation, we often need to estimate the L^s-norm $\|du\|_{s,B}$. In Theorem 6.2.12, we discuss the weak reverse Hölder inequality for du, which was established in [201]. In order to prove this result, we need the following lemma which is due to C. Nolder and his co-worker.

Lemma 6.2.11. *Suppose that $|v| \in L^s_{loc}(\Omega)$, $\sigma > 1$, and $0 < t < s$. If there exists a constant A such that*

$$\|v\|_{s,Q} \le A|Q|^{(t-s)/st}\|v\|_{t,2Q} \tag{6.2.29}$$

for all cubes Q with $2Q \subset \Omega$, then for all $r > 0$, there exists a constant B, depending only on σ, n, s, t, r, and A, such that

$$\|v\|_{s,Q} \le B|Q|^{(r-s)/sr}\|v\|_{r,\sigma Q}$$

for all cubes Q with $\sigma Q \subset \Omega$.

6.2.5 Estimates for du

Theorem 6.2.12. *Suppose that u is a solution of (1.2.10), $\sigma > 1$, and $p, q > 0$. Then, there exists a constant C, depending only on σ, n, p, a, b, and q, such that*

$$\|du\|_{p,Q} \leq C|Q|^{(q-p)/pq}\|du\|_{q,\sigma Q} \tag{6.2.30}$$

for all Q with $\sigma Q \subset \Omega$.

Proof. By Theorems 4.2.1, 6.2.1, and 3.2.3 with $p' = (p+1)/2$, we have

$$
\begin{aligned}
\|du\|_{p,Q} &\leq C_1|Q|^{1/n}\|u - u_{\sigma Q}\|_{p,\sqrt{\sigma}Q} \\
&\leq C_2|Q|^{(p'-p)/pp'}\|u - u_{\sigma Q}\|_{p',\sigma Q} \\
&\leq C_3|Q|^{(p'-p)/pp'}\|du\|_{p',\sigma Q}.
\end{aligned}
$$

Thus, du satisfies the reverse Hölder inequality (6.2.29), and hence (6.2.30) follows from Lemma 6.2.11. ∎

At this point, we should notice that, using Theorem 6.2.12 and the similar methods developed in the proofs of the $A_r(\Omega)$-weighted inequalities, all versions of the $A_r(\Omega)$-weighted weak reverse Hölder inequality for differential form u can be extended to du. For example, Theorem 6.2.12 implies that the proof holds if we replace u by du in Theorem 6.2.6. Therefore, we have the following $A_r(\Omega)$-weighted weak reverse Hölder inequality for du when u is an A-harmonic tensor in a domain $\Omega \subset \mathbf{R}^n$.

Theorem 6.2.13. *Let $u \in D'(\Omega, \wedge^l)$, $l = 0, 1, \ldots, n$, be an A-harmonic tensor in a domain $\Omega \subset \mathbf{R}^n$, $\sigma > 1$. Assume that $0 < s, t < \infty$, and $w \in A_r(\Omega)$ for some $r > 1$. Then, there exists a constant C, independent of u, such that*

$$\left(\frac{1}{|B|}\int_B |du|^s w^\alpha dx\right)^{1/s} \leq C\left(\frac{1}{|B|}\int_{\sigma B} |du|^t w^{\alpha t/s} dx\right)^{1/t} \tag{6.2.31}$$

for all balls B with $\sigma B \subset \Omega$ and any real number α with $0 < \alpha \leq 1$.

Now, if we choose $\alpha = 1$, $\alpha = s$ for $0 < s \leq 1$ and $\alpha = 1/t$ for $t \geq 1$ in Theorem 6.2.13, respectively, we obtain the following special cases for du.

Corollary 6.2.14. *Let $u \in D'(\Omega, \wedge^l)$, $l = 0, 1, \ldots, n$, be an A-harmonic tensor in a domain $\Omega \subset \mathbf{R}^n$, $\sigma > 1$. Assume that $0 < s, t < \infty$, and $w \in A_r(\Omega)$ for some $r > 1$. Then, there exists a constant C, independent of u, such that*

$$\left(\frac{1}{|B|}\int_B |du|^s w dx\right)^{1/s} \leq C\left(\frac{1}{|B|}\int_{\sigma B} |du|^t w^{t/s} dx\right)^{1/t}$$

for all balls B with $\sigma B \subset \Omega$.

Corollary 6.2.15. *Let $u \in D'(\Omega, \wedge^l)$, $l = 0, 1, \ldots, n$, be an A-harmonic tensor in a domain $\Omega \subset \mathbf{R}^n$, $\sigma > 1$. Assume that $0 < t < \infty$, $0 < s \leq 1$, and $w \in A_r(\Omega)$ for some $r > 1$. Then, there exists a constant C, independent of u, such that*

$$\left(\frac{1}{|B|} \int_B |du|^s w^s dx \right)^{1/s} \leq C \left(\frac{1}{|B|} \int_{\sigma B} |du|^t w^t dx \right)^{1/t}$$

for all balls B with $\sigma B \subset \Omega$.

Corollary 6.2.16. *Let $u \in D'(\Omega, \wedge^l)$, $l = 0, 1, \ldots, n$, be an A-harmonic tensor in a domain $\Omega \subset \mathbf{R}^n$, $\sigma > 1$. Assume that $t \geq 1$, $0 < s < \infty$, and $w \in A_r(\Omega)$ for some $r > 1$. Then, there exists a constant C, independent of u, such that*

$$\left(\frac{1}{|B|} \int_B |du|^s w^{1/t} dx \right)^{1/s} \leq C \left(\frac{1}{|B|} \int_{\sigma B} |du|^t w^{1/s} dx \right)^{1/t}$$

for all balls B with $\sigma B \subset \Omega$.

6.3 The second weighted case

Here we shall prove the following $A_r^\lambda(\Omega)$-weighted reverse Hölder inequalities which were established by S. Ding in 2001.

6.3.1 Inequalities with $A_r^\alpha(\Omega)$-weights

Theorem 6.3.1. *Let $u \in D'(\Omega, \wedge^l)$, $l = 0, 1, \ldots, n$, be an A-harmonic tensor in a domain $\Omega \subset \mathbf{R}^n$, $\sigma > 1$. Assume that $0 < s, t < \infty$, and $w \in A_r^\alpha(\Omega)$ for some $r > 1$ and $\alpha > 1$. Then, there exists a constant C, independent of u, such that*

$$\left(\int_B |u|^s w^{1/\alpha} dx \right)^{1/s} \leq C|B|^{(t-s)/st} \left(\int_{\sigma B} |u|^t w^{t/s} dx \right)^{1/t} \qquad (6.3.1)$$

for all balls B with $\sigma B \subset \Omega$.

Note that (6.3.1) has the following symmetric version:

$$\left(\frac{1}{|B|} \int_B |u|^s w^{1/\alpha} dx \right)^{1/s} \leq C \left(\frac{1}{|B|} \int_{\sigma B} |u|^t w^{t/s} dx \right)^{1/t}. \qquad (6.3.1)'$$

Proof. Choose $k = \alpha s/(\alpha - 1)$. It is easy to see that $s < k$ and $1/s = 1/k + (k-s)/ks$. From the Hölder inequality, we obtain

$$\left(\int_B |u|^s w^{1/\alpha} dx\right)^{1/s}$$
$$= \left(\int_B (|u|w^{1/\alpha s})^s dx\right)^{1/s}$$
$$\leq \left(\int_B |u|^k dx\right)^{1/k} \left(\int_B \left(w^{1/\alpha s}\right)^{ks/(k-s)} dx\right)^{(k-s)/ks} \qquad (6.3.2)$$
$$\leq \|u\|_{k,B} \cdot \left(\int_B w dx\right)^{1/\alpha s}$$

for all balls B with $B \subset \Omega$. Choosing $m = st/(s + t(r-1))$, by Theorem 6.2.1, we have
$$\|u\|_{k,B} \leq C_1 |B|^{(m-k)/km} \|u\|_{m,\sigma B}. \qquad (6.3.3)$$

Since $1/m = 1/t + (t-m)/mt$, using the Hölder inequality again, we obtain

$$\|u\|_{m,\sigma B}$$
$$= \left(\int_{\sigma B} \left(|u|w^{1/s}w^{-1/s}\right)^m dx\right)^{1/m}$$
$$\leq \left(\int_{\sigma B} |u|^t w^{t/s} dx\right)^{1/t} \left(\int_{\sigma B} \left(\frac{1}{w}\right)^{mt/s(t-m)} dx\right)^{(t-m)/mt} \qquad (6.3.4)$$
$$\leq \left(\int_{\sigma B} (1/w)^{1/(r-1)} dx\right)^{(r-1)/s} \left(\int_{\sigma B} |u|^t w^{t/s} dx\right)^{1/t}.$$

Combining (6.3.2), (6.3.3), and (6.3.4), we arrive at the estimate
$$\left(\int_B |u|^s w^{1/\alpha} dx\right)^{1/s}$$
$$\leq C_1 |B|^{(m-k)/km} \left(\int_B w dx\right)^{1/\alpha s}$$
$$\times \left(\int_{\sigma B} (1/w)^{1/(r-1)} dx\right)^{(r-1)/s} \left(\int_{\sigma B} |u|^t w^{t/s} dx\right)^{1/t}. \qquad (6.3.5)$$

Now, since $w \in A_r^\alpha(\Omega)$, it follows that

$$\left(\int_B w dx\right)^{1/\alpha s} \left(\int_{\sigma B} (1/w)^{1/(r-1)} dx\right)^{(r-1)/s}$$
$$\leq \left(\left(\int_B w dx\right)\left(\int_{\sigma B}(1/w)^{1/(r-1)}dx\right)^{\alpha(r-1)}\right)^{1/\alpha s}$$
$$\leq \left(|\sigma B|^{\alpha(r-1)+1}\left(\frac{1}{|\sigma B|}\int_B w dx\right)\right.$$
$$\left.\times \left(\frac{1}{|\sigma B|}\int_{\sigma B}\left(\frac{1}{w}\right)^{1/(r-1)}dx\right)^{\alpha(r-1)}\right)^{1/\alpha s} \qquad (6.3.6)$$
$$\leq C_2 |\sigma B|^{(r-1)/s+1/\alpha s}$$
$$\leq C_3 |B|^{(r-1)/s+1/\alpha s}.$$

Finally, substituting (6.3.6) into (6.3.5) and noticing

$$\frac{m-k}{km} = \frac{1}{k} - \frac{1}{m} = \frac{1}{s} - \frac{1}{\alpha s} - \frac{1}{t} - \frac{r-1}{s},$$

we obtain

$$\left(\int_B |u|^s w^{1/\alpha} dx\right)^{1/s} \le C|B|^{(t-s)/st} \left(\int_{\sigma B} |u|^t w^{t/s} dx\right)^{1/t}. \qquad \blacksquare$$

The constant α in Theorem 6.3.1 is arbitrary, and hence different values of α lead to different versions of the weak reverse Hölder inequality. For example, when $\alpha = t/s$, so that $t > s$, we have the following symmetric form of the weighted weak reverse Hölder inequality.

Corollary 6.3.2. *Let $u \in D'(\Omega, \wedge^l)$, $l = 0, 1, \dots, n$, be an A-harmonic tensor in a domain $\Omega \subset \mathbf{R}^n$, $\sigma > 1$. Assume that $0 < s < t < \infty$ and $w \in A_r^{t/s}(\Omega)$ for some $r > 1$. Then, there exists a constant C, independent of u, such that*

$$\left(\int_B |u|^s w^{s/t} dx\right)^{1/s} \le C|B|^{(t-s)/st} \left(\int_{\sigma B} |u|^t w^{t/s} dx\right)^{1/t} \qquad (6.3.7)$$

for all balls B with $\sigma B \subset \Omega$.

Now we shall prove the following interesting result.

Theorem 6.3.3. *Let $u \in D'(\Omega, \wedge^l)$, $l = 0, 1, \dots, n$, be an A-harmonic tensor in a domain $\Omega \subset \mathbf{R}^n$, $\sigma > 1$. Assume that $0 < s, t < \infty$, and $w \in A_r^{s/t}(\Omega)$ for some $r > 1$ and $0 < \alpha < 1$. Then, there exists a constant C, independent of u, such that*

$$\left(\int_B |u|^s w^\alpha dx\right)^{1/s} \le C|B|^{(t-s)/st} \left(\int_{\sigma B} |u|^t w^\alpha dx\right)^{1/t} \qquad (6.3.8)$$

for all balls B with $\sigma B \subset \Omega$.

Proof. Choose $k = s/(1-\alpha)$. Using the Hölder inequality, we obtain

$$\left(\int_B |u|^s w^\alpha dx\right)^{1/s}$$
$$= \left(\int_B (|u| w^{\alpha/s})^s dx\right)^{1/s}$$
$$\le \left(\int_B |u|^k dx\right)^{1/k} \left(\int_B w^{\alpha k/(k-s)} dx\right)^{(k-s)/ks} \qquad (6.3.9)$$
$$\le \|u\|_{k,B} \cdot \left(\int_B w dx\right)^{\alpha/s}.$$

Now, let $m = t/(1 + \alpha(r-1))$. From Theorem 6.2.1, we have

$$\|u\|_{k,B} \leq C_1 |B|^{(m-k)/km} \|u\|_{m,\sigma B}. \tag{6.3.10}$$

Using the Hölder inequality again, we find that

$$
\begin{aligned}
&\|u\|_{m,\sigma B} \\
&= \left(\int_{\sigma B} \left(|u| w^{\alpha/t} w^{-\alpha/t} \right)^m dx \right)^{1/m} \\
&\leq \left(\int_{\sigma B} |u|^t w^\alpha dx \right)^{1/t} \left(\int_{\sigma B} (1/w)^{\alpha m/(t-m)} dx \right)^{(t-m)/mt} \qquad (6.3.11) \\
&\leq \left(\int_{\sigma B} |u|^t w^\alpha dx \right)^{1/t} \left(\int_{\sigma B} (1/w)^{1/(r-1)} dx \right)^{\alpha(r-1)/t}.
\end{aligned}
$$

On the other hand, $w \in A_r^{s/t}(\Omega)$ yields

$$
\begin{aligned}
&\left(\int_B w \, dx \right)^{\alpha/s} \left(\int_{\sigma B} (1/w)^{1/(r-1)} dx \right)^{\alpha(r-1)/s} \\
&\leq \left(\left(\int_B w \, dx \right) \left(\int_{\sigma B} (1/w)^{1/(r-1)} dx \right)^{s(r-1)/t} \right)^{\alpha/s} \\
&\leq \left(|\sigma B|^{s(r-1)/t+1} \left(\frac{1}{|\sigma B|} \int_B w \, dx \right) \right. \\
&\qquad \left. \times \left(\frac{1}{|\sigma B|} \int_{\sigma B} \left(\frac{1}{w} \right)^{1/(r-1)} dx \right)^{s(r-1)/t} \right)^{\alpha/s} \qquad (6.3.12) \\
&\leq C_2 |\sigma B|^{\alpha(r-1)/t+\alpha/s} \\
&\leq C_3 |B|^{\alpha(r-1)/t+\alpha/s}.
\end{aligned}
$$

Finally, combining (6.3.9), (6.3.10), (6.3.11), and (6.3.12), we obtain

$$
\begin{aligned}
&\left(\int_B |u|^s w^\alpha dx \right)^{1/s} \\
&\leq C_1 |B|^{(m-k)/km} \left(\int_B w \, dx \right)^{\alpha/s} \left(\int_{\sigma B} (1/w)^{1/(r-1)} dx \right)^{\alpha(r-1)/t} \\
&\qquad \times \left(\int_{\sigma B} |u|^t w^\alpha dx \right)^{1/t} \\
&\leq C_4 |B|^{(t-s)/st} \left(\int_{\sigma B} |u|^t w^\alpha dx \right)^{1/t}
\end{aligned}
$$

which is equivalent to inequality (6.3.8). ∎

6.3.2 Inequalities with two parameters

In 2003, Y. Xing proved the following version of the $A_r^\lambda(\Omega)$-weighted weak reverse Hölder inequality for differential forms.

Theorem 6.3.4. *Let $u \in D'(\Omega, \wedge^l)$ be a differential form satisfying the A-harmonic equation (1.2.4) in a domain $\Omega \subset \mathbf{R}^n$, $l = 0, 1, \ldots, n$. Suppose that $0 < s, t < \infty, \sigma > 1$. If $w \in A_r^\lambda(\Omega)$ for some $r > 1$ and $\lambda > 0$, then there exists a constant C, independent of u, such that*

$$\left(\int_B |u|^s w^\beta dx \right)^{1/s} \le C|B|^{(t-s)/st} \left(\int_{\sigma B} |u|^t w^{\beta \lambda t/s} dx \right)^{1/t} \qquad (6.3.13)$$

for all balls B with $\sigma B \subset \Omega$ and any real number β with $0 < \beta < 1$.

Note that (6.3.13) has the following symmetric version:

$$\left(\frac{1}{|B|} \int_B |u|^s w^\beta dx \right)^{1/s} \le C \left(\frac{1}{|B|} \int_{\sigma B} |u|^t w^{\beta \lambda t/s} dx \right)^{1/t} \qquad (6.3.14)$$

or

$$\|u\|_{s,B,w^\beta} \le C|B|^{(t-s)/st} \|u\|_{t,\sigma B, w^{\beta \lambda t/s}}. \qquad (6.3.14)'$$

Proof. Choose $k = s/(1 - \beta)$, so that $s < k$ and $1/s = 1/k + (k - s)/ks$. Using Theorem 1.1.4, we obtain

$$\left(\int_B |u|^s w^\beta dx \right)^{1/s}$$
$$= \left(\int_B \left(|u| w^{\beta/s} \right)^s dx \right)^{1/s}$$
$$\le \left(\int_B |u|^k dx \right)^{1/k} \left(\int_B w^{k\beta/(k-s)} dx \right)^{(k-s)/ks} \qquad (6.3.15)$$
$$= \|u\|_{k,B} \left(\int_B w dx \right)^{\beta/s}$$

for all balls B with $B \subset \Omega$. Next, choose $m = st/(s + \beta \lambda t(r - 1))$, so that $m < t$. Using Lemma 3.1.1, we obtain

$$\|u\|_{k,B} \le C_1 |B|^{(m-k)/km} \|u\|_{m,\sigma B}. \qquad (6.3.16)$$

Since $1/m = 1/t + (t - m)/mt$, by Theorem 1.1.4 again, we have

$$\|u\|_{m,\sigma B}$$
$$= \left(\int_{\sigma B} \left(|u| w^{\beta \lambda/s} w^{-\beta \lambda/s} \right)^m dx \right)^{1/m}$$

$$\leq \left(\int_{\sigma B} |u|^t w^{\beta\lambda t/s} dx \right)^{1/t} \left(\int_{\sigma B} (1/w)^{\beta\lambda mt/s(t-m)} dx \right)^{(t-m)/mt}$$

$$= \left(\int_{\sigma B} |u|^t w^{\beta\lambda t/s} dx \right)^{1/t} \left(\int_{\sigma B} (1/w)^{1/(r-1)} dx \right)^{\beta\lambda(r-1)/s}. \qquad (6.3.17)$$

Combining (6.3.15), (6.3.16), and (6.3.17), we arrive at the following estimate:

$$\left(\int_B |u|^s w^\beta dx \right)^{1/s}$$

$$\leq C_1 |B|^{(m-k)/km} \left(\int_{\sigma B} |u|^t w^{\beta\lambda t/s} dx \right)^{1/t}$$

$$\times \left(\int_B w \, dx \right)^{\beta/s} \left(\int_{\sigma B} (1/w)^{1/(r-1)} dx \right)^{\beta\lambda(r-1)/s}. \qquad (6.3.18)$$

Now, since $w \in A_r^\lambda(\Omega)$, we find that

$$\left(\int_B w \, dx \right)^{\beta/s} \left(\int_{\sigma B} (1/w)^{1/(r-1)} dx \right)^{\beta\lambda(r-1)/s}$$

$$\leq \left(\left(\int_{\sigma B} w \, dx \right) \left(\int_{\sigma B} (1/w)^{1/(r-1)} dx \right)^{\lambda(r-1)} \right)^{\beta/s}$$

$$\leq \left(|\sigma B|^{\lambda(r-1)+1} \left(\frac{1}{|\sigma B|} \int_B w \, dx \right) \right.$$

$$\left. \times \left(\frac{1}{|\sigma B|} \int_{\sigma B} (1/w)^{1/(r-1)} dx \right)^{\lambda(r-1)} \right)^{\beta/s} \qquad (6.3.19)$$

$$\leq C_2 |\sigma B|^{\beta(\lambda(r-1)+1)/s}$$

$$= C_3 |B|^{\beta\lambda(r-1)/s+\beta/s}.$$

Finally, substituting (6.3.19) into (6.3.18) and using

$$\frac{m-k}{km} = \frac{1}{k} - \frac{1}{m} = \frac{1-\alpha}{s} - \frac{1}{t} - \frac{\beta\lambda(r-1)}{s},$$

we obtain

$$\left(\int_B |u|^s w^\beta dx \right)^{1/s} \leq C|B|^{(t-s)/st} \left(\int_{\sigma B} |u|^t w^{\beta\lambda t/s} dx \right)^{1/t}. \qquad \blacksquare$$

In Theorem 6.3.4, if we let $\beta = 1/p$, where $p > 1$ is a real number, then inequality (6.3.13) reduces to

$$\left(\int_B |u|^s w^{1/p} dx \right)^{1/s} \leq C|B|^{(t-s)/st} \left(\int_{\sigma B} |u|^t w^{\lambda t/ps} dx \right)^{1/t}. \qquad (6.3.20)$$

Similarly, when $\beta = s/(s+t)$ in Theorem 6.3.4, we have the following symmetric inequality:

$$\left(\frac{1}{|B|}\int_B |u|^s w^{s/(s+t)}dx\right)^{1/s} \leq C\left(\frac{1}{|B|}\int_{\sigma B}|u|^t w^{\lambda t/(s+t)}dx\right)^{1/t}. \quad (6.3.21)$$

Choosing $\lambda = 1$ in (6.3.21), we find the following version of the weak reverse Hölder inequality.

Corollary 6.3.5. *Let $u \in D'(\Omega, \wedge^l)$ be a differential form satisfying the A-harmonic equation (1.2.4) in a domain $\Omega \subset \mathbf{R}^n$, $l = 0, 1, \ldots, n$. Suppose that $0 < s, t < \infty, \sigma > 1$. If $w \in A_r(\Omega)$ for some $r > 1$, then there exists a constant C, independent of u, such that*

$$\left(\frac{1}{|B|}\int_B |u|^s w^{s/(s+t)}dx\right)^{1/s} \leq C\left(\frac{1}{|B|}\int_{\sigma B}|u|^t w^{t/(s+t)}dx\right)^{1/t}$$

for all balls B with $\sigma B \subset \Omega$.

Selecting $\beta = 1/\lambda$ with $\lambda > 1$ in Theorem 6.3.4, we obtain the following result.

Corollary 6.3.6. *Let $u \in D'(\Omega, \wedge^l)$ be a differential form satisfying the A-harmonic equation (1.2.4) in a domain $\Omega \subset \mathbf{R}^n$, $l = 0, 1, \ldots, n$. Suppose that $0 < s, t < \infty, \sigma > 1$. If $w \in A_r^\lambda(\Omega)$ for some $r > 1$ and $\lambda > 1$, then there exists a constant C, independent of u, such that*

$$\left(\frac{1}{|B|}\int_B |u|^s w^{1/\lambda}dx\right)^{1/s} \leq C\left(\frac{1}{|B|}\int_{\sigma B}|u|^t w^{t/s}dx\right)^{1/t}$$

for all balls B with $\sigma B \subset \Omega$.

Assuming $\beta = s$ with $0 < s < 1$ in Theorem 6.3.4, we have the following inequality.

Corollary 6.3.7. *Let $u \in D'(\Omega, \wedge^l)$ be a differential form satisfying the A-harmonic equation (1.2.4) in a domain $\Omega \subset \mathbf{R}^n$, $l = 0, 1, \ldots, n$. Suppose that $0 < t < \infty, \sigma > 1$, and $0 < s < 1$. If $w \in A_r^\lambda(\Omega)$ for some $r > 1$ and $\lambda > 0$, then there exists a constant C, independent of u, such that*

$$\left(\frac{1}{|B|}\int_B |u|^s w^s dx\right)^{1/s} \leq C\left(\frac{1}{|B|}\int_{\sigma B}|u|^t w^{\lambda t}dx\right)^{1/t}$$

for all balls B with $\sigma B \subset \Omega$.

Letting $\lambda = 1$ in Corollary 6.3.7, we find the following symmetric inequality.

Corollary 6.3.8. *Let $u \in D'(\Omega, \wedge^l)$ be a differential form satisfying the A-harmonic equation (1.2.4) in a domain $\Omega \subset \mathbf{R}^n$, $l = 0, 1, \ldots, n$. Suppose that $0 < t < \infty, \sigma > 1$, and $0 < s < 1$. If $w \in A_r(\Omega)$ for some $r > 1$, then there exists a constant C, independent of u, such that*

$$\left(\frac{1}{|B|} \int_B |u|^s w^s dx \right)^{1/s} \leq C \left(\frac{1}{|B|} \int_{\sigma B} |u|^t w^t dx \right)^{1/t} \tag{6.3.22}$$

for all balls B with $\sigma B \subset \Omega$.

Assuming $t > 1$ in Theorem 6.3.4 and letting $\beta = 1/t$ in (6.3.13), we have the following corollary.

Corollary 6.3.9. *Let $u \in D'(\Omega, \wedge^l)$ be a differential form satisfying the A-harmonic equation (1.2.4) in a domain $\Omega \subset \mathbf{R}^n$, $l = 0, 1, \ldots, n$. Suppose that $0 < s < \infty, \sigma > 1$, and $t > 1$. If $w \in A_r^\lambda(\Omega)$ for some $r > 1$ and $\lambda > 0$, then there exists a constant C, independent of u, such that*

$$\left(\frac{1}{|B|} \int_B |u|^s w^{1/t} dx \right)^{1/s} \leq C \left(\frac{1}{|B|} \int_{\sigma B} |u|^t w^{\lambda/s} dx \right)^{1/t} \tag{6.3.23}$$

for all balls B with $\sigma B \subset \Omega$.

Choosing $\lambda = 1$ in Corollary 6.3.9, inequality (6.3.23) reduces to

$$\left(\frac{1}{|B|} \int_B |u|^s w^{1/t} dx \right)^{1/s} \leq C \left(\frac{1}{|B|} \int_{\sigma B} |u|^t w^{1/s} dx \right)^{1/t} \tag{6.3.24}$$

or

$$\|u\|_{s, B, w^{1/t}} \leq C |B|^{(t-s)/st} \|u\|_{t, \sigma B, w^{1/s}} \tag{6.3.24'}$$

which has a nice symmetric property.

Applying Theorem 6.3.4 and the covering lemma, we can also prove the following global result.

Theorem 6.3.10. *Let $u \in D'(\Omega, \wedge^l)$, $l = 0, 1, \ldots, n$, be a differential form satisfying the A-harmonic equation (1.2.4) in a domain $\Omega \subset \mathbf{R}^n$ with $|\Omega| < \infty$. Assume that $0 < s \leq t < \infty$ and $w \in A_r^\lambda(\Omega)$ for some $r > 1$ and $\lambda > 0$. Then,*

$$\left(\frac{1}{|\Omega|} \int_\Omega |u|^s w^\beta dx \right)^{1/s} \leq \left(\frac{C}{|\Omega|} \int_\Omega |u|^t w^{\beta \lambda t/s} dx \right)^{1/t} \tag{6.3.25}$$

for any real number β with $0 < \beta < 1$.

6.3.3 $A_r^\lambda(\Omega)$-weighted inequalities for du

By the virtue of Theorem 6.2.12 and using methods similar to the proofs of Theorems 6.3.3 and 6.3.4, we can prove the following versions of the weighted weak reverse Hölder inequality for du.

Theorem 6.3.11. *Let $u \in D'(\Omega, \wedge^l)$, $l = 0, 1, \ldots, n$, be an A-harmonic tensor in a domain $\Omega \subset \mathbf{R}^n$, $\sigma > 1$. Assume that $0 < s, t < \infty$, and $w \in A_r^{s/t}(\Omega)$ for some $r > 1$ and $0 < \alpha < 1$. Then, there exists a constant C, independent of u, such that*

$$\left(\int_B |du|^s w^\alpha dx \right)^{1/s} \le C|B|^{(t-s)/st} \left(\int_{\sigma B} |du|^t w^\alpha dx \right)^{1/t} \qquad (6.3.26)$$

for all balls B with $\sigma B \subset \Omega$.

Theorem 6.3.12. *Let $u \in D'(\Omega, \wedge^l)$ be a differential form satisfying the A-harmonic equation (1.2.4) in a domain $\Omega \subset \mathbf{R}^n$, $l = 0, 1, \ldots, n$. Suppose that $0 < s, t < \infty, \sigma > 1$. If $w \in A_r^\lambda(\Omega)$ for some $r > 1$ and $\lambda > 0$, then there exists a constant C, independent of u, such that*

$$\left(\int_B |du|^s w^\beta dx \right)^{1/s} \le C|B|^{(t-s)/st} \left(\int_{\sigma B} |du|^t w^{\beta \lambda t/s} dx \right)^{1/t} \qquad (6.3.27)$$

for all balls B with $\sigma B \subset \Omega$ and any real number β with $0 < \beta < 1$.

Note that (6.3.27) has the following symmetric version:

$$\left(\frac{1}{|B|} \int_B |du|^s w^\beta dx \right)^{1/s} \le C \left(\frac{1}{|B|} \int_{\sigma B} |du|^t w^{\beta \lambda t/s} dx \right)^{1/t} \qquad (6.3.28)$$

or

$$\|du\|_{s, B, w^\beta} \le C|B|^{(t-s)/st} \|du\|_{t, \sigma B, w^{\beta \lambda t/s}}. \qquad (6.3.28)'$$

From the above results it is clear that all versions of the weak reverse Hölder inequality for differential forms u obtained in this section can be extended to differential forms du.

6.4 The third weighted case

In the last two sections, we have discussed $A_r(\Omega)$-weighted and $A_r^\lambda(\Omega)$-weighted weak reverse Hölder inequalities. In this section, we study $A_r(\lambda, \Omega)$-weighted weak reverse Hölder inequalities.

6.4.1 Local inequalities

In 2002, S. Ding and D. Sylvester [268] established the following $A_r(\lambda, \Omega)$-weighted weak reverse Hölder inequality.

Theorem 6.4.1. *Let $u \in D'(\Omega, \wedge^l)$ be a differential form satisfying the A-harmonic equation (1.2.4) in a domain $\Omega \subset \mathbf{R}^n$, $l = 0, 1, \ldots, n$. Suppose that $0 < s, t < \infty, \sigma > 1$. If $w \in A_r(\lambda, \Omega)$ for some $r > 1$ and $\lambda > 0$, then there exists a constant C, independent of u, such that*

$$\left(\int_B |u|^s w^{\alpha\lambda} dx\right)^{1/s} \leq C|B|^{(t-s)/st} \left(\int_{\sigma B} |u|^t w^{\alpha t/s} dx\right)^{1/t} \qquad (6.4.1)$$

for all balls B with $\sigma B \subset \Omega$ and any real number α with $0 < \alpha < 1$.

Note that (6.4.1) has the following symmetric version:

$$\left(\frac{1}{|B|} \int_B |u|^s w^{\alpha\lambda} dx\right)^{1/s} \leq C \left(\frac{1}{|B|} \int_{\sigma B} |u|^t w^{\alpha t/s} dx\right)^{1/t}. \qquad (6.4.2)$$

Proof. Choose $k = s/(1 - \alpha)$, so that $s < k$ and $1/s = 1/k + (k - s)/ks$. Applying Hölder's inequality, we find that

$$\begin{aligned}
\left(\int_B |u|^s w^{\alpha\lambda} dx\right)^{1/s} &= \left(\int_B \left(|u| w^{\alpha\lambda/s}\right)^s dx\right)^{1/s} \\
&\leq \left(\int_B |u|^k dx\right)^{1/k} \left(\int_B w^{k\alpha\lambda/(k-s)} dx\right)^{(k-s)/ks} \\
&= \|u\|_{k,B} \left(\int_B w^\lambda dx\right)^{\alpha/s}
\end{aligned} \qquad (6.4.3)$$

for all balls B with $B \subset \Omega$. Next, choose $m = st/(s + \alpha t(r - 1))$. Then, from Theorem 6.2.1, we obtain

$$\|u\|_{k,B} \leq C_1 |B|^{(m-k)/km} \|u\|_{m,\sigma B}. \qquad (6.4.4)$$

Since $1/m = 1/t + (t - m)/mt$, by the Hölder inequality again, we have

$$\|u\|_{m,\sigma B} = \left(\int_{\sigma B} \left(|u| w^{\alpha/s} w^{-\alpha/s}\right)^m dx\right)^{1/m}$$

$$\leq \left(\int_{\sigma B} |u|^t w^{\alpha t/s} dx\right)^{1/t} \left(\int_{\sigma B} (1/w)^{\alpha mt/s(t-m)} dx\right)^{(t-m)/mt}$$

$$= \left(\int_{\sigma B} |u|^t w^{\alpha t/s} dx\right)^{1/t} \left(\int_{\sigma B} (1/w)^{1/(r-1)} dx\right)^{\alpha(r-1)/s}. \tag{6.4.5}$$

Combining (6.4.3), (6.4.4), and (6.4.5), we arrive at the following estimate:

$$\left(\int_B |u|^s w^{\alpha\lambda} dx\right)^{1/s}$$

$$\leq C_1 |B|^{(m-k)/km} \left(\int_B w^\lambda dx\right)^{\alpha/s} \left(\int_{\sigma B} \left(\tfrac{1}{w}\right)^{1/(r-1)} dx\right)^{\alpha(r-1)/s} \tag{6.4.6}$$

$$\times \left(\int_{\sigma B} |u|^t w^{\alpha t/s} dx\right)^{1/t}.$$

Now since $w \in A_r(\lambda, \Omega)$, it follows that

$$\left(\int_B w^\lambda dx\right)^{\alpha/s} \left(\int_{\sigma B} (1/w)^{1/(r-1)} dx\right)^{\alpha(r-1)/s}$$

$$\leq \left(\left(\int_{\sigma B} w^\lambda dx\right)\left(\int_{\sigma B} (1/w)^{1/(r-1)} dx\right)^{(r-1)}\right)^{\alpha/s}$$

$$\leq \left(|\sigma B|^{(r-1)+1} \left(\tfrac{1}{|\sigma B|}\int_B w^\lambda dx\right)\right.$$

$$\left.\times \left(\tfrac{1}{|\sigma B|}\int_{\sigma B} (1/w)^{1/(r-1)} dx\right)^{(r-1)}\right)^{\alpha/s} \tag{6.4.7}$$

$$\leq C_2 |\sigma B|^{\alpha r/s}$$

$$= C_3 |B|^{\alpha r/s}.$$

Finally, substituting (6.4.7) into (6.4.6) and using $(m-k)/km = 1/k - 1/m = 1/s - 1/t - \alpha r/s$, we obtain

$$\left(\int_B |u|^s w^{\alpha\lambda} dx\right)^{1/s} \leq C|B|^{(t-s)/st} \left(\int_{\sigma B} |u|^t w^{\alpha t/s} dx\right)^{1/t}. \quad\blacksquare$$

If we choose $\alpha = 1/p$ where $p > 1$ in Theorem 6.4.1, then inequality (6.4.1) reduces to

$$\left(\int_B |u|^s w^{\lambda/p} dx\right)^{1/s} \leq C|B|^{(t-s)/st} \left(\int_{\sigma B} |u|^t w^{t/ps} dx\right)^{1/t}. \tag{6.4.8}$$

Setting $\alpha = s/(s+t)$ in Theorem 6.4.1, we have the following symmetric inequality:

$$\left(\frac{1}{|B|}\int_B |u|^s w^{\lambda s/(s+t)} dx\right)^{1/s} \le C\left(\frac{1}{|B|}\int_{\sigma B}|u|^t w^{t/(s+t)}dx\right)^{1/t}. \quad (6.4.9)$$

Choosing $\lambda = 1$ in (6.4.9), we obtain the following symmetric weak reverse Hölder inequality.

Corollary 6.4.2. *Let $u \in D'(\Omega, \wedge^l)$ be a differential form satisfying the A-harmonic equation (1.2.4) in a domain $\Omega \subset \mathbf{R}^n$, $l = 0, 1, \ldots, n$. Suppose that $0 < s, t < \infty, \sigma > 1$. If $w \in A_r(\Omega)$ for some $r > 1$, then there exists a constant C, independent of u, such that*

$$\left(\frac{1}{|B|}\int_B |u|^s w^{s/(s+t)} dx\right)^{1/s} \le C\left(\frac{1}{|B|}\int_{\sigma B}|u|^t w^{t/(s+t)}dx\right)^{1/t}$$

for all balls B with $\sigma B \subset \Omega$.

Selecting $\alpha = 1/\lambda$ with $\lambda > 1$ in Theorem 6.4.1, we have the following corollary.

Corollary 6.4.3. *Let $u \in D'(\Omega, \wedge^l)$ be a differential form satisfying the A-harmonic equation (1.2.4) in a domain $\Omega \subset \mathbf{R}^n$, $l = 0, 1, \ldots, n$. Suppose that $0 < s, t < \infty, \sigma > 1$. If $w \in A_r(\lambda, \Omega)$ for some $r > 1$ and $\lambda > 1$, then there exists a constant C, independent of u, such that*

$$\left(\frac{1}{|B|}\int_B |u|^s w dx\right)^{1/s} \le C\left(\frac{1}{|B|}\int_{\sigma B}|u|^t w^{t/\lambda s}dx\right)^{1/t}$$

for all balls B with $\sigma B \subset \Omega$.

6.4.2 Global inequality

As an application of the local result, Theorem 6.4.1, we shall prove the following global weak reverse Hölder inequality which can also be proved by using Theorem 1.1.4 directly.

Theorem 6.4.4. *Let $u \in D'(\Omega, \wedge^l)$ be a differential form satisfying the A-harmonic equation (1.2.4) in a bounded domain $\Omega \subset \mathbf{R}^n$, $l = 0, 1, \ldots, n$. Suppose that $1 < s \le t < \infty$ and $w \in A_r(\lambda, \Omega)$ for some $r > 1$ and $\lambda > 0$. Then, there exists a constant C, independent of u, such that*

$$\left(\frac{1}{|\Omega|}\int_\Omega |u|^s w^{\alpha\lambda} dx\right)^{1/s} \le C\left(\frac{1}{|\Omega|}\int_\Omega |u|^t w^{\alpha t/s}dx\right)^{1/t} \quad (6.4.10)$$

for any real number α with $0 < \alpha < 1$.

Proof. By (6.4.1) and Theorem 1.5.3, we obtain

$$\left(\int_\Omega |u|^s w^{\alpha\lambda} dx\right)^{1/s}$$

$$\leq \sum_{Q\in\mathcal{V}} \left(\int_Q |u|^s w^{\alpha\lambda} dx\right)^{1/s}$$

$$\leq \sum_{Q\in\mathcal{V}} \left(C_1 |Q|^{(t-s)/st} \int_{\sigma Q} |u|^t w^{\alpha t/s} dx\right)^{1/t}$$

$$\leq C_1 |\Omega|^{(t-s)/st} \sum_{Q\in\mathcal{V}} \left(\int_{\sigma Q} |u|^t w^{\alpha t/s} dx\right)^{1/t}$$

$$\leq C_1 |\Omega|^{(t-s)/st} \sum_{Q\in\mathcal{V}} \left(\int_\Omega |u|^t w^{\alpha t/s} dx\right)^{1/t}$$

$$\leq C_2 |\Omega|^{(t-s)/st} \left(\int_\Omega |u|^t w^{\alpha t/s} dx\right)^{1/t},$$

which is equivalent to (6.4.10). ∎

6.4.3 Analogies for du

Using Theorem 6.2.12 and the method we developed in the proof of Theorem 6.4.1, we obtain the following weak reverse Hölder inequality for du, where u satisfies the A-harmonic equation (1.2.4).

Theorem 6.4.5. *Let $u \in D'(\Omega, \wedge^l)$ be a differential form satisfying the A-harmonic equation (1.2.4) in a domain $\Omega \subset \mathbf{R}^n$, $l = 0, 1, \ldots, n$. Suppose that $0 < s, t < \infty, \sigma > 1$. If $w \in A_r(\lambda, \Omega)$ for some $r > 1$ and $\lambda > 0$, then there exists a constant C, independent of u, such that*

$$\left(\int_B |du|^s w^{\alpha\lambda} dx\right)^{1/s} \leq C|B|^{(t-s)/st} \left(\int_{\sigma B} |du|^t w^{\alpha t/s} dx\right)^{1/t} \qquad (6.4.11)$$

for all balls B with $\sigma B \subset \Omega$ and any real number α with $0 < \alpha < 1$.

Note that (6.4.11) has the following symmetric version:

$$\left(\frac{1}{|B|}\int_B |du|^s w^{\alpha\lambda} dx\right)^{1/s} \leq C \left(\frac{1}{|B|}\int_{\sigma B} |du|^t w^{\alpha t/s} dx\right)^{1/t}. \qquad (6.4.11)'$$

From Theorem 6.4.5 and the method to the proof of Theorem 6.4.4, we have the following global weak reverse Hölder inequality for du in a bounded domain $\Omega \subset \mathbf{R}^n$.

Theorem 6.4.6. *Let $u \in D'(\Omega, \wedge^l)$ be a differential form satisfying the A-harmonic equation (1.2.4) in a bounded domain $\Omega \subset \mathbf{R}^n$, $l = 0, 1, \ldots, n$. Suppose that $1 < s \le t < \infty$ and $w \in A_r(\lambda, \Omega)$ for some $r > 1$ and $\lambda > 0$. Then, there exists a constant C, independent of u, such that*

$$\left(\frac{1}{|\Omega|} \int_\Omega |du|^s w^{\alpha\lambda} dx \right)^{1/s} \le C \left(\frac{1}{|\Omega|} \int_\Omega |du|^t w^{\alpha t/s} dx \right)^{1/t} \qquad (6.4.12)$$

for any real number α with $0 < \alpha < 1$.

6.5 Two-weight inequalities

Many two-weight versions of the weak reverse Hölder inequality have been established recently. In this section, we discuss these results for $A_{r,\lambda}(\Omega)$-weights and $A_r(\lambda, \Omega)$-weights.

6.5.1 $A_{r,\lambda}(\Omega)$-weighted cases

In this section, we prove the $A_{r,\lambda}(\Omega)$-weighted weak reverse Hölder inequalities. The following two-weighted inequality is due to Y. Xing and G. Bao [84].

Theorem 6.5.1. *Let $u \in D'(\Omega, \wedge^l)$, $l = 0, 1, \ldots, n$, be an A-harmonic tensor in a domain $\Omega \subset \mathbf{R}^n$, $\rho > 1$. Assume that $0 < s, t < \infty$, and $(w_1, w_2) \in A_{r,\lambda}(\Omega)$ for some $1 < r < \infty$ and $\lambda \ge 1$. Then, there exists a constant C, independent of u, such that*

$$\left(\frac{1}{|B|} \int_B |u|^s w_1^\ell dx \right)^{1/s} \le C \left(\frac{1}{|B|} \int_{\rho B} |u|^t w_2^{\ell t/s} dx \right)^{1/t} \qquad (6.5.1)$$

for all balls B with $\rho B \subset \Omega$ and $0 < \ell < \lambda$.

Proof. Let $k = s\lambda/(\lambda - \ell)$ and $m = \frac{st\lambda r'}{tr\ell + s\lambda r'}$, so that $s < k$ and $m < t$. Using Theorem 1.1.4, we have

$$\left(\int_B |u|^s w_1^\ell dx \right)^{1/s}$$

$$= \left(\int_B \left(|u| w_1^{\ell/s} \right)^s dx \right)^{1/s}$$

$$\le \left(\int_B |u|^k dx \right)^{1/k} \left(\int_B \left(w_1^{\ell/s} \right)^{ks/(k-s)} dx \right)^{(k-s)/ks} \qquad (6.5.2)$$

$$= \|u\|_{k,B} \left(\int_B w_1^\lambda dx \right)^{\ell/s\lambda}.$$

From Theorem 6.2.1, it follows that

$$\|u\|_{k,B} \le C_1 |B|^{(m-k)/km} \|u\|_{m,\rho B}. \tag{6.5.3}$$

Using the generalized Hölder inequality with $1/m = 1/t + (t-m)/mt$ again, we obtain

$\|u\|_{m,\rho B}$

$$\le \left(\int_{\rho B} |u|^t w_2^{\ell t/s} dx \right)^{1/t} \left(\int_{\rho B} \left(\frac{1}{w_2} \right)^{\ell m t/(st-sm)} dx \right)^{(t-m)/mt}$$

$$= \left(\int_{\rho B} |u|^t w_2^{\ell t/s} dx \right)^{1/t} \left(\int_{\rho B} \left(\frac{1}{w_2} \right)^{\lambda r'/r} dx \right)^{\ell r/s \lambda r'}. \tag{6.5.4}$$

Combining (6.5.2), (6.5.3), and (6.5.4), we find that

$\left(\int_B |u|^s w_1^\ell dx \right)^{1/s}$

$$\le C_1 |B|^{(m-k)/km} \left(\int_B w_1^\lambda dx \right)^{\ell/s\lambda} \left(\int_{\rho B} |u|^t w_2^{\ell t/s} dx \right)^{1/t}$$

$$\times \left(\int_{\rho B} \left(\frac{1}{w_2} \right)^{\lambda r'/r} dx \right)^{\ell r/s\lambda r'}. \tag{6.5.5}$$

Now since $(w_1, w_2) \in A_{r,\lambda}(\Omega)$, we obtain

$\left(\int_B w_1^\lambda dx \right)^{\ell/\lambda s} \left(\int_{\rho B} \left(\frac{1}{w_2} \right)^{\lambda r'/r} dx \right)^{r\ell/s\lambda r'}$

$$\le |\rho B|^{r\ell/\lambda s} \left(\left(\frac{1}{|\rho B|} \int_{\rho B} w_1^\lambda dx \right)^{1/\lambda r} \left(\frac{1}{|\rho B|} \int_{\rho B} \left(\frac{1}{w_2} \right)^{\lambda r'/r} dx \right)^{1/\lambda r'} \right)^{r\ell/s}$$

$$\le C_2 |B|^{r\ell/\lambda s}. \tag{6.5.6}$$

Finally, substituting (6.5.6) into (6.5.5), we have

$\left(\int_B |u|^s w_1^\ell dx \right)^{1/s}$

$$\le C_3 |B|^{(m-k)/mk+r\ell/s\lambda} \left(\int_{\rho B} |u|^t w_2^{\ell t/s} dx \right)^{1/t}$$

$$= C_3 |B|^{(1/s-1/t)} \left(\int_{\rho B} |u|^t w_2^{\ell t/s} dx \right)^{1/t}. \qquad \blacksquare \tag{6.5.7}$$

In Theorem 6.5.1, particular values of the parameter ℓ lead to different versions of the weak reverse Hölder inequality. For example, choosing $\ell = s$ with $s < \lambda$, we obtain the following symmetric weak reverse Hölder inequality

$$\left(\frac{1}{|B|}\int_B |u|^s w_1^s dx\right)^{1/s} \leq C\left(\frac{1}{|B|}\int_{\rho B} |u|^t w_2^t dx\right)^{1/t} \tag{6.5.8}$$

for all balls B with $\rho B \subset \Omega$.

Similarly, if we select $\ell = 1/t$ with $\lambda t > 1$, inequality (6.5.1) reduces to

$$\left(\frac{1}{|B|}\int_B |u|^s w_1^{1/t} dx\right)^{1/s} \leq C\left(\frac{1}{|B|}\int_{\rho B} |u|^t w_2^{1/s} dx\right)^{1/t} \tag{6.5.9}$$

for all balls B with $\rho B \subset \Omega$.

Now, we prove the following global weak reverse Hölder inequality.

Theorem 6.5.2. *Let $u \in D'(\Omega, \wedge^l)$, $l = 0, 1, \ldots, n$, be an A-harmonic tensor in a domain $\Omega \subset \mathbf{R}^n$ with $|\Omega| < \infty$. Assume that $1 < s \leq t < \infty$, $0 < \ell < \lambda$, and $(w_1, w_2) \in A_{r,\lambda}(\Omega)$ for some $1 < r < \infty$ and $\lambda \geq 1$, then*

$$\left(\frac{1}{|\Omega|}\int_\Omega |u|^s w_1^\ell dx\right)^{1/s} \leq C\left(\frac{1}{|\Omega|}\int_\Omega |u|^t w_2^{\ell t/s} dx\right)^{1/t}. \tag{6.5.10}$$

Proof. Using Theorems 6.5.1 and 1.5.3, we obtain

$$\begin{aligned}
&\left(\int_\Omega |u|^s w_1^\ell dx\right)^{1/s} \\
&\leq \sum_{D \in \mathcal{V}} \left(\int_D |u|^s w_1^\ell dx\right)^{1/s} \\
&\leq \sum_{D \in \mathcal{V}} C_k |D|^{(t-s)/st} \left(\int_{\rho D} |u|^t w_2^{\ell t/s} dx\right)^{1/t} \\
&\leq \sum_{D \in \mathcal{V}} C_k |\Omega|^{(t-s)/st} \left(\int_\Omega |u|^t w_2^{\ell t/s} dx\right)^{1/t} \\
&\leq C|\Omega|^{(t-s)/st} \left(\int_\Omega |u|^t w_2^{\ell t/s} dx\right)^{1/t}. \quad \blacksquare
\end{aligned} \tag{6.5.11}$$

6.5.2 $A_r(\lambda, \Omega)$-weighted cases

Theorem 6.5.3. *Let $u \in D'(\Omega, \wedge^l)$, $l = 0, 1, \ldots, n$, be an A-harmonic tensor in a domain $\Omega \subset \mathbf{R}^n$, $\rho > 1$. Assume that $0 < s, t < \infty$, and $(w_1, w_2) \in A_r(\lambda, \Omega)$ for some $\lambda > 0$ and $1 < r < \infty$. Then, there exists a constant C, independent of u, such that*

$$\left(\int_B |u|^s w_1^{\alpha\lambda} dx \right)^{1/s} \leq C|B|^{(t-s)/st} \left(\int_{\rho B} |u|^t w_2^{\alpha t/s} dx \right)^{1/t} \tag{6.5.12}$$

for all balls B with $\rho B \subset \Omega$ and any real number α with $0 < \alpha < 1$.

Proof. Choose $k = s/(1-\alpha)$, so that $s < k$ and $1/s = 1/k + (k-s)/ks$. Applying Hölder's inequality, we have

$$\begin{aligned}
&\left(\int_B |u|^s w_1^{\alpha\lambda} dx \right)^{1/s} \\
&= \left(\int_B \left(|u| w_1^{\alpha\lambda/s} \right)^s dx \right)^{1/s} \\
&\leq \left(\int_B |u|^k dx \right)^{1/k} \left(\int_B w_1^{k\alpha\lambda/(k-s)} dx \right)^{(k-s)/ks} \\
&= \|u\|_{k,B} \left(\int_B w_1^\lambda dx \right)^{\alpha/s}
\end{aligned} \tag{6.5.13}$$

for all balls B with $B \subset \Omega$. Next, choose $m = st/(s + \alpha t(r-1))$. Using Theorem 6.2.1, we obtain

$$\|u\|_{k,B} \leq C_1 |B|^{(m-k)/km} \|u\|_{m,\sigma B}. \tag{6.5.14}$$

Since $1/m = 1/t + (t-m)/mt$, by the Hölder inequality again, we have

$$\begin{aligned}
&\|u\|_{m,\sigma B} \\
&= \left(\int_{\sigma B} \left(|u| w_2^{\alpha/s} w_2^{-\alpha/s} \right)^m dx \right)^{1/m} \\
&\leq \left(\int_{\sigma B} |u|^t w_2^{\alpha t/s} dx \right)^{1/t} \left(\int_{\sigma B} (1/w_2)^{\alpha m t/s(t-m)} dx \right)^{(t-m)/mt} \\
&= \left(\int_{\sigma B} |u|^t w_2^{\alpha t/s} dx \right)^{1/t} \left(\int_{\sigma B} (1/w_2)^{1/(r-1)} dx \right)^{\alpha(r-1)/s}.
\end{aligned} \tag{6.5.15}$$

Combining (6.5.13), (6.5.14), and (6.5.15), we arrive at the following estimate:

$$\begin{aligned}
&\left(\int_B |u|^s w_1^{\alpha\lambda} dx \right)^{1/s} \\
&\leq C_1 |B|^{(m-k)/km} \left(\int_B w_1^\lambda dx \right)^{\alpha/s} \left(\int_{\sigma B} \left(\frac{1}{w_2} \right)^{1/(r-1)} dx \right)^{\alpha(r-1)/s} \\
&\quad \times \left(\int_{\sigma B} |u|^t w_2^{\alpha t/s} dx \right)^{1/t}.
\end{aligned} \tag{6.5.16}$$

Now since $(w_1, w_2) \in A_r(\lambda, \Omega)$, we find that

$$\left(\int_B w_1^\lambda dx\right)^{\alpha/s} \left(\int_{\sigma B} (1/w_2)^{1/(r-1)} dx\right)^{\alpha(r-1)/s}$$

$$\le \left(\left(\int_{\sigma B} w_1^\lambda dx\right) \left(\int_{\sigma B} (1/w_2)^{1/(r-1)} dx\right)^{(r-1)}\right)^{\alpha/s}$$

$$\le \left(|\sigma B|^{(r-1)+1} \left(\frac{1}{|\sigma B|} \int_B w_1^\lambda dx\right)\right.$$

$$\left. \times \left(\frac{1}{|\sigma B|} \int_{\sigma B} (1/w_2)^{1/(r-1)} dx\right)^{(r-1)}\right)^{\alpha/s} \tag{6.5.17}$$

$$\le C_2 |\sigma B|^{\alpha r/s}$$

$$= C_3 |B|^{\alpha r/s}.$$

Finally, substituting (6.5.17) into (6.5.16) and using

$$\frac{m-k}{km} = \frac{1}{k} - \frac{1}{m} = \frac{1}{s} - \frac{1}{t} - \frac{\alpha r}{s},$$

we obtain

$$\left(\int_B |u|^s w_1^{\alpha\lambda} dx\right)^{1/s} \le C|B|^{(t-s)/st} \left(\int_{\sigma B} |u|^t w_2^{\alpha t/s} dx\right)^{1/t}. \quad \blacksquare$$

6.5.3 Two-weight inequalities for du

From Theorem 6.2.12 (weak reverse Hölder inequality for du), we can prove the following $A_{r,\lambda}(\Omega)$-weighted weak reverse Hölder inequality for du.

Theorem 6.5.4. *Let $u \in D'(\Omega, \wedge^l)$, $l = 0, 1, \ldots, n$, be an A-harmonic tensor in a domain $\Omega \subset \mathbf{R}^n$, $\rho > 1$. Assume that $0 < s, t < \infty$, and $(w_1, w_2) \in A_{r,\lambda}(\Omega)$ for some $1 < r < \infty$ and $\lambda \ge 1$. Then, there exists a constant C, independent of u, such that*

$$\left(\frac{1}{|B|} \int_B |du|^s w_1^\beta dx\right)^{1/s} \le C \left(\frac{1}{|B|} \int_{\rho B} |du|^t w_2^{\beta t/s} dx\right)^{1/t}$$

for all balls B with $\rho B \subset \Omega$ and $0 < \beta < \lambda$.

Similarly, following the proof of Theorem 6.5.3, we can prove $A_r(\lambda, \Omega)$-weighted weak reverse Hölder inequality for du.

Theorem 6.5.5.   *Let $u \in D'(\Omega, \wedge^l)$, $l = 0, 1, \ldots, n$, be an A-harmonic tensor in a domain $\Omega \subset \mathbf{R}^n$, $\rho > 1$. Assume that $0 < s, t < \infty$, and $(w_1, w_2) \in$*

$A_r(\lambda, \Omega)$ *for some* $\lambda > 0$ *and* $1 < r < \infty$. *Then, there exists a constant* C, *independent of* u, *such that*

$$\left(\int_B |du|^s w_1^{\alpha\lambda} dx \right)^{1/s} \leq C |B|^{(s-t)/st} \left(\int_{\rho B} |du|^t w_2^{\alpha t/s} dx \right)^{1/t}$$

for all balls B *with* $\rho B \subset \Omega$ *and any real number* α *with* $0 < \alpha < 1$.

6.6 Inequalities with Orlicz norms

In this section, we shall present the weak reverse Hölder inequality with $L^p(\log L)^\alpha$-norm. For this, we recall that the $L^p(\log L)^\alpha$-norm of a measurable function on Ω is defined as

$$\|f\|_{L^p \log^\alpha L} = \inf \left\{ k : \int_\Omega |f|^p \log^\alpha \left(e + \frac{|f|}{k} \right) dx \leq k^p \right\}, \qquad (6.6.1)$$

where $0 < p < \infty$ and $\alpha > 0$ are real numbers. Let E be any subset of $\mathbf{R}^n$. The functional of a measurable function f over E is defined by

$$[f]_{L^p(\log L)^\alpha(E)} = \left(\int_E |f|^p \log^\alpha \left(e + \frac{|f|}{\|f\|_p} \right) dx \right)^{\frac{1}{p}}, \qquad (6.6.2)$$

where $\|f\|_p = \left(\int_E |f(x)|^p dx \right)^{1/p}$. From Theorem 1.9.1, we know that the norm $\|f\|_{L^p \log^\alpha L}$ is equivalent to $[f]_{L^p(\log L)^\alpha}$ for $0 < p < 1$ and $\alpha \geq 0$. Hence, we do not distinguish the norm $\|f\|_{L^p \log^\alpha L}$ from the functional $[f]_{L^p(\log L)^\alpha}$ throughout this section, and for convenience write

$$\|f\|_{L^p \log^\alpha L} = \left(\int_E |f|^p \log^\alpha \left(e + \frac{|f|}{\|f\|_p} \right) dx \right)^{\frac{1}{p}}. \qquad (6.6.3)$$

6.6.1 Elementary inequalities

It is easy to see that for any constant k, there exist constants $m > 0$ and $M > 0$, such that

$$m \log(e + t) \leq \log \left(e + \frac{t}{k} \right) \leq M \log(e + t), \quad t > 0. \qquad (6.6.4)$$

From the weak reverse Hölder inequality (Lemma 3.1.1), we know that the norms $\|u\|_{s,B}$ and $\|u\|_{t,B}$ are comparable when $0 < d_1 \leq diam(B) \leq d_2 < \infty$. Hence, we may assume that $0 < m_1 \leq \|u\|_{s,B} \leq M_1 < \infty$ and $0 < m_2 \leq$

$\|u\|_{t,B} \leq M_2 < \infty$ for some constants m_i and M_i, $i = 1, 2$. Thus, we have

$$C_1 \log (e + |u|) \leq \log \left(e + \frac{|u|}{\|u\|_{s,B}} \right) \leq C_2 \log(e + |u|) \qquad (6.6.5)$$

and

$$C_3 \log(e + |u|) \leq \log \left(e + \frac{|u|}{\|u\|_{t,B}} \right) \leq C_4 \log(e + |u|) \qquad (6.6.6)$$

for any $s > 0$ and $t > 0$, where C_i are constants, $i = 1, 2, 3, 4$. Using (6.6.5) and (6.6.6), we obtain

$$C_5 \left(\int_B |u|^s \log^\alpha \left(e + \frac{|u|}{\|u\|_{t,B}} \right) dx \right)^{1/s}$$

$$\leq \|u\|_{L^s(\log L)^\alpha(B)} = \left(\int_B |u|^s \log^\alpha \left(e + \frac{|u|}{\|u\|_{s,B}} \right) dx \right)^{1/s} \qquad (6.6.7)$$

$$\leq C_6 \left(\int_B |u|^s \log^\alpha \left(e + \frac{|u|}{\|u\|_{t,B}} \right) dx \right)^{1/s}$$

and

$$C_7 \|u\|_{L^t(\log L)^\alpha(B)}$$

$$\leq \left(\int_B |u|^t \log^\alpha \left(e + \frac{|u|}{\|u\|_{s,B}} \right) dx \right)^{1/t} \qquad (6.6.8)$$

$$\leq C_8 \|u\|_{L^t(\log L)^\alpha(B)}$$

for any ball B and any $s > 0, t > 0$, and $\alpha > 0$. Consequently, $\|u\|_{L^s(\log L)^\alpha(B)} < \infty$ if and only if

$$\left(\int_B |u|^s \log^\alpha \left(e + \frac{|u|}{\|u\|_{t,B}} \right) dx \right)^{1/s} < \infty.$$

6.6.2 $L^p(\log L)^\alpha$-norm inequalities

Now, we shall prove the following weak reverse Hölder inequality with $L^p(\log L)^\alpha$-norm.

Theorem 6.6.1. *Let $u \in D'(\Omega, \wedge^l)$, $l = 0, 1, \ldots, n$, be an A-harmonic tensor in a domain $\Omega \subset \mathbf{R}^n$, $\sigma > 1$, and $0 < s, t < \infty$. Then, there exists a constant C, independent of u, such that*

$$\|u\|_{L^s(\log L)^\alpha(B)} \leq C|B|^{(t-s)/st} \|u\|_{L^t(\log L)^\beta(\sigma B)} \qquad (6.6.9)$$

for any constants $\alpha > 0$ *and* $\beta > 0$, *and all balls* B *with* $\sigma B \subset \Omega$ *and* $diam(B) \geq d_0 > 0$, *where* d_0 *is a fixed constant.*

Proof. For any ball $B \subset \Omega$ with $diam(B) \geq d_0 > 0$, we choose $\varepsilon > 0$ small enough and a constant C_1 such that

$$|B|^{-\varepsilon/st} \leq C_1. \tag{6.6.10}$$

By Lemma 3.1.1, we have

$$\|u\|_{s+\varepsilon,B} \leq C_2 |B|^{(t-(s+\varepsilon))/t(s+\varepsilon)} \|u\|_{t,\sigma_1 B} \tag{6.6.11}$$

for some $\sigma_1 > 1$. Now from (6.6.7), we have

$$C_3 \left(\int_B |u|^s \log^\alpha \left(e + \frac{|u|}{\|u\|_{t,B}} \right) dx \right)^{1/s}$$
$$\leq \|u\|_{L^s(\log L)^\alpha(B)} \tag{6.6.12}$$
$$\leq C_4 \left(\int_B |u|^s \log^\alpha \left(e + \frac{|u|}{\|u\|_{t,B}} \right) dx \right)^{1/s}.$$

Setting $B_1 = \{x \in B : |u|/\|u\|_{t,B} \geq 1\}$, $B_2 = \{x \in B : |u|/\|u\|_{t,B} < 1\}$ and using (6.6.12) and the elementary inequality $|a+b|^s \leq 2^s(|a|^s + |b|^s)$, where $s > 0$ is any constant, we find that

$$\|u\|_{L^s(\log L)^\alpha(B)}$$
$$= \left(\int_B |u|^s \log^\alpha \left(e + \frac{|u|}{\|u\|_{s,B}} \right) dx \right)^{1/s}$$
$$= \left(\int_{B_1} |u|^s \log^\alpha \left(e + \frac{|u|}{\|u\|_{s,B}} \right) dx \right.$$
$$\left. + \int_{B_2} |u|^s \log^\alpha \left(e + \frac{|u|}{\|u\|_{s,B}} \right) dx \right)^{1/s}$$
$$\leq 2^{1/s} \left(\left(\int_{B_1} |u|^s \log^\alpha \left(e + \frac{|u|}{\|u\|_{s,B}} \right) dx \right)^{\frac{1}{s}} \right. \tag{6.6.13}$$
$$\left. + \left(\int_{B_2} |u|^s \log^\alpha \left(e + \frac{|u|}{\|u\|_{s,B}} \right) dx \right)^{\frac{1}{s}} \right)$$
$$\leq 2^{1/s} \left(C_5 \left(\int_{B_1} |u|^s \log^\alpha \left(e + \frac{|u|}{\|u\|_{t,B}} \right) dx \right)^{\frac{1}{s}} \right.$$
$$\left. + C_6 \left(\int_{B_2} |u|^s \log^\alpha \left(e + \frac{|u|}{\|u\|_{t,B}} \right) dx \right)^{\frac{1}{s}} \right).$$

Next, since $|u|/\|u\|_{t,B} > 1$ on B_1, for $\varepsilon > 0$ appeared in (6.6.10), there exists $C_7 > 0$ such that

$$\log^{\alpha}\left(e+\frac{|u|}{\|u\|_{t,\sigma_1 B}}\right) \leq C_7\left(\frac{|u|}{\|u\|_{t,\sigma_1 B}}\right)^{\varepsilon}. \qquad (6.6.14)$$

Combining $(6.6.10), (6.6.11)$, and $(6.6.14)$, we obtain

$$\left(\int_{B_1}|u|^s \log^{\alpha}\left(e+\frac{|u|}{\|u\|_{t,B}}\right)dx\right)^{1/s}$$

$$\leq C_8\left(\frac{1}{\|u\|_{t,\sigma_1 B}^{\varepsilon}}\int_{B_1}|u|^{s+\varepsilon}dx\right)^{1/s}$$

$$\leq C_8\left(\frac{1}{\|u\|_{t,\sigma_1 B}^{\varepsilon}}\int_{B}|u|^{s+\varepsilon}dx\right)^{1/s} \qquad (6.6.15)$$

$$= \frac{C_8}{\|u\|_{t,\sigma_1 B}^{\varepsilon/s}}\left(\left(\int_{B}|u|^{s+\varepsilon}dx\right)^{\frac{1}{s+\varepsilon}}\right)^{(s+\varepsilon)/s}$$

$$\leq \frac{C_9}{\|u\|_{t,\sigma_1 B}^{\varepsilon/s}}\left(|B|^{(t-(s+\varepsilon))/t(s+\varepsilon)}\|u\|_{t,\sigma_1 B}\right)^{(s+\varepsilon)/s}$$

$$\leq C_{10}|B|^{(t-s)/st}\|u\|_{t,\sigma_1 B}.$$

Finally, since

$$\log^{\alpha}\left(e+\frac{|u|}{\|u\|_{t,B}}\right) \leq M_1\log^{\alpha}(e+1) \leq M_2, \quad x \in B_2$$

it follows that

$$\left(\int_{B_2}|u|^s\log^{\alpha}\left(e+\frac{|u|}{\|u\|_{t,B}}\right)dx\right)^{1/s}$$

$$\leq \left(\int_{B_2}M_2|u|^s dx\right)^{1/s} \qquad (6.6.16)$$

$$\leq C_{11}\|u\|_{s,B}$$

which in view of Lemma 3.1.1 gives

$$\|u\|_{s,B} \leq C_{12}|B|^{(t-s)/st}\|u\|_{t,\sigma_2 B}. \qquad (6.6.17)$$

Substituting $(6.6.17)$ into $(6.6.16)$ yields

$$\left(\int_{B_2}|u|^s\log^{\alpha}\left(e+\frac{|u|}{\|u\|_{t,B}}\right)dx\right)^{1/s} \leq C_{13}|B|^{(t-s)/st}\|u\|_{t,\sigma_2 B}. \qquad (6.6.18)$$

Combining $(6.6.13), (6.6.15)$, and $(6.6.18)$, we obtain

$$\|u\|_{L^s(\log L)^{\alpha}(B)} \leq C_{14}|B|^{(t-s)/st}\|u\|_{t,\sigma_3 B}, \qquad (6.6.19)$$

where $\sigma_3 = \max\{\sigma_1, \sigma_2\}$. Now since $\log^\beta \left(e + \frac{|u|}{\|u\|_{t,\sigma_3 B}} \right) \geq 1$ for $\beta > 0$, we obtain

$$\|u\|_{L^s (\log L)^\alpha (B)}$$

$$\leq C_{14} |B|^{(t-s)/st} \|u\|_{t,\sigma_3 B}$$

$$= C_{14} |B|^{(t-s)/st} \left(\int_{\sigma_3 B} |u|^t dx \right)^{1/t}$$

$$\leq C_{14} |B|^{(t-s)/st} \left(\int_{\sigma_3 B} |u|^t \log^\beta \left(e + \frac{|u|}{\|u\|_{t,\sigma_3 B}} \right) dx \right)^{1/t}$$

$$= C_{14} |B|^{(t-s)/st} \|u\|_{L^t (\log L)^\beta (\sigma_3 B)}. \quad \blacksquare$$

Using a similar method developed in the proof of Theorem 6.6.1 and Theorem 6.2.12, we can prove the following version of the weak reverse Hölder inequality with $L^p (\log L)^\alpha$-norm.

Theorem 6.6.2. *Let u be an A-harmonic tensor in a domain $\Omega \subset \mathbf{R}^n$, $\sigma > 1$, and $0 < s, t < \infty$. Then, there exists a constant C, independent of u, such that*

$$\|du\|_{L^s (\log L)^\alpha (B)} \leq C|B|^{(t-s)/st} \|du\|_{L^t (\log L)^\beta (\sigma B)} \qquad (6.6.20)$$

for all balls B with $\sigma B \subset \Omega$ and $diam(B) \geq d_0 > 0$. Here d_0 is a fixed constant, $\alpha > 0$ and $\beta > 0$ are any constants.

Note that the above version of the weak reverse Hölder inequality cannot be obtained by replacing u by du in Theorem 6.6.1 since du may not be a solution of the A-harmonic equation.

Notes to Chapter 6. In this chapter, we have developed the weak reverse Hölder inequalities for differential forms satisfying some version of the A-harmonic equation. However, several other different versions of the weak reverse Hölder inequality have been established during the recent years. For example, see [283] for reverse Hölder inequalities with boundary integrals and L^p-estimates for solutions of nonlinear elliptic and parabolic boundary-value problems. We encourage readers to see [284–297, 126, 268, 282, 171, 79, 84] for further interesting versions of the weak reverse Hölder inequality.

Chapter 7
Inequalities for operators

The purpose of this chapter is to present a series of the local and global estimates for some operators, including the homotopy operator T, the Laplace–Beltrami operator $\Delta = dd^\star + d^\star d$, Green's operator G, the gradient operator ∇, the Hardy–Littlewood maximal operator, and the differential operator, which act on the space of harmonic forms defined in a domain in $\mathbf{R}^n$, and the compositions of some of these operators. We introduce the Hardy–Littlewood maximal operator $\mathbb{M}_s$ and the sharp maximal operator $\mathbb{M}_s^\sharp$ applied to differential forms in Section 7.1. We develop some basic estimates for Green's operator $\nabla \circ T$ and $d \circ T$ in Section 7.2. We establish some L^s-estimates and imbedding inequalities for the compositions of homotopy operator T and Green's operator G in Section 7.3. In Section 7.4, we prove some Poincaré-type inequalities for $T \circ G$ and $G \circ T$. In Section 7.5, we obtain Poincaré-type inequalities for the homotopy operator T. In Section 7.6, we study various estimates for the composition $T \circ H$. In Section 7.7, we provide the estimates for the compositions of three operators. Finally, in Section 7.8, we offer some norm comparison theorems for the maximal operators.

7.1 Introduction

We begin this section by introducing the Hardy–Littlewood maximal operator. For a locally L^s-integrable form $u(y)$, the *Hardy–Littlewood maximal operator* $\mathbb{M}_s$ is defined by

$$\mathbb{M}_s(u) = \mathbb{M}_s u = \mathbb{M}_s u(x) = \sup_{r>0} \left(\frac{1}{|B(x,r)|} \int_{B(x,r)} |u(y)|^s dy \right)^{1/s}, \quad (7.1.1)$$

where $B(x,r)$ is the ball of radius r, centered at x, $1 \leq s < \infty$. We write $\mathbb{M}(u) = \mathbb{M}_1(u)$ if $s = 1$. Similarly, for a locally L^s-integrable form u, we define the *sharp maximal operator* $\mathbb{M}_s^\sharp$ by

R.P. Agarwal et al., *Inequalities for Differential Forms*,
DOI 10.1007/978-0-387-68417-8_7, © Springer Science+Business Media, LLC 2009

$$\mathbb{M}_s^\sharp(u) = \mathbb{M}_s^\sharp u = \mathbb{M}_s^\sharp u(x)$$

$$= \sup_{r>0} \left(\frac{1}{|B(x,r)|} \int_{B(x,r)} |u(y) - u_{B(x,r)}|^s dy \right)^{1/s}. \tag{7.1.2}$$

These operators and those mentioned above play an important role in many diverse fields, including partial differential equations and analysis. For example, the Laplace–Beltrami operator and Green's operator are widely used in physics, potential theory, and nonlinear elasticity, etc. The gradient operator and the homotopy operator are effective tools in analysis which have found many applications in different areas of mathematics, including topology and differential geometry. Some estimates for the Laplace–Beltrami operator Δ, Green's operator G, and projection operator have been established in previous chapters. Also, see [298–309, 60, 261, 127, 128, 113, 48, 163–165, 174, 133, 83, 86] for recent results related to these operators. In this chapter, we develop several other local and global estimates for these operators and their compositions. We also study the differential operator d and codifferential operator $d^\star$ and some other related operators.

7.2 Some basic estimates

The L^s-estimates for the compositions of Green's operator G, differential operator d, and codifferential operator $d^\star$ have been developed and used in Chapter 3. Now, we recall the following estimate which was established in [21].

7.2.1 Estimates related to Green's operator

Let $u \in C^\infty(\wedge^l M)$, $l = 1, 2, \ldots, n-1$. For $1 < s < \infty$, there exists a constant C, independent of u, such that

$$\|dd^\star G(u)\|_{s,M} + \|d^\star dG(u)\|_{s,M} + \|dG(u)\|_{s,M}$$

$$+\|d^\star G(u)\|_{s,M} + \|G(u)\|_{s,M} \tag{7.2.1}$$

$$\leq C\|u\|_{s,M}.$$

Proof of inequality (7.2.1). First, we shall prove (7.2.1) for the case $s \geq 2$ by using the closed graph theorem. From [144], we have the following L^2-estimate:

$$\|dd^\star G(u)\|_{2,M} + \|d^\star dG(u)\|_{2,M} + \|dG(u)\|_{2,M}$$

$$+\|d^\star G(u)\|_{2,M} + \|G(u)\|_{2,M} \tag{7.2.2}$$

$$\leq C_1 \|u\|_{2,M}.$$

Let $\|u_n\|_{s,M} + \|G(u_n) - v\|_{s,M} \to 0$ as $n \to \infty$. Since L^s is imbedded in L^2, it follows that $\|u_n\|_{2,M} + \|G(u_n) - v\|_{2,M} \to 0$ as $n \to \infty$. From (7.2.2), we have

$$\|G(u_n)\|_{2,M} = \|G(u_n - H(u_n))\|_{2,M}$$

$$\leq C_2 \|u_n - H(u_n)\|_{2,M}$$

$$\leq C_2 \|u_n\|_{2,M} + \|H(u_n)\|_{2,M} \tag{7.2.3}$$

$$\leq C_3 \|u_n\|_{2,M}$$

$$\to 0$$

as $n \to \infty$, where H is the projection operator. Thus, $v = 0$, and so by the closed graph theorem G is bounded. Repeating this procedure for the compositions $d^\star dG$, $dd^\star G$, dG, and $d^\star G$, the proof for the case $s \geq 2$ follows. Next, we assume that $1 < s < 2$. Notice that for smooth forms, Green's operator commutes with anything the Laplacian does (see [23]) and is self-adjoint. In particular, when $\eta, u \in C^\infty(\wedge^l M)$, we have

$$\begin{cases} (G(\eta), u) &= (\eta, G(u)), \\ dG(\eta) &= Gd(\eta), \\ d^\star G(u) &= Gd^\star(u). \end{cases} \tag{7.2.4}$$

Let $\eta_n = G(u)(|G(u)|^2 + 1/n)^{(s-2)/2}$ and observe that $\eta_n \in C^\infty$. By the Lebesgue dominated convergence theorem (LDCT), we obtain

$$\|\eta_n\|_{t,M}^t \to \|G(u)\|_{s,M}^s \tag{7.2.5}$$

as $n \to \infty$, where t is a positive number with $1/t + 1/s = 1$. Next notice that $|(G(u), \eta_n)|$ increases to $\|G(u)\|_{s,M}^s$ (again by the LDCT). Therefore, given $\varepsilon > 0$, we may choose a large positive integer N, so that for $n > N$ we have $\|G(u)\|_{s,M}^s \leq |(G(u), \eta)| + \varepsilon$. Now since $1 < s < 2$, it follows that $t > 2$. Hence, using Hölder inequality and the case for $s \geq 2$, we have

$$\|G(u)\|_{s,M}^s \leq |(G(u), \eta)| + \varepsilon$$

$$= |(u, G(\eta_n))| + \varepsilon$$

$$\leq \|u\|_{s,M} \|G(\eta_n)\|_{t,M} + \varepsilon \tag{7.2.6}$$

$$\leq C_4 \|u\|_{s,M} \|\eta_n\|_{t,M} + \varepsilon$$

$$\to C_5 \|u\|_{s,M} \|G(u)\|_{t,M}^{s/t} + \varepsilon.$$

Thus, letting $\varepsilon \to 0$, we obtain

$$\|G(u)\|_{s,M}^s \leq C_6 \|u\|_{s,M} \|G(u)\|_{s,M}^{s/t}. \tag{7.2.7}$$

Assuming $\|G(u)\|_{t,M} > 0$, and dividing (7.2.7) by $\|G(u)\|_{s,M}^{s/t}$, we obtain

$$\|G(u)\|_{s,M} \leq C_7 \|u\|_{s,M}. \tag{7.2.8}$$

Notice that (7.2.8) obviously holds for the case $\|G(u)\|_{t,M} = 0$. Next, we set $\eta_n = dG(u)(|dG(u)|^2 + 1/n)^{(s-2)/2}$. As above, $\eta_n \in C^\infty$ and by the LDCT,

$$\|\eta_n\|_{t,M}^t \to \|dG(u)\|_{s,M}^s \tag{7.2.9}$$

as $n \to \infty$. Again we see that $|(dG(u), \eta_n)|$ increases to $\|dG(u)\|_{s,M}^s$ by the LDCT. Therefore, for $\varepsilon > 0$, we may choose a large positive integer N, such that for $n > N$, we have

$$\|dG(u)\|_{s,M}^s \leq |(dG(u), \eta_n)| + \varepsilon.$$

Now we find that

$$\begin{aligned}
\|dG(u)\|_{s,M}^s &\leq |(dG(u), \eta_n)| + \varepsilon \\
&= |(u, G(d^\star \eta_n))| + \varepsilon \\
&\leq \|u\|_{s,M} \|G(d^\star \eta_n)\|_{t,M} + \varepsilon \\
&\leq C_8 \|u\|_{s,M} \|\eta_n\|_{t,M} + \varepsilon \\
&\to C_8 \|u\|_{s,M} \|dG(u)\|_{t,M}^{s/t} + \varepsilon.
\end{aligned} \tag{7.2.10}$$

Thus, as above, we obtain the following analog of (7.2.8):

$$\|dG(u)\|_{s,M} \leq C_9 \|u\|_{s,M}. \tag{7.2.11}$$

Finally, choosing

$$\begin{aligned}
\eta_n &= d^\star G(u)(|d^\star G(u)|^2 + 1/n)^{(s-2)/2}, \\
\eta_n &= dd^\star G(u)(|dd^\star G(u)|^2 + 1/n)^{(s-2)/2}, \\
\eta_n &= d^\star dG(u)(|d^\star dG(u)|^2 + 1/n)^{(s-2)/2},
\end{aligned}$$

and following the same method, it follows that

$$\begin{aligned}
\|d^\star G(u)\|_{s,M} &\leq C_{10} \|u\|_{s,M}, \\
\|dd^\star G(u)\|_{s,M} &\leq C_{11} \|u\|_{s,M}, \\
\|d^\star dG(u)\|_{s,M} &\leq C_{12} \|u\|_{s,M}.
\end{aligned}$$

Combining these inequalities with (7.2.8) and (7.2.11), we obtain the required inequality (7.2.1).   ∎

7.2.2 Estimates for $\nabla \circ T$

Now, we prove the following basic $A_r(\Omega)$-weighted estimate for $\nabla \circ T$.

Theorem 7.2.1. *Let $u \in L^s_{loc}(\Omega, \wedge^l)$, $l = 1, 2, \ldots, n$, $1 < s < \infty$, be a differential form satisfying (1.2.10) in a bounded, convex domain $\Omega \subset \mathbf{R}^n$ and $T : C^\infty(\Omega, \wedge^l) \to C^\infty(\Omega, \wedge^{l-1})$ be the homotopy operator defined in (1.5.1). Assume that $\rho > 1$ and $(w_1, w_2) \in A^\lambda_r(\Omega)$ for some $r > 1$ and $\lambda > 0$. Then, there exists a constant C, independent of u, such that*

$$\left(\int_B |\nabla(Tu)|^s w_1^\alpha dx \right)^{1/s} \leq C|B| \left(\int_{\rho B} |u|^s w_2^{\alpha\lambda} dx \right)^{1/s} \tag{7.2.12}$$

for any real number α with $0 < \alpha < 1$.

Note that (7.2.12) can be written as

$$\|\nabla(Tu)\|_{s,B,w_1^\alpha} \leq C|B|\|u\|_{s,\rho B,w_2^{\alpha\lambda}}. \tag{7.2.12$'$}$$

Proof. Let $t = s/(1-\alpha)$, so that $1 < s < t$. Using the Hölder inequality, we have

$$
\begin{aligned}
\left(\int_B |\nabla(Tu)|^s w_1^\alpha dx \right)^{1/s} \\
= \left(\int_B \left(|\nabla(Tu)| w_1^{\alpha/s} \right)^s dx \right)^{1/s} \\
\leq \|\nabla(Tu)\|_{t,B} \left(\int_B w_1^{t\alpha/(t-s)} dx \right)^{(t-s)/st} \\
= \|\nabla(Tu)\|_{t,B} \left(\int_B w_1 dx \right)^{\alpha/s}.
\end{aligned}
\tag{7.2.13}
$$

Thus, from inequality (1.5.5), we obtain

$$\|\nabla(Tu)\|_{t,B} \leq C_1|B|\|u\|_{t,B}. \tag{7.2.14}$$

Choose

$$m = \frac{s}{1 + \alpha\lambda(r-1)},$$

so that $m < s$. Substituting (7.2.14) into (7.2.13) and using Lemma 3.1.1, we find that

$$\left(\int_B |\nabla(Tu)|^s w_1^\alpha dx \right)^{1/s}$$

$$\leq C_1 |B| \|u\|_{t,B} \left(\int_B w_1 dx \right)^{\alpha/s} \tag{7.2.15}$$

$$\leq C_2 |B| |B|^{(m-t)/mt} \|u\|_{m,\rho B} \left(\int_B w_1 dx \right)^{\alpha/s} .$$

Now, using the Hölder inequality with $1/m = 1/s + (s-m)/sm$, we have

$$\|u\|_{m,\rho B}$$

$$= \left(\int_{\rho B} |u|^m dx \right)^{1/m}$$

$$= \left(\int_{\rho B} \left(|u| w_2^{\alpha\lambda/s} w_2^{-\alpha\lambda/s} \right)^m dx \right)^{1/m} \tag{7.2.16}$$

$$\leq \left(\int_{\rho B} |u|^s w_2^{\alpha\lambda} dx \right)^{1/s} \left(\int_{\rho B} \left(\frac{1}{w_2} \right)^{1/(r-1)} dx \right)^{\alpha\lambda(r-1)/s}$$

for all balls B with $\rho B \subset \Omega$. Substituting (7.2.16) into (7.2.15), we obtain

$$\left(\int_B |\nabla(Tu)|^s w_1^\alpha dx \right)^{1/s}$$

$$\leq C_2 |B| |B|^{(m-t)/mt} \left(\int_{\rho B} |u|^s w_2^\alpha dx \right)^{1/s}$$

$$\times \left(\int_B w_1 dx \right)^{\alpha/s} \left(\int_{\rho B} \left(\frac{1}{w_2} \right)^{1/(r-1)} dx \right)^{\alpha\lambda(r-1)/s} \tag{7.2.17}$$

$$= C_2 |B| |B|^{(m-t)/mt} \left(\int_{\rho B} |u|^s w_2^\alpha dx \right)^{1/s} \|w_1\|_{1,B}^{\alpha/s} \cdot \left\| \frac{1}{w_2} \right\|_{1/(r-1),\rho B}^{\alpha\lambda/s}.$$

Next, since $(w_1, w_2) \in A_r^\lambda(\Omega)$, we find

$$\|w_1\|_{1,B}^{\alpha/s} \cdot \left\| \frac{1}{w_2} \right\|_{1/(r-1),\rho B}^{\alpha\lambda/s}$$

$$= \left(\int_B w_1 dx \right)^{\alpha/s} \left(\int_{\rho B} \left(\frac{1}{w_2} \right)^{1/(r-1)} dx \right)^{\alpha\lambda(r-1)/s}$$

$$\leq \left[\left(\int_{\rho B} w_1 dx \right) \left(\int_{\rho B} \left(\frac{1}{w_2} \right)^{1/(r-1)} dx \right)^{\lambda(r-1)} \right]^{\alpha/s}$$

$$= \left[|\rho B|^{\lambda(r-1)+1} \left(\frac{1}{|\rho B|} \int_{\rho B} w_1 dx \right) \right. \tag{7.2.18}$$

$$\left. \times \left(\frac{1}{|\rho B|} \int_{\rho B} \left(\frac{1}{w_2} \right)^{1/(r-1)} dx \right)^{\lambda(r-1)} \right]^{\alpha/s}$$

$$\leq C_3 |\rho B|^{\alpha\lambda(r-1)/s + \alpha/s}$$

$$\leq C_4 |B|^{\alpha\lambda(r-1)/s + \alpha/s}.$$

Combining (7.2.18) and (7.2.17) and using

$$\frac{m-t}{mt} = -\frac{\alpha}{s} - \frac{\alpha\lambda(r-1)}{s},$$

we find

$$\left(\int_B |\nabla(Tu)|^s w_1^\alpha \, dx\right)^{1/s} \le C_5 |B| \left(\int_{\rho B} |u|^s w_2^{\alpha\lambda} \, dx\right)^{1/s} \qquad (7.2.19)$$

for all balls B with $\rho B \subset \Omega$. Thus, inequality (7.2.12) holds. $\blacksquare$

The method similar to the proof of Theorem 7.2.1 also leads to the following inequality:

$$\|Tu\|_{s,B,w_1^\alpha} \le C|B| diam(B) \|u\|_{s,\rho B, w_2^{\alpha\lambda}}, \qquad (7.2.20)$$

where α is any real number with $0 < \alpha \le 1$ and $\rho > 1$.

Next, we prove the following local weighted imbedding theorem for differential forms under the homotopy operator T.

Theorem 7.2.2. *Let $u \in L^s_{loc}(\Omega, \wedge^l)$, $l = 1, 2, \ldots, n$, $1 < s < \infty$, be a differential form satisfying (1.2.10) in a bounded, convex domain $\Omega \subset \mathbf{R}^n$ and $T : L^s(\Omega, \wedge^l) \to W^{1,s}(\Omega, \wedge^{l-1})$, $l = 1, 2, \ldots, n$, be the operator defined in (1.5.1). Assume that $\rho > 1$ and $(w_1, w_2) \in A_r^\lambda(\Omega)$ for some $r > 1$ and $\lambda > 0$. Then, there exists a constant C, independent of u, such that*

$$\|Tu\|_{W^{1,s}(B),\, w_1^\alpha} \le C|B| \|u\|_{s,\rho B, w_2^{\alpha\lambda}} \qquad (7.2.21)$$

for all balls B with $\rho B \subset \Omega$ and any real number α with $0 < \alpha < 1$.

Proof. From the definition of the $\|\cdot\|_{W^{1,s}(B),\, w_1^\alpha}$ norm, (7.2.12)′, and (7.2.20), we have

$$\|Tu\|_{W^{1,s}(B),\, w_1^\alpha}$$
$$= diam(B)^{-1} \|Tu\|_{s,B,w_1^\alpha} + \|\nabla(Tu)\|_{s,B,w_1^\alpha}$$
$$\le diam(B)^{-1} \left(C_1 |B| diam(B) \|u\|_{s,\rho B, w_2^{\alpha\lambda}}\right) + C_2 |B| \|u\|_{s,\rho B, w_2^{\alpha\lambda}}$$
$$\le C_1 |B| \|u\|_{s,\rho B, w_2^{\alpha\lambda}} + C_2 |B| \|u\|_{s,\rho B, w_2^{\alpha\lambda}}$$
$$\le C_3 |B| \|u\|_{s,\rho B, w_2^{\alpha\lambda}},$$

which is equivalent to (7.2.21). $\blacksquare$

In Theorems 7.2.1 and 7.2.2, the parameters α and λ are arbitrary real numbers with $0 < \alpha < 1$ and $\lambda > 0$. Therefore, for different values of α

and λ we obtain different versions of the weighted imbedding inequality. For example, let $\lambda = 1$ and $t = 1 - \alpha$ in Theorem 7.2.1 and write $d\mu = w_1(x)dx$ and $d\nu = w_2(x)dx$. Then, inequality (7.2.12) reduces to

$$\left(\int_B |\nabla(Tu)|^s w_1^{-t} d\mu \right)^{1/s} \leq C|B| \left(\int_{\rho B} |u|^s w_2^{-t} d\nu \right)^{1/s}. \qquad (7.2.22)$$

If we choose $\alpha = 1/r$ in Theorem 7.2.1, then (7.2.12) reduces to

$$\left(\int_B |\nabla(Tu)|^s w_1^{1/r} dx \right)^{1/s} \leq C|B| \left(\int_{\rho B} |u|^s w_2^{\lambda/r} dx \right)^{1/s}, \qquad (7.2.23)$$

where $r > 1$ and $\lambda > 0$.

If we choose $\alpha = 1/s$ in Theorem 7.2.1, then $0 < \alpha < 1$ since $1 < s < \infty$. Thus, (7.2.12) reduces to the following symmetric version:

$$\left(\int_B |\nabla(Tu)|^s w_1^{1/s} dx \right)^{1/s} \leq C|B| \left(\int_{\rho B} |u|^s w_2^{\lambda/s} dx \right)^{1/s}, \qquad (7.2.24)$$

where $\lambda > 0$.

Next, set $\lambda = s$ in (7.2.24), we have

$$\left(\int_B |\nabla(Tu)|^s w_1^{1/s} dx \right)^{1/s} \leq C|B| \left(\int_{\rho B} |u|^s w_2 dx \right)^{1/s}, \qquad (7.2.25)$$

where $s > 1$.

Finally, select $\lambda = r$ in (7.2.23), to obtain

$$\left(\int_B |\nabla(Tu)|^s w_1^{1/r} dx \right)^{1/s} \leq C|B| \left(\int_{\rho B} |u|^s w_2 dx \right)^{1/s}, \qquad (7.2.26)$$

where $r > 1$.

In addition to the condition $(w_1, w_2) \in A_r^\lambda(\Omega)$ for some $r > 1$ and $\lambda > 0$, if $w_1(x) \in A_r(\Omega)$ in Theorems 7.2.1 and 7.2.2, then using Theorem 6.1.4, we can also prove that inequalities (7.2.12), (7.2.20), and (7.2.21) hold, that is,

$$\|\nabla(Tu)\|_{s,B,w_1} \leq C|B| \|u\|_{s,\rho B,w_2^\lambda}, \qquad (7.2.27)$$

$$\|Tu\|_{s,B,w_1} \leq C|B| diam(B) \|u\|_{s,\rho B,w_2^\lambda}, \qquad (7.2.28)$$

and

$$\|Tu\|_{W^{1,s}(B),w_1} \leq C|B| \|u\|_{s,B,w_2^\lambda}. \qquad (7.2.29)$$

Using the local weighted inequalities developed above, we shall now prove the following global estimates.

Theorem 7.2.3. *Let $u \in L^s(D, \wedge^l)$, $l = 1, 2, \ldots, n$, $1 < s < \infty$, be a differential form satisfying (1.2.10) in a bounded, convex domain $D \subset \mathbf{R}^n$ and $T : L^s(D, \wedge^l) \to W^{1,s}(D, \wedge^{l-1})$, $l = 1, 2, \ldots, n$, be the homotopy operator defined in (1.5.1). Assume that $(w_1, w_2) \in A_r^\lambda(\Omega)$ for some $r > 1$ and $\lambda > 0$. Then, there exists a constant C, independent of u, such that*

$$\|\nabla(Tu)\|_{s,D,w_1^\alpha} \le C\|u\|_{s,D,w_2^{\alpha\lambda}} \tag{7.2.30}$$

and

$$\|Tu\|_{W^{1,s}(D),\, w_1^\alpha} \le C\|u\|_{s,D,w_2^{\alpha\lambda}} \tag{7.2.31}$$

for any real number α with $0 < \alpha < 1$.

Proof. Using (7.2.12) and Theorem 1.5.3, we find that

$$\begin{aligned}
\|\nabla(Tu)\|_{s,D,w_1^\alpha} &= \left(\int_D |\nabla(Tu)|^s w_1^\alpha dx\right)^{1/s}\\
&\le \sum_{Q\in\mathcal{V}} \left(C_1|Q| \left(\int_{\rho Q} |u|^s w_2^{\alpha\lambda}dx\right)^{1/s}\right)\\
&\le C_1|D| \sum_{Q\in\mathcal{V}} \left(\int_{\rho Q} |u|^s w_2^{\alpha\lambda}dx\right)^{1/s}\\
&\le C_1|D| \sum_{Q\in\mathcal{V}} \left(\int_D |u|^s w_2^{\alpha\lambda}dx\right)^{1/s}\\
&\le C_3 \left(\int_D |u|^s w_2^{\alpha\lambda}dx\right)^{1/s}\\
&= C_3\|u\|_{s,D,w_2^{\alpha\lambda}}
\end{aligned} \tag{7.2.32}$$

since D is bounded. Thus, (7.2.30) holds. Similarly, using Theorem 1.5.3 and (7.2.20), we have

$$\|Tu\|_{s,D,w_1^\alpha} \le C_4 diam(D)\|u\|_{s,D,w_2^{\alpha\lambda}}. \tag{7.2.33}$$

From the definition of the $\|\cdot\|_{W^{1,s}(D),\, w^\alpha}$ norm, (7.2.32), and (7.2.33), we obtain

$$\begin{aligned}
\|Tu\|_{W^{1,s}(D),\, w_1^\alpha} &= diam(D)^{-1}\|Tu\|_{s,D,w_1^\alpha} + \|\nabla(Tu)\|_{s,D,w_1^\alpha}\\
&\le C_4\|u\|_{s,D,w_2^{\alpha\lambda}} + C_3\|u\|_{s,D,w_2^{\alpha\lambda}}\\
&\le C_5\|u\|_{s,D,w_2^{\alpha\lambda}},
\end{aligned} \tag{7.2.34}$$

which implies that (7.2.31) holds. ∎

Remark. Choosing some particular values of α in (7.2.30) and (7.2.31) leads to global results similar to the local case. Further, in addition to the condition $(w_1, w_2) \in A_r^\lambda(\Omega)$, if $w_1 \in A_r(\Omega)$ in Theorems 7.2.3, we find that (7.2.30) and (7.2.31) become

$$\|\nabla(Tu)\|_{s,D,w_1} \leq C\|u\|_{s,D,w_2^\lambda}, \tag{7.2.35}$$

$$\|Tu\|_{W^{1,s}(D),w_1} \leq C\|u\|_{s,D,w_2^\lambda}, \tag{7.2.36}$$

respectively.

7.2.3 Estimates for $d \circ T$

We shall conclude this section with the local and global weighted estimates for $d \circ T$, which will be used later.

Theorem 7.2.4. *Let* $u \in L_{loc}^s(\Omega, \wedge^l)$, $l = 1, 2, \ldots, n$, $1 < s < \infty$, *be a differential form satisfying (1.2.10) in a bounded, convex domain* $\Omega \subset \mathbf{R}^n$ *and* $T : C^\infty(\Omega, \wedge^l) \to C^\infty(\Omega, \wedge^{l-1})$ *be the homotopy operator defined in (1.5.1). Assume that* $\sigma > 1$ *and* $w \in A_r(\Omega)$ *for some* $r > 1$. *Then,*

$$\left(\int_B |d(Tu)|^s w^\alpha dx\right)^{1/s} \leq C|B|^{(1-\alpha)/s} \left(\int_{\sigma B} |u|^s w^\alpha dx\right)^{1/s} \tag{7.2.37}$$

for all balls B *with* $\sigma B \subset \Omega$ *and any real number* α *with* $0 < \alpha \leq 1$.

Proof. First, we assume that $0 < \alpha < 1$. Let $t = s/(1-\alpha)$. From Caccioppoli-type estimates for the solutions of the A-harmonic equation, there exists a constant C_1, independent of u, such that

$$\|du\|_{t,B} \leq C_1 diam(B)^{-1}\|u\|_{t,\sigma B} \tag{7.2.38}$$

for any solution u of the A-harmonic equation in Ω and all balls or cubes B with $\sigma B \subset \Omega$, where $\sigma > 1$. Now let $m = s/(1 + \alpha(r-1))$. Using the Hölder inequality, (7.2.38), Lemma 3.1.1, and the basic estimate

$$\|d(Tu)\|_{s,B} \leq \|u\|_{s,B} + C_2 diam(B)\|du\|_{s,B}, \tag{7.2.39}$$

we have

$$\left(\int_B |d(Tu)|^s w^\alpha dx\right)^{1/s}$$

$$= \left(\int_B \left(|d(Tu)|w^{\alpha/s}\right)^s dx\right)^{1/s}$$

$$\leq \|d(Tu)\|_{t,B} \left(\int_B w^{t\alpha/(t-s)} dx\right)^{(t-s)/st}$$

$$= \|d(Tu)\|_{t,B} \left(\int_B w dx\right)^{\alpha/s}$$

$$\leq \left(\|u\|_{t,B} + C_3 diam(B)\|du\|_{t,B}\right) \left(\int_B w dx\right)^{\alpha/s} \qquad (7.2.40)$$

$$\leq \left(\|u\|_{t,B} + C_4\|u\|_{t,\sigma_1 B}\right) \left(\int_B w dx\right)^{\alpha/s}$$

$$\leq C_5\|u\|_{t,\sigma_1 B} \left(\int_B w dx\right)^{\alpha/s}$$

$$\leq C_6|B|^{(m-t)/mt}\|u\|_{m,\sigma_2 B} \left(\int_B w dx\right)^{\alpha/s},$$

where $\sigma_2 > \sigma_1 > 1$. Using the Hölder inequality again, we obtain

$$\|u\|_{m,\sigma_2 B}$$

$$= \left(\int_{\sigma_2 B} \left(|u|w^{\alpha/s}w^{-\alpha/s}\right)^m dx\right)^{1/m} \qquad (7.2.41)$$

$$\leq \left(\int_{\sigma_2 B} |u|^s w^\alpha dx\right)^{1/s} \left(\int_{\sigma_2 B} \left(\frac{1}{w}\right)^{1/(r-1)} dx\right)^{\alpha(r-1)/s}$$

for all balls B with $\sigma_2 B \subset \Omega$. Substituting (7.2.41) into (7.2.40), and then using the condition of $A_r(\Omega)$-weights, we find that

$$\left(\int_B |d(Tu)|^s w^\alpha dx\right)^{1/s} \leq C_6|B|^{(1-\alpha)/s} \left(\int_{\sigma_2 B} |u|^s w^\alpha dx\right)^{1/s}$$

which finishes the proof of Theorem 7.2.4 for the case $0 < \alpha < 1$. For the case $\alpha = 1$, the proof is similar to that of Theorem 7.2.1. ∎

Now, we shall prove the following global $A_r(\Omega)$-weighted imbedding inequality in a bounded domain Ω for A-harmonic tensors.

Theorem 7.2.5. *Let $u \in L^s(\Omega, \wedge^l)$, $l = 1, 2, \ldots, n$, $1 < s < \infty$, be an A-harmonic tensor in a bounded, convex domain $\Omega \subset \mathbf{R}^n$ and $T : C^\infty(\Omega, \wedge^l) \to C^\infty(\Omega, \wedge^{l-1})$ be a homotopy operator defined in (1.5.1). Assume that $w \in A_r(\Omega)$ for some $r > 1$. Then, there exists a constant C, independent of u, such that*

$$\left(\int_\Omega |d(Tu)|^s w^\alpha dx\right)^{1/s} \leq C \left(\int_\Omega |u|^s w^\alpha dx\right)^{1/s} \qquad (7.2.42)$$

for any real number α with $0 < \alpha \leq 1$.

Proof. Using (7.2.37) and the covering lemma, we have

$$\left(\int_\Omega |d(Tu)|^s w^\alpha dx\right)^{1/s}$$

$$\leq \sum_{Q\in\mathcal{V}}\left(C_1|Q|^{(1-\alpha)/s}\left(\int_{\rho Q}|u|^s w^\alpha dx\right)^{1/s}\right)$$

$$\leq C_1|\Omega|^{(1-\alpha)/s}\sum_{Q\in\mathcal{V}}\left(\int_{\rho Q}|u|^s w^\alpha dx\right)^{1/s}$$

$$\leq C_1|\Omega|^{(1-\alpha)/s}\sum_{Q\in\mathcal{V}}\left(\int_\Omega |u|^s w^\alpha dx\right)^{1/s}$$

$$\leq C_3\left(\int_\Omega |u|^s w^\alpha dx\right)^{1/s},$$

which implies that (7.2.42) holds. ∎

7.3 Compositions of operators

We prove the local Sobolev–Poincaré imbedding theorems for the compositions of the gradient operator ∇, the homotopy operator T, the Laplace–Beltrami operator Δ, and Green's operator G acting on a smooth form.

7.3.1 Estimates for $\nabla \circ T \circ G$ and $T \circ \Delta \circ G$

Lemma 7.3.1. *Let $u \in C^\infty(\wedge^l B)$, $l = 1, 2, \ldots, n$, and $1 < p < \infty$. Then, $\nabla(T(G(u))) \in L^p(\wedge^l B)$ and $T(\Delta(G(u))) \in W^{1,p}(B, \wedge^l)$. Moreover, there exists a constant C, independent of u, such that*

$$\|\nabla(T(G(u)))\|_{p,B} \leq C|B|\|u\|_{p,B} \qquad (7.3.1)$$

and

$$\|T(\Delta(G(u)))\|_{W^{1,p}(B)} \leq C|B|\|u\|_{p,B} \qquad (7.3.2)$$

for all balls B with $B \subset \mathbf{R}^n$.

Proof. We only need to prove (7.3.1) and (7.3.2). In fact, from these two inequalities, the remaining part of the theorem follows. From inequality (1.5.5), we have

$$\|\nabla(T(\omega))\|_{p,B} \leq C|B|\|\omega\|_{p,B} \qquad (7.3.3)$$

for any $\omega \in L^p_{loc}(\wedge^l B)$. Now, setting $\omega = G(u)$ in (7.3.3), we find that

$$\|\nabla(T(G(u)))\|_{p,B} \leq C|B|\|G(u)\|_{p,B}. \qquad (7.3.4)$$

Using (3.3.3) and (7.3.4), we obtain (7.3.1). Now, applying (3.3.14), (1.5.5), (1.5.6), and Theorem 3.3.2, we obtain

$$
\begin{aligned}
&\|T(\Delta(G(u)))\|_{W^{1,p}(B)} \\
&= diam(B)^{-1}\|T(\Delta(G(u)))\|_{p,B} + \|\nabla(T(\Delta(G(u))))\|_{p,B} \\
&\leq diam(B)^{-1} \cdot C_1|B|diam(B)\|\Delta(G(u))\|_{p,B} + C_2|B|\|\Delta(G(u))\|_{p,B} \\
&\leq C_3|B|\|\Delta(G(u))\|_{p,B} \\
&\leq C_3|B|\|u\|_{p,B}.
\end{aligned}
$$

Thus, inequality (7.3.2) holds.   ■

Theorem 7.3.2. *Let* $u \in C^\infty(\wedge^l\Omega)$, $l = 1, 2, \ldots, n$, $1 < p < \infty$, *be a differential form satisfying* (1.2.10) *in a bounded, convex domain* Ω *and* $T : C^\infty(\Omega, \wedge^l) \to C^\infty(\Omega, \wedge^{l-1})$ *be the homotopy operator defined in* (1.5.1). *Assume that* $\rho > 1$ *and* $(w_1, w_2) \in A_{r,\lambda}(\Omega)$ *for some* $\lambda \geq 1$ *and* $1 < r < \infty$ *with* $1/r + 1/r' = 1$. *Then,* $T(\Delta(G(u))) \in L^p(\wedge^l B, w_1^\alpha)$. *Moreover, there exists a constant* C, *independent of* u, *such that*

$$
\|T(\Delta(G(u)))\|_{p,B,w_1^\alpha} \leq C|B|diam(B)\|u\|_{p,\rho B,w_2^\alpha} \tag{7.3.5}
$$

for all balls B *with* $\rho B \subset \Omega$ *and any real number* α *with* $\alpha < \lambda$.

Proof. Similar to the proof of Lemma 7.3.1, we only need to prove (7.3.5). We choose $t = \lambda p/(\lambda - \alpha)$, then the Hölder inequality yields

$$
\begin{aligned}
&\|T(\Delta(G(u)))\|_{p,B,w_1^\alpha} \\
&= \left(\int_B \left(|T(\Delta(G(u)))|w_1^{\alpha/p}\right)^p dx\right)^{1/p} \\
&\leq \|T(\Delta(G(u)))\|_{t,B} \left(\int_B w_1^{t\alpha/(t-p)}dx\right)^{(t-p)/pt} \\
&= \|T(\Delta(G(u)))\|_{t,B} \left(\int_B w_1^\lambda dx\right)^{\alpha/\lambda p}.
\end{aligned}
\tag{7.3.6}
$$

Now, applying (1.5.6) and Theorem 3.3.2, we obtain

$$
\begin{aligned}
&\|T(\Delta(G(u)))\|_{t,B} \\
&\leq C_1|B|diam(B)\|\Delta(G(u))\|_{t,B} \\
&\leq C_2|B|diam(B)\|u\|_{t,B}.
\end{aligned}
\tag{7.3.7}
$$

Choose $m = \lambda p/(\lambda + \alpha(r - 1))$, so that $m < p$. By the weak reverse Hölder inequality, we have

$$
\|u\|_{t,B} \leq C_3|B|^{(m-t)/mt}\|u\|_{m,\rho B}. \tag{7.3.8}
$$

Combining $(7.3.6), (7.3.7)$, and $(7.3.8)$, we find that

$$\|T(\Delta(G(u)))\|_{p,B,w_1^\alpha}$$

$$\leq C_4|B|diam(B)|B|^{(m-t)/mt}\|u\|_{m,\rho B}\left(\int_B w_1^\lambda dx\right)^{\alpha/\lambda p}. \tag{7.3.9}$$

Using the Hölder inequality again with $1/m = 1/p + (p-m)/pm$, it follows that

$$\|u\|_{m,\rho B}$$

$$= \left(\int_{\rho B}|u|^m dx\right)^{1/m}$$

$$= \left(\int_{\rho B}\left(|u|w_2^{\alpha/p}w_2^{-\alpha/p}\right)^m dx\right)^{1/m} \tag{7.3.10}$$

$$\leq \left(\int_{\rho B}|u|^p w_2^\alpha dx\right)^{1/p}\left(\int_{\rho B}\left(\tfrac{1}{w_2}\right)^{\lambda/(r-1)}dx\right)^{\alpha(r-1)/\lambda p}$$

for all balls B with $\rho B \subset M$. Substituting $(7.3.10)$ into $(7.3.9)$, we obtain

$$\|T(\Delta(G(u)))\|_{p,B,w_1^\alpha}$$

$$\leq C_4|B|diam(B)|B|^{(m-t)/mt}\|u\|_{p,\rho B,w_2^\alpha}$$

$$\times \left(\int_B w_1^\lambda dx\right)^{\alpha/\lambda p}\left(\int_{\rho B}\left(\tfrac{1}{w_2}\right)^{\lambda/(r-1)}dx\right)^{\alpha(r-1)/\lambda p}. \tag{7.3.11}$$

From the condition $(w_1,w_2) \in A_{r,\lambda}(M)$, it is easy to find the estimate

$$\left(\int_B w_1^\lambda dx\right)^{\alpha/\lambda p}\left(\int_{\rho B}\left(\tfrac{1}{w_2}\right)^{\lambda/(r-1)}dx\right)^{\alpha(r-1)/\lambda p}$$

$$\leq C_5|B|^{\alpha r/\lambda p}\left(\tfrac{1}{|\rho B|}\int_{\rho B}w_1^\lambda dx\right)^{\alpha/\lambda p}$$

$$\times \left(\tfrac{1}{|\rho B|}\int_{\rho B}\left(\tfrac{1}{w_2}\right)^{\lambda/(r-1)}dx\right)^{\alpha(r-1)/\lambda p} \tag{7.3.12}$$

$$\leq C_6|B|^{\alpha r/\lambda p}.$$

Finally, substituting $(7.3.12)$ into $(7.3.11)$ and using $(m-t)/mt = -\alpha r/\lambda p$, we obtain

$$\|T(\Delta(G(u)))\|_{p,B,w_1^\alpha} \leq C|B|diam(B)\|u\|_{p,\rho B,w_2^\alpha}. \quad \blacksquare$$

Using the same method developed in the proof of Theorem 7.3.2, we shall prove the following two-weight estimate for the composition of operators.

Theorem 7.3.3. *Let $u \in C^\infty(\wedge^l \Omega)$, $l = 1, 2, \ldots, n$, $1 < p < \infty$, be a differential form satisfying (1.2.10) in a bounded, convex domain Ω and $T : C^\infty(\Omega, \wedge^l) \to C^\infty(\Omega, \wedge^{l-1})$ be the homotopy operator defined in (1.5.1). Assume that $\rho > 1$ and $(w_1, w_2) \in A_{r,\lambda}(\Omega)$ for some $\lambda \geq 1$ and $1 < r < \infty$ with $1/r + 1/r' = 1$. Then, $\nabla(T(\Delta(G(u)))) \in L^p(\wedge^l B, w_1^\alpha)$. Moreover, there exists a constant C, independent of u, such that*

$$\|\nabla(T(\Delta(G(u))))\|_{p,B,w_1^\alpha} \leq C|B|\|u\|_{p,\rho B,w_2^\alpha} \tag{7.3.13}$$

for all balls B with $\rho B \subset \Omega$ and any real number α with $\alpha < \lambda$.

Proof. We only need to prove (7.3.13). Choose $t = \lambda p/(\lambda - \alpha)$, then from the Hölder inequality, we find that

$$\|\nabla(T(\Delta(G(u))))\|_{p,B,w_1^\alpha}$$
$$= \left(\int_B \left(|\nabla(T(\Delta(G(u))))| w_1^{\alpha/p} \right)^p dx \right)^{1/p}$$
$$\leq \|\nabla(T(\Delta(G(u))))\|_{t,B} \left(\int_B w_1^{t\alpha/(t-p)} dx \right)^{(t-p)/pt} \tag{7.3.14}$$
$$= \|\nabla(T(\Delta(G(u))))\|_{t,B} \left(\int_B w_1^\lambda dx \right)^{\alpha/\lambda p}.$$

From inequality (1.5.6) and Theorem 3.3.2, we obtain

$$\|\nabla(T(\Delta(G(u))))\|_{t,B} \leq C_1|B|\|\Delta(G(u))\|_{t,B} \leq C_2|B|\|u\|_{t,B}. \tag{7.3.15}$$

Substituting (7.3.15) into (7.3.14), we find that

$$\|\nabla(T(\Delta(G(u))))\|_{p,B,w_1^\alpha} \leq C_3|B|\|u\|_{t,B} \left(\int_B w_1^\lambda dx \right)^{\alpha/\lambda p}. \tag{7.3.16}$$

Choosing $m = \lambda p/(\lambda + \alpha(r-1))$ and using the weak reverse Hölder inequality, we have

$$\|u\|_{t,B} \leq C_4|B|^{(m-t)/mt}\|u\|_{m,\rho B}. \tag{7.3.17}$$

Combining (7.3.16) and (7.3.17), it follows that

$$\|\nabla(T(\Delta(G(u))))\|_{p,B,w_1^\alpha}$$
$$\leq C_5|B||B|^{(m-t)/mt}\|u\|_{m,\rho B} \left(\int_B w_1^\lambda dx \right)^{\alpha/\lambda p}. \tag{7.3.18}$$

Now, it is easy to see that inequality (7.3.18) plays the same role as inequality (7.3.9) in the proof of Theorem 7.3.2. Using the same strategy, we establish (7.3.13). ∎

Now, we are ready to prove one of our main results, the local two-weight Sobolev–Poincaré imbedding theorem for the composition of operators applied to A-harmonic tensors in a bounded and convex domain.

Theorem 7.3.4. *Let* $u \in C^\infty(\wedge^l \Omega)$, $l = 1, 2, \ldots, n$, $1 < p < \infty$, *be a differential form satisfying* (1.2.10) *in a bounded, convex domain* Ω *and* $T : C^\infty(\Omega, \wedge^l) \to C^\infty(\Omega, \wedge^{l-1})$ *be the homotopy operator defined in* (1.5.1). *Assume that* $\rho > 1$ *and* $(w_1, w_2) \in A_{r,\lambda}(\Omega)$ *for some* $\lambda \geq 1$ *and* $1 < r < \infty$ *with* $1/r + 1/r' = 1$. *Then,* $T(\Delta(G(u))) \in W^{1,p}(B, \wedge^l, w_1^\alpha)$. *Moreover, there exists a constant* C, *independent of* u, *such that*

$$\|T(\Delta(G(u)))\|_{W^{1,p}(B),w_1^\alpha} \leq C|B| \|u\|_{p,\rho B,w_2^\alpha} \qquad (7.3.19)$$

for all balls B *with* $\rho B \subset \Omega$ *and any real number* α *with* $\alpha < \lambda$.

Proof. We only need to prove (7.3.19). From the definition of the $\| \cdot \|_{W^{1,p}(B),w_1^\alpha}$ norm, inequalities (7.3.5) and (7.3.13), we have

$$\|T(\Delta(G(u)))\|_{W^{1,p}(B),w_1^\alpha}$$

$$= diam(B)^{-1}\|T(\Delta(G(u)))\|_{p,B,w_1^\alpha} + \|\nabla(T(\Delta(G(u))))\|_{p,B,w_1^\alpha}$$

$$\leq diam(B)^{-1} \cdot C_1|B|diam(B)\|u\|_{p,\rho B,w_2^\alpha} + C_2|B| \|u\|_{p,\rho B,w_2^\alpha}$$

$$\leq C_1|B| \|u\|_{p,\rho B,w_2^\alpha} + C_2|B| \|u\|_{p,\rho B,w_2^\alpha}$$

$$\leq C_3|B| \|u\|_{p,\rho B,w_2^\alpha}.$$

Hence, inequality (7.3.19) follows.　∎

Note that we have proved rather general versions of two-weighted L^p-estimates in Theorems 7.3.2, 7.3.3, and 7.3.4, where the parameters λ and α are any real numbers with $\lambda \geq 1$ and $\alpha < \lambda$. Thus, for example, choosing $\alpha = 1$ in these results, we obtain Corollaries 7.3.5, ,7.3.6, and 7.3.7, respectively.

Corollary 7.3.5. *Let* $u \in C^\infty(\wedge^l \Omega)$, $l = 1, 2, \ldots, n$, $1 < p < \infty$, *be a differential form satisfying* (1.2.10) *in a bounded, convex domain* Ω *and* $T : C^\infty(\Omega, \wedge^l) \to C^\infty(\Omega, \wedge^{l-1})$ *be the homotopy operator defined in* (1.5.1). *Assume that* $\rho > 1$ *and* $(w_1, w_2) \in A_{r,\lambda}(\Omega)$ *for some* $\lambda > 1$ *and* $1 < r < \infty$ *with* $1/r + 1/r' = 1$. *Then, there exists a constant* C, *independent of* u, *such that*

$$\|T(\Delta(G(u)))\|_{p,B,w_1} \leq C|B|diam(B)\|u\|_{p,\rho B,w_2} \qquad (7.3.20)$$

for all balls B *with* $\rho B \subset \Omega$.

Corollary 7.3.6. *Let* $u \in C^\infty(\wedge^l \Omega)$, $l = 1, 2, \ldots, n$, $1 < p < \infty$, *be a differential form satisfying* (1.2.10) *in a bounded, convex domain* Ω *and* $T : C^\infty(\Omega, \wedge^l) \to C^\infty(\Omega, \wedge^{l-1})$ *be the homotopy operator defined in* (1.5.1). *Assume that* $\rho > 1$ *and* $(w_1, w_2) \in A_{r,\lambda}(\Omega)$ *for some* $\lambda > 1$ *and* $1 < r < \infty$ *with* $1/r + 1/r' = 1$. *Then, there exists a constant* C, *independent of* u, *such* *that*

$$\|\nabla(T(\Delta(G(u))))\|_{p,B,w_1} \leq C|B|\|u\|_{p,\rho B,w_2} \qquad (7.3.21)$$

for all balls B with $\rho B \subset \Omega$.

Corollary 7.3.7. *Let $u \in C^{\infty}(\wedge^l \Omega)$, $l = 1, 2, \ldots, n$, $1 < p < \infty$, be a differential form satisfying (1.2.10) in a bounded, convex domain Ω and $T : C^{\infty}(\Omega, \wedge^l) \to C^{\infty}(\Omega, \wedge^{l-1})$ be the homotopy operator defined in (1.5.1). Assume that $\rho > 1$ and $(w_1, w_2) \in A_{r,\lambda}(\Omega)$ for some $\lambda > 1$ and $1 < r < \infty$ with $1/r + 1/r' = 1$. Then, there exists a constant C, independent of u, such that*

$$\|T(\Delta(G(u)))\|_{W^{1,p}(B),w_1} \leq C|B|\|u\|_{p,\rho B,w_2} \qquad (7.3.22)$$

for all balls B with $\rho B \subset \Omega$.

If we let $w_1(x) = w_2(x) = w(x)$ in (7.3.5), (7.3.13), and (7.3.19), respectively, we obtain

$$\|T(\Delta(G(u)))\|_{p,B,w^\alpha} \leq C|B|diam(B)\|u\|_{p,\rho B,w^\alpha},$$

$$\|\nabla(T(\Delta(G(u))))\|_{p,B,w^\alpha} \leq C|B|\|u\|_{p,\rho B,w^\alpha},$$

$$\|T(\Delta(G(u)))\|_{W^{1,p}(B),w^\alpha} \leq C|B|\|u\|_{p,\rho B,w^\alpha},$$

where the weight $w(x)$ satisfies

$$\sup_{B \subset E} \left(\frac{1}{|B|} \int_B (w)^\lambda dx \right)^{1/\lambda r} \left(\frac{1}{|B|} \int_B \left(\frac{1}{w}\right)^{\lambda r'/r} dx \right)^{1/\lambda r'} < \infty, \qquad (7.3.23)$$

which is a generalization of the usual A_r-weights.

It is easy to see that if $w_1(x) = w_2(x) = w(x)$ and $\lambda = 1$, then the pair of weights (w_1, w_2) satisfy

$$\sup_{B \subset E} \left(\frac{1}{|B|} \int_B w dx \right)^{1/r} \left(\frac{1}{|B|} \int_B \left(\frac{1}{w}\right)^{r'/r} dx \right)^{1/r'} < \infty,$$

that is

$$\sup_{B \subset E} \left(\left(\frac{1}{|B|} \int_B w dx \right) \left(\frac{1}{|B|} \int_B \left(\frac{1}{w}\right)^{1/(r-1)} dx \right)^{r-1} \right)^{1/r} < \infty$$

since $r'/r = 1/(r-1)$, and hence the $A_{r,\lambda}(\Omega)$-weight reduces to the usual $A_r(\Omega)$-weight. Hence, setting $w_1(x) = w_2(x) = w(x)$ and $\lambda = 1$ in Theorem 7.3.4, we obtain the following local $A_r(\Omega)$-weighted imbedding theorem.

Theorem 7.3.8. *Let $u \in C^\infty(\wedge^l \Omega)$, $l = 1, 2, \ldots, n$, $1 < p < \infty$, be a differential form satisfying (1.2.10) in a bounded, convex domain Ω and $T : C^\infty(\Omega, \wedge^l) \to C^\infty(\Omega, \wedge^{l-1})$ be the homotopy operator defined in (1.5.1). Assume that $w \in A_r(\Omega)$ for some $1 < r < \infty$. Then, $T(\Delta(G(u))) \in W^{1,p}(B, \wedge^l, w^\alpha)$. Moreover, there exists a constant C, independent of u, such that*

$$\|T(\Delta(G(u)))\|_{W^{1,p}(B), w^\alpha} \leq C|B| \|u\|_{p, \rho B, w^\alpha} \qquad (7.3.24)$$

for all balls B with $\rho B \subset \Omega$ and any real number α with $\alpha < 1$. Here $\rho > 1$ is some constant.

7.3.2 Global estimates on manifolds

Now, we shall prove the following global $A_{r,\lambda}(\Omega)$-weighted L^p-estimates for operators in a bounded, convex domain M.

Theorem 7.3.9. *Let $u \in C^\infty(\wedge^l \Omega)$, $l = 1, 2, \ldots, n$, be an A-harmonic form in a bounded, convex domain Ω. Assume that $1 < p < \infty$ and $T : C^\infty(\Omega, \wedge^l) \to C^\infty(\Omega, \wedge^{l-1})$ is the homotopy operator defined in (1.5.1) and $(w_1, w_2) \in A_{r,\lambda}(\Omega)$ for some $\lambda \geq 1$ and $1 < r < \infty$ with $1/r + 1/r' = 1$. Then, $T(\Delta(G(u))) \in L^p(\wedge^l \Omega, w_1^\alpha)$ and $\nabla(T(\Delta(G(u)))) \in L^p(\wedge^l \Omega, w_1^\alpha)$. Moreover, there exists a constant C, independent of u, such that*

$$\|T(\Delta(G(u)))\|_{p, \Omega, w_1^\alpha} \leq C|\Omega| diam(\Omega) \|u\|_{p, \Omega, w_2^\alpha} \qquad (7.3.25)$$

and

$$\|\nabla(T(\Delta(G(u))))\|_{p, \Omega, w_1^\alpha} \leq C|\Omega| \|u\|_{p, \Omega, w_2^\alpha} \qquad (7.3.26)$$

for any real number α with $\alpha < \lambda$.

Proof. Applying Theorems 7.3.2 and 1.5.3, we find that

$$\begin{aligned}
&\|T(\Delta(G(u)))\|_{p, \Omega, w_1^\alpha} \\
&\leq \sum_{B \in \mathcal{V}} \|T(\Delta(G(u)))\|_{p, B, w_1^\alpha} \\
&\leq \sum_{B \in \mathcal{V}} \left(C_1 |B| diam(B) \|u\|_{p, \rho B, w_2^\alpha} \right) \\
&\leq C_1 |\Omega| diam(\Omega) N \|u\|_{p, \Omega, w_2^\alpha} \\
&\leq C_2 |\Omega| diam(\Omega) \|u\|_{p, \Omega, w_2^\alpha},
\end{aligned} \qquad (7.3.27)$$

which is equivalent to inequality (7.3.25). Similarly, we can prove (7.3.26). ∎

Now, we are ready to prove the following global two-weight Sobolev–Poincaré imbedding theorem for the composition of operators applied to A-harmonic tensors in a bounded and convex domain.

Theorem 7.3.10. *Let $u \in C^\infty(\wedge^l \Omega)$, $l = 1, 2, \ldots, n$, $1 < p < \infty$, be a differential form satisfying (1.2.10) in a bounded, convex domain Ω. Assume that $1 < p < \infty$, $T: C^\infty(\Omega, \wedge^l) \to C^\infty(\Omega, \wedge^{l-1})$ is the homotopy operator defined in (1.5.1), and $(w_1, w_2) \in A_{r,\lambda}(\Omega)$ for some $\lambda \geq 1$ and $1 < r < \infty$ with $1/r + 1/r' = 1$. Then, $T(\Delta(G(u))) \in W^{1,p}(\Omega, \wedge^l, w_1^\alpha)$. Moreover, there exists a constant C, independent of u, such that*

$$\|T(\Delta(G(u)))\|_{W^{1,p}(\Omega), w_1^\alpha} \leq C\|u\|_{p, \Omega, w_2^\alpha} \qquad (7.3.28)$$

for any real number α with $\alpha < \lambda$.

Proof. Applying the definition of the $\|\cdot\|_{W^{1,p}(\Omega), w_1^\alpha}$ norm, (7.3.25), and (7.3.26), we have

$$\|T(\Delta(G(u)))\|_{W^{1,p}(\Omega), w_1^\alpha}$$

$$= diam(\Omega)^{-1}\|T(\Delta(G(u)))\|_{p, \Omega, w_1^\alpha} + \|\nabla(T(\Delta(G(u))))\|_{p, \Omega, w_1^\alpha}$$

$$\leq C_1 diam(\Omega)^{-1} \cdot |\Omega| diam(\Omega)\|u\|_{p, \Omega, w_2^\alpha} + C_2|\Omega|\|u\|_{p, \Omega, w_2^\alpha} \qquad (7.3.29)$$

$$\leq C_1|\Omega|\|u\|_{p, \Omega, w_2^\alpha} + C_2|\Omega|\|u\|_{p, \Omega, w_2^\alpha}$$

$$\leq C_3|\Omega|\|u\|_{p, \Omega, w_2^\alpha}.$$

Thus, inequality (7.3.28) holds. ∎

Choosing $\alpha = 1$ in Theorem 7.3.10, we have the following version of the imbedding theorem.

Theorem 7.3.11. *Let $u \in C^\infty(\wedge^l \Omega)$, $l = 1, 2, \ldots, n$, $1 < p < \infty$, be a differential form satisfying (1.2.10) in a bounded, convex domain Ω. Assume that $1 < p < \infty$ and $T : C^\infty(\Omega, \wedge^l) \to C^\infty(\Omega, \wedge^{l-1})$ is the homotopy operator defined in (1.5.1), and $(w_1, w_2) \in A_{r,\lambda}(\Omega)$ for some $\lambda > 1$ and $1 < r < \infty$ with $1/r + 1/r' = 1$. Then, $T(\Delta(G(u))) \in W^{1,p}(B, \wedge^l, w_1)$. Moreover, there exists a constant C, independent of u, such that*

$$\|T(\Delta(G(u)))\|_{W^{1,p}(\Omega), w_1} \leq C\|u\|_{p, \Omega, w_2}. \qquad (7.3.30)$$

If we select $w_1 = w_2 = w$ and $\lambda = 1$ in Theorem 7.3.10, we obtain the following $A_r(\Omega)$-weighted Sobolev–Poincaré imbedding theorem.

Theorem 7.3.12. *Let $u \in C^\infty(\wedge^l \Omega)$, $l = 1, 2, \ldots, n$, $1 < p < \infty$, be a differential form satisfying (1.2.10) in a bounded, convex domain Ω. Assume that $1 < p < \infty$ and $T : C^\infty(\Omega, \wedge^l) \to C^\infty(\Omega, \wedge^{l-1})$ is the homotopy operator defined in (1.5.1) and $w \in A_r(\Omega)$ for some $1 < r < \infty$. Then, $T(\Delta(G(u))) \in W^{1,p}(\Omega, \wedge^l, w^\alpha)$. Moreover, there exists a constant C, independent of u, such* *that*

$$\|T(\Delta(G(u)))\|_{W^{1,r}(\Omega),w^\alpha} \le C\|u\|_{p,\Omega,w^\alpha} \qquad (7.3.31)$$

for any real number α with $\alpha < 1$.

Similarly, if we choose $w_1 = w_2 = w$ and $\lambda = 1$ in Theorem 7.3.9, we obtain the following $A_r(\Omega)$-weighted L^p-estimates which are very useful.

Theorem 7.3.13. *Let $u \in C^\infty(\wedge^l\Omega)$, $l = 1,2,\ldots,n$, $1 < p < \infty$, be a differential form satisfying (1.2.10) in a bounded, convex domain Ω. Assume that $1 < p < \infty$ and $T : C^\infty(\Omega,\wedge^l) \to C^\infty(\Omega,\wedge^{l-1})$ is the homotopy operator defined in (1.5.1), and $w \in A_r(\Omega)$ for some $1 < r < \infty$. Then, $T(\Delta(G(u))) \in L^p(\wedge^l\Omega,w^\alpha)$ and $\nabla(T(\Delta(G(u)))) \in L^p(\wedge^l\Omega,w^\alpha)$. Moreover, there exists a constant C, independent of u, such that*

$$\|T(\Delta(G(u)))\|_{p,\Omega,w^\alpha} \le C|\Omega|diam(\Omega)\|u\|_{p,\Omega,w^\alpha},$$

$$\|\nabla(T(\Delta(G(u))))\|_{p,\Omega,w^\alpha} \le C|\Omega|\|u\|_{p,\Omega,w^\alpha} \qquad (7.3.32)$$

for any real number α with $\alpha < 1$.

Remark. Similar to the local case, for different choices of α and λ in Theorems 7.3.9 and 7.3.10, we have different versions of the global $A_{r,\lambda}(\Omega)$ Sobolev–Poincaré imbedding theorems.

7.3.3 L^s-estimates for $T \circ G$

Now, we prove the following local estimate for the composition of operators T and G acting on a differential form in a domain Ω, which can be used to develop the imbedding theorems for the composition of operators.

Theorem 7.3.14. *Let $u \in L^s_{loc}(\Omega,\wedge^l)$, $l = 1,2,\ldots,n$, $1 < s < \infty$, be a smooth solution of the A-harmonic equation (1.2.10) in a bounded, convex domain Ω and $T : C^\infty(\Omega,\wedge^l) \to C^\infty(\Omega,\wedge^{l-1})$ be the homotopy operator defined in (1.5.1). Assume that $\rho > 1$ and $w \in A_r(\Omega)$ for some $1 < r < \infty$. Then, $T(G(u)) \in L^s_{loc}(\Omega,\wedge^l)$. Moreover, there exists a constant C, independent of u, such that*

$$\|T(G(u))\|_{s,B,w^\alpha} \le C|B|diam(B)\|u\|_{s,\rho B,w^\alpha} \qquad (7.3.33)$$

for all balls B with $\rho B \subset \Omega$ and any real number α with $0 < \alpha \le 1$.

Proof. We only need to prove that inequality (7.3.33) holds. From inequalities (1.5.6) and (7.2.1), we have

$$\|T(G(u))\|_{s,B}$$

$$\leq C_1|B|diam(B)\|G(u)\|_{s,B} \qquad (7.3.34)$$

$$\leq C_2|B|diam(B)\|u\|_{s,B}.$$

We first show that (7.3.33) holds for $0 < \alpha < 1$. Let $t = s/(1-\alpha)$. Using the Hölder inequality, we obtain

$$\|T(G(u))\|_{s,B,w^\alpha}$$

$$= \left(\int_B \left(|T(G(u))|w^{\alpha/s}\right)^s dx\right)^{1/s}$$

$$\leq \|T(G(u))\|_{t,B} \left(\int_B w^{t\alpha/(t-s)}dx\right)^{(t-s)/st} \qquad (7.3.35)$$

$$= \|T(G(u))\|_{t,B} \left(\int_B wdx\right)^{\alpha/s}.$$

By (7.3.34), we have

$$\|T(G(u))\|_{t,B} \leq C_1|B|diam(B)\|u\|_{t,B}. \qquad (7.3.36)$$

Choose $m = s/(1 + \alpha(r-1))$, so that $m < s$. Substituting (7.3.36) into (7.3.35) and using Lemma 3.1.1, we find that

$$\|T(G(u))\|_{s,B,w^\alpha}$$

$$\leq C_1|B|diam(B)\|u\|_{t,B} \left(\int_B wdx\right)^{\alpha/s} \qquad (7.3.37)$$

$$\leq C_2|B|diam(B)|B|^{(m-t)/mt}\|u\|_{m,\rho B} \left(\int_B wdx\right)^{\alpha/s}.$$

Now using Hölder's inequality with $1/m = 1/s + (s-m)/sm$, we obtain

$$\|u\|_{m,\rho B}$$

$$= \left(\int_{\rho B} |u|^m dx\right)^{1/m}$$

$$= \left(\int_{\rho B} \left(|u|w^{\alpha/s}w^{-\alpha/s}\right)^m dx\right)^{1/m} \qquad (7.3.38)$$

$$\leq \left(\int_{\rho B} |u|^s w^\alpha dx\right)^{1/s} \left(\int_{\rho B} \left(\frac{1}{w}\right)^{1/(r-1)} dx\right)^{\alpha(r-1)/s}$$

for all balls B with $\rho B \subset \Omega$. Substituting (7.3.38) into (7.3.37), we have

$$\|T(G(u))\|_{s,B,w^{\alpha}}$$

$$\leq C_2|B|diam(B)|B|^{(m-t)/mt}\left(\int_{\rho B}|u|^s w^{\alpha}dx\right)^{1/s}$$

$$\times\left(\int_B wdx\right)^{\alpha/s}\left(\int_{\rho B}\left(\frac{1}{w}\right)^{1/(r-1)}dx\right)^{\alpha(r-1)/s}.$$

(7.3.39)

Since $w \in A_r(\Omega)$, it follows that

$$\|w\|_{1,B}^{\alpha/s}\cdot\|1/w\|_{1/(r-1),\rho B}^{\alpha/s}$$

$$\leq\left(\left(\int_{\rho B}wdx\right)\left(\int_{\rho B}(1/w)^{1/(r-1)}dx\right)^{r-1}\right)^{\alpha/s}$$

$$=\left(|\rho B|^r\left(\frac{1}{|\rho B|}\int_{\rho B}wdx\right)\left(\frac{1}{|\rho B|}\int_{\rho B}\left(\frac{1}{w}\right)^{1/(r-1)}dx\right)^{r-1}\right)^{\alpha/s}$$

(7.3.40)

$$\leq C_3|B|^{\alpha r/s}.$$

Combining (7.3.40) and (7.3.39), we find that

$$\|T(G(u))\|_{s,B,w^{\alpha}}\leq C_4|B|diam(B)\|u\|_{s,\rho B,w^{\alpha}}\qquad(7.3.41)$$

for all balls B with $\rho B \subset \Omega$. Thus, (7.3.33) holds if $0 < \alpha < 1$.

Next, we prove (7.3.33) for $\alpha = 1$, that is, we need to show that

$$\|T(G(u))\|_{s,B,w}\leq C|B|diam(B)\|u\|_{s,\rho B,w}.\qquad(7.3.42)$$

By Lemma 1.4.7, there exist constants $\beta > 1$ and $C_5 > 0$, such that

$$\|\,w\,\|_{\beta,B}\leq C_5|B|^{(1-\beta)/\beta}\,\|\,w\,\|_{1,B}\,.\qquad(7.3.43)$$

Choose $t = s\beta/(\beta - 1)$, so that $1 < s < t$ and $\beta = t/(t - s)$. Since $1/s = 1/t + (t - s)/st$, by Hölder's inequality, (7.3.34), and (7.3.43), we have

$$\left(\int_B |T(G(u))|^s w dx\right)^{1/s}$$

$$= \left(\int_B \left(|T(G(u))|w^{1/s}\right)^s dx\right)^{1/s}$$

$$\leq \left(\int_B |T(G(u))|^t dx\right)^{1/t} \left(\int_B \left(w^{1/s}\right)^{st/(t-s)} dx\right)^{(t-s)/st}$$

$$\leq C_6 \|T(G(u))\|_{t,B} \cdot \|w\|_{\beta,B}^{1/s} \tag{7.3.44}$$

$$\leq C_6 |B| diam(B) \|u\|_{t,B} \cdot \|w\|_{\beta,B}^{1/s}$$

$$\leq C_7 |B| diam(B) |B|^{(1-\beta)/\beta s} \|w\|_{1,B}^{1/s} \cdot \|u\|_{t,B}$$

$$\leq C_7 |B| diam(B) |B|^{-1/t} \|w\|_{1,B}^{1/s} \cdot \|u\|_{t,B}.$$

Let $m = s/r$. From Lemma 3.1.1, we find that

$$\|u\|_{t,B} \leq C_8 |B|^{(m-t)/mt} \|u\|_{m,\rho B}. \tag{7.3.45}$$

Now, using Hölder's inequality again, we obtain

$$\|u\|_{m,\rho B}$$

$$= \left(\int_{\rho B} \left(|u|w^{1/s} w^{-1/s}\right)^m dx\right)^{1/m} \tag{7.3.46}$$

$$\leq \left(\int_{\rho B} |u|^s w dx\right)^{1/s} \left(\int_{\rho B} \left(\frac{1}{w}\right)^{1/(r-1)} dx\right)^{(r-1)/s}.$$

Next, since $w \in A_r(\Omega)$, it follows that

$$\|w\|_{1,B}^{1/s} \cdot \|1/w\|_{1/(r-1),\rho B}^{1/s}$$

$$\leq \left(\left(\int_{\rho B} w dx\right) \left(\int_{\rho B} (1/w)^{1/(r-1)} dx\right)^{r-1}\right)^{1/s}$$

$$= \left(|\rho B|^r \left(\frac{1}{|\rho B|} \int_{\rho B} w dx\right) \left(\frac{1}{|\rho B|} \int_{\rho B} \left(\frac{1}{w}\right)^{1/(r-1)} dx\right)^{r-1}\right)^{1/s} \tag{7.3.47}$$

$$\leq C_9 |B|^{r/s}.$$

Combining (7.3.44), (7.3.45), (7.3.46), and (7.3.47), we find that

$$\|T(G(u))\|_{s,B,w}$$

$$\leq C_{10}|B|diam(B)|B|^{-1/t}\|w\|_{1,B}^{1/s}|B|^{(m-t)/mt}\|u\|_{m,\rho B}$$

$$\leq C_{10}|B|diam(B)|B|^{-1/m}\|w\|_{1,B}^{1/s}\cdot\|1/w\|_{1/(r-1),\rho B}^{1/s}\|u\|_{s,\rho B,w}$$

$$\leq C_{11}|B|diam(B)\|u\|_{s,\rho B,w}$$

for all balls B with $\rho B \subset \Omega$. Hence, inequality (7.3.42) holds. ∎

7.3.4 Local imbedding theorems for $T \circ G$

First, we prove an unweighted imbedding inequality for $T \circ G$, and then extend it into the weighted case.

Theorem 7.3.15. *Let $u \in L_{loc}^s(\Omega, \wedge^l)$, $l = 1, 2, \ldots, n$, $1 < s < \infty$, be a smooth differential form satisfying (1.2.10) in a bounded, convex domain Ω and $T : C^\infty(\Omega, \wedge^l) \to C^\infty(\Omega, \wedge^{l-1})$ be the homotopy operator defined in (1.5.1). Assume that $w \in A_r(\Omega)$ for some $1 < r < \infty$. Then, there exists a constant C, independent of u, such that*

$$\|T(G(u))\|_{W^{1,s}(B)} \leq C|B|\|u\|_{s,B} \qquad (7.3.48)$$

for all balls B with $B \subset \Omega$.

Proof. Using the definition of the $\|\cdot\|_{W^{1,s}(B)}$-norm, inequalities (1.5.5) and (1.5.6), and Lemma 3.3.1, we find that

$$\|T(G(u))\|_{W^{1,s}(B)}$$

$$= diam(B)^{-1}\|T(G(u))\|_{s,B} + \|\nabla(T(G(u)))\|_{s,B}$$

$$\leq diam(B)^{-1}\cdot C_1|B|diam(B)\|G(u)\|_{s,B} + C_2|B|\|G(u)\|_{s,B}$$

$$\leq C_3|B|\|G(u)\|_{s,B}$$

$$\leq C_4|B|\|u\|_{s,B}. \qquad \blacksquare$$

Theorem 7.3.16. *Let $u \in L_{loc}^s(\Omega, \wedge^l)$, $l = 1, 2, \ldots, n$, $1 < s < \infty$, be a smooth differential form satisfying (1.2.10) in a bounded, convex domain Ω and $T : C^\infty(\Omega, \wedge^l) \to C^\infty(\Omega, \wedge^{l-1})$ be the homotopy operator defined in (1.5.1). Assume that $\rho > 1$ and $(w_1, w_2) \in A_{r,\lambda}(\Omega)$ for some $\lambda \geq 1$ and $1 < r < \infty$. Then, $\nabla(T(G(u))) \in L_{loc}^s(\Omega, \wedge^l)$. Moreover, there exists a constant C, independent of u, such that*

$$\|\nabla(T(G(u)))\|_{s,B,w_1^\alpha} \le C|B|\|u\|_{s,\rho B,w_2^\alpha} \qquad (7.3.49)$$

for all balls B with $\rho B \subset \Omega$ and any real number α with $0 < \alpha < \lambda$.

Proof. If (7.3.49) holds, then $\nabla(T(G(u))) \in L_{loc}^s(\Omega, \wedge^l)$ follows. Hence, we only need to prove (7.3.49). From (1.5.5) and Lemma 3.3.1, we know that

$$\|\nabla(T(G(u)))\|_{s,B} \le C|B|\|G(u)\|_{s,B} \le C|B|\|u\|_{s,B}.$$

Using the method developed in the proof of Theorem 5.7.3, we can extend the above inequality to the two-weighted case

$$\|\nabla(T(G(u)))\|_{s,B,w_1^\alpha} \le C|B|\|u\|_{s,\rho B,w_2^\alpha}. \qquad \blacksquare$$

Choosing $w_1(x) = w_2(x) = w(x)$ and setting $\lambda = 1$ in Theorem 7.3.16, we obtain the following $A_r(\Omega)$-weighted inequality:

$$\|\nabla(T(G(u)))\|_{s,B,w^\alpha} \le C|B|\|u\|_{s,\rho B,w^\alpha}, \qquad (7.3.50)$$

where $w \in A_r(\Omega)$ and $0 < \alpha < 1$.

Theorem 7.3.17 *Let $u \in L_{loc}^s(\Omega, \wedge^l)$, $l = 1, 2, \ldots, n$, $1 < s < \infty$, be a smooth differential form satisfying (1.2.10) in a bounded, convex domain Ω and $T : C^\infty(\Omega, \wedge^l) \to C^\infty(\Omega, \wedge^{l-1})$ be the homotopy operator defined in (1.5.1). Assume that $\rho > 1$ and $w \in A_r(\Omega)$ for some $1 < r < \infty$. Then, there exists a constant C, independent of u, such that*

$$\|T(G(u))\|_{W^{1,s}(B),w^\alpha} \le C|B|\|u\|_{s,\rho B,w^\alpha} \qquad (7.3.51)$$

for all balls B with $\rho B \subset \Omega$ and any real number α with $0 < \alpha \le 1$.

Proof. For $0 < \alpha < 1$, using inequalities (3.3.15), (7.3.50), and Theorem 7.3.14, we obtain

$$\begin{aligned}
&\|T(G(u))\|_{W^{1,s}(B),w^\alpha} \\
&= diam(B)^{-1}\|T(G(u))\|_{s,B,w^\alpha} + \|\nabla(T(G(u)))\|_{s,B,w^\alpha} \\
&\le diam(B)^{-1}C_1 diam(B)|B|\|u\|_{s,\rho_1 B,w^\alpha} + C_2|B|\|u\|_{s,\rho_2 B,w^\alpha} \\
&\le C_3|B|\|u\|_{s,\rho B,w^\alpha},
\end{aligned} \qquad (7.3.52)$$

where $\rho = max\{\rho_1, \rho_2\}$. For the case $\alpha = 1$, using the method similar to the proof of Theorem 3.2.10, we can also prove that (7.3.51) holds. $\blacksquare$

7.3.5 Global imbedding theorems for $T \circ G$

Next, we prove the global $A_r(\Omega)$-weighted estimates for the composition of operators and the $A_r(\Omega)$-weighted imbedding theorem for $T \circ G$.

Theorem 7.3.18. *Let $u \in L^s(\Omega, \wedge^l)$, $l = 1, 2, \ldots, n$, $1 < s < \infty$, be an A-harmonic tensor in a bounded, convex domain Ω and $T : C^\infty(\Omega, \wedge^l) \to C^\infty(\Omega, \wedge^{l-1})$ be the homotopy operator defined in (1.5.1). Assume that $w \in A_r(\Omega)$ for some $r > 1$. Then, there exists a constant C, independent of u, such that*

$$\|T(G(u))\|_{s,\Omega,w^\alpha} \le C|\Omega| diam(\Omega) \|u\|_{s,\Omega,w^\alpha} \qquad (7.3.53)$$

for any real number α with $0 < \alpha \le 1$.

Proof. Let $\mathcal{V} = \{Q_i\}$ be a Whitney cover of Ω. Applying Theorems 1.5.3 and 7.3.14, we obtain

$$\|T(G(u))\|_{s,\Omega,w^\alpha}$$

$$\le \sum_{Q \in \mathcal{V}} \|T(G(u))\|_{s,Q,w^\alpha}$$

$$\le \sum_{Q \in \mathcal{V}} \left(C_1 |Q| diam(Q) \|u\|_{s,\rho Q,w^\alpha} \right)$$

$$\le C_2 |\Omega| diam(\Omega) N \|u\|_{s,\Omega,w^\alpha}$$

$$\le C_3 |\Omega| diam(\Omega) \|u\|_{s,\Omega,w^\alpha}. \qquad \blacksquare$$

Similar to the proof of Theorem 7.3.18, using Theorems 1.5.3 and 7.3.16, we obtain the following global weighted estimate for the composition of the operators ∇, T, and G.

Theorem 7.3.19. *Let $u \in L^s(\Omega, \wedge^l)$, $l = 1, 2, \ldots, n$, $1 < s < \infty$, be an A-harmonic tensor in a bounded, convex domain Ω and $T : C^\infty(\Omega, \wedge^l) \to C^\infty(\Omega, \wedge^{l-1})$ be the homotopy operator defined in (1.5.1). Assume that and $(w_1, w_2) \in A_{r,\lambda}(\Omega)$ for some $\lambda \ge 1$ and $1 < r < \infty$. Then, there exists a constant C, independent of u, such that*

$$\|\nabla(T(G(u)))\|_{s,\Omega,w_1^\alpha} \le C|\Omega| \|u\|_{s,\Omega,w_2^\alpha}$$

for any real number α with $0 < \alpha < \lambda$.

Setting $w_1(x) = w_2(x) = w(x)$ and $\lambda = 1$ in Theorem 7.3.19, we have the following $A_r(\Omega)$-weighted inequality:

$$\|\nabla(T(G(u)))\|_{s,\Omega,w^\alpha} \le C|\Omega| \|u\|_{s,\Omega,w^\alpha}, \qquad (7.3.54)$$

where $w \in A_r(\Omega)$.

Now, we are ready to prove the following global $A_r(\Omega)$-weighted Sobolev–Poincaré imbedding theorem for the composition of operators applied to A-harmonic tensors.

Theorem 7.3.20. *Let $u \in L^s(\Omega, \wedge^l)$, $l = 1, 2, \ldots, n$, $1 < s < \infty$, be an A-harmonic tensor in a bounded, convex domain Ω. Assume that $T : C^\infty(\Omega, \wedge^l) \to C^\infty(\Omega, \wedge^{l-1})$ is the homotopy operator defined in (1.5.1) and $w \in A_r(\Omega)$ for some $1 < r < \infty$. Then, $T(G(u)) \in W^{1,s}(\Omega, \wedge^l)$. Moreover, there exists a constant C, independent of u, such that*

$$\|T(G(u))\|_{W^{1,s}(\Omega), w^\alpha} \le C|\Omega|\|u\|_{s, \Omega, w^\alpha} \tag{7.3.55}$$

for any real number α with $0 < \alpha \le 1$.

Proof. Applying (3.3.15), (7.3.54), and Theorem 7.3.18, we find that

$$\|T(G(u))\|_{W^{1,s}(\Omega), w^\alpha}$$
$$= diam(\Omega)^{-1}\|T(G(u))\|_{s, \Omega, w^\alpha} + \|\nabla(T(G(u)))\|_{s, \Omega, w^\alpha}$$
$$\le C_1 diam(\Omega)^{-1} \cdot |\Omega| diam(\Omega)\|u\|_{s, \Omega, w^\alpha} + C_2|\Omega|\|u\|_{s, \Omega, w^\alpha}$$
$$\le C_1|\Omega|\|u\|_{s, \Omega, w^\alpha} + C_2|\Omega|\|u\|_{s, \Omega, w^\alpha}$$
$$\le C_3|\Omega|\|u\|_{s, \Omega, w^\alpha}. \qquad \blacksquare$$

7.3.6 Some special cases

We should notice that the parameter α in the above theorems makes our results more applicable and powerful. In fact, particular values of α lead to several interesting inequalities. For example, choosing $\alpha = 1$ in Theorems 7.3.18 and 7.3.20, and (7.3.54), we have the following corollaries, respectively.

Corollary 7.3.21. *Let $u \in L^s(\Omega, \wedge^l)$, $l = 1, 2, \ldots, n$, $1 < s < \infty$, be an A-harmonic tensor in a bounded, convex domain Ω and $T : C^\infty(\Omega, \wedge^l) \to C^\infty(\Omega, \wedge^{l-1})$ be the homotopy operator defined by (1.5.1). Assume that $w \in A_r(\Omega)$ for some $r > 1$. Then, there exists a constant C, independent of u, such that*

$$\|T(G(u))\|_{s, \Omega, w} \le C|\Omega| diam(\Omega)\|u\|_{s, \Omega, w}. \tag{7.3.56}$$

Corollary 7.3.22. *Let $u \in L^s(\Omega, \wedge^l)$, $l = 1, 2, \ldots, n$, $1 < s < \infty$, be an A-harmonic tensor in a bounded, convex domain Ω and $T : C^\infty(\Omega, \wedge^l) \to C^\infty(\Omega, \wedge^{l-1})$ be the homotopy operator defined in (1.5.1). Assume that $w \in A_r(\Omega)$ for some $r > 1$. Then, there exists a constant C, independent of u, such that*

$$\|\nabla(T(G(u)))\|_{s,\Omega,w} \le C|\Omega| \|u\|_{s,\Omega,w}. \qquad (7.3.57)$$

Corollary 7.3.23. *Let* $u \in L^s(\Omega, \wedge^l)$, $l = 1, 2, \ldots, n$, $1 < s < \infty$, *be an A-harmonic tensor in a bounded, convex domain* Ω. *Assume that* $T : C^\infty(\Omega, \wedge^l) \to C^\infty(\Omega, \wedge^{l-1})$ *is the homotopy operator defined in* (1.5.1) *and* $w \in A_r(\Omega)$ *for some* $1 < r < \infty$. *Then,* $T(G(u)) \in W^{1,s}(\Omega, \wedge^l)$. *Moreover, there exists a constant* C, *independent of* u, *such that*

$$\|T(G(u))\|_{W^{1,s}(\Omega),w} \le C|\Omega| \|u\|_{s,\Omega,w}. \qquad (7.3.58)$$

Next, since $1 < s < \infty$, letting $\alpha = 1/s$ in (7.3.53), (7.3.54), and (7.3.55), respectively, we obtain

$$\|T(G(u))\|_{s,\Omega,w^{1/s}} \le C|\Omega| diam(\Omega) \|u\|_{s,\Omega,w^{1/s}},$$

$$\|\nabla(T(G(u)))\|_{s,\Omega,w^{1/s}} \le C|\Omega| \|u\|_{s,\Omega,w^{1/s}},$$

$$\|T(G(u))\|_{W^{1,s}(\Omega),w^{1/s}} \le C|\Omega| \|u\|_{s,\Omega,w^{1/s}}.$$

Similarly, selecting $\alpha = s/(r+s)$ in (7.3.53), (7.3.54), and (7.3.55), respectively, we find that

$$\|T(G(u))\|_{s,\Omega,w^{s/(r+s)}} \le C|\Omega| diam(\Omega) \|u\|_{s,\Omega,w^{s/(r+s)}},$$

$$\|\nabla(T(G(u)))\|_{s,\Omega,w^{s/(r+s)}} \le C|\Omega| \|u\|_{s,\Omega,w^{s/(r+s)}},$$

$$\|T(G(u))\|_{W^{1,s}(\Omega),w^{s/(r+s)}} \le C|\Omega| \|u\|_{s,\Omega,w^{s/(r+s)}}.$$

Finally, from Theorems 7.3.18 and 7.3.19, we have the following property for the compositions of operators.

Corollary 7.3.24. *The compositions* $T \circ G$ *and* $\nabla \circ T \circ G$ *are bounded operators if* Ω *is a bounded, convex domain.*

7.3.7 L^s-estimates for $\Delta \circ G \circ d$

In the following theorem, we study the composition of the operators Δ, G, and d.

Theorem 7.3.25. *Let* $u \in C^\infty(\wedge^l \Omega)$, $l = 1, 2, \ldots, n-1$, *be an A-harmonic tensor on a bounded, convex* Ω. *Assume that* $\rho > 1$, $1 < s < \infty$, *and* $w \in A_r(\Omega)$ *for some* $r > 1$. *Then, there exists a constant* C, *independent of* u, *such that*

$$\|\Delta(G(du))\|_{s,B,w^\alpha} \le C\|du\|_{s,\rho B,w^\alpha} \qquad (7.3.59)$$

for any ball B with $\rho B \subset \Omega$ and any real number α with $0 < \alpha \le 1$.

Proof. From inequality (7.2.1), for any smooth l-form ω, we have

$$\|dd^\star G(\omega)\|_{s,B} + \|d^\star dG(\omega)\|_{s,B} \le C_1\|\omega\|_{s,B}, \qquad (7.3.60)$$

where $1 < s < \infty$ and C_1 is a constant. Choosing $\omega = du$, we find that

$$\|dd^\star G(du)\|_{s,B} + \|d^\star dG(du)\|_{s,B} \le C_1\|du\|_{s,B}. \qquad (7.3.61)$$

By the definition of the Laplace–Beltrami operator Δ, Minkowski's inequality, and (7.3.61), we obtain

$$
\begin{aligned}
\|\Delta G(du)\|_{s,B} &= \|(d^\star d + dd^\star)G(du)\|_{s,B} \\
&\le \|d^\star dG(du)\|_{s,B} + \|dd^\star G(du)\|_{s,B} \\
&\le C_1\|du\|_{s,B}.
\end{aligned}
\qquad (7.3.62)
$$

We shall first show that (7.3.59) holds for $0 < \alpha < 1$. Let $t = s/(1 - \alpha)$. Using the Hölder inequality and (7.3.62), we have

$$
\begin{aligned}
\|\Delta(G(du))\|_{s,B,w^\alpha} \\
= \left(\int_B \left(|\Delta(G(du))|w^{\alpha/s}\right)^s dx\right)^{1/s} \\
\le \|\Delta(G(du))\|_{t,B} \left(\int_B w^{t\alpha/(t-s)}dx\right)^{(t-s)/st} \\
= \|\Delta(G(du))\|_{t,B} \left(\int_B wdx\right)^{\alpha/s} \\
\le C_1\|du\|_{t,B} \left(\int_B wdx\right)^{\alpha/s}.
\end{aligned}
\qquad (7.3.63)
$$

Let $m = s/(1 + \alpha(r - 1))$, so that $m < s$. Applying Theorem 6.2.12, we find that

$$\|du\|_{t,B} \le C_2|B|^{(m-t)/mt}\|du\|_{m,\rho B}. \qquad (7.3.64)$$

Now substituting (7.3.64) into (7.3.63), we have

$$\|\Delta(G(du))\|_{s,B,w^\alpha} \le C_3|B|^{(m-t)/mt}\|du\|_{m,\rho B} \left(\int_B wdx\right)^{\alpha/s}. \qquad (7.3.65)$$

Using the Hölder inequality again with $1/m = 1/s + (s - m)/sm$, we obtain

$$\|du\|_{m,\rho B} = \left(\int_{\rho B} |du|^m dx\right)^{1/m}$$

$$= \left(\int_{\rho B} \left(|du| w^{\alpha/s} w^{-\alpha/s}\right)^m dx\right)^{1/m} \tag{7.3.66}$$

$$\leq \|du\|_{s,\rho B, w^\alpha} \left(\int_{\rho B} \left(\tfrac{1}{w}\right)^{1/(r-1)} dx\right)^{\alpha(r-1)/s}$$

for all balls B with $\rho B \subset \Omega$. Next, substitution of (7.3.66) into (7.3.65) gives

$$\|\Delta(G(du))\|_{s,B,w^\alpha}$$

$$\leq C_3 |B|^{(m-t)/mt} \|du\|_{s,\rho B, w^\alpha} \left(\int_B w\,dx\right)^{\alpha/s} \tag{7.3.67}$$

$$\times \left(\int_{\rho B} \left(\tfrac{1}{w}\right)^{1/(r-1)} dx\right)^{\alpha(r-1)/s}.$$

Since $w \in A_r(\Omega)$, it follows that

$$\|w\|_{1,B}^{\alpha/s} \cdot \|1/w\|_{1/(r-1),\rho B}^{\alpha/s}$$

$$\leq \left(\left(\int_{\rho B} w\,dx\right) \left(\int_{\rho B} \left(\tfrac{1}{w}\right)^{1/(r-1)} dx\right)^{r-1}\right)^{\alpha/s}$$

$$= \left(|\rho B|^r \left(\tfrac{1}{|\rho B|} \int_{\rho B} w\,dx\right) \left(\tfrac{1}{|\rho B|} \int_{\rho B} \left(\tfrac{1}{w}\right)^{1/(r-1)} dx\right)^{r-1}\right)^{\alpha/s} \tag{7.3.68}$$

$$\leq C_4 |B|^{\alpha r/s}.$$

Finally, combining (7.3.68) and (7.3.67), we obtain

$$\|\Delta(G(du))\|_{s,B,w^\alpha} \leq C_5 \|du\|_{s,\rho B, w^\alpha} \tag{7.3.69}$$

for all balls B with $\rho B \subset \Omega$. Hence (7.3.59) holds if $0 < \alpha < 1$.

Next, we prove (7.3.59) for $\alpha = 1$, that is, we show that

$$\|\Delta(G(du))\|_{s,B,w} \leq C \|du\|_{s,\rho B, w}. \tag{7.3.70}$$

By Lemma 1.4.7, there exist constants $\beta > 1$ and $C_6 > 0$, such that

$$\| w \|_{\beta,B} \leq C_6 |B|^{(1-\beta)/\beta} \| w \|_{1,B} \tag{7.3.71}$$

for any cube or ball $B \subset \mathbf{R}^n$. Set $t = s\beta/(\beta - 1)$, so that $1 < s < t$ and $\beta = t/(t-s)$. Note that $1/s = 1/t + (t-s)/st$. Using Theorem 1.1.4, (7.3.62), and (7.3.71), we have

$$\left(\int_B |\Delta(G(du))|^s w\,dx\right)^{1/s}$$

$$= \left(\int_B \left(|\Delta(G(du))|w^{1/s}\right)^s dx\right)^{1/s}$$

$$\leq \left(\int_B |\Delta(G(du))|^t dx\right)^{1/t} \left(\int_B \left(w^{1/s}\right)^{st/(t-s)} dx\right)^{(t-s)/st}$$

$$\leq C_7 \|\Delta(G(du))\|_{t,B} \cdot \|w\|_{\beta,B}^{1/s} \tag{7.3.72}$$

$$\leq C_7 \|du\|_{t,B} \cdot \|w\|_{\beta,B}^{1/s}$$

$$\leq C_8 |B|^{(1-\beta)/\beta s} \|w\|_{1,B}^{1/s} \cdot \|du\|_{t,B}$$

$$\leq C_8 |B|^{-1/t} \|w\|_{1,B}^{1/s} \cdot \|du\|_{t,B}.$$

Next, choose $m = s/r$. Then, from Theorem 6.2.12, it follows that

$$\|du\|_{t,B} \leq C_9 |B|^{(m-t)/mt} \|du\|_{m,\rho B}. \tag{7.3.73}$$

Thus, Theorem 1.1.4 yields

$$\|du\|_{m,\rho B}$$

$$= \left(\int_{\rho B} \left(|du|w^{1/s}w^{-1/s}\right)^m dx\right)^{1/m} \tag{7.3.74}$$

$$\leq \left(\int_{\rho B} |du|^s w\,dx\right)^{1/s} \left(\int_{\rho B} \left(\frac{1}{w}\right)^{1/(r-1)} dx\right)^{(r-1)/s}$$

for all balls B with $\rho B \subset \Omega$. Now, since $w \in A_r(\Omega)$, (7.3.68) with $\alpha = 1$ gives

$$\|w\|_{1,B}^{1/s} \cdot \|1/w\|_{1/(r-1),\rho B}^{1/s} \leq C_{10} |B|^{r/s}. \tag{7.3.74'}$$

Finally, combining $(7.3.72), (7.3.73), (7.3.74)$, and $(7.3.74)'$, we have

$$\|\Delta(G(du))\|_{s,B,w}$$

$$\leq C_{11} |B|^{-1/t} \|w\|_{1,B}^{1/s} |B|^{(m-t)/mt} \|du\|_{m,\rho B}$$

$$\leq C_{11} |B|^{-1/m} \|w\|_{1,B}^{1/s} \cdot \|1/w\|_{1/(r-1),\rho B}^{1/s} \|du\|_{s,\rho B,w}$$

$$\leq C_{12} \|du\|_{s,\rho B,w}$$

for all balls B with $\rho B \subset \Omega$. Hence, (7.3.70) holds. ∎

7.4 Poincaré-type inequalities for operators

Recently, S. Ding, Y. Xing, and their co-workers investigated various estimates for different compositions of operators, such as the homotopy operator, Green's operator, the projection operator, and the maximal operator. We
present some of their contributions in the remaining sections of this chapter.
These results have applications in the L^p-theory of differential forms and related areas. We first present the Poincaré-type estimates for the compositions
of operators, including $T \circ G$, $G \circ T$, and $T \circ H$.

7.4.1 Poincaré-type inequalities for $T \circ G$

We begin with the Poincaré-type estimates for the composition $T \circ G$ of
the homotopy operator T and Green's operator G. We will discuss single as
well as two-weighted cases. Our first result provides the local estimate in a
bounded, convex domain Ω.

Theorem 7.4.1. *Let $u \in L^s_{loc}(\Omega, \wedge^l)$, $l = 1, 2, \ldots, n$, $1 < s < \infty$,
be a smooth differential form in a bounded, convex domain Ω and $T :
C^\infty(\Omega, \wedge^l) \to C^\infty(\Omega, \wedge^{l-1})$ be the homotopy operator defined in (1.5.1).
Then, there exists a constant C, independent of u, such that*

$$\|T(G(u)) - (T(G(u)))_B\|_{s,B} \le C|B|diam(B)\|u\|_{s,B} \qquad (7.4.1)$$

for all balls $B \subset \Omega$.

Note that since $B \subset \Omega$ and Ω is bounded, inequality (7.4.1) can be written
as

$$\|T(G(u)) - (T(G(u)))_B\|_{s,B} \le C\|u\|_{s,B}. \qquad (7.4.1)'$$

However, we include the factor $|B|diam(B)$ on the right-hand side of (7.4.1),
so that it is more flexible, particularly, it helps in establishing the global
estimates.

Proof. It is well known that for any differential form u, we have

$$\|Tu\|_{s,B} \le C_1|B|diam(B)\|u\|_{s,B} \qquad (7.4.2)$$

for all balls $B \subset \Omega$. Recall the decomposition

$$u = d(Tu) + T(du) = u_B + T(du) \qquad (7.4.3)$$

for any differential form u, where $u_B = d(Tu)$. Since Ω is bounded, there
exists a constant C_2 such that $|B|diam(B) \le C_2$. From (7.4.2) and (7.4.3),
it follows that

$$\|d(Tu)\|_{s,B} = \|u - T(du)\|_{s,B}$$

$$\leq \|u\|_{s,B} + \|T(du)\|_{s,B} \qquad (7.4.4)$$

$$\leq \|u\|_{s,B} + C_3|B|diam(B)\|du\|_{s,B}.$$

Now from Lemma 3.3.1, we find that

$$\|G(u)\|_{s,B} \leq C_4\|u\|_{s,B} \quad \text{and} \quad \|d(G(u))\|_{s,B} \leq C_4\|u\|_{s,B} \qquad (7.4.5)$$

for some constant $C_4 > 0$. Next, from $(7.4.2), (7.4.3), (7.4.4),$ and $(7.4.5)$, we obtain

$$\|T(G(u)) - (T(G(u)))_B\|_{s,B}$$

$$= \|Td(T(G(u)))\|_{s,B}$$

$$\leq C_5|B|diam(B)\|dT(G(u))\|_{s,B}$$

$$\leq C_6|B|diam(B)(\|G(u)\|_{s,B} + C_7|B|diam(B)\|d(G(u))\|_{s,B}) \qquad (7.4.6)$$

$$\leq C_6|B|diam(B)(\|G(u)\|_{s,B} + C_8\|d(G(u))\|_{s,B})$$

$$\leq C_6|B|diam(B)(\|u\|_{s,B} + C_9\|u\|_{s,B})$$

$$\leq C_{10}|B|diam(B)\|u\|_{s,B},$$

and hence inequality (7.4.1) holds. ∎

Theorem 7.4.2. *Suppose that $u \in L_{loc}^s(\Omega, \wedge^l)$, $l = 1, 2, \ldots, n$, $1 < s < \infty$, is a solution of the nonhomogeneous A-harmonic equation (1.2.10) in a bounded, convex domain Ω and $T : C^\infty(\Omega, \wedge^l) \to C^\infty(\Omega, \wedge^{l-1})$ is the homotopy operator defined in (1.5.1). Assume that $\rho > 1$ and $w \in A_r(\Omega)$ for some $1 < r < \infty$. Then, there exists a constant C, independent of u, such that*

$$\|T(G(u)) - (T(G(u)))_B\|_{s,B,w^\alpha} \leq C|B|diam(B)\|u\|_{s,\rho B, w^\alpha} \qquad (7.4.7)$$

for all balls B with $\rho B \subset \Omega$ and any real number α with $0 < \alpha \leq 1$.

The above L^s-norm inequality can also be written in the integral form as

$$\left(\int_B |T(G(u)) - (T(G(u)))_B|^s w^\alpha dx \right)^{1/s} \leq C \left(\int_{\rho B} |u|^s w^\alpha dx \right)^{1/s}. \qquad (7.4.7)'$$

Proof. First, we prove (7.4.7) for $0 < \alpha < 1$. Set $t = s/(1 - \alpha)$. Using the Hölder inequality and (7.4.1), we have

$$\|T(G(u)) - (T(G(u)))_B\|_{s,B,w^\alpha}$$

$$= \left(\int_B \left(|TG(u) - (TG(u))_B| w^{\alpha/s} \right)^s dx \right)^{1/s}$$

$$\leq \|TG(u) - (TG(u))_B\|_{t,B} \left(\int_B w^{t\alpha/(t-s)} dx \right)^{(t-s)/st} \qquad (7.4.8)$$

$$= \|TG(u) - (TG(u))_B\|_{t,B} \left(\int_B w dx \right)^{\alpha/s}$$

$$\leq C_3 |B| diam(B) \|u\|_{t,B} \left(\int_B w dx \right)^{\alpha/s}.$$

Select $m = s/(1+\alpha(r-1))$, so that $m < s$. Recalling the weak reverse Hölder inequality

$$\|u\|_{t,B} \leq C_4 |B|^{(m-t)/mt} \|u\|_{m,\rho B} \qquad (7.4.9)$$

and using (7.4.8), we obtain

$$\|TG(u) - (TG(u))_B\|_{s,B,w^\alpha}$$
$$\leq C_5 |B|^{1+(m-t)/mt} diam(B) \|u\|_{m,\rho B} \left(\int_B w dx \right)^{\alpha/s}. \qquad (7.4.10)$$

Now using the Hölder inequality again with $1/m = 1/s + (s-m)/sm$, we have

$$\|u\|_{m,\rho B} = \left(\int_{\rho B} |u|^m dx \right)^{1/m}$$

$$= \left(\int_{\rho B} \left(|u| w^{\alpha/s} w^{-\alpha/s} \right)^m dx \right)^{1/m} \qquad (7.4.11)$$

$$\leq \|u\|_{s,\rho B,w^\alpha} \left(\int_{\rho B} \left(\tfrac{1}{w} \right)^{1/(r-1)} dx \right)^{\alpha(r-1)/s}$$

for all balls B with $\rho B \subset \Omega$. Substituting (7.4.11) into (7.4.10), we find that

$$\|TG(u) - (TG(u))_B\|_{s,B,w^\alpha}$$

$$\leq C_6 |B|^{1+\frac{m-t}{mt}} diam(B) \|u\|_{s,\rho B,w^\alpha} \qquad (7.4.12)$$

$$\times \left(\int_B w dx \right)^{\alpha/s} \left(\int_{\rho B} \left(\tfrac{1}{w} \right)^{1/(r-1)} dx \right)^{\alpha(r-1)/s}.$$

Next, since $w \in A_r(\Omega)$, it follows that

$$\|w\|_{1,B}^{\alpha/s} \cdot \|1/w\|_{1/(r-1),\rho B}^{\alpha/s}$$

$$\leq \left(\left(\int_{\rho B} w dx \right) \left(\int_{\rho B} \left(\tfrac{1}{w} \right)^{1/(r-1)} dx \right)^{r-1} \right)^{\alpha/s}$$

$$= \left(|\rho B|^r \left(\tfrac{1}{|\rho B|} \int_{\rho B} w dx \right) \left(\tfrac{1}{|\rho B|} \int_{\rho B} \left(\tfrac{1}{w} \right)^{1/(r-1)} dx \right)^{r-1} \right)^{\alpha/s}$$

$$\leq C_7 |B|^{\alpha r/s}.$$

(7.4.13)

Combining (7.4.13) and (7.4.12), we obtain

$$\|TG(u) - (TG(u))_B\|_{s,B,w^\alpha} \leq C_8 |B| diam(B) \|u\|_{s,\rho B,w^\alpha} \tag{7.4.14}$$

for all balls B with $\rho B \subset \Omega$. Thus, (7.4.7) holds if $0 < \alpha < 1$. Next, we shall prove (7.4.7) for $\alpha = 1$, that is, we will show that

$$\|TG(u) - (TG(u))_B\|_{s,B,w} \leq C |B| diam(B) \|u\|_{s,\rho B,w}. \tag{7.4.15}$$

By Lemma 1.4.7, there exist constants $\beta > 1$ and $C_9 > 0$, such that

$$\| w \|_{\beta,B} \leq C_9 |B|^{(1-\beta)/\beta} \| w \|_{1,B} \tag{7.4.16}$$

for any cube or ball $B \subset \mathbf{R}^n$. Set $t = s\beta/(\beta - 1)$, so that $1 < s < t$ and $\beta = t/(t - s)$. Note that $1/s = 1/t + (t - s)/st$. Using Theorem 1.1.4, (7.4.1), and (7.4.16), we find that

$$\left(\int_B |TG(u) - (TG(u))_B|^s w dx \right)^{1/s}$$

$$= \left(\int_B \left(|TG(u) - (TG(u))_B| w^{1/s} \right)^s dx \right)^{1/s}$$

$$\leq \left(\int_B |TG(u) - (TG(u))_B|^t dx \right)^{1/t} \left(\int_B \left(w^{1/s} \right)^{st/(t-s)} dx \right)^{(t-s)/st}$$

$$\leq C_{10} \|TG(u) - (TG(u))_B\|_{t,B} \cdot \|w\|_{\beta,B}^{1/s}$$

$$\leq C_{11} |B| diam(B) \|u\|_{t,B} \cdot \|w\|_{\beta,B}^{1/s}$$

$$\leq C_{12} |B|^{1+(1-\beta)/\beta s} diam(B) \|w\|_{1,B}^{1/s} \cdot \|u\|_{t,B}$$

$$\leq C_{12} |B|^{1-1/t} diam(B) \|w\|_{1,B}^{1/s} \cdot \|u\|_{t,B}.$$

(7.4.17)

Next, choose $m = s/r$. Using the weak reverse Hölder inequality again for the new exponent, we have

$$\|u\|_{t,B} \le C_{13}|B|^{(m-t)/mt}\|u\|_{m,\rho B}. \tag{7.4.18}$$

Now, from Lemma 1.1.4, we find that

$$\|u\|_{m,\rho B} = \left(\int_{\rho B} \left(|u|w^{1/s}w^{-1/s}\right)^m dx\right)^{1/m}$$

$$\le \left(\int_{\rho B} |u|^s w dx\right)^{1/s} \left(\int_{\rho B} \left(\tfrac{1}{w}\right)^{1/(r-1)} dx\right)^{(r-1)/s} \tag{7.4.19}$$

for all balls B with $\rho B \subset \Omega$. Finally, since $w \in A_r(\Omega)$, as for (7.4.13), we have

$$\|w\|_{1,B}^{1/s} \cdot \|1/w\|_{1/(r-1),\rho B}^{1/s} \le C_{14}|B|^{r/s}. \tag{7.4.19}'$$

Combining $(7.4.17), (7.4.18), (7.4.19)$, and $(7.4.19)'$, we obtain

$$\|TG(u) - (TG(u))_B\|_{s,B,w}$$

$$\le C_{15}|B|^{1-1/t}diam(B)\|w\|_{1,B}^{1/s}|B|^{(m-t)/mt}\|u\|_{m,\rho B}$$

$$\le C_{15}|B|^{1-1/m}diam(B)\|w\|_{1,B}^{1/s} \cdot \|1/w\|_{1/(r-1),\rho B}^{1/s}\|u\|_{s,\rho B,w}$$

$$\le C_{16}|B|diam(B)\|u\|_{s,\rho B,w}$$

for all balls B with $\rho B \subset \Omega$. Hence, (7.4.15) follows. ∎

Theorem 7.4.2 for $\alpha = 1/s$ gives the inequality

$$\left(\int_B |T(G(u)) - (T(G(u)))_B|^s w^{1/s} dx\right)^{1/s} \le C \left(\int_{\rho B} |u|^s w^{1/s} dx\right)^{1/s}.$$

Let $\alpha = 1 - t, 0 < t < 1$ and the measure μ be defined by $d\mu = w(x)dx$. Then, from Theorem 7.4.2, we have

$$\left(\int_B |T(G(u)) - (T(G(u)))_B|^s w^{-t} d\mu\right)^{1/s} \le C \left(\int_{\rho B} |u|^s w^{-t} d\mu\right)^{1/s}.$$

If we choose $\alpha = 1 - 1/s$ and $d\mu = w(x)dx$, then from Theorem 7.4.2, we obtain

$$\left(\int_B |T(G(u)) - (T(G(u)))_B|^s w^{-1/s} d\mu\right)^{1/s} \le C \left(\int_{\rho B} |u|^s w^{-1/s} d\mu\right)^{1/s}.$$

Let $\alpha = 1$ in Theorem 7.4.2. Then, we find that

$$\left(\int_B |T(G(u)) - (T(G(u)))_B|^s w dx\right)^{1/s} \leq C\left(\int_{\rho B} |u|^s w dx\right)^{1/s}.$$

Next, we extend our local weighted Poincaré-type inequality to the global case.

Theorem 7.4.3. *Let $u \in D'(\Omega, \wedge^1)$ be a solution of the nonhomogeneous A-harmonic equation (1.2.10), $T : C^\infty(\Omega, \wedge^1) \to C^\infty(\Omega, \wedge^0)$ be the homotopy operator defined in (1.5.1), and G be Green's operator. Assume that $w \in A_r(\Omega)$ for some $1 < r < \infty$ and s is a fixed exponent associated with the A-harmonic equation (1.2.10). Then, there exists a constant C, independent of u, such that*

$$\left(\int_\Omega |T(G(u)) - (T(G(u)))_{Q_0}|^s w dx\right)^{1/s} \leq C\left(\int_\Omega |u|^s w dx\right)^{1/s} \qquad (7.4.20)$$

for any bounded, convex δ-John domain $\Omega \subset \mathbf{R}^n$. Here $Q_0 \subset \Omega$ is a fixed cube.

Proof. Letting $\alpha = 1$ in (7.4.7). Then, we can write (7.4.7) as

$$\int_Q |TG(u) - (TG(u))_Q|^s d\mu(x) \leq C_1 |Q| diam(Q) \int_{\sigma Q} |u|^s d\mu(x), \qquad (7.4.21)$$

where the measure $\mu(x)$ is defined by $d\mu(x) = w(x)dx$. We use the notation and the covering $\mathcal{V}$ described in Theorem 1.5.3 (covering lemma) and the properties of the measure $\mu(x)$: if $w \in A_r$, then

$$\mu(NQ) \leq MN^{nr}\mu(Q) \qquad (7.4.22)$$

for each cube Q with $NQ \subset \mathbf{R}^n$, and

$$\max(\mu(Q_i), \mu(Q_{i+1})) \leq MN^{nr}\mu(Q_i \cap Q_{i+1}) \qquad (7.4.23)$$

for the sequence of cubes $Q_i, Q_{i+1}, i = 0, 1, \ldots, k-1$. Now, by the elementary inequality $|a + b|^s \leq 2^s(|a|^s + |b|^s)$, $s > 0$, we find that

$$\int_\Omega |TG(u) - (TG(u))_{Q_0}|^s w dx$$

$$= \int_\Omega |TG(u) - (TG(u))_{Q_0}|^s d\mu(x)$$

$$\leq 2^s \sum_{Q \in \mathcal{V}} \int_Q |TG(u) - (TG(u))_Q|^s d\mu(x) \qquad (7.4.24)$$

$$+ 2^s \sum_{Q \in \mathcal{V}} \int_Q |(TG(u))_{Q_0} - (TG(u))_Q|^s d\mu(x).$$

From (7.4.21) and the covering lemma, we estimate the first sum as follows:

$$\sum_{Q \in \mathcal{V}} \int_Q |TG(u) - (TG(u))_Q|^s d\mu(x)$$

$$\leq C_1 \sum_{Q \in \mathcal{V}} |Q| diam(Q) \left(\int_{\sigma Q} |u|^s w dx \right)$$

$$\leq C_1 \sum_{Q \in \mathcal{V}} \left(\int_{\sigma Q} |u|^s w dx \right) \tag{7.4.25}$$

$$\leq C_1 N \left(\int_\Omega |u|^s w dx \right).$$

Next, we shall estimate the second sum in (7.4.24). Fix a cube $Q \in \mathcal{V}$ and let $Q_0, Q_1, \ldots, Q_k = Q$ be the chain in the covering lemma. Clearly,

$$|(TG(u))_{Q_0} - (TG(u))_Q| \leq \sum_{i=0}^{k-1} |(TG(u))_{Q_i} - (TG(u))_{Q_{i+1}}|. \tag{7.4.26}$$

Using (7.4.21) and (7.4.23), we obtain

$$|(TG(u))_{Q_i} - (TG(u))_{Q_{i+1}}|^s$$

$$= \frac{1}{\mu(Q_i \cap Q_{i+1})} \int_{Q_i \cap Q_{i+1}} |(TG(u))_{Q_i} - (TG(u))_{Q_{i+1}}|^s d\mu(x)$$

$$\leq \frac{M N^{n r}}{\max(\mu(Q_i), \mu(Q_{i+1}))} \int_{Q_i \cap Q_{i+1}} |(TG(u))_{Q_i} - (TG(u))_{Q_{i+1}}|^s d\mu(x)$$

$$\leq C_2 \sum_{j=i}^{i+1} \frac{1}{\mu(Q_j)} \int_{Q_j} |TG(u) - (TG(u))_{Q_j}|^s d\mu(x)$$

$$\leq C_3 \sum_{j=i}^{i+1} \frac{diam(Q_j)}{\mu(Q_j)} \int_{\sigma Q_j} |u|^s w dx.$$

Now since $Q \subset N Q_j$ for $j = i, i+1, 0 \leq i \leq k-1$ (see the covering lemma), it follows that

$$|(TG(u))_{Q_i} - (TG(u))_{Q_{i+1}}|^s \chi_Q(x)$$

$$\leq C_3 \sum_{j=i}^{i+1} \frac{\chi_{N Q_j}(x) diam(\Omega)}{\mu(Q_j)} \left(\int_{\sigma Q_j} |u|^s w dx \right).$$

By (7.4.26) and $diam(\Omega) < \infty$, and

$$|a + b|^{1/s} \leq 2^{1/s} (|a|^{1/s} + |b|^{1/s}),$$

we obtain

$$|(TG(u))_{Q_0} - (TG(u))_Q| \chi_Q(x)$$

$$\leq C_4 \sum_{R \in \mathcal{V}} \left(\frac{1}{\mu(R)} \left(\int_{\sigma R} |u|^s w dx \right) \right)^{1/s} \cdot \chi_{N R}(x)$$

for every $x \in \mathbf{R}^n$. Hence, we have

$$\sum_{Q \in \mathcal{V}} \int_Q |(TG(u))_{Q_0} - (TG(u))_Q|^s d\mu(x)$$

$$\leq C_5 \int_{\mathbf{R}^n} \left| \sum_{R \in \mathcal{V}} \left(\frac{1}{\mu(R)} \left(\int_{\sigma R} |u|^s w dx \right) \right)^{1/s} \chi_{NR}(x) \right|^s d\mu(x). \qquad (7.4.27)$$

From (7.4.27) and Theorem 1.4.15, it follows that

$$\sum_{Q \in \mathcal{V}} \int_Q |(TG(u))_{Q_0} - (TG(u))_Q|^s d\mu(x)$$

$$\leq C_6 \int_{\mathbf{R}^n} \left| \sum_{R \in \mathcal{V}} \left(\frac{1}{\mu(R)} \left(\int_{\sigma R} |u|^s w dx \right) \right)^{1/s} \chi_R(x) \right|^s d\mu(x).$$

Finally, since

$$\sum_{R \in \mathcal{V}} \chi_R(x) \leq \sum_{R \in \mathcal{V}} \chi_{\sigma R}(x) \leq N \chi_\Omega(x),$$

by the elementary inequality

$$\left| \sum_{i=1}^N t_i \right|^s \leq N^{s-1} \sum_{i=1}^N |t_i|^s,$$

and the covering lemma, we find that

$$\sum_{Q \in \mathcal{V}} \int_Q |(TG(u))_{Q_0} - (TG(u))_Q|^s d\mu(x)$$

$$\leq C_7 \int_{\mathbf{R}^n} \left(\sum_{R \in \mathcal{V}} \frac{1}{\mu(R)} \left(\int_{\sigma R} |u|^s w dx \right) \chi_R(x) \right) d\mu(x)$$

$$= C_7 \sum_{R \in \mathcal{V}} \left(\int_{\sigma R} |u|^s w dx \right) \qquad (7.4.28)$$

$$\leq C_8 \left(\int_\Omega |u|^s w dx \right).$$

Combination of (7.4.24), (7.4.25), and (7.4.28) immediately gives (7.4.20).
∎

Now, we are ready to prove the local and global imbedding inequalities.

Theorem 7.4.4. *Suppose that* $u \in L^s_{loc}(\Omega, \wedge^l)$, $l = 1, 2, \ldots, n$, $1 < s < \infty$, *is a smooth differential form in a bounded, convex domain* Ω *and* $T : C^\infty(\Omega, \wedge^l) \to C^\infty(\Omega, \wedge^{l-1})$ *is the homotopy operator defined in (1.5.1). Assume that* $\rho > 1$ *and* $w \in A_r(\Omega)$ *for some* $1 < r < \infty$. *Then, there exists* a constant C, independent of u, such that

$$\|T(G(u)) - (T(G(u)))_B\|_{W^{1,s}(B),w^\alpha} \leq C|B|\|u\|_{s,\rho B,w^\alpha} \qquad (7.4.29)$$

for all balls B with $\rho B \subset \Omega$ and any real number α with $0 < \alpha \leq 1$.

Proof. For any differential form, we have the following basic inequality:

$$\|d(T(u))\|_{s,B} \leq \|u\|_{s,B} + C_1 diam(B)\|du\|_{s,B}.$$

Replacing u by $G(u)$ in above inequality yields

$$\|d(T(G(u)))\|_{s,B} \leq \|G(u)\|_{s,B} + C_1 diam(B)\|dG(u)\|_{s,B}.$$

Since Ω is bounded, using inequality (7.2.1), we have

$$\|d(T(G(u)))\|_{s,B} \leq C_2\|u\|_{s,B} + C_3 diam(B)\|u\|_{s,B}$$

$$\leq C_4\|u\|_{s,B}.$$

From (1.5.5), (1.5.6), (7.2.1), and the above inequality, we have

$$\|\nabla((T(G(u))_B)\|_{s,B}$$

$$= \|\nabla((T(G(u)) - Td(T(G(u)))))\|_{s,B}$$

$$\leq \|\nabla(T(G(u))\|_{s,B} + \|\nabla(Td(T(G(u))))\|_{s,B}$$

$$\leq C_6|B|\|T(G(u)\|_{s,B} + C_7|B|\|d(T(G(u)))\|_{s,B}$$

$$\leq C_8 diam(B)|B|^2\|G(u)\|_{s,B} + C_9|B|\|u\|_{s,B}$$

$$\leq C_{10} diam(B)|B|^2\|u\|_{s,B} + C_9|B|\|u\|_{s,B}$$

$$\leq C_{11}|B|\|u\|_{s,B}$$

since $|B| \subset \Omega$ and Ω is bounded. We can extend the above inequality into the weighted version

$$\|\nabla((T(G(u))_B)\|_{s,B,w^\alpha} \leq C_{12}|B|\|u\|_{s,\rho B,w^\alpha}. \qquad (7.4.30)$$

Applying (3.3.15), (7.4.30), and Theorems 7.4.2 and 5.6.3, we obtain

$$\|T(G(u)) - (T(G(u)))_B\|_{W^{1,s}(B),w^\alpha}$$

$$= diam(B)^{-1}\|T(G(u)) - (T(G(u)))_B\|_{s,B,w^\alpha}$$

$$+\|\nabla(T(G(u)) - (T(G(u)))_B)\|_{s,B,w^\alpha}$$

$$= diam(B)^{-1}\|T(G(u)) - (T(G(u)))_B\|_{s,B,w^\alpha}$$

$$+\|\nabla(T(G(u))) - \nabla((T(G(u)))_B)\|_{s,B,w^\alpha}$$

$$\leq diam(B)^{-1}\|T(G(u)) - (T(G(u)))_B\|_{s,B,w^\alpha}$$

$$+\|\nabla(T(G(u)))\|_{s,B,w^\alpha} + \|\nabla((T(G(u)))_B)\|_{s,B,w^\alpha}$$

$$\leq C_{13}|B|\|u\|_{s,\rho B,w^\alpha}. \qquad\blacksquare$$

We should notice that all steps before inequality (7.4.30) in the proof of Theorem 7.4.4 will still hold if the ball B is replaced by a bounded, convex domain $\Omega \subset \mathbf{R}^n$. Thus, we have the following global inequality without weights.

Theorem 7.4.4′. *Suppose that* $u \in L^s_{loc}(\Omega, \wedge^l)$, $l = 1, 2, \ldots, n$, $1 < s < \infty$, *is a smooth differential form in a bounded, convex domain* Ω *and* $T : C^\infty(\Omega, \wedge^l) \to C^\infty(\Omega, \wedge^{l-1})$ *is the homotopy operator defined in* (1.5.1). *Then, there exists a constant* C, *independent of* u, *such that*

$$\|T(G(u)) - (T(G(u)))_\Omega\|_{W^{1,s}(\Omega)} \leq C|\Omega|\|u\|_{s,\Omega}.$$

Next, using Theorem 7.4.3, we prove the following global weighted imbedding inequality.

Theorem 7.4.5. *Let* $u \in D'(\Omega, \wedge^1)$ *be a solution of the nonhomogeneous A-harmonic equation* (1.2.10), $T : C^\infty(\Omega, \wedge^1) \to C^\infty(\Omega, \wedge^0)$ *be the homotopy operator defined in* (1.5.1), *and* G *be Green's operator. Assume that* $w \in A_r(\Omega)$ *for some* $1 < r < \infty$ *and* s *is a fixed exponent associated with the A-harmonic equation* (1.2.10). *Then, there exists a constant* C, *independent of* u, *such that*

$$\|T(G(u)) - (T(G(u)))_{Q_0}\|_{W^{1,s}(\Omega),w} \leq C\|u\|_{s,\Omega,w} \qquad (7.4.31)$$

for any bounded δ-*John domain* $\Omega \subset \mathbf{R}^n$. *Here* $Q_0 \subset \Omega$ *is a fixed cube.*

Proof. Since u is a 1-form, then $(T(G(u)))_{Q_0}$ is a closed 0-form (function). Thus, we have

$$\|\nabla((T(G(u)))_{Q_0})\|_{s,\Omega,w} = \|d((T(G(u)))_{Q_0})\|_{s,\Omega,w} = 0. \qquad (7.4.32)$$

Applying (3.3.15), (7.4.32), and Theorems 7.4.3 and 7.3.18, we obtain

$$\|T(G(u)) - (T(G(u)))_{Q_0}\|_{W^{1,s}(\Omega),w}$$

$$= diam(\Omega)^{-1}\|T(G(u)) - (T(G(u)))_{Q_0}\|_{s,\Omega,w}$$

$$+ \|\nabla(T(G(u)) - (T(G(u)))_{Q_0})\|_{s,\Omega,w}$$

$$= diam(\Omega)^{-1}\|T(G(u)) - (T(G(u)))_{Q_0}\|_{s,\Omega,w}$$

$$+ \|\nabla(T(G(u))) - \nabla((T(G(u)))_{Q_0})\|_{s,\Omega,w}$$

$$\leq diam(\Omega)^{-1}\|T(G(u)) - (T(G(u)))_{Q_0}\|_{s,\Omega,w}$$

$$+ \|\nabla(T(G(u)))\|_{s,\Omega,w} + \|\nabla((T(G(u)))_{Q_0})\|_{s,\Omega,w}$$

$$= diam(\Omega)^{-1}\|T(G(u)) - (T(G(u)))_{Q_0}\|_{s,\Omega,w} + \|\nabla(T(G(u)))\|_{s,\Omega,w}$$

$$\leq diam(\Omega)^{-1}C_1 diam(\Omega)|\Omega|\|u\|_{s,\Omega,w} + C_2|\Omega|\|u\|_{s,\Omega,w}$$

$$\leq C_3|\Omega|\|u\|_{s,\Omega,w}$$

$$\leq C_4\|u\|_{s,\Omega,w},$$

that is,

$$\|T(G(u)) - (T(G(u)))_{Q_0}\|_{W^{1,s}(\Omega),w} \leq C_4\|u\|_{s,\Omega,w}. \qquad \blacksquare$$

In our earlier results, we have developed the $A_r(\Omega)$-weighted Poincaré-type estimates for the composition $T \circ G$. Now, we shall present estimates with different weights, such as $A_r(\lambda, \Omega)$-weights, $A_r^\lambda(\Omega)$-weights.

The proof of the following inequalities with $A_r(\lambda, \Omega)$-weights is similar to the proof of Theorem 7.4.4.

Theorem 7.4.6. *Let $u \in L_{loc}^s(\Omega, \wedge^l)$, $l = 1, 2, \ldots, n$, $1 < s < \infty$, be a differential form satisfying the nonhomogeneous A-harmonic equation (1.2.10) in a bounded, convex domain $\Omega \subset \mathbf{R}^n$ and $T : C^\infty(\Omega, \wedge^l) \to C^\infty(\Omega, \wedge^{l-1})$ be a homotopy operator defined in (1.5.1). Assume that $w \in A_r(\lambda, \Omega)$ for some $r > 1$ and $\lambda > 0$. Then, there exists a constant C, independent of u, such that*

$$\|T(G(u)) - (T(G(u)))_B\|_{s,B,w^{\alpha\lambda}} \leq C|B|diam(B)\|u\|_{s,\rho B,w^\alpha} \qquad (7.4.33)$$

and

$$\|T(G(u)) - (T(G(u)))_B\|_{W^{1,s}(B),\ w^{\alpha\lambda}} \leq C|B|\|u\|_{s,\rho B,w^\alpha} \qquad (7.4.34)$$

for all balls B with $\rho B \subset \Omega$ and arbitrary real number α with $0 < \alpha < 1$. Here $\rho > 1$ is some constant.

Note that inequality (7.4.33) can be written as

$$\left(\int_B |T(G(u)) - (T(G(u)))_B|^s w^{\alpha\lambda} dx\right)^{\frac{1}{s}} \leq C|B| diam(B) \left(\int_{\rho B} |u|^s w^\alpha dx\right)^{\frac{1}{s}}.$$

Similarly, we have the analog of inequalities (7.4.33) and (7.4.34) with $A_r^\lambda(\Omega)$-weights.

Theorem 7.4.7. *Let $u \in L_{loc}^s(\Omega, \wedge^l)$, $l = 1, 2, \ldots, n$, $1 < s < \infty$, be a differential form satisfying (1.2.10) in a bounded, convex domain $\Omega \subset \mathbf{R}^n$ and $T : C^\infty(\Omega, \wedge^l) \rightarrow C^\infty(\Omega, \wedge^{l-1})$ be a homotopy operator defined in (1.5.1). Assume that $\rho > 1$ and $w \in A_r^\lambda(\Omega)$ for some $r > 1$ and $\lambda > 0$. Then, there exists a constant C, independent of u, such that*

$$\|T(G(u)) - (T(G(u)))_B\|_{s,B,w^\alpha} \leq C|B| diam(B)\|u\|_{s,\rho B,w^{\alpha\lambda}} \qquad (7.4.35)$$

and

$$\|T(G(u)) - (T(G(u)))_B\|_{W^{1,s}(B),\ w^\alpha} \leq C|B|\|u\|_{s,\rho B,w^{\alpha\lambda}} \qquad (7.4.36)$$

for all balls B with $\rho B \subset \Omega$ and any real number α with $0 < \alpha < 1$.

Each of the above two inequalities has an integral version. For example, inequality (7.4.35) can be written as

$$\left(\int_B |T(G(u)) - (T(G(u)))_B|^s w^\alpha dx\right)^{\frac{1}{s}} \leq C|B| diam(B) \left(\int_{\rho B} |u|^s w^{\alpha\lambda} dx\right)^{\frac{1}{s}}.$$

Now, we extend the above estimates into the following two-weight case.

Theorem 7.4.8. *Let $u \in L_{loc}^s(\Omega, \wedge^l)$, $l = 1, 2, \ldots, n$, $1 < s < \infty$, be a solution of the nonhomogeneous A-harmonic equation (1.2.10) in a bounded, convex domain $\Omega \subset \mathbf{R}^n$ and $T : C^\infty(\Omega, \wedge^l) \rightarrow C^\infty(\Omega, \wedge^{l-1})$ be the homotopy operator defined in (1.5.1). Suppose that $\rho > 1$ and $(w_1, w_2) \in A_r(\lambda, \Omega)$ for some $\lambda > 0$ and $1 < r < \infty$. Then, there exists a constant C, independent of u, such that*

$$\left(\int_B |T(G(u)) - (T(G(u)))_B|^s w_1^{\alpha\lambda} dx \right)^{\frac{1}{s}} \le C|B|diam(B) \left(\int_{\rho B} |u|^s w_2^\alpha dx \right)^{\frac{1}{s}}$$

$$(7.4.37)$$

and

$$\|T(G(u)) - (T(G(u)))_B\|_{W^{1,s}(B),\, w_1^{\alpha\lambda}} \le C|B|\|u\|_{s,\rho B, w_2^\alpha} \qquad (7.4.37)'$$

for all balls B with $\rho B \subset \Omega$ and any real number α with $0 < \alpha < 1$.

Note that (7.4.37) can be written as

$$\|T(G(u)) - (T(G(u)))_B\|_{s,B,w_1^{\alpha\lambda}} \le C|B|diam(B)\|u\|_{s,\rho B,w_2^\alpha}. \qquad (7.4.37)''$$

Proof. Select $t = s/(1-\alpha)$, so that $1 < s < t$. Noticing $1/s = 1/t + (t-s)/st$ and using the Hölder inequality, we obtain

$$\left(\int_B |T(G(u)) - (T(G(u)))_B|^s w_1^{\alpha\lambda} dx \right)^{1/s}$$

$$= \left(\int_B \left(|T(G(u)) - (T(G(u)))_B| w_1^{\alpha\lambda/s} \right)^s dx \right)^{1/s}$$

$$\le \left(\int_B |T(G(u)) - (T(G(u)))_B|^t dx \right)^{1/t} \qquad (7.4.38)$$

$$\times \left(\int_B \left(w_1^{\alpha\lambda/s} \right)^{st/(t-s)} dx \right)^{(t-s)/st}$$

$$\le \|T(G(u)) - (T(G(u)))_B\|_{t,B} \cdot \left(\int_B w_1^\lambda dx \right)^{\alpha/s}$$

for all balls B with $B \subset \Omega$. Applying Theorem 7.4.1, we have

$$\|T(G(u)) - (T(G(u)))_B\|_{t,B} \le C_1|B|diam(B)\|u\|_{t,B}. \qquad (7.4.39)$$

Choose $m = s/(1 + \alpha(r-1))$, so that $m < s < t$. Substituting (7.4.39) into (7.4.38) and using Lemma 3.1.1, we obtain

$$\left(\int_B |T(G(u)) - (T(G(u)))_B|^s w_1^{\alpha\lambda} dx \right)^{1/s}$$

$$\le C_1|B|diam(B)\|u\|_{t,B} \left(\int_B w_1^\lambda dx \right)^{\alpha/s} \qquad (7.4.40)$$

$$\le C_2|B|diam(B)|B|^{(m-t)/mt}\|u\|_{m,\rho B} \left(\int_B w_1^\lambda dx \right)^{\alpha/s}.$$

Using the Hölder inequality with $1/m = 1/s + (s-m)/sm$, we find that

$$\|u\|_{m,\rho B}$$

$$= \left(\int_{\rho B} |u|^m dx \right)^{1/m}$$

$$= \left(\int_{\rho B} \left(|u| w_2^{\alpha/s} w_2^{-\alpha/s} \right)^m dx \right)^{1/m} \tag{7.4.41}$$

$$\leq \left(\int_{\rho B} |u|^s w_2^\alpha dx \right)^{1/s} \left(\int_{\rho B} \left(\frac{1}{w_2} \right)^{1/(r-1)} dx \right)^{\alpha(r-1)/s}$$

for all balls B with $\rho B \subset \Omega$. Substituting (7.4.41) into (7.4.40), it follows that

$$\left(\int_B |T(G(u)) - (T(G(u)))_B|^s w_1^{\alpha\lambda} dx \right)^{1/s}$$

$$\leq C_3 |B| diam(B) |B|^{(m-t)/mt} \left(\int_{\rho B} |u|^s w_2^\alpha dx \right)^{1/s} \tag{7.4.42}$$

$$\times \left(\int_B w_1^\lambda dx \right)^{\alpha/s} \left(\int_{\rho B} \left(\frac{1}{w_2} \right)^{1/(r-1)} dx \right)^{\alpha(r-1)/s}.$$

Now, using the condition $(w_1, w_2) \in A_r(\lambda, \Omega)$, we obtain

$$\left(\int_B w_1^\lambda dx \right)^{\alpha/s} \left(\int_{\rho B} \left(\frac{1}{w_2} \right)^{1/(r-1)} dx \right)^{\alpha(r-1)/s}$$

$$\leq \left(\left(\int_{\rho B} w_1^\lambda dx \right) \left(\int_{\rho B} (1/w_2)^{1/(r-1)} dx \right)^{r-1} \right)^{\alpha/s} \tag{7.4.43}$$

$$= \left(|\rho B|^r \left(\frac{1}{|\rho B|} \int_{\rho B} w_1^\lambda dx \right) \left(\frac{1}{|\rho B|} \int_{\rho B} \left(\frac{1}{w_2} \right)^{1/(r-1)} dx \right)^{r-1} \right)^{\alpha/s}$$

$$\leq C_4 |B|^{\alpha r/s}.$$

Finally, combining (7.4.42) and (7.4.43), we find that

$$\left(\int_B |T(G(u)) - (T(G(u)))_B|^s w_1^{\alpha\lambda} dx \right)^{\frac{1}{s}}$$

$$\leq C |B| diam(B) \left(\int_{\rho B} |u|^s w_2^\alpha dx \right)^{\frac{1}{s}}$$

for all balls B with $\rho B \subset \Omega$. The proof of inequality (7.4.37)' is similar to that of Theorem 7.4.4. ∎

Theorem 7.4.9. *Suppose that $u \in L^s_{loc}(\Omega, \wedge^l)$, $l = 1, 2, \ldots, n$, $1 < s < \infty$, is a differential form satisfying (1.2.10) in a bounded, convex domain*

$\Omega \subset \mathbf{R}^n$ and $T : C^\infty(\Omega, \wedge^l) \to C^\infty(\Omega, \wedge^{l-1})$ *is a homotopy operator defined in (1.5.1). Assume that $\rho > 1$ and $(w_1, w_2) \in A_{r,\lambda}(\Omega)$ for some $\lambda \geq 1$ and $1 < r < \infty$. Then, there exists a constant C, independent of u, such that*

$$\left(\int_B |T(G(u)) - (T(G(u)))_B|^s w_1^\alpha dx \right)^{\frac{1}{s}}$$

$$\leq C|B|diam(B) \left(\int_{\rho B} |u|^s w_2^\alpha dx \right)^{\frac{1}{s}} \tag{7.4.44}$$

and

$$\|T(G(u)) - (T(G(u)))_B\|_{W^{1,s}(B),\, w_1^\alpha} \leq C|B|diam(B)\|u\|_{s,\rho B, w_2^{\alpha\lambda}} \tag{7.4.44}'$$

for all balls B with $\rho B \subset \Omega$ and any real number α with $0 < \alpha < \lambda$.

Note that (7.4.44) can be written as

$$\|T(G(u)) - (T(G(u)))_B\|_{s,B,w_1^\alpha} \leq C|B|diam(B)\|u\|_{s,\rho B, w_2^\alpha}. \tag{7.4.44}''$$

Proof. Choose $t = \lambda s/(\lambda - \alpha)$. Since $1/s = 1/t + (t-s)/st$, from the Hölder inequality, we obtain

$$\left(\int_B |T(G(u)) - (T(G(u)))_B|^s w_1^\alpha dx \right)^{1/s}$$

$$= \left(\int_B \left(|T(G(u)) - (T(G(u)))_B| w_1^{\alpha/s} \right)^s dx \right)^{1/s}$$

$$\leq \left(\int_B |T(G(u)) - (T(G(u)))_B|^t dx \right)^{1/t}$$

$$\times \left(\int_B \left(w_1^{\alpha/s} \right)^{st/(t-s)} dx \right)^{(t-s)/st} \tag{7.4.45}$$

$$\leq \|T(G(u)) - (T(G(u)))_B\|_{t,B} \cdot \left(\int_B w_1^\lambda dx \right)^{\alpha/\lambda s}$$

for all balls B with $B \subset \Omega$. Using Theorem 7.4.1, we find that

$$\|T(G(u)) - (T(G(u)))_B\|_{t,B} \leq C_1 |B|diam(B)\|u\|_{t,B}. \tag{7.4.46}$$

Set $m = \lambda s/(\lambda + \alpha(r-1))$, so that $m < s < t$. Substituting (7.4.46) into (7.4.45) and using Lemma 3.1.1, we obtain

$$\left(\int_B |T(G(u)) - (T(G(u)))_B|^s w_1^\alpha dx\right)^{1/s}$$

$$\leq C_1 |B| diam(B) \|u\|_{t,B} \left(\int_B w_1^\lambda dx\right)^{\alpha/\lambda s} \tag{7.4.47}$$

$$\leq C_2 |B| diam(B) |B|^{(m-t)/mt} \|u\|_{m,\rho B} \left(\int_B w_1^\lambda dx\right)^{\alpha/\lambda s}.$$

Now, using Theorem 1.1.4 with $1/m = 1/s + (s-m)/sm$, we have

$$\|u\|_{m,\rho B}$$

$$= \left(\int_{\rho B} |u|^m dx\right)^{1/m}$$

$$= \left(\int_{\rho B} \left(|u| w_2^{\alpha/s} w_2^{-\alpha/s}\right)^m dx\right)^{1/m} \tag{7.4.48}$$

$$\leq \left(\int_{\rho B} |u|^s w_2^\alpha dx\right)^{1/s} \left(\int_{\rho B} \left(\frac{1}{w_2}\right)^{\lambda/(r-1)} dx\right)^{\alpha(r-1)/\lambda s}$$

for all balls B with $\rho B \subset \Omega$. Substituting (7.4.48) into (7.4.47), we obtain

$$\left(\int_B |T(G(u)) - (T(G(u)))_B|^s w_1^\alpha dx\right)^{1/s}$$

$$\leq C_3 |B| |B|^{(m-t)/mt} \left(\int_{\rho B} |u|^s w_2^\alpha dx\right)^{1/s} \tag{7.4.49}$$

$$\times \left(\int_B w_1^\lambda dx\right)^{\alpha/\lambda s} \left(\int_{\rho B} \left(\frac{1}{w_2}\right)^{\lambda/(r-1)} dx\right)^{\alpha(r-1)/\lambda s}.$$

Next, since $(w_1, w_2) \in A_{r,\lambda}(\Omega)$, it follows that

$$\left(\int_B w_1^\lambda dx\right)^{\alpha/\lambda s} \left(\int_{\rho B} \left(\frac{1}{w_2}\right)^{\lambda/(r-1)} dx\right)^{\alpha(r-1)/\lambda s}$$

$$\leq \left(\left(\int_{\rho B} w_1^\lambda dx\right) \left(\int_{\rho B} (1/w_2)^{\lambda/(r-1)} dx\right)^{r-1}\right)^{\alpha/\lambda s} \tag{7.4.50}$$

$$= \left(|\rho B|^r \left(\frac{1}{|\rho B|} \int_{\rho B} w_1^\lambda dx\right) \left(\frac{1}{|\rho B|} \int_{\rho B} \left(\frac{1}{w_2^\lambda}\right)^{\frac{1}{r-1}} dx\right)^{r-1}\right)^{\alpha/\lambda s}$$

$$\leq C_4 |B|^{\alpha r/\lambda s}.$$

Combining (7.4.49) and (7.4.50), we have

$$\left(\int_B |T(G(u)) - (T(G(u)))_B|^s w_1^\alpha dx\right)^{\frac{1}{s}} \leq C|B| diam(B) \left(\int_{\rho B} |u|^s w_2^\alpha dx\right)^{\frac{1}{s}}$$

for all balls B with $\rho B \subset \Omega$. This completes the proof of (7.4.44). The proof of (7.4.44)' is similar to that of Theorem 7.4.4.   ∎

Next, we shall discuss the following version of two-weight Poincaré inequalities for the composition of the operators acting on differential forms.

Theorem 7.4.10. *Let $u \in L^s_{loc}(\Omega, \wedge^l)$, $l = 1, 2, \ldots, n$, $1 < s < \infty$, be a differential form satisfying (1.2.10) in a bounded, convex domain $\Omega \subset \mathbf{R}^n$ and $T : C^\infty(\Omega, \wedge^l) \to C^\infty(\Omega, \wedge^{l-1})$ be a homotopy operator defined in (1.5.1). Suppose that $(w_1, w_2) \in A_r^\lambda(\Omega)$ for some $r > 1$ and $\lambda > 0$. If $0 < \alpha < 1$ and $\sigma > 1$, then there exists a constant C, independent of u, such that*

$$\left(\int_B |T(G(u)) - (T(G(u)))_B|^s w_1^\alpha dx \right)^{\frac{1}{s}}$$
$$\leq C|B|diam(B) \left(\int_{\sigma B} |u|^s w_2^{\alpha\lambda} dx \right)^{\frac{1}{s}} \tag{7.4.51}$$

and

$$\|T(G(u)) - (T(G(u)))_B\|_{W^{1,s}(B), \, w_1^\alpha} \leq C|B|\|u\|_{s,\sigma B, w_2^{\alpha\lambda}} \tag{7.4.51'}$$

for all balls B with $\sigma B \subset \Omega$.

Proof. Set $t = s/(1 - \alpha)$, so that $1 < s < t$. Noticing that $1/s = 1/t + (t - s)/st$, by the Hölder inequality, we find that

$$\left(\int_B |T(G(u)) - (T(G(u)))_B|^s w_1^\alpha dx \right)^{1/s}$$
$$\leq \left(\int_B (|T(G(u)) - (T(G(u)))_B| w_1^{\alpha/s})^s dx \right)^{1/s}$$
$$\leq \left(\int_B |T(G(u)) - (T(G(u)))_B|^t dx \right)^{1/t}$$
$$\times \left(\int_B w_1^{\alpha t/(t-s)} dx \right)^{(t-s)/st} \tag{7.4.52}$$
$$= \|T(G(u)) - (T(G(u)))_B\|_{t,B} \left(\int_B w_1 dx \right)^{\alpha/s}.$$

Next, choose

$$m = \frac{s}{\alpha\lambda(r - 1) + 1}.$$

Using Theorem 7.4.1 and Lemma 3.1.1, we obtain

$$\|T(G(u)) - (T(G(u)))_B\|_{t,B}$$
$$\leq C_1|B|diam(B)\|u\|_{t,B} \tag{7.4.53}$$
$$\leq C_2|B|diam(B)|B|^{(m-t)/mt}\|u\|_{m,\sigma B}$$

for all balls B with $\sigma B \subset \Omega$. Now since $1/m = 1/s + (s-m)/sm$, by the Hölder inequality again, we find that

$$\|u\|_{m,\sigma B}$$

$$= \left(\int_{\sigma B} \left(|u| w_2^{\alpha\lambda/s} w_2^{-\alpha\lambda/s} \right)^m dx \right)^{1/m}$$

$$\leq \left(\int_{\sigma B} |u|^s w_2^{\alpha\lambda} dx \right)^{1/s} \left(\int_{\sigma B} \left(\frac{1}{w_2} \right)^{\alpha\lambda m/(s-m)} dx \right)^{(s-m)/sm} \qquad (7.4.54)$$

$$= \left(\int_{\sigma B} |u|^s w_2^{\alpha\lambda} dx \right)^{1/s} \left(\int_{\sigma B} \left(\frac{1}{w_2} \right)^{1/(r-1)} dx \right)^{\alpha\lambda(r-1)/s}.$$

From $(7.4.52), (7.4.53),$ and $(7.4.54),$ we have

$$\left(\int_B |T(G(u)) - (T(G(u)))_B|^s w_1^\alpha dx \right)^{1/s}$$

$$\leq C_2 |B|^{(m-t)/tm} |B| \operatorname{diam}(B) \left(\int_B w_1 dx \right)^{\alpha/s} \qquad (7.4.55)$$

$$\times \left(\int_{\sigma B} \left(\frac{1}{w_2} \right)^{1/(r-1)} dx \right)^{\alpha\lambda(r-1)/s} \left(\int_{\sigma B} |u|^s w_2^{\alpha\lambda} dx \right)^{1/s}.$$

Since $(w_1, w_2) \in A_r^\lambda(\Omega)$, it follows that

$$\left(\int_B w_1 dx \right)^{\alpha/s} \left(\int_{\sigma B} \left(\frac{1}{w_2} \right)^{1/(r-1)} dx \right)^{\alpha\lambda(r-1)/s}$$

$$= \left[\left(\int_B w_1 dx \right) \left(\int_{\sigma B} \left(\frac{1}{w_2} \right)^{1/(r-1)} dx \right)^{\lambda(r-1)} \right]^{\alpha/s}$$

$$\leq \left[|\sigma B|^{\lambda(r-1)+1} \left(\frac{1}{|\sigma B|} \int_{\sigma B} w_1 dx \right) \right. \qquad (7.4.56)$$

$$\times \left. \left(\frac{1}{|\sigma B|} \int_{\sigma B} \left(\frac{1}{w_2} \right)^{1/(r-1)} dx \right)^{\lambda(r-1)} \right]^{\alpha/s}$$

$$\leq C_3 |\sigma B|^{\alpha\lambda(r-1)/s + \alpha/s}$$

$$\leq C_4 |B|^{\alpha\lambda(r-1)/s + \alpha/s}.$$

Finally, substituting $(7.4.56)$ into $(7.4.55)$ and using

$$\frac{m-t}{mt} = \frac{-\alpha}{s} - \frac{\alpha\lambda(r-1)}{s},$$

we obtain

$$\left(\int_B |T(G(u)) - (T(G(u)))_B|^s w_1^\alpha dx \right)^{\frac{1}{s}} \le C|B|diam(B) \left(\int_{\sigma B} |u|^s w_2^{\alpha\lambda} dx \right)^{\frac{1}{s}}.$$

Thus, inequality (7.4.51) holds. By the same method used in the proof of Theorem 7.4.4, we can prove inequality (7.4.51)'.　∎

For $\lambda = 1/\alpha$, Theorem 7.4.10 reduces to the following corollary.

Corollary 7.4.11. *Let $u \in L^s_{loc}(\Omega, \wedge^l)$, $l = 1, 2, \ldots, n$, $1 < s < \infty$, be a differential form satisfying (1.2.10) in a bounded, convex domain $\Omega \subset \mathbf{R}^n$ and $T : C^\infty(\Omega, \wedge^l) \to C^\infty(\Omega, \wedge^{l-1})$ be a homotopy operator defined in (1.5.1). Suppose that $(w_1, w_2) \in A_r^{1/\alpha}(\Omega)$ for some $r > 1$. If $0 < \alpha < 1$ and $\sigma > 1$, then there exists a constant C, independent of u, such that*

$$\left(\int_B |T(G(u)) - (T(G(u)))_B|^s w_1^\alpha dx \right)^{\frac{1}{s}} \le C|B|diam(B) \left(\int_{\sigma B} |u|^s w_2 dx \right)^{\frac{1}{s}} \tag{7.4.57}$$

and

$$\|T(G(u)) - (T(G(u)))_B\|_{W^{1,s}(B),\, w_1^\alpha} \le C|B|\|u\|_{s,\sigma B, w_2}$$

for all balls B with $\sigma B \subset \Omega$.

Selecting $\alpha = 1/s$ in Theorem 7.4.10, we have the following two-weighted Poincaré inequalities.

Corollary 7.4.12. *Let $u \in L^s_{loc}(\Omega, \wedge^l)$, $l = 1, 2, \ldots, n$, $1 < s < \infty$, be a differential form satisfying (1.2.10) in a bounded, convex domain $\Omega \subset \mathbf{R}^n$ and $T : C^\infty(\Omega, \wedge^l) \to C^\infty(\Omega, \wedge^{l-1})$ be a homotopy operator defined in (1.5.1). Suppose that $(w_1, w_2) \in A_r^\lambda(\Omega)$ for some $r > 1$, $\lambda > 0$, and $\sigma > 1$, then there exists a constant C, independent of u, such that*

$$\left(\int_B |T(G(u)) - (T(G(u)))_B|^s w_1^{\frac{1}{s}} dx \right)^{1/s} \le C|B|diam(B) \left(\int_{\sigma B} |u|^s w_2^{\lambda/s} dx \right)^{1/s} \tag{7.4.58}$$

and

$$\|T(G(u)) - (T(G(u)))_B\|_{W^{1,s}(B),\, w_1^{1/s}} \le C|B|\|u\|_{s,\sigma B, w_2^{\lambda/s}}$$

for all balls B with $\sigma B \subset \Omega$.

The choice $\lambda = 1$ in Corollary 7.4.12 gives the following symmetric two-weighted inequalities.

Corollary 7.4.13. *Let* $u \in L^s_{loc}(\Omega, \wedge^l)$, $l = 1, 2, \ldots, n$, $1 < s < \infty$, *be a differential form satisfying* (1.2.10) *in a bounded, convex domain* $\Omega \subset \mathbf{R}^n$ *and* $T : C^\infty(\Omega, \wedge^l) \to C^\infty(\Omega, \wedge^{l-1})$ *be a homotopy operator defined in* (1.5.1). *Suppose that* $(w_1, w_2) \in A_r(\Omega)$ *for some* $r > 1$ *and* $\sigma > 1$, *then there exists a constant* C, *independent of* u, *such that*

$$\left(\int_B |T(G(u)) - (T(G(u)))_B|^s w_1^{\frac{1}{s}} dx \right)^{1/s}$$
$$\leq C|B| \mathrm{diam}(B) \left(\int_{\sigma B} |u|^s w_2^{1/s} dx \right)^{1/s} \tag{7.4.59}$$

and

$$\|T(G(u)) - (T(G(u)))_B\|_{W^{1,s}(B),\, w_1^{1/s}} \leq C|B| \|u\|_{s, \sigma B, w_2^{1/s}}$$

for all balls B *with* $\sigma B \subset \Omega$.

Finally, when $\lambda = s$ in Theorem 7.4.10, we obtain the following two-weighted inequalities.

Corollary 7.4.14. *Let* $u \in L^s_{loc}(\Omega, \wedge^l)$, $l = 1, 2, \ldots, n$, $1 < s < \infty$, *be a differential form satisfying* (1.2.10) *in a bounded, convex domain* $\Omega \subset \mathbf{R}^n$ *and* $T : C^\infty(\Omega, \wedge^l) \to C^\infty(\Omega, \wedge^{l-1})$ *be a homotopy operator defined in* (1.5.1). *Suppose that* $(w_1, w_2) \in A_r^s(\Omega)$ *for some* $r > 1$. *If* $0 < \alpha < 1$ *and* $\sigma > 1$, *then there exists a constant* C, *independent of* u, *such that*

$$\left(\int_B |T(G(u)) - (T(G(u)))_B|^s w_1^\alpha dx \right)^{\frac{1}{s}}$$
$$\leq C|B| \mathrm{diam}(B) \left(\int_{\sigma B} |u|^s w_2^{\alpha s} dx \right)^{\frac{1}{s}} \tag{7.4.60}$$

and

$$\|T(G(u)) - (T(G(u)))_B\|_{W^{1,s}(B),\, w_1^\alpha} \leq C|B| \|u\|_{s, \sigma B, w_2^{\alpha s}}$$

for all balls B *with* $\sigma B \subset \Omega$.

7.4.2 Poincaré-type inequalities for $G \circ T$

In the previous section, we have obtained various estimates for the composition $T \circ G$. In this section, we present local and global Poincaré-type estimates for $G \circ T$ with single weight and two-weights. We begin with the following basic Poincaré-type estimates for $G \circ T$ applied to the smooth solutions of the nonhomogeneous A-harmonic equation.

Theorem 7.4.15. *Let* $u \in L^s_{loc}(\Omega, \wedge^l)$, $l = 1, 2, \ldots, n$, $1 < s < \infty$, *be a smooth differential form in a bounded, convex domain* Ω *and* $T : C^\infty(\Omega, \wedge^l) \to C^\infty(\Omega, \wedge^{l-1})$ *be the homotopy operator defined in* (1.5.1).

Then, there exists a constant C, independent of u, such that

$$\|G(T(u)) - (G(T(u)))_B\|_{s,B} \le C|B|diam(B)\|u\|_{s,B} \qquad (7.4.61)$$

for all balls $B \subset \Omega$.

As in Theorem 7.4.1, Poincaré-type estimate (7.4.61) can also be written as

$$\|G(T(u)) - (G(T(u)))_B\|_{s,B} \le C\|u\|_{s,B}, \qquad (7.4.61)'$$

but because of the same reason as in (7.4.1), we keep the factor $|B|diam(B)$ on the right-hand side of (7.4.61).

Proof. Replacing u by Tu in the second inequality in (7.4.5) and using (7.4.2), we have

$$\|d(G(Tu))\|_{s,B} \le C_1\|Tu\|_{s,B}$$

$$\le C_2|B|diam(B)\|u\|_{s,B} \qquad (7.4.62)$$

$$\le C_3\|u\|_{s,B}.$$

From (7.4.2) and replacing u by $G(u)$ in (7.4.4), then using (7.4.62), we obtain

$$\|G(T(u)) - (G(T(u)))_B\|_{s,B}$$

$$= \|Td(G(T(u)))\|_{s,B}$$

$$\le C_4|B|diam(B)\|d(G(T(u)))\|_{s,B} \qquad (7.4.63)$$

$$\le C_5 B|diam(B)\|u\|_{s,B}.$$

Thus, inequality (7.4.61) holds. ∎

Theorem 7.4.16. *Let $u \in L^s_{loc}(\Omega, \wedge^l)$, $l = 1, 2, \ldots, n$, $1 < s < \infty$, be a solution of the nonhomogeneous A-harmonic equation (1.2.10) in a bounded, convex domain Ω and $T : C^\infty(\Omega, \wedge^l) \to C^\infty(\Omega, \wedge^{l-1})$ be the homotopy operator defined in (1.5.1). Assume that $\rho > 1$ and $w \in A_r(\Omega)$ for some $1 < r < \infty$. Then, there exists a constant C, independent of u, such that*

$$\|G(T(u)) - (G(T(u)))_B\|_{s,B,w^\alpha} \le C|B|diam(B)\|u\|_{s,\rho B,w^\alpha} \qquad (7.4.64)$$

for all balls B with $\rho B \subset \Omega$ and any real number α with $0 < \alpha \le 1$.

The above L^s-norm inequality can also be written in the integral form as

$$\left(\int_B |G(T(u)) - (G(T(u)))_B|^s w^\alpha dx \right)^{1/s} \leq C \left(\int_{\rho B} |u|^s w^\alpha dx \right)^{1/s}.$$

$$(7.4.64)'$$

Theorem 7.4.17. *Let* $u \in D'(\Omega, \wedge^1)$ *be a solution of the nonhomogeneous A-harmonic equation (1.2.10),* $T : C^\infty(\Omega, \wedge^l) \to C^\infty(\Omega, \wedge^{l-1})$, $l = 1, 2, \ldots, n$, *be the homotopy operator defined in (1.5.1), and G be Green's operator. Assume that* $w \in A_r(\Omega)$ *for some* $1 < r < \infty$ *and s is a fixed exponent associated with the A-harmonic equation (1.2.10). Then, there exists a constant C, independent of u, such that*

$$\left(\int_\Omega |G(T(u)) - (G(T(u)))_{Q_0}|^s w dx \right)^{1/s} \leq C \left(\int_\Omega |u|^s w dx \right)^{1/s} \qquad (7.4.65)$$

for any bounded, convex δ-*John domain* $\Omega \subset \mathbf{R}^n$. *Here* $Q_0 \subset \Omega$ *is a fixed cube.*

Next, we present the local and global imbedding inequalities.

Theorem 7.4.18. *Let* $u \in L^s_{loc}(\Omega, \wedge^l)$, $l = 1, 2, \ldots, n$, $1 < s < \infty$, *be a smooth differential form in a bounded, convex domain* Ω *and* $T : C^\infty(\Omega, \wedge^l) \to C^\infty(\Omega, \wedge^{l-1})$ *be the homotopy operator defined in (1.5.1). Assume that* $\rho > 1$ *and* $w \in A_r(\Omega)$ *for some* $1 < r < \infty$. *Then, there exists a constant C, independent of u, such that*

$$\|G(T(u)) - (G(T(u)))_B\|_{W^{1,s}(B), w^\alpha} \leq C|B| \|u\|_{s, \rho B, w^\alpha} \qquad (7.4.66)$$

for all balls B with $\rho B \subset \Omega$ *and any real number* α *with* $0 < \alpha \leq 1$.

Theorem 7.4.19. *Let* $u \in D'(\Omega, \wedge^1)$ *be a solution of the nonhomogeneous A-harmonic equation (1.2.10),* $T : C^\infty(\Omega, \wedge^l) \to C^\infty(\Omega, \wedge^{l-1})$, $l = 1, 2, \ldots, n$, *be the homotopy operator defined in (1.5.1), and G be Green's operator. Assume that* $w \in A_r(\Omega)$ *for some* $1 < r < \infty$ *and s is a fixed exponent associated with the A-harmonic equation (1.2.10). Then, there exists a constant C, independent of u, such that*

$$\|G(T(u)) - (G(T(u)))_{Q_0}\|_{W^{1,s}(\Omega), w} \leq C\|u\|_{s, \Omega, w} \qquad (7.4.67)$$

for any bounded, convex δ-*John domain* $\Omega \subset \mathbf{R}^n$. *Here* $Q_0 \subset \Omega$ *is a fixed cube.*

We have presented above the $A_r(\Omega)$-weighted Poincaré-type estimates for the composition $G \circ T$. Now, we present some other estimates with different weights, such as $A_r(\lambda, \Omega)$-weights, $A_r^\lambda(\Omega)$-weights. The proofs of the following results are similar to those provided in earlier chapters, and hence are omitted.

Theorem 7.4.20. *Let $u \in L^s_{loc}(\Omega, \wedge^l)$, $l = 1, 2, \ldots, n$, $1 < s < \infty$, be a differential form satisfying the nonhomogeneous A-harmonic equation (1.2.10) in a bounded, convex domain $\Omega \subset \mathbf{R}^n$ and $T : C^\infty(\Omega, \wedge^l) \to C^\infty(\Omega, \wedge^{l-1})$ be a homotopy operator defined in (1.5.1). Assume that $w \in A_r(\lambda, \Omega)$ for some $r > 1$ and $\lambda > 0$. Then, there exists a constant C, independent of u, such that*

$$\|G(T(u)) - (G(T(u)))_B\|_{s,B,w^{\alpha\lambda}} \leq C|B|diam(B)\|u\|_{s,\rho B,w^\alpha} \qquad (7.4.68)$$

and

$$\|G(T(u)) - (G(T(u)))_B\|_{W^{1,s}(B),\ w^{\alpha\lambda}} \leq C|B|\|u\|_{s,\rho B,w^\alpha} \qquad (7.4.69)$$

for all balls B with $\rho B \subset \Omega$ and any real number α with $0 < \alpha < 1$. Here $\rho > 1$ is some constant.

Note that inequality (7.4.68) can be written as

$$\left(\int_B |G(T(u)) - (G(T(u)))_B|^s w^{\alpha\lambda} dx \right)^{\frac{1}{s}} \leq C|B|diam(B) \left(\int_{\rho B} |u|^s w^\alpha dx \right)^{\frac{1}{s}}.$$

Theorem 7.4.21. *Let $u \in L^s_{loc}(\Omega, \wedge^l)$, $l = 1, 2, \ldots, n$, $1 < s < \infty$, be a differential form satisfying (1.2.10) in a bounded, convex domain $\Omega \subset \mathbf{R}^n$ and $T : C^\infty(\Omega, \wedge^l) \to C^\infty(\Omega, \wedge^{l-1})$ be a homotopy operator defined in (1.5.1). Assume that $\rho > 1$ and $w \in A^\lambda_r(\Omega)$ for some $r > 1$ and $\lambda > 0$. Then, there exists a constant C, independent of u, such that*

$$\|G(T(u)) - (G(T(u)))_B\|_{s,B,w^\alpha} \leq C|B|diam(B)\|u\|_{s,\rho B,w^{\alpha\lambda}} \qquad (7.4.70)$$

and

$$\|G(T(u)) - (G(T(u)))_B\|_{W^{1,s}(B),\ w^\alpha} \leq C|B|diam(B)\|u\|_{s,\rho B,w^{\alpha\lambda}} \qquad (7.4.71)$$

for all balls B with $\rho B \subset \Omega$ and any real number α with $0 < \alpha < 1$.

It should be noticed that each of the above two inequalities has an integral version. For example, inequality (7.4.70) can be written as

$$\left(\int_B |G(T(u)) - (G(T(u)))_B|^s w^\alpha dx \right)^{\frac{1}{s}} \leq C|B|diam(B) \left(\int_{\rho B} |u|^s w^{\alpha\lambda} dx \right)^{\frac{1}{s}}.$$

We also notice that the above estimates can be extended into the following two-weight case.

Theorem 7.4.22. *Let $u \in L^s_{loc}(\Omega, \wedge^l)$, $l = 1, 2, \ldots, n$, $1 < s < \infty$, be a solution of the nonhomogeneous A-harmonic equation (1.2.10) in a bounded,* *convex domain* $\Omega \subset \mathbf{R}^n$ *and* $T : C^\infty(\Omega, \wedge^l) \to C^\infty(\Omega, \wedge^{l-1})$ *be the homotopy operator defined in (1.5.1). Suppose that* $\rho > 1$ *and* $(w_1, w_2) \in A_r(\lambda, \Omega)$ *for some* $\lambda > 0$ *and* $1 < r < \infty$. *Then, there exists a constant* C, *independent of* u, *such that*

$$\left(\int_B |G(T(u)) - (G(T(u)))_B|^s w_1^{\alpha\lambda} dx \right)^{\frac{1}{s}} \leq C|B| diam(B) \left(\int_{\rho B} |u|^s w_2^\alpha dx \right)^{\frac{1}{s}}$$

$$(7.4.72)$$

and

$$\|G(T(u)) - (G(T(u)))_B\|_{W^{1,s}(B), w_1^{\alpha\lambda}} \leq C|B| \|u\|_{s, \rho B, w_2^\alpha} \qquad (7.4.72)'$$

for all balls B *with* $\rho B \subset \Omega$ *and any real number* α *with* $0 < \alpha < 1$.

Note that (7.4.72) can be written as

$$\|G(T(u)) - (G(T(u)))_B\|_{s, B, w_1^{\alpha\lambda}} \leq C|B| diam(B) \|u\|_{s, \rho B, w_2^\alpha}. \qquad (7.4.72)''$$

Theorem 7.4.23. *Let* $u \in L_{loc}^s(\Omega, \wedge^l)$, $l = 1, 2, \ldots, n$, $1 < s < \infty$, *be a differential form satisfying (1.2.10) in a bounded, convex domain* $\Omega \subset \mathbf{R}^n$ *and* $T : C^\infty(\Omega, \wedge^l) \to C^\infty(\Omega, \wedge^{l-1})$ *be the homotopy operator defined in (1.5.1). Suppose that* $\rho > 1$ *and* $(w_1, w_2) \in A_{r,\lambda}(\Omega)$ *for some* $\lambda \geq 1$ *and* $1 < r < \infty$. *Then, there exists a constant* C, *independent of* u, *such that*

$$\left(\int_B |G(T(u)) - (G(T(u)))_B|^s w_1^\alpha dx \right)^{\frac{1}{s}}$$
$$\leq C|B| diam(B) \left(\int_{\rho B} |u|^s w_2^\alpha dx \right)^{\frac{1}{s}}$$

$$(7.4.73)$$

and

$$\|G(T(u)) - (G(T(u)))_B\|_{W^{1,s}(B), w_1^\alpha} \leq C|B| \|u\|_{s, \rho B, w_2^\alpha} \qquad (7.4.73)'$$

for all balls B *with* $\rho B \subset \Omega$ *and any real number* α *with* $0 < \alpha < \lambda$.

Note that (7.4.73) can be written as

$$\|G(T(u)) - (G(T(u)))_B\|_{s, B, w_1^\alpha} \leq C|B| diam(B) \|u\|_{s, \rho B, w_2^\alpha}. \qquad (7.4.73)''$$

Next, we discuss the following version of two-weight Poincaré inequalities for the compositions of the operators acting on differential forms.

Theorem 7.4.24. *Let* $u \in L_{loc}^s(\Omega, \wedge^l)$, $l = 1, 2, \ldots, n$, $1 < s < \infty$, *be a differential form satisfying (1.2.10) in a bounded, convex domain* $\Omega \subset \mathbf{R}^n$ *and* $T : C^\infty(\Omega, \wedge^l) \to C^\infty(\Omega, \wedge^{l-1})$ *be the homotopy operator defined in (1.5.1).*

Suppose that $(w_1, w_2) \in A_r^\lambda(\Omega)$ for some $r > 1$ and $\lambda > 0$. If $0 < \alpha < 1$ and $\sigma > 1$, then there exists a constant C, independent of u, such that

$$\left(\int_B |G(T(u)) - (G(T(u)))_B|^s w_1^\alpha dx \right)^{\frac{1}{s}}$$
$$\leq C|B| diam(B) \left(\int_{\sigma B} |u|^s w_2^{\alpha\lambda} dx \right)^{\frac{1}{s}} \tag{7.4.74}$$

and

$$\|G(T(u)) - (G(T(u)))_B\|_{W^{1,s}(B), w_1^\alpha} \leq C|B|\|u\|_{s,\sigma B, w_2^{\alpha\lambda}} \tag{7.4.74}'$$

for all balls B with $\sigma B \subset \Omega$.

If we choose $\lambda = 1/\alpha$ in Theorem 7.4.24, we have the following version of the A_r^λ-weighted Poincaré inequality.

Corollary 7.4.25. *Let $u \in L_{loc}^s(\Omega, \wedge^l)$, $l = 1, 2, \ldots, n$, $1 < s < \infty$, be a differential form satisfying (1.2.10) in a bounded, convex domain $\Omega \subset \mathbf{R}^n$ and $T : C^\infty(\Omega, \wedge^l) \to C^\infty(\Omega, \wedge^{l-1})$ be the homotopy operator defined in (1.5.1). Suppose that $(w_1, w_2) \in A_r^{1/\alpha}(\Omega)$ for some $r > 1$. If $0 < \alpha < 1$ and $\sigma > 1$, then there exists a constant C, independent of u, such that*

$$\left(\int_B |G(T(u)) - (G(T(u)))_B|^s w_1^\alpha dx \right)^{\frac{1}{s}}$$
$$\leq C|B| diam(B) \left(\int_{\sigma B} |u|^s w_2 dx \right)^{\frac{1}{s}} \tag{7.4.75}$$

for all balls B with $\sigma B \subset \Omega$.

Choosing $\alpha = 1/s$ in Theorem 7.4.24, we have the following two-weighted Poincaré inequality.

Corollary 7.4.26. *Let $u \in L_{loc}^s(\Omega, \wedge^l)$, $l = 1, 2, \ldots, n$, $1 < s < \infty$, be a differential form satisfying (1.2.10) in a bounded, convex domain $\Omega \subset \mathbf{R}^n$ and $T : C^\infty(\Omega, \wedge^l) \to C^\infty(\Omega, \wedge^{l-1})$ be the homotopy operator defined in (1.5.1). Suppose that $(w_1, w_2) \in A_r^\lambda(\Omega)$ for some $r > 1$, $\lambda > 0$, and $\sigma > 1$, then there exists a constant C, independent of u, such that*

$$\left(\int_B |G(T(u)) - (G(T(u)))_B|^s w_1^{\frac{1}{s}} dx \right)^{\frac{1}{s}}$$
$$\leq C|B| diam(B) \left(\int_{\sigma B} |u|^s w_2^{\lambda/s} dx \right)^{\frac{1}{s}} \tag{7.4.76}$$

for all balls B with $\sigma B \subset \Omega$.

When $\lambda = 1$ in Corollary 7.4.26, we obtain the following symmetric two-weight inequality.

Corollary 7.4.27. *Let* $u \in L^s_{loc}(\Omega, \wedge^l)$, $l = 1, 2, \ldots, n$, $1 < s < \infty$, *be a differential form satisfying* (1.2.10) *in a bounded, convex domain* $\Omega \subset \mathbf{R}^n$ *and* $T : C^\infty(\Omega, \wedge^l) \to C^\infty(\Omega, \wedge^{l-1})$ *be the homotopy operator defined in* (1.5.1). *Suppose that* $(w_1, w_2) \in A_r(\Omega)$ *for some* $r > 1$ *and* $\sigma > 1$, *then there exists a constant* C, *independent of* u, *such that*

$$\left(\int_B |G(T(u)) - (G(T(u)))_B|^s w_1^{\frac{1}{s}} dx \right)^{\frac{1}{s}}$$
$$\leq C|B| diam(B) \left(\int_{\sigma B} |u|^s w_2^{\frac{1}{s}} dx \right)^{\frac{1}{s}} \tag{7.4.77}$$

for all balls B *with* $\sigma B \subset \Omega$.

Finally, choose $\lambda = s$ in Theorem 7.4.24, to obtain the following two-weight inequalities.

Corollary 7.4.28. *Let* $u \in L^s_{loc}(\Omega, \wedge^l)$, $l = 1, 2, \ldots, n$, $1 < s < \infty$, *be a differential form satisfying* (1.2.10) *in a bounded, convex domain* $\Omega \subset \mathbf{R}^n$ *and* $T : C^\infty(\Omega, \wedge^l) \to C^\infty(\Omega, \wedge^{l-1})$ *be the homotopy operator defined in* (1.5.1). *Suppose that* $(w_1, w_2) \in A^s_r(\Omega)$ *for some* $r > 1$. *If* $0 < \alpha < 1$ *and* $\sigma > 1$, *then there exists a constant* C, *independent of* u, *such that*

$$\left(\int_B |G(T(u)) - (G(T(u)))_B|^s w_1^\alpha dx \right)^{\frac{1}{s}}$$
$$\leq C|B| diam(B) \left(\int_{\sigma B} |u|^s w_2^{\alpha s} dx \right)^{\frac{1}{s}} \tag{7.4.78}$$

and

$$\|G(T(u)) - (G(T(u)))_B\|_{W^{1,s}(B), w_1^\alpha} \leq C|B| \|u\|_{s, \sigma B, w_2^{\alpha s}} \tag{7.4.78$'$}$$

for all balls B *with* $\sigma B \subset \Omega$.

7.5 The homotopy operator

We have developed various Poincaré-type estimates for the compositions $T \circ G$ and $G \circ T$ in the previous sections. Now, we obtain Poincaré-type estimates for the homotopy operator T applied to the differential forms satisfying the nonhomogeneous A-harmonic equation.

7.5.1 Basic estimates for T

First, we prove the following basic local Poincaré-type estimate for the homotopy operator T. Then, we extend it to the different weighted cases and the global cases. We also obtain some weighted imbedding theorems.

Theorem 7.5.1. *Let* $u \in L_{loc}^s(\Omega, \wedge^l)$, $l = 1, 2, \ldots, n$, $1 < s < \infty$, *be a solution of the A-harmonic equation* (1.2.10) *in a bounded, convex domain* Ω *and* $T : C^\infty(\Omega, \wedge^l) \rightarrow C^\infty(\Omega, \wedge^{l-1})$ *be the homotopy operator defined in* (1.5.1). *Then, there exists a constant* C, *independent of* u, *such that*

$$\|T(u) - (T(u))_B\|_{s,B} \leq C|B| diam(B) \|u\|_{s,\rho B} \qquad (7.5.1)$$

for all balls B *with* $\rho B \subset \Omega$, *where* $\rho > 1$ *is a constant.*

Proof. Choosing the closed form $c = 0$ in the Caccioppoli inequality, we have

$$\|du\|_{s,B} \leq C_1 (diam(B))^{-1} \|u\|_{s,\rho B} \qquad (7.5.2)$$

for any solution u of the A-harmonic equation. Using the decomposition for the form $T(u)$ yields

$$\begin{aligned} T(u) &= d(T(T(u))) + T(d(T(u))) \\ &= (T(u))_B + T(d(T(u))). \end{aligned} \qquad (7.5.3)$$

Now, since $B \subset \Omega$ and Ω is bounded, we have $|B| \leq K$ for some constant $K > 0$. Thus, from (7.4.2), (7.4.4), and (7.5.2), it follows that

$$\begin{aligned} &\|T(u) - (T(u))_B\|_{s,B} \\ &= \|Td(T(u))\|_{s,B} \\ &\leq C_2|B| diam(B) \|d(T(u))\|_{s,B} \\ &\leq C_2|B| diam(B)(\|u\|_{s,B} + C_3|B| diam(B) \|du\|_{s,B}) \\ &\leq C_2|B| diam(B)(\|u\|_{s,B} + C_4|B| \|u\|_{s,\rho B}) \\ &\leq C_2|B| diam(B)(\|u\|_{s,B} + C_5 \|u\|_{s,\rho B}) \\ &\leq C_6|B| diam(B) \|u\|_{s,\rho B}, \end{aligned} \qquad (7.5.4)$$

that is

$$\|T(u) - (T(u))_B\|_{s,B} \leq C_7|B| diam(B) \|u\|_{s,\rho B}. \qquad \blacksquare \qquad (7.5.5)$$

From the proof of Theorem 7.5.1, it is easy to see that the following result still holds if the condition that u is a solution of the A-harmonic equation is dropped. We state it as the following theorem.

Theorem 7.5.1'. *Let* $u \in L_{loc}^s(\Omega, \wedge^l)$, $l = 1, 2, \ldots, n$, $1 < s < \infty$, *be a smooth differential form in a bounded, convex domain* Ω *and* $T :*

$C^\infty(\Omega, \wedge^l) \to C^\infty(\Omega, \wedge^{l-1})$ *be the homotopy operator defined in* (1.5.1). *Then, there exists a constant C, independent of u, such that*

$$\|T(u) - (T(u))_B\|_{s,B} \le C|B|diam(B)(\|u\|_{s,B} + \|du\|_{s,B}) \qquad (7.5.5)'$$

for all balls $B \subset \Omega$.

7.5.2 $A_r(\Omega)$-weighted estimates for T

Based on the basic Poincaré-type estimate for the homotopy operator T established above, we can state the following $A_r(\Omega)$-weighted inequality.

Theorem 7.5.2. *Let $u \in L^s_{loc}(\Omega, \wedge^l)$, $l = 1, 2, \ldots, n$, $1 < s < \infty$, be a solution of the nonhomogeneous A-harmonic equation* (1.2.10) *in a bounded, convex domain Ω and $T : C^\infty(\Omega, \wedge^l) \to C^\infty(\Omega, \wedge^{l-1})$ be the homotopy operator defined in* (1.5.1). *Assume that $\rho > 1$ and $w(x) \in A_r(\Omega)$ for some $1 < r < \infty$. Then, there exists a constant C, independent of u, such that*

$$\|T(u) - (T(u))_B\|_{s,B,w^\alpha} \le C|B|diam(B)\|u\|_{s,\rho B,w^\alpha} \qquad (7.5.6)$$

for all balls B with $\rho B \subset \Omega$ and any real number α with $0 < \alpha \le 1$.

The above L^s-norm inequality can also be written in the integral form as follows:

$$\left(\int_B |T(u) - (T(u))_B|^s w^\alpha dx \right)^{1/s} \le C \left(\int_{\rho B} |u|^s w^\alpha dx \right)^{1/s}. \qquad (7.5.6)'$$

Also, using the procedure developed to extend the local inequalities into the John domains, we have the following global Poincaré-type inequality.

Theorem 7.5.3. *Let $u \in D'(\Omega, \wedge^1)$ be a solution of the nonhomogeneous A-harmonic equation* (1.2.10) *and $T : C^\infty(\Omega, \wedge^l) \to C^\infty(\Omega, \wedge^{l-1})$, $l = 1, 2, \ldots, n$, be the homotopy operator defined in* (1.5.1). *Assume that $w \in A_r(\Omega)$ for some $1 < r < \infty$ and s is a fixed exponent associated with the A-harmonic equation* (1.2.10). *Then, there exists a constant C, independent of u, such that*

$$\left(\int_\Omega |T(u) - (T(u))_{Q_0}|^s w dx \right)^{1/s} \le C \left(\int_\Omega |u|^s w dx \right)^{1/s} \qquad (7.5.7)$$

for any bounded, convex δ-John domain $\Omega \subset \mathbf{R}^n$. Here $Q_0 \subset \Omega$ is a fixed cube.

7.5.3 Poincaré-type imbedding for T

By the same method used to prove the imbedding inequalities, we can prove the following local and global imbedding inequalities, Theorems 7.5.4 and 7.5.5, respectively.

Theorem 7.5.4. *Let $u \in L_{loc}^s(\Omega, \wedge^l)$, $l = 1, 2, \ldots, n$, $1 < s < \infty$, be a smooth differential form in a bounded, convex domain Ω and $T : C^\infty(\Omega, \wedge^l) \to C^\infty(\Omega, \wedge^{l-1})$ be the homotopy operator defined in (1.5.1). Assume that $\rho > 1$ and $w(x) \in A_r(\Omega)$ for some $1 < r < \infty$. Then, there exists a constant C, independent of u, such that*

$$\|T(u) - (T(u))_B\|_{W^{1,s}(B),w^\alpha} \leq C|B|\|u\|_{s,\rho B, w^\alpha} \tag{7.5.8}$$

for all balls B with $\rho B \subset \Omega$ and any real number α with $0 < \alpha \leq 1$.

Theorem 7.5.5. *Let $u \in D'(\Omega, \wedge^1)$ be a solution of the nonhomogeneous A-harmonic equation (1.2.10) and $T : C^\infty(\Omega, \wedge^l) \to C^\infty(\Omega, \wedge^{l-1})$, $l = 1, 2, \ldots, n$, be the homotopy operator defined in (1.5.1). Assume that $w \in A_r(\Omega)$ for some $1 < r < \infty$ and s is a fixed exponent associated with the A-harmonic equation (1.2.10). Then, there exists a constant C, independent of u, such that*

$$\|T(u) - (T(u))_{Q_0}\|_{W^{1,s}(\Omega),w} \leq C\|u\|_{s,\Omega,w} \tag{7.5.9}$$

for any bounded, convex δ-John domain $\Omega \subset \mathbf{R}^n$. Here $Q_0 \subset \Omega$ is a fixed cube.

So far, we have presented the $A_r(\Omega)$-weighted Poincaré-type estimates for the homotopy operator T. Now, we state other estimates with different weights, such as $A_r(\lambda, \Omega)$-weights, $A_r^\lambda(\Omega)$-weights.

Theorem 7.5.6. *Let $u \in L_{loc}^s(\Omega, \wedge^l)$, $l = 1, 2, \ldots, n$, $1 < s < \infty$, be a differential form satisfying the nonhomogeneous A-harmonic equation (1.2.10) in a bounded, convex domain $\Omega \subset \mathbf{R}^n$ and $T : C^\infty(\Omega, \wedge^l) \to C^\infty(\Omega, \wedge^{l-1})$ be the homotopy operator defined in (1.5.1). Assume that $w \in A_r(\lambda, \Omega)$ for some $r > 1$ and $\lambda > 0$. Then, there exists a constant C, independent of u, such that*

$$\|T(u) - (T(u))_B\|_{s,B,w^{\alpha\lambda}} \leq C|B|diam(B)\|u\|_{s,\rho B,w^\alpha} \tag{7.5.10}$$

and

$$\|T(u) - (T(u))_B\|_{W^{1,s}(B),\ w^{\alpha\lambda}} \leq C|B|\|u\|_{s,\rho B,w^\alpha} \tag{7.5.11}$$

for all balls B with $\rho B \subset \Omega$ and any real number α with $0 < \alpha < 1$. Here $\rho > 1$ is some constant.

Note that inequality (7.5.10) can be written as

$$\left(\int_B |T(u)-(T(u))_B|^s w^{\alpha\lambda} dx\right)^{\frac{1}{s}} \le C|B|diam(B)\left(\int_{\rho B} |u|^s w^{\alpha} dx\right)^{\frac{1}{s}}.$$

Theorem 7.5.7. *Let $u \in L_{loc}^s(\Omega, \wedge^l)$, $l = 1, 2, \ldots, n$, $1 < s < \infty$, be a differential form satisfying (1.2.10) in a bounded, convex domain $\Omega \subset \mathbf{R}^n$ and $T : C^\infty(\Omega, \wedge^l) \to C^\infty(\Omega, \wedge^{l-1})$ be the homotopy operator defined in (1.5.1). Assume that $\rho > 1$ and $w \in A_r^\lambda(\Omega)$ for some $r > 1$ and $\lambda > 0$. Then, there exists a constant C, independent of u, such that*

$$\|T(u)-(T(u))_B\|_{s,B,w^\alpha} \le C|B|diam(B)\|u\|_{s,\rho B,w^{\alpha\lambda}} \qquad (7.5.12)$$

and

$$\|T(u)-(T(u))_B\|_{W^{1,s}(B),\ w^\alpha} \le C|B|\|u\|_{s,\rho B,w^{\alpha\lambda}} \qquad (7.5.13)$$

for all balls B with $\rho B \subset \Omega$ and any real number α with $0 < \alpha < 1$.

The above inequalities have integral representations, for example, inequality (7.5.12) can be written as

$$\left(\int_B |T(u)-(T(u))_B|^s w^\alpha dx\right)^{\frac{1}{s}} \le C|B|diam(B)\left(\int_{\rho B} |u|^s w^{\alpha\lambda} dx\right)^{\frac{1}{s}}.$$

7.5.4 Two-weight Poincaré-type imbedding for T

The above estimates can be extended into the following two-weight case.

Theorem 7.5.8. *Let $u \in L_{loc}^s(\Omega, \wedge^l)$, $l = 1, 2, \ldots, n$, $1 < s < \infty$, be a solution of the nonhomogeneous A-harmonic equation (1.2.10) in a bounded, convex domain $\Omega \subset \mathbf{R}^n$ and $T : C^\infty(\Omega, \wedge^l) \to C^\infty(\Omega, \wedge^{l-1})$ be the homotopy operator defined in (1.5.1). Suppose that $\rho > 1$ and $(w_1, w_2) \in A_r(\lambda, \Omega)$ for some $\lambda > 0$ and $1 < r < \infty$. Then, there exists a constant C, independent of u, such that*

$$\left(\int_B |T(u)-(T(u))_B|^s w_1^{\alpha\lambda} dx\right)^{\frac{1}{s}} \le C|B|diam(B)\left(\int_{\rho B} |u|^s w_2^\alpha dx\right)^{\frac{1}{s}} \qquad (7.5.14)$$

and

$$\|T(u)-(T(u))_B\|_{W^{1,s}(B),w_1^{\alpha\lambda}} \le C|B|\|u\|_{s,\rho B,w_2^\alpha} \qquad (7.5.14)'$$

for all balls B with $\rho B \subset \Omega$ and any real number α with $0 < \alpha < 1$.

Note that inequality (7.5.14) can be written as

$$\|T(u)-(T(u))_B\|_{s,B,w_1^{\alpha\lambda}} \le C|B|diam(B)\|u\|_{s,\rho B,w_2^\alpha}. \qquad (7.5.14)''$$

In Theorem 7.5.8, we have assumed that $(w_1, w_2) \in A_r(\lambda, \Omega)$. If the weights w_1 and w_2 satisfy some other condition, say $(w_1, w_2) \in A_{r,\lambda}(\Omega)$, we have the following version of Poincaré-type inequality.

Theorem 7.5.9. *Let $u \in L_{loc}^s(\Omega, \wedge^l)$, $l = 1, 2, \ldots, n$, $1 < s < \infty$, be a differential form satisfying (1.2.10) in a bounded, convex domain $\Omega \subset \mathbf{R}^n$ and $T : C^\infty(\Omega, \wedge^l) \to C^\infty(\Omega, \wedge^{l-1})$ be the homotopy operator defined in (1.5.1). Suppose that $\rho > 1$ and $(w_1, w_2) \in A_{r,\lambda}(\Omega)$ for some $\lambda \geq 1$ and $1 < r < \infty$. Then, there exists a constant C, independent of u, such that*

$$\left(\int_B |T(u) - (T(u))_B|^s w_1^\alpha dx \right)^{\frac{1}{s}} \leq C|B| diam(B) \left(\int_{\rho B} |u|^s w_2^\alpha dx \right)^{\frac{1}{s}}$$
(7.5.15)

and

$$\|T(u) - (T(u))_B\|_{W^{1,s}(B), w_1^\alpha} \leq C|B| \|u\|_{s, \rho B, w_2^\alpha}$$
(7.5.16)

for all balls B with $\rho B \subset \Omega$ and any real number α with $0 < \alpha < \lambda$.

Note that inequality (7.5.15) can be written as

$$\|T(u) - (T(u))_B\|_{s, B, w_1^\alpha} \leq C|B| diam(B) \|u\|_{s, \rho B, w_2^\alpha}.$$
(7.5.16)'

Similarly, if $(w_1, w_2) \in A_r^\lambda(\Omega)$, we have the following version of two-weight Poincaré inequality for differential forms.

Theorem 7.5.10. *Let $u \in L_{loc}^s(\Omega, \wedge^l)$, $l = 1, 2, \ldots, n$, $1 < s < \infty$, be a differential form satisfying (1.2.10) in a bounded, convex domain $\Omega \subset \mathbf{R}^n$ and $T : C^\infty(\Omega, \wedge^l) \to C^\infty(\Omega, \wedge^{l-1})$ be the homotopy operator defined in (1.5.1). Suppose that $(w_1, w_2) \in A_r^\lambda(\Omega)$ for some $r > 1$ and $\lambda > 0$. If $0 < \alpha < 1$ and $\sigma > 1$, then there exists a constant C, independent of u, such that*

$$\left(\int_B |T(u) - (T(u))_B|^s w_1^\alpha dx \right)^{\frac{1}{s}} \leq C|B| diam(B) \left(\int_{\sigma B} |u|^s w_2^{\alpha\lambda} dx \right)^{\frac{1}{s}}$$
(7.5.17)

and

$$\|T(u) - (T(u))_B\|_{W^{1,s}(B), w_1^\alpha} \leq C|B| \|u\|_{s, \sigma B, w_2^{\alpha\lambda}}$$
(7.5.18)

for all balls B with $\sigma B \subset \Omega$.

If we choose $\lambda = 1/\alpha$ in Theorem 7.5.10, we have the following version of the Poincaré inequality with $(w_1, w_2) \in A_r^{1/\alpha}(\Omega)$.

Corollary 7.5.11. *Let $u \in L_{loc}^s(\Omega, \wedge^l)$, $l = 1, 2, \ldots, n$, $1 < s < \infty$, be a differential form satisfying (1.2.10) in a bounded, convex domain $\Omega \subset \mathbf{R}^n$ and $T : C^\infty(\Omega, \wedge^l) \to C^\infty(\Omega, \wedge^{l-1})$ be the homotopy operator defined in (1.5.1). Suppose that $(w_1, w_2) \in A_r^{1/\alpha}(\Omega)$ for some $r > 1$. If $0 < \alpha < 1$ and $\sigma > 1$, then there exists a constant C, independent of u, such that*

$$\left(\int_B |T(u) - (T(u))_B|^s w_1^\alpha dx\right)^{\frac{1}{s}} \leq C|B|diam(B) \left(\int_{\sigma B} |u|^s w_2 dx\right)^{\frac{1}{s}}$$
(7.5.19)

for all balls B with $\sigma B \subset \Omega$.

Choosing $\alpha = 1/s$ in Theorem 7.5.10, we obtain the following two-weighted Poincaré inequality.

Corollary 7.5.12. *Let $u \in L^s_{loc}(\Omega, \wedge^l)$, $l = 1, 2, \ldots, n$, $1 < s < \infty$, be a differential form satisfying (1.2.10) in a bounded, convex domain $\Omega \subset \mathbf{R}^n$ and $T : C^\infty(\Omega, \wedge^l) \to C^\infty(\Omega, \wedge^{l-1})$ be the homotopy operator defined in (1.5.1). Suppose that $(w_1, w_2) \in A_r^\lambda(\Omega)$ for some $r > 1$, $\lambda > 0$, and $\sigma > 1$, then there exists a constant C, independent of u, such that*

$$\left(\int_B |T(u) - (T(u))_B|^s w_1^{\frac{1}{s}} dx\right)^{\frac{1}{s}} \leq C|B|diam(B) \left(\int_{\sigma B} |u|^s w_2^{\lambda/s} dx\right)^{\frac{1}{s}}$$
(7.5.20)

for all balls B with $\sigma B \subset \Omega$.

Letting $\lambda = 1$ in Corollary 7.5.12, we find the following symmetric two-weighted inequality.

Corollary 7.5.13. *Let $u \in L^s_{loc}(\Omega, \wedge^l)$, $l = 1, 2, \ldots, n$, $1 < s < \infty$, be a differential form satisfying (1.2.10) in a bounded, convex domain $\Omega \subset \mathbf{R}^n$ and $T : C^\infty(\Omega, \wedge^l) \to C^\infty(\Omega, \wedge^{l-1})$ be the homotopy operator defined in (1.5.1). Suppose that $(w_1, w_2) \in A_r(\Omega)$ for some $r > 1$ and $\sigma > 1$, then there exists a constant C, independent of u, such that*

$$\left(\int_B |T(u) - (T(u))_B|^s w_1^{\frac{1}{s}} dx\right)^{\frac{1}{s}} \leq C|B|diam(B) \left(\int_{\sigma B} |u|^s w_2^{\frac{1}{s}} dx\right)^{\frac{1}{s}}$$
(7.5.21)

for all balls B with $\sigma B \subset \Omega$.

Finally, when $\lambda = s$ in Theorem 7.5.10, we have the following two-weighted inequality.

Corollary 7.5.14. *Let $u \in L^s_{loc}(\Omega, \wedge^l)$, $l = 1, 2, \ldots, n$, $1 < s < \infty$, be a differential form satisfying (1.2.10) in a bounded, convex domain $\Omega \subset \mathbf{R}^n$ and $T : C^\infty(\Omega, \wedge^l) \to C^\infty(\Omega, \wedge^{l-1})$ be the homotopy operator defined in (1.5.1). Suppose that $(w_1, w_2) \in A_r^s(\Omega)$ for some $r > 1$. If $0 < \alpha < 1$ and $\sigma > 1$, then there exists a constant C, independent of u, such that*

$$\left(\int_B |T(u) - (T(u))_B|^s w_1^\alpha dx\right)^{\frac{1}{s}} \leq C|B|diam(B) \left(\int_{\sigma B} |u|^s w_2^{\alpha s} dx\right)^{\frac{1}{s}}$$
(7.5.22)

and

$$\|T(u) - (T(u))_B\|_{W^{1,s}(B),w_1^\alpha} \le C|B| \|u\|_{s,\sigma B, w_2^{\alpha s}}$$

for all balls B with $\sigma B \subset \Omega$.

7.6 Homotopy and projection operators

In the previous sections, we have discussed Poincaré-type estimates for some compositions of operators. In this section, we shall develop Poincaré-type estimates for the composition of the homotopy operator T and the projection operator H which are applied to the differential forms or the solutions of the nonhomogeneous A-harmonic equation.

7.6.1 Basic estimates for $T \circ H$

As we have done in Section 7.5, here also, first, we shall prove the following basic local Poincaré-type estimate for the composition $T \circ H$. Then, we shall discuss the different weighted cases, the local cases and the global cases. As applications of the L^s-norm inequalities, we will also develop some weighted imbedding inequalities.

Theorem 7.6.1. *Let $u \in L^s_{loc}(\Omega, \wedge^l)$, $l = 1, 2, \ldots, n$, $1 < s < \infty$, be a solution of the A-harmonic equation (1.2.10) in a bounded, convex domain Ω, H be the projection operator, and $T : C^\infty(\Omega, \wedge^l) \to C^\infty(\Omega, \wedge^{l-1})$ be the homotopy operator defined in (1.5.1). Then, there exists a constant C, independent of u, such that*

$$\|T(H(u)) - (T(H(u)))_B\|_{s,B} \le C|B| \, diam(B) \|u\|_{s,\rho B} \qquad (7.6.1)$$

for all balls B with $\rho B \subset \Omega$, where $\rho > 1$ is a constant.

Proof. Let H be the projection operator and T be the homotopy operator. Then, for any differential form u, we have

$$H(u) = u - \Delta G(u). \qquad (7.6.2)$$

Replacing u by du in (7.6.2), we obtain

$$H(du) = du - \Delta G(du). \qquad (7.6.3)$$

Now, since the differential operator d commutes with Δ and G, using (7.6.3) we find that

$$d(H(u)) = d(u - \Delta G(u))$$
$$= du - d\Delta G(u)$$
$$= du - \Delta G(du) \qquad (7.6.4)$$
$$= H(du).$$

Thus, H commutes with the differential operator d. Next, from (7.6.3) and Theorem 3.3.2, we have

$$\|d(H(u))\|_{s,B} = \|H(du)\|_{s,B}$$
$$= \|du - \Delta G(du)\|_{s,B}$$
$$\leq \|du\|_{s,B} + \|\Delta G(du)\|_{s,B} \qquad (7.6.5)$$
$$\leq \|du\|_{s,B} + C_1\|du\|_{s,B}$$
$$\leq C_2\|du\|_{s,B}.$$

Using (7.4.2), (7.4.4), and (7.6.5), we find that

$$\|T(H(u)) - (T(H(u)))_B\|_{s,B}$$
$$= \|Td(T(H(u)))\|_{s,B}$$
$$\leq C_3|B|diam(B)\|d(T(H(u)))\|_{s,B}$$
$$\leq C_3|B|diam(B)(\|H(u)\|_{s,B}$$
$$+ C_4|B|diam(B)\|d(H(u))\|_{s,B})$$
$$\leq C_3|B|diam(B)(\|H(u)\|_{s,B} \qquad (7.6.6)$$
$$+ C_5diam(B)\|d(H(u))\|_{s,B})$$
$$\leq C_3|B|diam(B)(\|H(u)\|_{s,B} + C_5diam(B)\|du\|_{s,B})$$
$$\leq C_3|B|diam(B)(\|H(u)\|_{s,B} + C_6\|u\|_{s,\rho B}).$$

Applying Theorem 3.3.2 and (7.6.2), it follows that

$$\|H(u)\|_{s,B} = \|u - \Delta G(u)\|_{s,B}$$
$$\leq \|u\|_{s,B} + \|\Delta G(u)\|_{s,B}$$
$$\leq \|u\|_{s,B} + C_7\|u\|_{s,B} \qquad (7.6.7)$$
$$\leq C_8\|u\|_{s,B}.$$

Finally, substituting (7.6.7) into (7.6.6), we obtain

$$\|T(H(u)) - (T(H(u)))_B\|_{s,B}$$
$$\leq C_3|B|diam(B)(\|H(u)\|_{s,B} + C_6\|u\|_{s,\rho B})$$
$$\leq C_9|B|diam(B)\|u\|_{s,\rho B}$$

which is equivalent to inequality (7.6.1).   ■

Note that in the proof of Theorem 7.6.1, we have used Caccioppoli inequality to obtain (7.6.6), which requires that u to be a solution of the A-harmonic equation. If we drop this condition, then the above result can be stated as follows.

Theorem 7.6.1′.   *Let $u \in L^s_{loc}(\Omega, \wedge^l)$, $l = 1, 2, \ldots, n$, $1 < s < \infty$, be a smooth differential form in a bounded, convex domain Ω, H be the projection operator, and $T : C^\infty(\Omega, \wedge^l) \to C^\infty(\Omega, \wedge^{l-1})$ be the homotopy operator defined in (1.5.1). Then, there exists a constant C, independent of u, such that*

$$\|T(H(u)) - (T(H(u)))_B\|_{s,B} \leq C|B|diam(B)(\|u\|_{s,B} + \|du\|_{s,B}) \quad (7.6.1)′$$

for all balls $B \subset \Omega$.

7.6.2 $A_r(\Omega)$-weighted inequalities for $T \circ H$

Using the unweighted Poincaré-type inequality obtained above, we can easily establish the $A_r(\Omega)$-weighted inequalities for $T \circ H$. We state these results in the following theorems.

Theorem 7.6.2.   *Let $u \in L^s_{loc}(\Omega, \wedge^l)$, $l = 1, 2, \ldots, n$, $1 < s < \infty$, be a solution of the nonhomogeneous A-harmonic equation (1.2.10) in a bounded, convex domain Ω and $T : C^\infty(\Omega, \wedge^l) \to C^\infty(\Omega, \wedge^{l-1})$ be the homotopy operator defined in (1.5.1). Assume that $\rho > 1$ and $w \in A_r(\Omega)$ for some $1 < r < \infty$. Then, there exists a constant C, independent of u, such that*

$$\|T(H(u)) - (T(H(u)))_B\|_{s,B,w^\alpha} \leq C|B|diam(B)\|u\|_{s,\rho B,w^\alpha} \quad (7.6.8)$$

for all balls B with $\rho B \subset \Omega$ and any real number α with $0 < \alpha \leq 1$.

Note that the above L^s-norm inequality can also be written as

$$\left(\int_B |T(H(u)) - (T(H(u)))_B|^s w^\alpha dx\right)^{1/s} \leq C \left(\int_{\rho B} |u|^s w^\alpha dx\right)^{1/s}. \quad (7.6.8)′$$

Theorem 7.6.3. *Let* $u \in D'(\Omega, \wedge^1)$ *be a solution of the nonhomogeneous A-harmonic equation* (1.2.10) *and* $T : C^\infty(\Omega, \wedge^l) \to C^\infty(\Omega, \wedge^{l-1})$, $l = 1, 2, \ldots, n$, *be the homotopy operator defined in* (1.5.1). *Assume that* $w \in A_r(\Omega)$ *for some* $1 < r < \infty$ *and* s *is a fixed exponent associated with the A-harmonic equation* (1.2.10). *Then, there exists a constant* C, *independent of* u, *such that*

$$\left(\int_\Omega |T(H(u)) - (T(H(u)))_{Q_0}|^s w dx \right)^{1/s} \leq C \left(\int_\Omega |u|^s w dx \right)^{1/s} \quad (7.6.9)$$

for any bounded, convex δ-*John domain* $\Omega \subset \mathbf{R}^n$. *Here* $Q_0 \subset \Omega$ *is a fixed cube.*

Using Theorem 7.6.2 and the definition of the $\| \cdot \|_{W^{1,s}(E),w^\alpha}$-norm, we can prove the following local and global imbedding inequalities.

Theorem 7.6.4. *Let* $u \in L^s_{loc}(\Omega, \wedge^l)$, $l = 1, 2, \ldots, n$, $1 < s < \infty$, *be a solution of the nonhomogeneous A-harmonic equation* (1.2.10) *in a bounded, convex domain* Ω *and* $T : C^\infty(\Omega, \wedge^l) \to C^\infty(\Omega, \wedge^{l-1})$ *be the homotopy operator defined in* (1.5.1). *Assume that* $\rho > 1$ *and* $w \in A_r(\Omega)$ *for some* $1 < r < \infty$. *Then, there exists a constant* C, *independent of* u, *such that*

$$\|T(H(u)) - (T(H(u)))_B\|_{W^{1,s}(B),w^\alpha} \leq C|B| \|u\|_{s, \rho B, w^\alpha} \quad (7.6.10)$$

for all balls B *with* $\rho B \subset \Omega$ *and any real number* α *with* $0 < \alpha \leq 1$.

Theorem 7.6.5. *Let* $u \in D'(\Omega, \wedge^1)$ *be a solution of the nonhomogeneous A-harmonic equation* (1.2.10) *and* $T : C^\infty(\Omega, \wedge^l) \to C^\infty(\Omega, \wedge^{l-1})$, $l = 1, 2, \ldots, n$, *be the homotopy operator defined in* (1.5.1). *Assume that* $w \in A_r(\Omega)$ *for some* $1 < r < \infty$ *and* s *is a fixed exponent associated with the A-harmonic equation* (1.2.10). *Then, there exists a constant* C, *independent of* u, *such that*

$$\|T(H(u)) - (T(H(u)))_{Q_0}\|_{W^{1,s}(\Omega),w} \leq C \|u\|_{s, \Omega, w} \quad (7.6.11)$$

for any bounded, convex δ-*John domain* $\Omega \subset \mathbf{R}^n$. *Here* $Q_0 \subset \Omega$ *is a fixed cube.*

7.6.3 Other single weighted cases

In the previous several theorems, we have presented the $A_r(\Omega)$-weighted Poincaré-type estimates for the composition $T \circ H$. Now, we state results with different weights, namely $A_r(\lambda, \Omega)$-weights and $A_r^\lambda(\Omega)$-weights.

Theorem 7.6.6. *Let $u \in L^s_{loc}(\Omega, \wedge^l)$, $l = 1, 2, \ldots, n$, $1 < s < \infty$, be a differential form satisfying the nonhomogeneous A-harmonic equation (1.2.10) in a bounded, convex domain $\Omega \subset \mathbf{R}^n$ and $T : C^\infty(\Omega, \wedge^l) \to C^\infty(\Omega, \wedge^{l-1})$ be the homotopy operator defined in (1.5.1). Assume that $w \in A_r(\lambda, \Omega)$ for some $r > 1$ and $\lambda > 0$. Then, there exists a constant C, independent of u, such that*

$$\|T(H(u)) - (T(H(u)))_B\|_{s,B,w^{\alpha\lambda}} \le C|B|diam(B)\|u\|_{s,\rho B,w^\alpha} \qquad (7.6.12)$$

and

$$\|T(H(u)) - (T(H(u)))_B\|_{W^{1,s}(B),\ w^{\alpha\lambda}} \le C|B|\|u\|_{s,\rho B,w^\alpha} \qquad (7.6.13)$$

for all balls B with $\rho B \subset \Omega$ and any real number α with $0 < \alpha < 1$. Here $\rho > 1$ is some constant.

The above inequality (7.6.12) can be written as

$$\left(\int_B |T(H(u)) - (T(H(u)))_B|^s w^{\alpha\lambda}dx\right)^{\frac{1}{s}} \le C|B|diam(B)\left(\int_{\rho B} |u|^s w^\alpha dx\right)^{\frac{1}{s}}.$$

Theorem 7.6.7. *Let $u \in L^s_{loc}(\Omega, \wedge^l)$, $l = 1, 2, \ldots, n$, $1 < s < \infty$, be a differential form satisfying (1.2.10) in a bounded, convex domain $\Omega \subset \mathbf{R}^n$ and $T : C^\infty(\Omega, \wedge^l) \to C^\infty(\Omega, \wedge^{l-1})$ be the homotopy operator defined in (1.5.1). Assume that $\rho > 1$ and $w \in A^\lambda_r(\Omega)$ for some $r > 1$ and $\lambda > 0$. Then, there exists a constant C, independent of u, such that*

$$\|T(H(u)) - (T(H(u)))_B\|_{s,B,w^\alpha} \le C|B|diam(B)\|u\|_{s,\rho B,w^{\alpha\lambda}} \qquad (7.6.14)$$

and

$$\|T(H(u)) - (T(H(u)))_B\|_{W^{1,s}(B),\ w^\alpha} \le C|B|\|u\|_{s,\rho B,w^{\alpha\lambda}} \qquad (7.6.15)$$

for all balls B with $\rho B \subset \Omega$ and any real number α with $0 < \alpha < 1$.

The integral version of inequality (7.6.14) can be written as

$$\left(\int_B |T(H(u)) - (T(H(u)))_B|^s w^\alpha dx\right)^{\frac{1}{s}} \le C|B|diam(B)\left(\int_{\rho B} |u|^s w^{\alpha\lambda}dx\right)^{\frac{1}{s}}.$$

7.6.4 Inequalities with two-weights in $A_r(\lambda, \Omega)$

Now, we present the following versions of two-weight Poincaré inequalities for differential forms satisfying the nonhomogeneous A-harmonic equation.

Theorem 7.6.8. *Let* $u \in L^s_{loc}(\Omega, \wedge^l)$, $l = 1, 2, \ldots, n$, $1 < s < \infty$, *be a solution of the nonhomogeneous A-harmonic equation (1.2.10) in a bounded, convex domain* $\Omega \subset \mathbf{R}^n$ *and* $T : C^\infty(\Omega, \wedge^l) \to C^\infty(\Omega, \wedge^{l-1})$ *be the homotopy operator defined in (1.5.1). Suppose that* $\rho > 1$ *and* $(w_1, w_2) \in A_r(\lambda, \Omega)$ *for some* $\lambda > 0$ *and* $1 < r < \infty$. *Then, there exists a constant* C, *independent of* u, *such that*

$$\left(\int_B |T(H(u)) - (T(H(u)))_B|^s w_1^{\alpha \lambda} dx \right)^{\frac{1}{s}} \leq C|B| diam(B) \left(\int_{\rho B} |u|^s w_2^{\alpha} dx \right)^{\frac{1}{s}}$$
(7.6.16)

and

$$\|T(H(u)) - (T(H(u)))_B\|_{W^{1,s}(B), w_1^{\alpha \lambda}} \leq C|B| \|u\|_{s, \rho B, w_2^{\alpha}}$$
(7.6.17)

for all balls B *with* $\rho B \subset \Omega$ *and any real number* α *with* $0 < \alpha < 1$.

Inequality (7.6.16) can be written as

$$\|T(H(u)) - (T(H(u)))_B\|_{s, B, w_1^{\alpha \lambda}} \leq C|B| diam(B) \|u\|_{s, \rho B, w_2^{\alpha}}.$$
(7.6.17)′

Theorem 7.6.9. *Let* $u \in L^s_{loc}(\Omega, \wedge^l)$, $l = 1, 2, \ldots, n$, $1 < s < \infty$, *be a differential form satisfying (1.2.10) in a bounded, convex domain* $\Omega \subset \mathbf{R}^n$ *and* $T : C^\infty(\Omega, \wedge^l) \to C^\infty(\Omega, \wedge^{l-1})$ *be the homotopy operator defined in (1.5.1). Suppose that* $\rho > 1$ *and* $(w_1, w_2) \in A_{r,\lambda}(\Omega)$ *for some* $\lambda \geq 1$ *and* $1 < r < \infty$. *Then, there exists a constant* C, *independent of* u, *such that*

$$\left(\int_B |T(H(u)) - (T(H(u)))_B|^s w_1^{\alpha} dx \right)^{\frac{1}{s}} \leq C|B| diam(B) \left(\int_{\rho B} |u|^s w_2^{\alpha} dx \right)^{\frac{1}{s}}$$
(7.6.18)

and

$$\|T(H(u)) - (T(H(u)))_B\|_{W^{1,s}(B), w_1^{\alpha}} \leq C|B| \|u\|_{s, \rho B, w_2^{\alpha}}$$
(7.6.19)

for all balls B *with* $\rho B \subset \Omega$ *and any real number* α *with* $0 < \alpha < \lambda$.

Note that (7.6.18) is equivalent to

$$\|T(H(u)) - (T(H(u)))_B\|_{s, B, w_1^{\alpha}} \leq C|B| diam(B) \|u\|_{s, \rho B, w_2^{\alpha}}.$$
(7.6.19)′

7.6.5 Inequalities with two-weights in $A_r^\lambda(\Omega)$

In the following results, we will state other two-weight cases.

Theorem 7.6.10. *Let* $u \in L^s_{loc}(\Omega, \wedge^l)$, $l = 1, 2, \ldots, n$, $1 < s < \infty$, *be a differential form satisfying (1.2.10) in a bounded, convex domain* $\Omega \subset \mathbf{R}^n$ *and*

$T : C^\infty(\Omega, \wedge^l) \to C^\infty(\Omega, \wedge^{l-1})$ be the homotopy operator defined in (1.5.1). Suppose that $(w_1, w_2) \in A_r^\lambda(\Omega)$ for some $r > 1$ and $\lambda > 0$. If $0 < \alpha < 1$ and $\sigma > 1$, then there exists a constant C, independent of u, such that

$$\left(\int_B |T(H(u)) - (T(H(u)))_B|^s w_1^\alpha dx\right)^{1/s}$$

$$\leq C|B| diam(B) \left(\int_{\sigma B} |u|^s w_2^{\alpha\lambda} dx\right)^{1/s} \qquad (7.6.20)$$

and

$$\|T(H(u)) - (T(H(u)))_B\|_{W^{1,s}(B), w_1^\alpha} \leq C|B| \|u\|_{s, \sigma B, w_2^{\alpha\lambda}} \qquad (7.6.21)$$

for all balls B with $\sigma B \subset \Omega$.

If we choose $\lambda = 1/\alpha$ in Theorem 7.6.10, we have the following version of the A_r^λ-weighted Poincaré inequality.

Corollary 7.6.11. Let $u \in L^s_{loc}(\Omega, \wedge^l)$, $l = 1, 2, \ldots, n$, $1 < s < \infty$, be a differential form satisfying (1.2.10) in a bounded, convex domain $\Omega \subset \mathbf{R}^n$ and $T : C^\infty(\Omega, \wedge^l) \to C^\infty(\Omega, \wedge^{l-1})$ be the homotopy operator defined in (1.5.1). Suppose that $(w_1, w_2) \in A_r^{1/\alpha}(\Omega)$ for some $r > 1$. If $0 < \alpha < 1$ and $\sigma > 1$, then there exists a constant C, independent of u, such that

$$\left(\int_B |T(H(u)) - (T(H(u)))_B|^s w_1^\alpha dx\right)^{1/s}$$

$$\leq C|B| diam(B) \left(\int_{\sigma B} |u|^s w_2 dx\right)^{1/s} \qquad (7.6.22)$$

for all balls B with $\sigma B \subset \Omega$.

Choosing $\alpha = 1/s$ in Theorem 7.6.10, we find the following two-weight Poincaré inequality.

Corollary 7.6.12. Let $u \in L^s_{loc}(\Omega, \wedge^l)$, $l = 1, 2, \ldots, n$, $1 < s < \infty$, be a differential form satisfying (1.2.10) in a bounded, convex domain $\Omega \subset \mathbf{R}^n$ and $T : C^\infty(\Omega, \wedge^l) \to C^\infty(\Omega, \wedge^{l-1})$ be the homotopy operator defined in (1.5.1). Suppose that $(w_1, w_2) \in A_r^\lambda(\Omega)$ for some $r > 1$, $\lambda > 0$, and $\sigma > 1$, then there exists a constant C, independent of u, such that

$$\left(\int_B |T(H(u)) - (T(H(u)))_B|^s w_1^{\frac{1}{s}} dx\right)^{1/s}$$

$$\leq C|B| diam(B) \left(\int_{\sigma B} |u|^s w_2^{\lambda/s} dx\right)^{1/s} \qquad (7.6.23)$$

for all balls B with $\sigma B \subset \Omega$.

Letting $\lambda = 1$ in Corollary 7.6.12, we obtain the following symmetric two-weight inequality.

Corollary 7.6.13. *Let* $u \in L^s_{loc}(\Omega, \wedge^l)$, $l = 1, 2, \ldots, n$, $1 < s < \infty$, *be a differential form satisfying* (1.2.10) *in a bounded, convex domain* $\Omega \subset \mathbf{R}^n$ *and* $T : C^\infty(\Omega, \wedge^l) \to C^\infty(\Omega, \wedge^{l-1})$ *be the homotopy operator defined in* (1.5.1). *Suppose that* $(w_1, w_2) \in A_r(\Omega)$ *for some* $r > 1$ *and* $\sigma > 1$, *then there exists a constant* C, *independent of* u, *such that*

$$\left(\int_B |T(H(u)) - (T(H(u)))_B|^s w_1^{\frac{1}{s}} dx \right)^{1/s}$$
$$\leq C|B| diam(B) \left(\int_{\sigma B} |u|^s w_2^{\frac{1}{s}} dx \right)^{1/s} \tag{7.6.24}$$

for all balls B *with* $\sigma B \subset \Omega$.

Finally, when $\lambda = s$ in Theorem 7.6.10, we have the following two-weighted inequality.

Corollary 7.6.14. *Let* $u \in L^s_{loc}(\Omega, \wedge^l)$, $l = 1, 2, \ldots, n$, $1 < s < \infty$, *be a differential form satisfying* (1.2.10) *in a bounded, convex domain* $\Omega \subset \mathbf{R}^n$ *and* $T : C^\infty(\Omega, \wedge^l) \to C^\infty(\Omega, \wedge^{l-1})$ *be the homotopy operator defined in* (1.5.1). *Suppose that* $(w_1, w_2) \in A^s_r(\Omega)$ *for some* $r > 1$. *If* $0 < \alpha < 1$ *and* $\sigma > 1$, *then there exists a constant* C, *independent of* u, *such that*

$$\left(\int_B |T(H(u)) - (T(H(u)))_B|^s w_1^\alpha dx \right)^{1/s}$$
$$\leq C|B| diam(B) \left(\int_{\sigma B} |u|^s w_2^{\alpha s} dx \right)^{1/s} \tag{7.6.25}$$

and

$$\|T(H(u)) - (T(H(u)))_B\|_{W^{1,s}(B), w_1^\alpha} \leq C|B| \|u\|_{s, \sigma B, w_2^{\alpha s}} \tag{7.6.26}$$

for all balls B *with* $\sigma B \subset \Omega$.

7.6.6 Basic estimates for $H \circ T$

Similar to the case of $T \circ H$, Poincaré-type inequalities also hold for $H \circ T$. We list them as follows.

Theorem 7.6.15. *Let* $u \in L^s_{loc}(\Omega, \wedge^l)$, $l = 1, 2, \ldots, n$, $1 < s < \infty$, *be a solution of the A-harmonic equation* (1.2.10) *in a bounded, convex domain* Ω, H *be the projection operator, and* $T : C^\infty(\Omega, \wedge^l) \to C^\infty(\Omega, \wedge^{l-1})$ *be* *the homotopy operator defined in (1.5.1). Then, there exists a constant C, independent of u, such that*

$$\|H(T(u)) - (H(T(u)))_B\|_{s,B} \leq C|B|diam(B)\|u\|_{s,\rho B} \tag{7.6.27}$$

for all balls B with $\rho B \subset \Omega$, where $\rho > 1$ is a constant.

Proof. For any differential form ω, we have

$$\|\omega - (\omega)_B\|_{s,B} = \|Td(\omega)\|_{s,B} \tag{7.6.28}$$

for any ball B. Now, replacing ω by $H(T(u))$ in (7.6.28), and using (7.4.2), we obtain

$$
\begin{aligned}
\|H(T(u)) &- (H(T(u)))_B\|_{s,B} \\
&= \|Td(H(T(u)))\|_{s,B} \\
&\leq C_1|B|diam(B)\|d(H(T(u)))\|_{s,B}.
\end{aligned}
\tag{7.6.29}
$$

In the proof of Theorem 7.6.1, we have shown that H commutes with d. Thus, inequality (7.6.29) can be written as

$$\|H(T(u)) - (H(T(u)))_B\|_{s,B} \leq C_1|B|diam(B)\|H(d(T(u)))\|_{s,B}. \tag{7.6.30}$$

Replacing u by $d(T(u))$ in (7.6.2), we find that

$$H(d(T(u))) = d(T(u)) - \Delta G(d(T(u))). \tag{7.6.31}$$

For any differential form u (note that u need not be a solution of the A-harmonic equation), using Theorem 3.3.2, we have

$$\|\Delta G(u)\|_{s,B} \leq C_2\|u\|_{s,B}. \tag{7.6.32}$$

Next, replacing u by $d(T(u))$ in (7.6.32), we find that

$$\|\Delta G(d(T(u)))\|_{s,B} \leq C_2\|d(T(u))\|_{s,B}. \tag{7.6.33}$$

Finally, applying inequality (3.3.8) and Caccioppoli's inequality, and noticing that $|B| \leq |\Omega| \leq C_3$ since Ω is bounded, we obtain

$$
\begin{aligned}
\|d(T(u))\|_{s,B} &\leq \|u\|_{s,B} + C_4|B|diam(B)\|du\|_{s,B} \\
&\leq \|u\|_{s,B} + C_5|B|\|u - c\|_{s,\rho B} \\
&\leq \|u\|_{s,B} + C_6|B|\|u\|_{s,\rho B} \\
&\leq \|u\|_{s,B} + C_7\|u\|_{s,\rho B} \\
&\leq C_8\|u\|_{s,\rho B}
\end{aligned}
\tag{7.6.34}
$$

for some $\rho > 1$, where we have chosen $c = 0$ when we used Caccioppoli's inequality. Combination of (7.6.33) and (7.6.34) yields

$$\|\Delta G(d(T(u)))\|_{s,B} \leq C_9 \|u\|_{s,\rho B}. \tag{7.6.35}$$

Hence, from (7.6.30), (7.6.31), (7.6.34), and (7.6.35), it follows that

$$
\begin{aligned}
&\|H(T(u)) - (H(T(u)))_B\|_{s,B} \\
&\quad \leq C_1|B|diam(B)\|H(d(T(u)))\|_{s,B} \\
&\quad \leq C_1|B|diam(B)\|d(T(u)) - \Delta G(d(T(u)))\|_{s,B} \\
&\quad \leq C_1|B|diam(B)\left(\|d(T(u))\|_{s,B} + \|\Delta G(d(T(u)))\|_{s,B}\right) \\
&\quad \leq C_1|B|diam(B)\left(C_8\|u\|_{s,B} + C_9\|u\|_{s,\rho B}\right) \\
&\quad \leq C_{10}|B|diam(B)\|u\|_{s,\rho B},
\end{aligned}
\tag{7.6.36}
$$

that is

$$\|H(T(u)) - (H(T(u)))_B\|_{s,B} \leq C_{10}|B|diam(B)\|u\|_{s,\rho B}. \qquad \blacksquare$$

The weak version of Theorem 7.6.15 can be stated as follows.

Theorem 7.6.15′. *Let $u \in L^s_{loc}(\Omega, \wedge^l)$, $l = 1, 2, \ldots, n$, $1 < s < \infty$, be a smooth differential form in a bounded, convex domain Ω, H be the projection operator, and $T : C^\infty(\Omega, \wedge^l) \to C^\infty(\Omega, \wedge^{l-1})$ be the homotopy operator defined in (1.5.1). Then, there exists a constant C, independent of u, such that*

$$\|H(T(u)) - (H(T(u)))_B\|_{s,B} \leq C|B|diam(B)(\|u\|_{s,B} + \|du\|_{s,B}) \tag{7.6.36′}$$

for all balls $B \subset \Omega$.

7.6.7 Weighted inequalities for $H \circ T$

Theorem 7.6.16. *Let $u \in L^s_{loc}(\Omega, \wedge^l)$, $l = 1, 2, \ldots, n$, $1 < s < \infty$, be a solution of the nonhomogeneous A-harmonic equation (1.2.10) in a bounded, convex domain Ω and $T : C^\infty(\Omega, \wedge^l) \to C^\infty(\Omega, \wedge^{l-1})$ be the homotopy operator defined in (1.5.1). Assume that $\rho > 1$ and $w \in A_r(\Omega)$ for some $1 < r < \infty$. Then, there exists a constant C, independent of u, such that*

$$\|H(T(u)) - (H(T(u)))_B\|_{s,B,w^\alpha} \leq C|B|diam(B)\|u\|_{s,\rho B,w^\alpha} \tag{7.6.37}$$

for all balls B with $\rho B \subset \Omega$ and any real number α with $0 < \alpha \leq 1$.

Note that the above L^s-norm inequality can also be written as

$$\left(\int_B |H(T(u)) - (H(T(u)))_B|^s w^\alpha dx\right)^{1/s} \leq C \left(\int_{\rho B} |u|^s w^\alpha dx\right)^{1/s}.$$
$$(7.6.37)'$$

Theorem 7.6.17. *Let $u \in D'(\Omega, \wedge^1)$ be a solution of the nonhomogeneous A-harmonic equation (1.2.10) and $T : C^\infty(\Omega, \wedge^l) \rightarrow C^\infty(\Omega, \wedge^{l-1})$ be the homotopy operator defined in (1.5.1). Assume that $w \in A_r(\Omega)$ for some $1 < r < \infty$ and s is a fixed exponent associated with the A-harmonic equation (1.2.10). Then, there exists a constant C, independent of u, such that*

$$\left(\int_\Omega |H(T(u)) - (H(T(u)))_{Q_0}|^s w dx\right)^{1/s} \leq C \left(\int_\Omega |u|^s w dx\right)^{1/s} \qquad (7.6.38)$$

for any bounded, convex δ-John domain $\Omega \subset \mathbf{R}^n$. Here $Q_0 \subset \Omega$ is a fixed cube.

From Theorem 7.6.16 and the definition of the $\| \cdot \|_{W^{1,s}(E),w^\alpha}$-norm, we can prove the following local and global imbedding inequalities.

Theorem 7.6.18. *Let $u \in L^s_{loc}(\Omega, \wedge^l)$, $l = 1, 2, \ldots, n$, $1 < s < \infty$, be a solution of the nonhomogeneous A-harmonic equation (1.2.10) in a bounded, convex domain Ω and $T : C^\infty(\Omega, \wedge^l) \rightarrow C^\infty(\Omega, \wedge^{l-1})$ be the homotopy operator defined in (1.5.1). Assume that $\rho > 1$ and $w \in A_r(\Omega)$ for some $1 < r < \infty$. Then, there exists a constant C, independent of u, such that*

$$\|H(T(u)) - (H(T(u)))_B\|_{W^{1,s}(B),w^\alpha} \leq C|B|\|u\|_{s,\rho B,w^\alpha} \qquad (7.6.39)$$

for all balls B with $\rho B \subset \Omega$ and any real number α with $0 < \alpha \leq 1$.

Theorem 7.6.19. *Let $u \in D'(\Omega, \wedge^1)$ be a solution of the nonhomogeneous A-harmonic equation (1.2.10) and $T : C^\infty(\Omega, \wedge^l) \rightarrow C^\infty(\Omega, \wedge^{l-1})$ be the homotopy operator defined in (1.5.1). Assume that $w \in A_r(\Omega)$ for some $1 < r < \infty$ and s is a fixed exponent associated with the A-harmonic equation (1.2.10). Then, there exists a constant C, independent of u, such that*

$$\|H(T(u)) - (H(T(u)))_{Q_0}\|_{W^{1,s}(\Omega),w} \leq C\|u\|_{s,\Omega,w} \qquad (7.6.40)$$

for any bounded, convex δ-John domain $\Omega \subset \mathbf{R}^n$. Here $Q_0 \subset \Omega$ is a fixed cube.

In the above discussion, we have presented the $A_r(\Omega)$-weighted Poincaré-type estimates for the composition $H \circ T$. Next, we state results with different weights, namely $A_r(\lambda, \Omega)$-weights and $A_r^\lambda(\Omega)$-weights.

Theorem 7.6.20. *Let* $u \in L^s_{loc}(\Omega, \wedge^l)$, $l = 1, 2, \ldots, n$, $1 < s < \infty$, *be a differential form satisfying the nonhomogeneous A-harmonic equation (1.2.10) in a bounded, convex domain* $\Omega \subset \mathbf{R}^n$ *and* $T : C^\infty(\Omega, \wedge^l) \rightarrow C^\infty(\Omega, \wedge^{l-1})$ *be the homotopy operator defined in (1.5.1). Assume that* $w \in A_r(\lambda, \Omega)$ *for some* $r > 1$ *and* $\lambda > 0$. *Then, there exists a constant* C, *independent of* u, *such that*

$$\|H(T(u)) - (H(T(u)))_B\|_{s,B,w^{\alpha\lambda}} \leq C|B|diam(B)\|u\|_{s,\rho B,w^\alpha} \qquad (7.6.41)$$

and

$$\|H(T(u)) - (H(T(u)))_B\|_{W^{1,s}(B),\ w^{\alpha\lambda}} \leq C|B|\|u\|_{s,\rho B,w^\alpha} \qquad (7.6.42)$$

for all balls B *with* $\rho B \subset \Omega$ *and any real number* α *with* $0 < \alpha < 1$. *Here* $\rho > 1$ *is some constant.*

The above inequality (7.6.41) can be written as

$$\left(\int_B |H(T(u)) - (H(T(u)))_B|^s w^{\alpha\lambda} dx \right)^{\frac{1}{s}} \leq C|B|diam(B) \left(\int_{\rho B} |u|^s w^\alpha dx \right)^{\frac{1}{s}}.$$

Theorem 7.6.21. *Let* $u \in L^s_{loc}(\Omega, \wedge^l)$, $l = 1, 2, \ldots, n$, $1 < s < \infty$, *be a differential form satisfying (1.2.10) in a bounded, convex domain* $\Omega \subset \mathbf{R}^n$ *and* $T : C^\infty(\Omega, \wedge^l) \rightarrow C^\infty(\Omega, \wedge^{l-1})$ *be the homotopy operator defined in (1.5.1). Assume that* $\rho > 1$ *and* $w \in A^\lambda_r(\Omega)$ *for some* $r > 1$ *and* $\lambda > 0$. *Then, there exists a constant* C, *independent of* u, *such that*

$$\|H(T(u)) - (H(T(u)))_B\|_{s,B,w^\alpha} \leq C|B|diam(B)\|u\|_{s,\rho B,w^{\alpha\lambda}} \qquad (7.6.43)$$

and

$$\|H(T(u)) - (H(T(u)))_B\|_{W^{1,s}(B),\ w^\alpha} \leq C|B|\|u\|_{s,\rho B,w^{\alpha\lambda}} \qquad (7.6.44)$$

for all balls B *with* $\rho B \subset \Omega$ *and any real number* α *with* $0 < \alpha < 1$.

The integral version of inequality (7.6.43) can be written as

$$\left(\int_B |H(T(u)) - (H(T(u)))_B|^s w^\alpha dx \right)^{\frac{1}{s}} \leq C|B|diam(B) \left(\int_{\rho B} |u|^s w^{\alpha\lambda} dx \right)^{\frac{1}{s}}.$$

7.6.8 Two-weight inequalities for $H \circ T$

In the following theorems, we state the two-weight Poincaré-type inequalities for $H \circ T$. We begin with the following estimate with weights $(w_1, w_2) \in A_r(\lambda, \Omega)$.

Theorem 7.6.22. *Let* $u \in L^s_{loc}(\Omega, \wedge^l)$, $l = 1, 2, \ldots, n$, $1 < s < \infty$, *be a solution of the nonhomogeneous A-harmonic equation (1.2.10) in a bounded, convex domain* $\Omega \subset \mathbf{R}^n$ *and* $T : C^\infty(\Omega, \wedge^l) \to C^\infty(\Omega, \wedge^{l-1})$ *be the homotopy operator defined in (1.5.1). Suppose that* $\rho > 1$ *and* $(w_1, w_2) \in A_r(\lambda, \Omega)$ *for some* $\lambda > 0$ *and* $1 < r < \infty$. *Then, there exists a constant* C, *independent of* u, *such that*

$$\left(\int_B |H(T(u)) - (H(T(u)))_B|^s w_1^{\alpha \lambda} dx \right)^{1/s}$$
$$\leq C|B| diam(B) \left(\int_{\rho B} |u|^s w_2^\alpha dx \right)^{1/s} \tag{7.6.45}$$

and

$$\|H(T(u)) - (H(T(u)))_B\|_{W^{1,s}(B), w_1^{\alpha\lambda}} \leq C|B| \|u\|_{s, \rho B, w_2^\alpha} \tag{7.6.46}$$

for all balls B *with* $\rho B \subset \Omega$ *and any real number* α *with* $0 < \alpha < 1$.

Inequality (7.6.45) can be written as

$$\|H(T(u)) - (H(T(u)))_B\|_{s, B, w_1^{\alpha\lambda}} \leq C|B| diam(B) \|u\|_{s, \rho B, w_2^\alpha}. \tag{7.6.46}'$$

Theorem 7.6.23. *Let* $u \in L^s_{loc}(\Omega, \wedge^l)$, $l = 1, 2, \ldots, n$, $1 < s < \infty$, *be a differential form satisfying (1.2.10) in a bounded, convex domain* $\Omega \subset \mathbf{R}^n$ *and* $T : C^\infty(\Omega, \wedge^l) \to C^\infty(\Omega, \wedge^{l-1})$ *be the homotopy operator defined in (1.5.1). Suppose that* $\rho > 1$ *and* $(w_1, w_2) \in A_{r,\lambda}(\Omega)$ *for some* $\lambda \geq 1$ *and* $1 < r < \infty$. *Then, there exists a constant* C, *independent of* u, *such that*

$$\left(\int_B |H(T(u)) - (H(T(u)))_B|^s w_1^\alpha dx \right)^{1/s}$$
$$\leq C|B| diam(B) \left(\int_{\rho B} |u|^s w_2^\alpha dx \right)^{1/s} \tag{7.6.47}$$

and

$$\|H(T(u)) - (H(T(u)))_B\|_{W^{1,s}(B), w_1^\alpha} \leq C|B| \|u\|_{s, \rho B, w_2^\alpha} \tag{7.6.48}$$

for all balls B *with* $\rho B \subset \Omega$ *and any real number* α *with* $0 < \alpha < \lambda$.

Note that inequality (7.6.47) is equivalent to

$$\|H(T(u)) - (H(T(u)))_B\|_{s, B, w_1^\alpha} \leq C|B| diam(B) \|u\|_{s, \rho B, w_2^\alpha}. \tag{7.6.48}'$$

Now, we discuss the following version of two-weight Poincaré inequalities for differential forms.

Theorem 7.6.24. *Let* $u \in L^s_{loc}(\Omega, \wedge^l)$, $l = 1, 2, \ldots, n$, $1 < s < \infty$, *be a differential form satisfying (1.2.10) in a bounded, convex domain* $\Omega \subset \mathbf{R}^n$ *and*

$T : C^\infty(\Omega, \wedge^l) \to C^\infty(\Omega, \wedge^{l-1})$ *be the homotopy operator defined in (1.5.1).*
Suppose that $(w_1, w_2) \in A_r^\lambda(\Omega)$ *for some* $r > 1$ *and* $\lambda > 0$. *If* $0 < \alpha < 1$ *and*
$\sigma > 1$, *then there exists a constant* C, *independent of* u, *such that*

$$\left(\int_B |H(T(u)) - (H(T(u)))_B|^s w_1^\alpha dx\right)^{\frac{1}{s}}$$
$$\leq C|B|diam(B)\left(\int_{\sigma B} |u|^s w_2^{\alpha\lambda} dx\right)^{\frac{1}{s}} \tag{7.6.49}$$

and

$$\|H(T(u)) - (H(T(u)))_B\|_{W^{1,s}(B), w_1^\alpha} \leq C|B|\|u\|_{s,\sigma B, w_2^{\alpha\lambda}} \tag{7.6.50}$$

for all balls B *with* $\sigma B \subset \Omega$.

Selecting $\lambda = 1/\alpha$ in Theorem 7.6.24, we find the following version of the
A_r^λ-weighted Poincaré inequality.

Corollary 7.6.25. *Let* $u \in L^s_{loc}(\Omega, \wedge^l)$, $l = 1, 2, \ldots, n$, $1 < s < \infty$, *be a*
differential form satisfying (1.2.10) in a bounded, convex domain $\Omega \subset \mathbf{R}^n$ *and*
$T : C^\infty(\Omega, \wedge^l) \to C^\infty(\Omega, \wedge^{l-1})$ *be the homotopy operator defined in (1.5.1).*
Suppose that $(w_1, w_2) \in A_r^{1/\alpha}(\Omega)$ *for some* $r > 1$. *If* $0 < \alpha < 1$ *and* $\sigma > 1$,
then there exists a constant C, *independent of* u, *such that*

$$\left(\int_B |H(T(u)) - (H(T(u)))_B|^s w_1^\alpha dx\right)^{\frac{1}{s}}$$
$$\leq C|B|diam(B)\left(\int_{\sigma B} |u|^s w_2 dx\right)^{\frac{1}{s}} \tag{7.6.51}$$

for all balls B *with* $\sigma B \subset \Omega$.

Now setting $\alpha = 1/s$ in Theorem 7.6.24, we obtain the following two-weight
Poincaré inequality.

Corollary 7.6.26. *Let* $u \in L^s_{loc}(\Omega, \wedge^l)$, $l = 1, 2, \ldots, n$, $1 < s < \infty$, *be a*
differential form satisfying (1.2.10) in a bounded, convex domain $\Omega \subset \mathbf{R}^n$ *and*
$T : C^\infty(\Omega, \wedge^l) \to C^\infty(\Omega, \wedge^{l-1})$ *be the homotopy operator defined in (1.5.1).*
Suppose that $(w_1, w_2) \in A_r^\lambda(\Omega)$ *for some* $r > 1$, $\lambda > 0$, *and* $\sigma > 1$, *then there*
exists a constant C, *independent of* u, *such that*

$$\left(\int_B |H(T(u)) - (H(T(u)))_B|^s w_1^{\frac{1}{s}} dx\right)^{\frac{1}{s}}$$
$$\leq C|B|diam(B)\left(\int_{\sigma B} |u|^s w_2^{\lambda/s} dx\right)^{\frac{1}{s}} \tag{7.6.52}$$

for all balls B *with* $\sigma B \subset \Omega$.

Letting $\lambda = 1$ in Corollary 7.6.26, we have the following symmetric two-weight inequality.

Corollary 7.6.27. *Let $u \in L^s_{loc}(\Omega, \wedge^l)$, $l = 1, 2, \ldots, n$, $1 < s < \infty$, be a differential form satisfying (1.2.10) in a bounded, convex domain $\Omega \subset \mathbf{R}^n$ and $T : C^\infty(\Omega, \wedge^l) \to C^\infty(\Omega, \wedge^{l-1})$ be the homotopy operator defined in (1.5.1). Suppose that $(w_1, w_2) \in A_r(\Omega)$ for some $r > 1$ and $\sigma > 1$, then there exists a constant C, independent of u, such that*

$$\left(\int_B |H(T(u)) - (H(T(u)))_B|^s w_1^{\frac{1}{s}} dx \right)^{\frac{1}{s}}$$

$$\leq C|B| diam(B) \left(\int_{\sigma B} |u|^s w_2^{\frac{1}{s}} dx \right)^{\frac{1}{s}} \tag{7.6.53}$$

for all balls B with $\sigma B \subset \Omega$.

Finally, when $\lambda = s$ in Theorem 7.6.24, we find the following two-weight inequality.

Corollary 7.6.28. *Let $u \in L^s_{loc}(\Omega, \wedge^l)$, $l = 1, 2, \ldots, n$, $1 < s < \infty$, be a differential form satisfying (1.2.10) in a bounded, convex domain $\Omega \subset \mathbf{R}^n$ and $T : C^\infty(\Omega, \wedge^l) \to C^\infty(\Omega, \wedge^{l-1})$ be the homotopy operator defined in (1.5.1). Suppose that $(w_1, w_2) \in A^s_r(\Omega)$ for some $r > 1$. If $0 < \alpha < 1$ and $\sigma > 1$, then there exists a constant C, independent of u, such that*

$$\left(\int_B |H(T(u)) - (H(T(u)))_B|^s w_1^\alpha dx \right)^{\frac{1}{s}}$$

$$\leq C|B| diam(B) \left(\int_{\sigma B} |u|^s w_2^{\alpha s} dx \right)^{\frac{1}{s}} \tag{7.6.54}$$

and

$$\|H(T(u)) - (H(T(u)))_B\|_{W^{1,s}(B), w_1^\alpha} \leq C|B| \|u\|_{s, \sigma B, w_2^{\alpha s}} \tag{7.6.55}$$

for all balls B with $\sigma B \subset \Omega$.

7.6.9 Some global inequalities

From the covering lemma, we can extend the local inequalities stated in this section to the global cases. For example, Theorems 7.6.20, 7.6.21, 7.6.22, and 7.6.23 can be extended to the following Theorems 7.6.29, 7.6.30, 7.6.31, and 7.6.32, respectively.

Theorem 7.6.29. *Let $u \in L^s_{loc}(\Omega, \wedge^1)$, $1 < s < \infty$, be a differential form satisfying the nonhomogeneous A-harmonic equation (1.2.10) in a bounded* domain $\Omega \subset \mathbf{R}^n$ and $T : C^\infty(\Omega, \wedge^l) \to C^\infty(\Omega, \wedge^{l-1})$, $l = 1, 2, \ldots, n$, be the homotopy operator defined in (1.5.1). Assume that $w \in A_r(\lambda, \Omega)$ for some $r > 1$ and $\lambda > 0$. Then, there exists a constant C, independent of u, such that

$$\|H(T(u)) - (H(T(u)))_{B_0}\|_{s,\Omega,w^{\alpha\lambda}} \leq C\|u\|_{s,\Omega,w^\alpha} \tag{7.6.56}$$

and

$$\|H(T(u)) - (H(T(u)))_{B_0}\|_{W^{1,s}(\Omega),\ w^{\alpha\lambda}} \leq C\|u\|_{s,\Omega,w^\alpha} \tag{7.6.57}$$

for all bounded, convex domains Ω and any real number α with $0 < \alpha < 1$. Here $B_0 \subset \Omega$ is a fixed ball.

Theorem 7.6.30. Let $u \in L^s_{loc}(\Omega, \wedge^1)$, $1 < s < \infty$, be a differential form satisfying (1.2.10) in a bounded, convex domain $\Omega \subset \mathbf{R}^n$ and $T : C^\infty(\Omega, \wedge^l) \to C^\infty(\Omega, \wedge^{l-1})$, $l = 1, 2, \ldots, n$, be the homotopy operator defined in (1.5.1). Assume that $w \in A^\lambda_r(\Omega)$ for some $r > 1$ and $\lambda > 0$. Then, there exists a constant C, independent of u, such that

$$\|H(T(u)) - (H(T(u)))_{B_0}\|_{s,\Omega,w^\alpha} \leq C\|u\|_{s,\Omega,w^{\alpha\lambda}} \tag{7.6.58}$$

and

$$\|H(T(u)) - (H(T(u)))_{B_0}\|_{W^{1,s}(\Omega),\ w^\alpha} \leq C\|u\|_{s,\Omega,w^{\alpha\lambda}} \tag{7.6.59}$$

for all bounded domains Ω and any real number α with $0 < \alpha < 1$. Here $B_0 \subset \Omega$ is a fixed ball.

Theorem 7.6.31. Let $u \in L^s_{loc}(\Omega, \wedge^1)$, $1 < s < \infty$, be a solution of the nonhomogeneous A-harmonic equation (1.2.10) in a bounded, convex domain $\Omega \subset \mathbf{R}^n$ and $T : C^\infty(\Omega, \wedge^l) \to C^\infty(\Omega, \wedge^{l-1})$, $l = 1, 2, \ldots, n$, be the homotopy operator defined in (1.5.1). Suppose that $(w_1, w_2) \in A_r(\lambda, \Omega)$ for some $\lambda > 0$ and $1 < r < \infty$. Then, there exists a constant C, independent of u, such that

$$\left(\int_\Omega |H(T(u)) - (H(T(u)))_{B_0}|^s w_1^{\alpha\lambda} dx \right)^{\frac{1}{s}} \leq C \left(\int_\Omega |u|^s w_2^\alpha dx \right)^{\frac{1}{s}} \tag{7.6.60}$$

and

$$\|H(T(u)) - (H(T(u)))_{B_0}\|_{W^{1,s}(\Omega),w_1^{\alpha\lambda}} \leq C\|u\|_{s,\Omega,w_2^\alpha} \tag{7.6.61}$$

for all bounded domains Ω and any real number α with $0 < \alpha < 1$. Here $B_0 \subset \Omega$ is a fixed ball.

Theorem 7.6.32. Let $u \in L^s_{loc}(\Omega, \wedge^1)$, $1 < s < \infty$, be a differential form satisfying (1.2.10) in a bounded, convex domain $\Omega \subset \mathbf{R}^n$ and $T : C^\infty(\Omega, \wedge^l) \to C^\infty(\Omega, \wedge^{l-1})$, $l = 1, 2, \ldots, n$, be the homotopy operator defined in (1.5.1). Suppose that $(w_1, w_2) \in A_{r,\lambda}(\Omega)$ for some $\lambda \geq 1$ and

$1 < r < \infty$. Then, there exists a constant C, independent of u, such that

$$\left(\int_\Omega |H(T(u)) - (H(T(u)))_{B_0}|^s w_1^\alpha dx \right)^{\frac{1}{s}} \le C \left(\int_\Omega |u|^s w_2^\alpha dx \right)^{\frac{1}{s}} \quad (7.6.62)$$

and

$$\|H(T(u)) - (H(T(u)))_{B_0}\|_{W^{1,s}(\Omega), w_1^\alpha} \le C\|u\|_{s,\Omega,w_2^\alpha}$$

for all bounded domains Ω and any real number α with $0 < \alpha < 1$. Here $B_0 \subset \Omega$ is a fixed ball.

7.7 Compositions of three operators

In this section, we discuss the different compositions generated by three basic operators, the homotopy operator T, the differential operator d, and the projection operator H. For a smooth differential form u, it is easy to see that $d(H(T(u)))$ and $d(T(H(u)))$ are closed. Thus, the Poincaré-type estimates for these compositions are trivial and hence we will not discuss them here.

7.7.1 Basic estimates for $T \circ d \circ H$

First, we prove the following unweighted Poincaré-type inequality for $T \circ d \circ H$. This elementary result forms the bases for the weighted Poincaré inequalities for $T \circ d \circ H$ which will be developed later in this chapter.

Theorem 7.7.1. Let $u \in L_{loc}^s(\Omega, \wedge^l)$, $l = 1, 2, \ldots, n-1$, $1 < s < \infty$, be a smooth differential form in a bounded, convex domain Ω, H be the projection operator, and $T : C^\infty(\Omega, \wedge^l) \to C^\infty(\Omega, \wedge^{l-1})$ be the homotopy operator defined in (1.5.1). Then, there exists a constant C, independent of u, such that

$$\|T(d(H(u))) - (T(d(H(u))))_B\|_{s,B} \le C|B|diam(B)\|du\|_{s,B} \quad (7.7.1)$$

for all balls B with $B \subset \Omega$.

Proof. From (7.6.4) and Theorem 3.3.2, we have

$$\|d\Delta G(u)\|_{s,B} \le C_1\|du\|_{s,B}. \quad (7.7.2)$$

Also, for any differential form u, $d(du) = 0$, and hence $d(d(H(u))) = 0$. Thus, from (7.6.4) and (7.7.2), we find that

$$\|T(d(H(u))) - (T(d(H(u))))_B\|_{s,B}$$

$$= \|Td(TdH(u))\|_{s,B}$$

$$\leq C_2|B|diam(B)\|dTdH(u)\|_{s,B}$$

$$\leq C_2|B|diam(B)(\|dH(u)\|_{s,B}$$

$$+C_3|B|diam(B)\|d(d(H(u)))\|_{s,B})$$

$$\leq C_2|B|diam(B)(\|dH(u)\|_{s,B} + 0)$$

$$\leq C_2|B|diam(B)\|dH(u)\|_{s,B} \qquad (7.7.3)$$

$$\leq C_2|B|diam(B)\|H(du)\|_{s,B}$$

$$= C_2|B|diam(B)\|du - \Delta G(du)\|_{s,B}$$

$$\leq C_2|B|diam(B)(\|du\|_{s,B} + \|\Delta G(du)\|_{s,B})$$

$$\leq C_2|B|diam(B)(\|du\|_{s,B} + C_3\|du\|_{s,B})$$

$$\leq C_4|B|diam(B)\|du\|_{s,B},$$

which implies that inequality (7.7.1) holds.   ∎

Note that the differential form u in Theorem 7.7.1 is not necessarily a solution of the A-harmonic equation. If u is a solution of the A-harmonic equation, then we can employ the Caccioppoli inequality, to obtain the following corollary.

Corollary 7.7.2. *Let $u \in L^s_{loc}(\Omega, \wedge^l)$, $l = 1, 2, \ldots, n$, $1 < s < \infty$, be a solution of the A-harmonic equation (1.2.10) in a bounded, convex domain Ω, H be the projection operator, and $T : C^\infty(\Omega, \wedge^l) \rightarrow C^\infty(\Omega, \wedge^{l-1})$ be the homotopy operator defined in (1.5.1). Then, there exists a constant C, independent of u, such that*

$$\|T(d(H(u))) - (T(d(H(u))))_B\|_{s,B} \leq C|B|diam(B)\|u\|_{s,\rho B} \qquad (7.7.4)$$

for all balls B with $\rho B \subset \Omega$, where $\rho > 1$ is a constant.

Since d is commutative with G and H, respectively, it follows that $HdT(u)$ and $GdT(u)$ are closed. Specifically, we have the following corollary.

Corollary 7.7.3. *Let $u \in L^s_{loc}(\Omega, \wedge^l)$, $l = 1, 2, \ldots, n$, be any smooth differential form in a bounded, convex domain Ω, H be the projection operator, G be Green's operator, and $T : C^\infty(\Omega, \wedge^l) \rightarrow C^\infty(\Omega, \wedge^{l-1})$ be the homotopy operator defined in (1.5.1). Then, $HdT(u)$ and $GdT(u)$ are closed forms.*

7.7.2 $A_r(\Omega)$-weighted inequalities for $T \circ d \circ H$

Based on the elementary inequality established above, we can now state the following $A_r(\Omega)$-weighted Poincaré inequalities.

Theorem 7.7.4. *Let $u \in L^s_{loc}(\Omega, \wedge^l)$, $l = 1, 2, \ldots, n$, $1 < s < \infty$, be a solution of the nonhomogeneous A-harmonic equation (1.2.10) in a bounded, convex domain Ω and $T : C^\infty(\Omega, \wedge^l) \to C^\infty(\Omega, \wedge^{l-1})$ be the homotopy operator defined in (1.5.1). Assume that $\rho > 1$ and $w \in A_r(\Omega)$ for some $1 < r < \infty$. Then, there exists a constant C, independent of u, such that*

$$\|T(d(H(u)) - (T(d(H(u))))_B\|_{s,B,w^\alpha} \le C|B|diam(B)\|u\|_{s,\rho B,w^\alpha} \quad (7.7.5)$$

for all balls B with $\rho B \subset \Omega$ and any real number α with $0 < \alpha \le 1$.

Note that the above L^s-norm inequality can also be written as

$$\left(\int_B |T(d(H(u))) - (T(d(H(u))))_B|^s w^\alpha dx \right)^{1/s} \le C \left(\int_{\rho B} |u|^s w^\alpha dx \right)^{1/s}.$$
$$(7.7.6)$$

Theorem 7.7.5. *Let $u \in D'(\Omega, \wedge^1)$ be a solution of the nonhomogeneous A-harmonic equation (1.2.10) and $T : C^\infty(\Omega, \wedge^l) \to C^\infty(\Omega, \wedge^{l-1})$, $l = 1, 2, \ldots, n$, be the homotopy operator defined in (1.5.1). Assume that $w \in A_r(\Omega)$ for some $1 < r < \infty$ and s is a fixed exponent associated with the A-harmonic equation (1.2.10). Then, there exists a constant C, independent of u, such that*

$$\left(\int_\Omega |T(d(H(u))) - (T(d(H(u))))_{Q_0}|^s w dx \right)^{1/s} \le C \left(\int_\Omega |u|^s w dx \right)^{1/s}$$
$$(7.7.7)$$

for any bounded, convex δ-John domain $\Omega \subset \mathbf{R}^n$. Here $Q_0 \subset \Omega$ is a fixed cube.

Using Theorem 7.7.4 and the definition of the $\|\cdot\|_{W^{1,s}(E),w^\alpha}$-norm, we can prove the following local and global imbedding inequalities.

Theorem 7.7.6. *Let $u \in L^s_{loc}(\Omega, \wedge^l)$, $l = 1, 2, \ldots, n$, $1 < s < \infty$, be a solution of the nonhomogeneous A-harmonic equation (1.2.10) in a bounded, convex domain Ω and $T : C^\infty(\Omega, \wedge^l) \to C^\infty(\Omega, \wedge^{l-1})$ be the homotopy operator defined in (1.5.1). Assume that $\rho > 1$ and $w \in A_r(\Omega)$ for some $1 < r < \infty$. Then, there exists a constant C, independent of u, such that*

$$\|T(d(H(u))) - (T(d(H(u))))_B\|_{W^{1,s}(B),w^\alpha}$$

$$\leq C|B|\|u\|_{s,\rho B,w^\alpha} \qquad (7.7.8)$$

for all balls B with $\rho B \subset \Omega$ and any real number α with $0 < \alpha \leq 1$.

Theorem 7.7.7. *Let $u \in D'(\Omega, \wedge^1)$ be a solution of the nonhomogeneous A-harmonic equation (1.2.10) and $T : C^\infty(\Omega, \wedge^l) \to C^\infty(\Omega, \wedge^{l-1})$, $l = 1, 2, \ldots, n$, be the homotopy operator defined in (1.5.1). Assume that $w \in A_r(\Omega)$ for some $1 < r < \infty$ and s is a fixed exponent associated with the A-harmonic equation (1.2.10), $r < s < \infty$. Then, there exists a constant C, independent of u, such that*

$$\|T(d(H(u))) - (T(d(H(u))))_{Q_0}\|_{W^{1,s}(\Omega),w} \leq C\|u\|_{s,\Omega,w} \qquad (7.7.9)$$

for any bounded, convex δ-John domain $\Omega \subset \mathbf{R}^n$. Here $Q_0 \subset \Omega$ is a fixed cube.

7.7.3 Cases of other weights

So far, we have presented the $A_r(\Omega)$-weighted Poincaré-type estimates for the composition $T \circ d \circ H$. Now, we state other estimates with different weights, namely $A_r(\lambda, \Omega)$-weights and $A_r^\lambda(\Omega)$-weights.

Theorem 7.7.8. *Let $u \in L^s_{loc}(\Omega, \wedge^l)$, $l = 1, 2, \ldots, n$, $1 < s < \infty$, be a differential form satisfying the nonhomogeneous A-harmonic equation (1.2.10) in a bounded, convex domain $\Omega \subset \mathbf{R}^n$ and $T : C^\infty(\Omega, \wedge^l) \to C^\infty(\Omega, \wedge^{l-1})$ be the homotopy operator defined in (1.5.1). Assume that $w \in A_r(\lambda, \Omega)$ for some $r > 1$ and $\lambda > 0$. Then, there exists a constant C, independent of u, such that*

$$\|T(d(H(u))) - (T(d(H(u))))_B\|_{s,B,w^{\alpha\lambda}}$$

$$\leq C|B|diam(B)\|u\|_{s,\rho B,w^\alpha} \qquad (7.7.10)$$

and

$$\|T(d(H(u))) - (T(d(H(u))))_B\|_{W^{1,s}(B),\ w^{\alpha\lambda}}$$

$$\leq C|B|\|u\|_{s,\rho B,w^\alpha} \qquad (7.7.11)$$

for all balls B with $\rho B \subset \Omega$ and any real number α with $0 < \alpha < 1$. Here $\rho > 1$ is some constant.

The above inequality (7.6.10) can be written as

$$\left(\int_B |T(d(H(u))) - (T(d(H(u))))_B|^s w^{\alpha\lambda} dx \right)^{\frac{1}{s}}$$

$$\leq C|B| diam(B) \left(\int_{\rho B} |u|^s w^\alpha dx \right)^{\frac{1}{s}}. \tag{7.7.12}$$

Theorem 7.7.9. *Let* $u \in L^s_{loc}(\Omega, \wedge^l)$, $l = 1, 2, \ldots, n$, $1 < s < \infty$, *be a differential form satisfying* (1.2.10) *in a bounded, convex domain* $\Omega \subset \mathbf{R}^n$ *and* $T : C^\infty(\Omega, \wedge^l) \to C^\infty(\Omega, \wedge^{l-1})$ *be the homotopy operator defined in* (1.5.1). *Assume that* $\rho > 1$ *and* $w \in A_r^\lambda(\Omega)$ *for some* $r > 1$ *and* $\lambda > 0$. *Then, there exists a constant* C, *independent of* u, *such that*

$$\|T(d(H(u))) - (T(d(H(u))))_B\|_{s,B,w^\alpha} \leq C|B| diam(B) \|u\|_{s,\rho B,w^{\alpha\lambda}} \tag{7.7.13}$$

and

$$\|T(d(H(u))) - (T(d(H(u))))_B\|_{W^{1,s}(B),\ w^\alpha}$$

$$\leq C|B| \|u\|_{s,\rho B,w^{\alpha\lambda}} \tag{7.7.14}$$

for all balls B *with* $\rho B \subset \Omega$ *and any real number* α *with* $0 < \alpha < 1$.

The integral version of inequality (7.7.13) can be written as

$$\left(\int_B |T(d(H(u))) - (T(d(H(u))))_B|^s w^\alpha dx \right)^{\frac{1}{s}}$$

$$\leq C|B| diam(B) \left(\int_{\rho B} |u|^s w^{\alpha\lambda} dx \right)^{\frac{1}{s}}. \tag{7.7.15}$$

7.7.4 Cases of two-weights

In our next several theorems, we provide the two-weight estimates for this composition of the operators. First, we state the following estimates with weights $(w_1, w_2) \in A_r(\lambda, \Omega)$.

Theorem 7.7.10. *Let* $u \in L^s_{loc}(\Omega, \wedge^l)$, $l = 1, 2, \ldots, n$, $1 < s < \infty$, *be a solution of the nonhomogeneous A-harmonic equation* (1.2.10) *in a bounded, convex domain* $\Omega \subset \mathbf{R}^n$ *and* $T : C^\infty(\Omega, \wedge^l) \to C^\infty(\Omega, \wedge^{l-1})$ *be the homotopy operator defined in* (1.5.1). *Assume that* $\rho > 1$ *and* $(w_1, w_2) \in A_r(\lambda, \Omega)$ *for some* $r > 1$ *and* $\lambda > 0$. *Then, there exists a constant* C, *independent of* u, *such that*

$$\left(\int_B |T(d(H(u))) - (T(d(H(u))))_B|^s w_1^{\alpha\lambda} dx \right)^{1/s}$$

$$\leq C|B| diam(B) \left(\int_{\rho B} |u|^s w_2^\alpha dx \right)^{1/s} \tag{7.7.16}$$

and

$$\|T(d(H(u))) - (T(d(H(u))))_B\|_{W^{1,s}(B),w_1^{\alpha\lambda}}$$

$$\leq C|B|\|u\|_{s,\rho B,w_2^{\alpha}} \tag{7.7.17}$$

for all balls B with $\rho B \subset \Omega$ and any real number α with $0 < \alpha < 1$.

Inequality (7.7.16) can be written as

$$\|T(d(H(u))) - (T(d(H(u))))_B\|_{s,B,w_1^{\alpha\lambda}} \leq C|B|diam(B)\|u\|_{s,\rho B,w_2^{\alpha}}. \tag{7.7.17$'$}$$

Theorem 7.7.11. *Let $u \in L_{loc}^s(\Omega, \wedge^l)$, $l = 1, 2, \ldots, n$, $1 < s < \infty$, be a differential form satisfying (1.2.10) in a bounded, convex domain $\Omega \subset \mathbf{R}^n$ and $T : C^\infty(\Omega, \wedge^l) \to C^\infty(\Omega, \wedge^{l-1})$ be the homotopy operator defined in (1.5.1). Suppose that $\rho > 1$ and $(w_1, w_2) \in A_{r,\lambda}(\Omega)$ for some $\lambda \geq 1$ and $1 < r < \infty$. Then, there exists a constant C, independent of u, such that*

$$\left(\int_B |T(d(H(u))) - (T(d(H(u))))_B|^s w_1^\alpha dx\right)^{\frac{1}{s}}$$

$$\leq C|B|diam(B) \left(\int_{\rho B} |u|^s w_2^\alpha dx\right)^{\frac{1}{s}} \tag{7.7.18}$$

and

$$\|T(d(H(u))) - (T(d(H(u))))_B\|_{W^{1,s}(B),w_1^{\alpha}}$$

$$\leq C|B|\|u\|_{s,\rho B,w_2^{\alpha}} \tag{7.7.19}$$

for all balls B with $\rho B \subset \Omega$ and any real number α with $0 < \alpha < \lambda$.

Note that (7.7.18) is equivalent to

$$\|T(d(H(u))) - (T(d(H(u))))_B\|_{s,B,w_1^{\alpha}} \leq C|B|diam(B)\|u\|_{s,\rho B,w_2^{\alpha}}. \tag{7.7.19$'$}$$

Now, we state the following versions of two-weight Poincaré inequalities for differential forms.

Theorem 7.7.12. *Let $u \in L_{loc}^s(\Omega, \wedge^l)$, $l = 1, 2, \ldots, n$, $1 < s < \infty$, be a differential form satisfying (1.2.10) in a bounded, convex domain $\Omega \subset \mathbf{R}^n$ and $T : C^\infty(\Omega, \wedge^l) \to C^\infty(\Omega, \wedge^{l-1})$ be the homotopy operator defined in (1.5.1). Suppose that $(w_1, w_2) \in A_r^\lambda(\Omega)$ for some $r > 1$ and $\lambda > 0$. If $0 < \alpha < 1$ and $\sigma > 1$, then there exists a constant C, independent of u, such that*

$$\left(\int_B |T(d(H(u))) - (T(d(H(u))))_B|^s w_1^\alpha dx\right)^{\frac{1}{s}}$$

$$\leq C|B|diam(B) \left(\int_{\sigma B} |u|^s w_2^{\alpha\lambda} dx\right)^{\frac{1}{s}} \tag{7.7.20}$$

and

$$\|T(d(H(u))) - (T(d(H(u))))_B\|_{W^{1,s}(B),w_1^\alpha}$$

$$\leq C|B|\|u\|_{s,\sigma B,w_2^{\alpha\lambda}} \tag{7.7.21}$$

for all balls B with $\sigma B \subset \Omega$.

If we choose $\lambda = 1/\alpha$ in Theorem 7.7.12, we have the following version of the A_r^λ-weighted Poincaré inequality.

Corollary 7.7.13. *Let $u \in L_{loc}^s(\Omega, \wedge^l)$, $l = 1, 2, \ldots, n$, $1 < s < \infty$, be a differential form satisfying (1.2.10) in a bounded, convex domain $\Omega \subset \mathbf{R}^n$ and $T : C^\infty(\Omega, \wedge^l) \to C^\infty(\Omega, \wedge^{l-1})$ be the homotopy operator defined in (1.5.1). Suppose that $(w_1, w_2) \in A_r^{1/\alpha}(\Omega)$ for some $r > 1$. If $0 < \alpha < 1$ and $\sigma > 1$, then there exists a constant C, independent of u, such that*

$$\left(\int_B |T(d(H(u))) - (T(d(H(u))))_B|^s w_1^\alpha dx\right)^{\frac{1}{s}}$$

$$\leq C|B|diam(B)\left(\int_{\sigma B} |u|^s w_2 dx\right)^{\frac{1}{s}} \tag{7.7.22}$$

for all balls B with $\sigma B \subset \Omega$.

Choosing $\alpha = 1/s$ in Theorem 7.7.12, we find the following two-weight Poincaré inequality.

Corollary 7.7.14. *Let $u \in L_{loc}^s(\Omega, \wedge^l)$, $l = 1, 2, \ldots, n$, $1 < s < \infty$, be a differential form satisfying (1.2.10) in a bounded, convex domain $\Omega \subset \mathbf{R}^n$ and $T : C^\infty(\Omega, \wedge^l) \to C^\infty(\Omega, \wedge^{l-1})$ be the homotopy operator defined in (1.5.1). Suppose that $(w_1, w_2) \in A_r^\lambda(\Omega)$ for some $r > 1$, $\lambda > 0$, and $\sigma > 1$, then there exists a constant C, independent of u, such that*

$$\left(\int_B |T(d(H(u))) - (T(d(H(u))))_B|^s w_1^{\frac{1}{s}} dx\right)^{\frac{1}{s}}$$

$$\leq C|B|diam(B)\left(\int_{\sigma B} |u|^s w_2^{\lambda/s} dx\right)^{\frac{1}{s}} \tag{7.7.23}$$

for all balls B with $\sigma B \subset \Omega$.

Let $\lambda = 1$ in Corollary 7.7.14, to obtain the following symmetric two-weight inequality.

Corollary 7.7.15. *Let $u \in L_{loc}^s(\Omega, \wedge^l)$, $l = 1, 2, \ldots, n$, $1 < s < \infty$, be a differential form satisfying (1.2.10) in a bounded, convex domain $\Omega \subset \mathbf{R}^n$ and $T : C^\infty(\Omega, \wedge^l) \to C^\infty(\Omega, \wedge^{l-1})$ be the homotopy operator defined in (1.5.1).*

Suppose that $(w_1, w_2) \in A_r(\Omega)$ for some $r > 1$ and $\sigma > 1$, then there exists a constant C, independent of u, such that

$$\left(\int_B |T(d(H(u))) - (T(d(H(u))))_B|^s w_1^{\frac{1}{s}} dx\right)^{\frac{1}{s}}$$

$$\leq C|B|diam(B) \left(\int_{\sigma B} |u|^s w_2^{\frac{1}{s}} dx\right)^{\frac{1}{s}}$$

(7.7.24)

for all balls B with $\sigma B \subset \Omega$.

Finally, when $\lambda = s$ in Theorem 7.6.12, we obtain the following two-weight inequality.

Corollary 7.7.16. Let $u \in L^s_{loc}(\Omega, \wedge^l)$, $l = 1, 2, \ldots, n$, $1 < s < \infty$, be a differential form satisfying (1.2.10) in a bounded, convex domain $\Omega \subset \mathbf{R}^n$ and $T : C^\infty(\Omega, \wedge^l) \to C^\infty(\Omega, \wedge^{l-1})$ be the homotopy operator defined in (1.5.1). Suppose that $(w_1, w_2) \in A_r^s(\Omega)$ for some $r > 1$. If $0 < \alpha < 1$ and $\sigma > 1$, then there exists a constant C, independent of u, such that

$$\left(\int_B |T(d(H(u))) - (T(d(H(u))))_B|^s w_1^\alpha dx\right)^{\frac{1}{s}}$$

$$\leq C|B|diam(B) \left(\int_{\sigma B} |u|^s w_2^{\alpha s} dx\right)^{\frac{1}{s}}$$

(7.7.25)

and

$$\|T(d(H(u))) - (T(d(H(u))))_B\|_{W^{1,s}(B), w_1^\alpha}$$

$$\leq C|B|\|u\|_{s, \sigma B, w_2^{\alpha s}}$$

(7.7.26)

for all balls B with $\sigma B \subset \Omega$.

7.7.5 Estimates for $T \circ H \circ d$

Theorem 7.7.17. Let $u \in L^s_{loc}(\Omega, \wedge^l)$, $l = 1, 2, \ldots, n-1$, $1 < s < \infty$, be a smooth differential form in a bounded, convex domain Ω, H be the projection operator, and $T : C^\infty(\Omega, \wedge^l) \to C^\infty(\Omega, \wedge^{l-1})$ be the homotopy operator defined in (1.5.1). Then, there exists a constant C, independent of u, such that

$$\|T(H(du)) - (T(H(du)))_B\|_{s,B} \leq C|B|diam(B)\|du\|_{s,B}$$

(7.7.27)

for all balls B with $B \subset \Omega$.

Proof. Let H be the projection operator and T be the homotopy operator. Then, for any differential form u, we have

$$\|T(H(du)) - (T(H(du)))_B\|_{s,B}$$

$$= \|Td(THd(u))\|_{s,B}$$

$$\leq C_1|B|diam(B)\|d(THd(u))\|_{s,B}$$

$$\leq C_1|B|diam(B)(\|Hd(u)\|_{s,B}$$

$$+C_2|B|diam(B)\|d(Hd(u))\|_{s,B}) \tag{7.7.28}$$

$$\leq C_1|B|diam(B)(\|Hd(u)\|_{s,B} + 0)$$

$$\leq C_1|B|diam(B)\|Hd(u)\|_{s,B}$$

$$\leq C_2|B|diam(B)\|du\|_{s,B},$$

and hence inequality (7.7.27) holds. ∎

Using the Caccioppoli inequality and Theorem 7.7.17, we have the following corollary immediately.

Corollary 7.7.18. *Let $u \in L_{loc}^s(\Omega, \wedge^l)$, $l = 1, 2, \ldots, n$, $1 < s < \infty$, be a differential form satisfying (1.2.10) in a bounded, convex domain Ω, H be the projection operator, and $T : C^\infty(\Omega, \wedge^l) \to C^\infty(\Omega, \wedge^{l-1})$ be the homotopy operator defined in (1.5.1). Then, there exists a constant C, independent of u, such that*

$$\|T(H(du)) - (T(H(du)))_B\|_{s,B} \leq C|B|\|u - c\|_{s,B} \tag{7.7.29}$$

for all balls B with $B \subset \Omega$, where c is any closed form.

7.7.6 Estimates for $H \circ T \circ d$

Theorem 7.7.19. *Let $u \in L_{loc}^s(\Omega, \wedge^l)$, $l = 1, 2, \ldots, n$, $1 < s < \infty$, be a smooth differential form in a bounded, convex domain Ω, H be the projection operator, and $T : C^\infty(\Omega, \wedge^l) \to C^\infty(\Omega, \wedge^{l-1})$ be the homotopy operator defined in (1.5.1). Then, there exists a constant C, independent of u, such that*

$$\|HT(du) - (HT(du))_B\|_{s,B} \leq C|B|diam(B)\|du\|_{s,B} \tag{7.7.30}$$

for all balls B with $B \subset \Omega$.

Proof. By the decomposition theorem and Theorem 3.3.2, we have

$$\|HT(du) - (HT(du))_B\|_{s,B}$$

$$= \|Td(HT(du))\|_{s,B}$$

$$\leq C_1|B|diam(B)\|d(HT(du))\|_{s,B}$$

$$\leq C_1|B|diam(B)\|HdT(du)\|_{s,B}$$

$$\leq C_1|B|diam(B)\|dT(du) - \Delta GdT(du)\|_{s,B} \qquad (7.7.31)$$

$$\leq C_1|B|diam(B)(\|dT(du)\|_{s,B} + \|\Delta GdT(du)\|_{s,B})$$

$$\leq C_1|B|diam(B)(\|dT(du)\|_{s,B} + \|dT(du)\|_{s,B})$$

$$\leq C_2|B|diam(B)\|dT(du)\|_{s,B}.$$

Also, from (3.3.8), we have

$$\|dT(du)\|_{s,B}$$

$$\leq \|du\|_{s,B} + C_3|B|diam(B)\|d(du)\|_{s,B}$$

$$\leq \|du\|_{s,B} + 0 \qquad (7.7.32)$$

$$\leq \|du\|_{s,B}.$$

Combining (7.7.31) and (7.7.32), we obtain

$$\|HT(du) - (HT(du))_B\|_{s,B} \leq C_2|B|diam(B)\|du\|_{s,B}. \qquad \blacksquare$$

7.8 The maximal operators

In this section, we first develop some estimates related to the Hardy–Littlewood maximal operator $\mathbb{M}_s$ and the sharp maximal operator $\mathbb{M}_s^\sharp$, and then study the relationship between $\|\mathbb{M}_s^\sharp\|_{s,\Omega}$ and $\|\mathbb{M}_s\|_{s,\Omega}$. We also discuss the compositions of these operators with other operators.

7.8.1 Global L^s-estimates

We begin this section with the following global estimate for the composition of the sharp maximal operator and the homotopy operator.

Theorem 7.8.1. *Let* $u \in L^s(\Omega, \wedge^l)$, $l = 1, 2, \ldots, n$, $1 < s < \infty$, *be a differential form satisfying* (1.2.10) *in a bounded, convex domain* Ω, $\mathbb{M}_s^\sharp$ *be the sharp maximal operator defined in* (7.1.2), *and* T *be the homotopy operator defined in* (1.5.1). *Then,*

$$\|\mathbb{M}_s^\sharp(Tu)\|_{s,\Omega} \le C\|u\|_{s,\Omega} \tag{7.8.1}$$

for some constant C, independent of u.

Proof. Using Theorem 7.5.1 over the ball $B(x,r)$, we have

$$\left(\tfrac{1}{|B(x,r)|} \int_{B(x,r)} |T(u) - (T(u))_{B(x,r)}|^s dy\right)^{1/s}$$

$$\le C_1 |B(x,r)|^{1-1/s} diam(B(x,r)) \left(\int_{\rho B(x,r)} |u|^s dy\right)^{1/s} \tag{7.8.2}$$

$$\le C_2 |\Omega|^{1-1/s} diam(\Omega) \|u\|_{s,\Omega}$$

$$\le C_3 \|u\|_{s,\Omega}$$

since $1 - 1/s > 0$ and Ω is bounded. Thus, it follows that

$$\sup_{r>0} \left(\frac{1}{|B(x,r)|} \int_{B(x,r)} |T(u) - (T(u))_{B(x,r)}|^s dy\right)^{1/s} \le C_3 \|u\|_{s,\Omega}. \tag{7.8.3}$$

Now, from (7.8.2) and (7.8.3), and using the definition of $\mathbb{M}_s^\sharp$, we obtain

$$\|\mathbb{M}_s^\sharp(Tu)\|_{s,\Omega}$$

$$= \left(\int_\Omega |\mathbb{M}_s^\sharp(Tu)|^s dx\right)^{1/s}$$

$$= \left(\int_\Omega \left|\sup_{r>0} \left(\tfrac{1}{|B(x,r)|} \int_{B(x,r)} |T(u) - (T(u))_{B(x,r)}|^s dy\right)^{1/s}\right|^s dx\right)^{1/s}$$

$$\le \left(\int_\Omega |C_3\|u\|_{s,\Omega}|^s dx\right)^{1/s}$$

$$\le C_4 \|u\|_{s,\Omega},$$

and hence

$$\|\mathbb{M}_s^\sharp(Tu)\|_{s,\Omega} \le C_4 \|u\|_{s,\Omega}. \qquad \blacksquare$$

Using Poincaré-type estimates obtained in the previous sections, we can obtain some other similar estimates. For example, applying Theorem 7.4.1 and the same method developed in the proof of Theorem 7.8.1, we have

$$\|\mathbb{M}_s^\sharp(T(G(u)))\|_{s,\Omega} \le C\|u\|_{s,\Omega}. \tag{7.8.4}$$

From [204], we know that if $u \in L^s(\Omega, \wedge^l)$, $1 < s < \infty$, then $\mathbb{M}(u) \in L^s(\Omega)$; specifically, we have the following theorem.

Theorem 7.8.2. Let $u \in L^s(\Omega, \wedge^l)$, $l = 0, 1, 2, \ldots, n$, $1 < s < \infty$, be a differential form in a bounded, convex domain Ω, $\mathbb{M}$ be the Hardy–Littlewood maximal operator defined in (7.1.1), and T be the homotopy operator defined in (1.5.1). Then,

$$\|\mathbb{M}u\|_{s,\Omega} \leq C\|u\|_{s,\Omega} \tag{7.8.5}$$

for some constant C, independent of u.

Note that (7.8.5) holds for any differential form in a bounded, convex domain Ω. Thus, we can replace u in (7.8.5) by $G(u)$, $H(u)$, and $T(u)$, respectively. Then, using the basic estimates for these operators, we find that

$$\|\mathbb{M}G(u)\|_{s,\Omega} \leq C_1\|u\|_{s,\Omega}, \tag{7.8.6}$$

$$\|\mathbb{M}H(u)\|_{s,\Omega} \leq C_2\|u\|_{s,\Omega}, \tag{7.8.7}$$

and

$$\|\mathbb{M}T(u)\|_{s,\Omega} \leq C_3\|u\|_{s,\Omega} \tag{7.8.8}$$

hold for any bounded domain Ω.

7.8.2 The norm comparison theorem

Theorem 7.8.3. Let $u \in L^s_{loc}(\Omega, \wedge^l)$, $l = 0, 1, 2, \ldots, n-1$, $1 < s < \infty$, be a smooth differential form in a bounded domain Ω, $\mathbb{M}$ be the Hardy–Littlewood maximal operator defined in (7.1.1), and $\mathbb{M}^\sharp_s$ be the sharp maximal operator defined in (7.1.2). Then,

$$\|\mathbb{M}^\sharp_s u\|_{s,\Omega} \leq C\|\mathbb{M}_s du\|_{s,\Omega} \tag{7.8.9}$$

for some constant C, independent of u.

Proof. For any ball $B(x,r) \subset \Omega$ with radius r, centered at $x \in \Omega$, using the basic Poincaré inequality (Theorem 3.2.3), we have

$$\left(\int_{B(x,r)} |u(y) - (u(y))_{B(x,r)}|^s dy\right)^{1/s} \leq C_1 |B(x,r)|^{1/n} \left(\int_{B(x,r)} |du|^s dy\right)^{1/s}$$

that is,

$$\left(\frac{1}{|B(x,r)|} \int_{B(x,r)} |u(y) - (u(y))_{B(x,r)}|^s dy\right)^{1/s}$$
$$\leq C_1 |B(x,r)|^{1/n} \left(\frac{1}{|B(x,r)|} \int_{B(x,r)} |du|^s dy\right)^{1/s}. \tag{7.8.10}$$

Now, by definitions of the Hardy–Littlewood maximal operator $\mathbb{M}_s$ and the sharp maximal operator $\mathbb{M}_s^\sharp$, and (7.8.10), we obtain

$$\|\mathbb{M}_s^\sharp(u)\|_{s,\Omega}$$

$$= \left(\int_\Omega |\mathbb{M}_s^\sharp(u)|^s dx\right)^{1/s}$$

$$= \left(\int_\Omega \left|\sup_{r>0}\left(\frac{1}{|B(x,r)|}\int_{B(x,r)}|u(y)-(u(y))_{B(x,r)}|^s dy\right)^{1/s}\right|^s dx\right)^{1/s}$$

$$\leq \left(\int_\Omega \left|\sup_{r>0} C_1|B(x,r)|^{1/n}\left(\frac{1}{|B(x,r)|}\int_{B(x,r)}|du(y)|^s dy\right)^{1/s}\right|^s dx\right)^{1/s}$$

$$\leq \left(\int_\Omega \left|\sup_{r>0} C_1|\Omega|^{1/n}\left(\frac{1}{|B(x,r)|}\int_{B(x,r)}|du(y)|^s dy\right)^{1/s}\right|^s dx\right)^{1/s}$$

$$\leq C_2\left(\int_\Omega \left|\sup_{r>0}\left(\frac{1}{|B(x,r)|}\int_{B(x,r)}|du(y)|^s dy\right)^{1/s}\right|^s dx\right)^{1/s}$$

$$\leq C_2\left(\int_\Omega |\mathbb{M}_s du(x)|^s dx\right)^{1/s}$$

$$= C_3\|\mathbb{M}_s du\|_{s,\Omega},$$

which is equivalent to

$$\|\mathbb{M}_s^\sharp u\|_{s,\Omega} \leq C_3\|\mathbb{M}_s du\|_{s,\Omega}. \quad \blacksquare$$

7.8.3 The fractional maximal operator

Next, following the idea given in [126], we introduce the fractional maximal operator of order α. Let $u(y)$ be a locally L^s-integrable form, $1 \leq s < \infty$, and α be a real number. We define the *fractional maximal operator* $\mathbb{M}_{s,\alpha}$ of order α by

$$\mathbb{M}_{s,\alpha}u(x) = \sup_{r>0}\left(\frac{1}{|B(x,r)|^{1+\alpha/n}}\int_{B(x,r)}|u(y)|^s dy\right)^{1/s}. \qquad (7.8.11)$$

Clearly, (7.8.11) for $\alpha = 0$ reduces to the Hardy–Littlewood maximal operator, and hence, we write $\mathbb{M}_s(u) = \mathbb{M}_{s,0}(u)$.

Theorem 7.8.4. *Let $u \in L^s(\Omega, \wedge^l)$, $l = 0,1,2,\ldots,n$, $1 < s < \infty$, be a smooth differential form satisfying equation (1.2.10) in a bounded domain Ω, $\mathbb{M}_s$ be the Hardy–Littlewood maximal operator defined in (7.1.1), and $\mathbb{M}_{s,\alpha}$ be the fractional maximal operator of order α. Then,*

$$\|\mathbf{M}_s du(x)\|_{s,\Omega} \leq C \|\mathbf{M}_{s,\alpha}(u(x) - c)\|_{s,\Omega} \tag{7.8.12}$$

for some constant C, independent of u, where $\alpha = s$ and c is any closed form.

Proof. Let $B(x,r) \subset \Omega$ be a ball with radius r and center at $x \in \Omega$. From Theorem 4.3.1 (Caccioppoli inequality), we have

$$\left(\int_{B(x,r)} |du(y)|^s dy \right)^{1/s} \leq C_1 |B(x,r)|^{-1/n} \left(\int_{B(x,r)} |u(y) - c|^s dy \right)^{1/s},$$

that is,

$$\begin{aligned}
&\left(\tfrac{1}{|B(x,r)|} \int_{B(x,r)} |du(y)|^s dy \right)^{1/s} \\
&\leq C_1 |B(x,r)|^{-1/n} \left(\tfrac{1}{|B(x,r)|} \int_{B(x,r)} |u(y) - c|^s dy \right)^{1/s}.
\end{aligned} \tag{7.8.13}$$

Now, by definitions of the maximal operators and (7.8.13), we obtain

$$\|\mathbf{M}_s du(x)\|_{s,\Omega}$$

$$= \left(\int_\Omega |\mathbf{M}_s du(x)|^s dx \right)^{1/s}$$

$$= \left(\int_\Omega \left| \sup_{r>0} \left(\tfrac{1}{|B(x,r)|} \int_{B(x,r)} |du(y)|^s dy \right)^{1/s} \right|^s dx \right)^{1/s}$$

$$\leq \left(\int_\Omega \left| \sup_{r>0} C_1 |B(x,r)|^{-1/n} \left(\tfrac{1}{|B(x,r)|} \int_{B(x,r)} |u(y) - c|^s dy \right)^{1/s} \right|^s dx \right)^{1/s}$$

$$\leq \left(\int_\Omega \left| \sup_{r>0} C_1 \left(\tfrac{1}{|B(x,r)|^{1+s/n}} \int_{B(x,r)} |u(y) - c|^s dy \right)^{1/s} \right|^s dx \right)^{1/s}$$

$$\leq C_2 \left(\int_\Omega |\mathbf{M}_{s,\alpha}(u(x) - c)|^s dx \right)^{1/s}$$

$$= C_2 \|\mathbf{M}_{s,\alpha}(u(x) - c)\|_{s,\Omega},$$

which can be written as

$$\|\mathbf{M}_s du(x)\|_{s,\Omega} \leq C_2 \|\mathbf{M}_{s,\alpha}(u(x) - c)\|_{s,\Omega}. \qquad \blacksquare$$

Note that in Theorem 7.8.4, c is any closed form. Thus, we can choose $c = 0$ in Theorem 7.8.4, to obtain the following corollary.

Corollary 7.8.5. *Let $u \in L^s(\Omega, \wedge^l)$, $l = 0, 1, 2, \ldots, n$, $1 < s < \infty$, be a smooth differential form satisfying equation (1.2.10) in a bounded domain Ω,*

$\mathbb{M}_s$ *be the Hardy–Littlewood maximal operator defined in* (7.1.1), *and* $\mathbb{M}_{s,\alpha}$ *be the fractional maximal operator of order* α. *Then,*

$$\|\mathbb{M}_s du(x)\|_{s,\Omega} \le C\|\mathbb{M}_{s,\alpha} u(x)\|_{s,\Omega}$$

for some constant C, independent of u.

We conclude this chapter with the following estimate for the composition of the sharp maximal operator, Green's operator, and the homotopy operator.

Theorem 7.8.6. *Let* $u \in L^s(\Omega, \wedge^l)$, $l = 0, 1, 2, \ldots, n$, $1 < s < \infty$, *be a differential form in a bounded, convex domain* Ω, $\mathbb{M}_s^\sharp$ *be the sharp maximal operator defined in* (7.1.2), *G be Green's operator, and T be the homotopy operator defined in* (1.5.1). *Then,*

$$\|\mathbb{M}_s^\sharp(TG(u))\|_{s,\Omega} \le C\|u\|_{s,\Omega} \tag{7.8.14}$$

for some constant C, independent of u.

7.9 Singular integrals

The purpose of this section is to estimate the singular integral of the composition of the homotopy operator T and the projection operator H. We can also discuss the integrals of other composite operators with a singular factor using the similar method. The consideration was motivated from physics. For instance, when calculating an electric field, we will deal with the integral

$$E(r) = \frac{1}{4\pi\epsilon_0} \int_D \rho(x) \frac{r-x}{\|r-x\|^3} dv,$$

where $\rho(x)$ is a charge density and x is the integral variable. It is singular if $r \in D$. Obviously, the singular integrals are more interesting to us because of their wide applications in different fields of mathematics and physics.

Theorem 7.9.1. *Let* $u \in L_{loc}^s(\Omega, \wedge^l)$, $l = 1, 2, \ldots, n$, $1 < s < \infty$, *be a solution of the nonhomogeneous A-harmonic equation in a bounded and convex domain* Ω, *H be the projection operator, and T be the homotopy operator. Then, there exists a constant C, independent of u, such that*

$$\left(\int_B |T(H(u)) - (T(H(u)))_B|^s \frac{1}{|x-x_B|^\alpha} dx\right)^{1/s}$$
$$\le C|B|^\gamma \left(\int_{\rho B} |u|^s \frac{1}{|x-x_B|^\lambda} dx\right)^{1/s} \tag{7.9.1}$$

for all balls B with $\rho B \subset \Omega$ and any real number α and λ with $\alpha > \lambda \geq 0$, where $\gamma = 1 + \frac{1}{n} - \frac{\alpha - \lambda}{ns}$ and x_B is the center of ball B.

Proof. Let $\varepsilon \in (0,1)$ be small enough such that $\varepsilon n < \alpha - \lambda$ and $B \subset \Omega$ be any ball with center x_B and radius r_B. Choose $t = s/(1-\varepsilon)$ and $\beta = t/(t-s)$, then $t > s$. Using the Hölder inequality and Theorem 7.6.1, we have

$$\left(\int_B (|TH(u) - (TH(u))_B|)^s \frac{1}{|x - x_B|^\alpha} dx \right)^{1/s}$$

$$= \left(\int_B \left(|TH(u) - (TH(u))_B| \frac{1}{|x - x_B|^{\alpha/s}} \right)^s dx \right)^{1/s}$$

$$\leq \|TH(u) - (TH(u))_B\|_{t,B} \left(\int_B \left(\frac{1}{|x - x_B|} \right)^{t\alpha/(t-s)} dx \right)^{(t-s)/st} \quad (7.9.2)$$

$$= \|TH(u) - (TH(u))_B\|_{t,B} \left(\int_B |x - x_B|^{-\alpha\beta} dx \right)^{1/\beta s}$$

$$\leq C_1 |B| diam(B) \|u\|_{t,B} \||x - x_B|^{-\alpha}\|_{\beta,B}^{1/s}.$$

We may assume that $x_B = 0$. Otherwise, we can move the center to the origin by a simple transformation. Then for any $x \in B$, $|x - x_B| \geq |x| - |x_B| = |x|$. By using the polar coordinate substitution, we have

$$\int_B |x - x_B|^{-\alpha\beta} dx \leq C \int_0^{r_B} \rho^{-\alpha\beta} \rho^{n-1} d\rho \leq \frac{C}{n - \alpha\beta} (r_B)^{n - \alpha\beta}. \quad (7.9.3)$$

Choose $m = nst/(ns + \alpha t - \lambda t)$, then $0 < m < s$. By the reverse Hölder inequality, we find that

$$\|u\|_{t,B} \leq C_2 |B|^{\frac{m-t}{mt}} \|u\|_{m,\sigma B}. \quad (7.9.4)$$

By the Hölder inequality again, we obtain

$$\|u\|_{m,\sigma B}$$

$$= \left(\int_{\sigma B} \left(|u||x - x_B|^{-\lambda/s} |x - x_B|^{\lambda/s} \right)^m dx \right)^{1/m}$$

$$\leq \left(\int_{\sigma B} \left(|u||x - x_B|^{-\lambda/s} \right)^s dx \right)^{1/s} \left(\int_{\sigma B} \left(|x - x_B|^{\lambda/s} \right)^{\frac{ms}{s-m}} dx \right)^{\frac{s-m}{ms}}$$

$$\leq \left(\int_{\sigma B} |u|^s |x - x_B|^{-\lambda} dx \right)^{1/s} C_3 (\sigma r_B)^{\lambda/s + n(s-m)/ms}$$

$$\leq C_4 \left(\int_{\sigma B} |u|^s |x - x_B|^{-\lambda} dx \right)^{1/s} (r_B)^{\lambda/s + n(s-m)/ms}. \quad (7.9.5)$$

Note that

$$diam(B) \cdot |B|^{1 + \frac{1}{t} - \frac{1}{m}} = |B|^{1 + \frac{1}{n} + \frac{1}{t} - \frac{ns + \alpha t - \lambda t}{nst}} = |B|^{1 + \frac{1}{n} - \frac{\alpha - \lambda}{ns}}. \quad (7.9.6)$$

Substituting (7.9.3), (7.9.4), and (7.9.5) into (7.9.2) and using (7.9.6), we have

$$\left(\int_B \left(|TH(u) - (TH(u))_B| \right)^s \frac{1}{|x-x_B|^\alpha} dx \right)^{1/s}$$

$$\leq C_5 |B|^\gamma \left(\int_{\sigma B} |u|^s |x - x_B|^{-\lambda} dx \right)^{1/s}. \qquad \blacksquare$$

Remark. 1) Replacing α by 2α and λ by α in Theorem 7.9.1, we have

$$\left(\int_B |T(H(u)) - (T(H(u)))_B|^s \frac{1}{|x-x_B|^{2\alpha}} dx \right)^{1/s}$$

$$\leq C |B|^\gamma \left(\int_{\rho B} |u|^s \frac{1}{|x-x_B|^\alpha} dx \right)^{1/s}.$$

2) If $\lambda = 0$, inequality (7.9.1) reduces to

$$\left(\int_B |T(H(u)) - (T(H(u)))_B|^s \frac{1}{|x-x_B|^\alpha} dx \right)^{1/s}$$

$$\leq C |B|^\gamma \left(\int_{\rho B} |u|^s dx \right)^{1/s}, \qquad (7.9.7)$$

which does not contain a singular factor in the integral on the right-hand side of the inequality.

Theorem 7.9.2. *Let $u \in D'(\Omega, \wedge^1)$ be a solution of the nonhomogeneous A-harmonic equation, H be the projection operator, and T be the homotopy operator. Assume that s is a fixed exponent associated with the nonhomogeneous A-harmonic equation. Then, there exists a constant C, independent of u, such that*

$$\left(\int_\Omega |T(H(u)) - (T(H(u)))_{B_0}|^s \frac{1}{d(x,\partial\Omega)^\alpha} dx \right)^{1/s}$$

$$\leq C \left(\int_\Omega |u|^s \frac{1}{d(x,\partial\Omega)^\lambda} dx \right)^{1/s} \qquad (7.9.8)$$

for any bounded and convex $L^s(\mu)$-averaging domain Ω. Here $B_0 \subset \Omega$ is a fixed ball and α and λ are constants with $0 \leq \lambda < \alpha < \min\{n, s+\lambda+n(s-1)\}$.

Proof. Let r_B be the radius of a ball $B \subset \Omega$. We may assume the center of B is 0. Then, $d(x, \partial\Omega) \geq r_B - |x|$ for any $x \in B$. Therefore, $d^{-1}(x, \partial\Omega) \leq \frac{1}{r_B - |x|}$ for any $x \in B$. Similar to the proof of Theorem 7.9.1, we have

$$\left(\int_B |TH(u) - (TH(u))_B|^s \frac{1}{d(x,\partial\Omega)^\alpha} dx\right)^{1/s}$$

$$\leq C_1 |B|^\gamma \left(\int_{\rho B} |u|^s \frac{1}{d(x,\partial\Omega)^\lambda} dx\right)^{1/s} \qquad (7.9.9)$$

for all balls B with $\rho B \subset \Omega$ and any real number α and λ with $\alpha > \lambda \geq 0$ where $\gamma = 1 + \frac{1}{n} - \frac{\alpha-\lambda}{ns}$. Write $d\mu = \frac{1}{d(x,\partial\Omega)^\alpha} dx$. Then,

$$\mu(B) = \int_B d\mu = \int_B \frac{1}{d(x,\partial\Omega)^\alpha} dx \geq \int_B \frac{1}{(diam(\Omega))^\alpha} dx = C_1 |B|,$$

and hence $\frac{1}{\mu(B)} \leq \frac{C_2}{|B|}$. Since Ω is an $L^s(\mu)$-averaging domain, using (7.9.9) and noticing that $\gamma - 1/s = (1 - 1/s) + (s + \lambda - \alpha)/ns > 0$, we have

$$\left(\frac{1}{\mu(\Omega)} \int_\Omega |T(H(u)) - (T(H(u)))_{B_0}|^s \frac{1}{d(x,\partial\Omega)^\alpha} dx\right)^{1/s}$$

$$= \left(\frac{1}{\mu(\Omega)} \int_\Omega |T(H(u)) - (T(H(u)))_{B_0}|^s d\mu\right)^{1/s}$$

$$\leq C_3 \sup_{4B \subset \Omega} \left(\frac{1}{\mu(B)} \int_B |T(H(u)) - (T(H(u)))_B|^s d\mu\right)^{1/s}$$

$$\leq C_4 \sup_{4B \subset \Omega} \left(\frac{1}{|B|} \int_B |T(H(u)) - (T(H(u)))_B|^s d\mu\right)^{1/s}$$

$$\leq C_5 \sup_{4B \subset \Omega} |B|^{\gamma-1/s} \left(\int_{\rho B} |u|^s \frac{1}{d(x,\partial\Omega)^\lambda} dx\right)^{1/s}$$

$$\leq C_5 |\Omega|^{\gamma-1/s} \left(\int_\Omega |u|^s \frac{1}{d(x,\partial\Omega)^\lambda} dx\right)^{1/s}$$

$$\leq C_6 \left(\int_\Omega |u|^s \frac{1}{d(x,\partial\Omega)^\lambda} dx\right)^{1/s},$$

which is equivalent to

$$\left(\int_\Omega |T(H(u)) - (T(H(u)))_{B_0}|^s \frac{1}{d(x,\partial\Omega)^\alpha} dx\right)^{1/s}$$

$$\leq C \left(\int_\Omega |u|^s \frac{1}{d(x,\partial\Omega)^\lambda} dx\right)^{1/s}. \qquad \blacksquare$$

Chapter 8
Estimates for Jacobians

In this chapter, we first present some integral inequalities related to a strictly increasing convex function on $[0, \infty)$ and then improve the Hölder inequality with $L^p(\log L)^\alpha(\Omega)$-norms to the case $0 < p, q < \infty$. Next, we prove $L^p(\log L)^\alpha(\Omega)$-integrability of Jacobians. We also show that the integrability exponents described in Theorem 8.2.7 are the best possible.

8.1 Introduction

Although, in Section 3.6 we have introduced some basic concepts and notations about Jacobians, for convenience, we recollect some of those here. We assume that Ω is an open subset of $\mathbf{R}^n$, $n \geq 2$, and $f : \Omega \to \mathbf{R}^n$, $f = (f^1, \ldots, f^n)$, is a mapping of Sobolev class $W^{1,p}_{loc}(\Omega, \mathbf{R}^n)$, $1 \leq p < \infty$, whose distributional differential $Df = [\partial f^i / \partial x_j] : \Omega \to GL(n)$ is a locally integrable function on Ω with values in the space $GL(n)$ of all $n \times n$-matrices. A homeomorphism $f : \Omega \to \mathbf{R}^n$ of Sobolev class $W^{1,n}_{loc}(\Omega, \mathbf{R}^n)$ is said to be K-quasiconformal, $1 \leq K < \infty$, if its differential matrix $Df(x)$ and the Jacobian determinant

$$J = J(x, f) = \det Df(x)$$

$$= \begin{vmatrix} f^1_{x_1} & f^1_{x_2} & f^1_{x_3} & \cdots & f^1_{x_n} \\ f^2_{x_1} & f^2_{x_2} & f^2_{x_3} & \cdots & f^2_{x_n} \\ \vdots & \vdots & \vdots & \ddots & \vdots \\ f^n_{x_1} & f^n_{x_2} & f^n_{x_3} & \cdots & f^n_{x_n} \end{vmatrix}$$

R.P. Agarwal et al., *Inequalities for Differential Forms*,
DOI 10.1007/978-0-387-68417-8_8, © Springer Science+Business Media, LLC 2009

satisfy

$$|Df(x)|^n \le KJ(x,f),$$

where

$$|Df(x)| = \max\{|Df(x)h| : |h| = 1\}$$

denotes the norm of the Jacobi matrix $Df(x)$. Now, since the Jacobian $J(x, f)$ is an n-form, specifically,

$$J(x, f)dx = df^1 \wedge \cdots \wedge df^n,$$

where $dx = dx_1 \wedge dx_2 \wedge \cdots \wedge dx_n$, many results about differential forms carry over for Jacobians. Jacob Jacobi (1804–1851), one of the nineteenth century Germany's most accomplished scientists, developed the theory of determinants and transformations for evaluating multiple integrals and solving differential equations. Since then, the Jacobian has played a crucial role in multidimensional analysis and related fields, including nonlinear elasticity, weakly differentiable mappings, continuum mechanics, nonlinear PDEs, and calculus of variations. The integrability of Jacobians has become a rather important topic, see [310, 311], for example. In 1973, F.W. Gehring [6] first proved the higher integrability properties of the Jacobians; he invented reverse Hölder inequalities and used these inequalities to obtain the $L^{1+\varepsilon}$-integrability of the Jacobians of quasiconformal mappings, $\varepsilon > 0$. In 1990, S. Müller [147] proved that the Jacobian of an orientation-preserving mapping $f \in W_{loc}^{1,n}(\Omega, \mathbf{R}^n)$ belongs to the Zygmund class $L \log L(E)$ for each compact $E \subset \Omega$. This result can be stated as the following theorem [312].

Theorem 8.1.1. *Let $\Omega \subset \mathbf{R}^n$ be a bounded domain. If $f \in W_{loc}^{1,n}(\Omega, \mathbf{R}^n)$ is an orientation-preserving mapping and $|Df| \in L^n(\Omega)$, then $J(x, f) \in L \log L(E)$ and*

$$\int_E J(x, f) \log \left(e + \frac{J(x, f)}{\int_\Omega J(y, f)dy} \right) dx \le C \int_\Omega |Df(x)|^n dx \qquad (8.1.1)$$

for each compact $E \subset \Omega$.

8.2 Global integrability

In this section, we shall develop some basic estimates and the weighted estimates for Jacobians of the orientation-preserving mappings. These estimates will provide the higher integrability of Jacobians.

8.2.1 Preliminary lemmas

We first establish some preliminary integral inequalities which will be used to study the $L^\varphi(\Omega)$-integrability of differential forms or functions. These results may also find applications in the study of the L^s-theory and related topics, such as $L^\varphi(\mu)$-averaging domains and Orlicz norm estimates.

Lemma 8.2.1. *Let φ be a strictly increasing convex function on $[0, \infty)$ with $\varphi(0) = 0$ and D be a domain in $\mathbf{R}^n$. Assume that u is a function in D such that $\varphi(|u|) \in L^1(D; \mu)$ and $\mu(\{x \in D : |u - c| > 0\}) > 0$ for any constant c. Then, for any positive constant a, we have*

$$\int_D \varphi(\frac{a}{2}|u - u_{D,\mu}|)d\mu \le \int_D \varphi(a|u|)d\mu. \tag{8.2.1}$$

Proof. First, we note that

$$|u_{D,\mu}| = \left|\frac{1}{\mu(D)} \int_D u \, d\mu\right| \le \frac{1}{\mu(D)} \int_D |u| \, d\mu. \tag{8.2.2}$$

Applying φ to the both sides of (8.2.2), and then using Jensen inequality, we obtain

$$\varphi(a|u_{D,\mu}|) \le \frac{1}{\mu(D)} \int_D \varphi(a|u|)d\mu,$$

where a is any positive constant. This inequality is equivalent to

$$\int_D \varphi(a|u_{D,\mu}|)d\mu \le \int_D \varphi(a|u|)d\mu. \tag{8.2.3}$$

Now, by the fact that φ is increasing and convex, from (8.2.3), we have

$$\int_D \varphi\left(\frac{a}{2}|u - u_{D,\mu}|\right) d\mu \le \int_D \varphi\left(\frac{a}{2}|u| + \frac{a}{2}|u_{D,\mu}|\right) d\mu$$
$$\le \frac{1}{2}\int_D \varphi(a|u|)d\mu + \frac{1}{2}\int_D \varphi(a|u_{D,\mu}|)d\mu$$
$$\le \int_D \varphi(a|u|)d\mu.$$

Thus, inequality (8.2.1) holds. ∎

Lemma 8.2.2. *Let φ be a strictly increasing convex function on $[0, \infty)$ with $\varphi(0) = 0$ and D be a domain in $\mathbf{R}^n$. Assume that u is a function in D such that $\varphi(|u|) \in L^1(D; \mu)$ and $\mu(\{x \in D : |u - c| > 0\}) > 0$ for any constant c. Then, for any positive constant a, we have*

$$\int_D \varphi(a|u|)d\mu \le C \int_D \varphi(2a|u-c|)d\mu, \tag{8.2.4}$$

where C is a positive constant.

Proof. For any constant c, let

$$C_1 = \int_D \varphi(2a|c|)d\mu = \varphi(2a|c|)\mu(D).$$

Note that $\mu(\{x \in D : 2a|u-c| > 0\}) = \mu(\{x \in D : |u-c| > 0\}) > 0$. Then, there exists a constant C_2 such that

$$C_1 \le C_2 \int_D \varphi(2a|u-c|)d\mu,$$

that is

$$\int_D \varphi(2a|c|)d\mu \le C_2 \int_D \varphi(2a|u-c|)d\mu.$$

Now, since φ is an increasing convex function, we obtain

$$\int_D \varphi(a|u|)d\mu \le \int_D \varphi\left(\tfrac{1}{2}(2a|u-c|) + \tfrac{1}{2}(2a|c|)\right) d\mu$$

$$\le \tfrac{1}{2}\int_D \varphi(2a|u-c|)d\mu + \tfrac{1}{2}\int_D \varphi(2a|c|)d\mu$$

$$\le \tfrac{1}{2}\int_D \varphi(2a|u-c|)d\mu + \tfrac{C_2}{2}\int_D \varphi(2a|u-c|)d\mu \tag{8.2.5}$$

$$\le C_3 \int_D \varphi(2a|u-c|)d\mu.$$

Hence, inequality (8.2.4) holds. ∎

Since c is any constant in Lemma 8.2.2, we may select $c = u_{D,\mu}$. Thus, combining Lemmas 8.2.1 and 8.2.2, we obtain the following corollary immediately.

Corollary 8.2.3. *Assume that φ is as above and u is a function in $D \subset \mathbf{R}^n$ such that $\varphi(|u|) \in L^1(D;\mu)$ and $\mu(\{x \in D : |u - u_{D,\mu}| > 0\}) > 0$. Then, for any positive constant a, we have*

$$\int_D \varphi\left(\frac{1}{2}a|u - u_{D,\mu}|\right) d\mu \le \int_D \varphi(a|u|)d\mu \le C \int_D \varphi(2a|u-u_{D,\mu}|)d\mu, \tag{8.2.6}$$

where C is a positive constant.

8.2.2 $L^p(\log L)^\alpha(\Omega)$-integrability

The purpose of this section is to study the $L^p(\log L)^\alpha(\Omega)$-integrability of the Jacobian of a composite mapping. For this, we keep using the symbols and notations introduced in Section 1.9. From Theorem 1.9.1, we find that the Orlicz space $L^\psi(\Omega)$ with $\psi(t) = t^p \log^\alpha(e + t)$ can be denoted by $L^p(\log L)^\alpha(\Omega)$ and the corresponding norm can also be written as $[f]_{L^p(\log L)^\alpha(\Omega)}$. The following version of Hölder's inequality appears in [312, Proposition 2.2].

Theorem 8.2.4. *Let* $1 < p, q < \infty$, $\alpha, \beta > 0$, $\frac{1}{p} + \frac{1}{q} = \frac{1}{r}$, $\frac{\alpha}{p} + \frac{\beta}{q} = \frac{\gamma}{r}$, *and* $f \in L^p(\log L)^\alpha(\Omega)$, $g \in L^q(\log L)^\beta(\Omega)$. *Then,* $fg \in L^r(\log L)^\gamma(\Omega)$ *and*

$$||fg||_{L^r \log^\gamma L} \le C ||f||_{L^p \log^\alpha L} ||g||_{L^q \log^\beta L}. \tag{8.2.7}$$

In a recent paper [194], we have improved the condition $1 < p, q < \infty$ to $0 < p, q < \infty$. We state and prove this result in the following theorem.

Theorem 8.2.5. *Let* $m, n, \alpha, \beta > 0$, $1/s = 1/m + 1/n$, $\alpha/m + \beta/n = \gamma/s$. *Assume that* $f \in L^m(\log L)^\alpha(\Omega)$ *and* $g \in L^n(\log L)^\beta(\Omega)$. *Then,* $fg \in L^s(\log L)^\gamma(\Omega)$ *and*

$$\left(\int_\Omega |fg|^s \log^\gamma \left(e + \frac{|fg|}{||fg||_s} \right) dx \right)^{1/s}$$
$$\le C \left(\int_\Omega |f|^m \log^\alpha \left(e + \frac{|f|}{||f||_m} \right) dx \right)^{1/m} \left(\int_\Omega |g|^n \log^\beta \left(e + \frac{|g|}{||g||_n} \right) dx \right)^{1/n}, \tag{8.2.8}$$

where C *is a positive constant.*

Note that inequality (8.2.8) can be written as

$$[fg]_{L^s(\log L)^\gamma(\Omega)} \le C \, [f]_{L^m(\log L)^\alpha(\Omega)} \, [g]_{L^n(\log L)^\beta(\Omega)}. \tag{8.2.9}$$

Proof. Using the elementary inequality

$$\log(e + x^a) \le \log(e + x)^{a+1}$$

for $a > 0, x > 0$, we obtain

$$\log \left(e + \frac{(|f|^s |g|^s)^{1/s}}{\||f|^s |g|^s\|_1^{1/s}} \right)$$

$$\leq \log \left(e + \frac{|f|^s |g|^s}{\||f|^s |g|^s\|_1} \right)^{1/s+1} \tag{8.2.10}$$

$$= (1/s + 1) \log \left(e + \frac{|f|^s |g|^s}{\||f|^s |g|^s\|_1} \right)$$

and

$$\log \left(e + \left(\frac{|f|}{\|f\|_m} \right)^s \right) \leq \log \left(e + \frac{|f|}{\|f\|_m} \right)^{s+1} \leq (s+1) \log \left(e + \frac{|f|}{\|f\|_m} \right), \tag{8.2.11}$$

$$\log \left(e + \left(\frac{|g|}{\|g\|_n} \right)^s \right) \leq \log \left(e + \frac{|g|}{\|g\|_n} \right)^{s+1} \leq (s+1) \log \left(e + \frac{|g|}{\|g\|_n} \right). \tag{8.2.12}$$

Now, from the Hölder inequality (8.2.7) with $1 = \frac{1}{m/s} + \frac{1}{n/s}$ (note that $m/s > 1, n/s > 1$ since $1/s = 1/m + 1/n$) and (8.2.10), (8.2.11), and (8.2.12), we have

$$\int_\Omega |fg|^s \log^\gamma \left(e + \frac{|fg|}{\|fg\|_s} \right) dx$$

$$= \int_\Omega (|f|^s |g|^s) \log^\gamma \left(e + \frac{\||f|^s |g|^s\|^{1/s}}{\||f|^s |g|^s\|_1^{1/s}} \right) dx$$

$$\leq C_1 \int_\Omega (|f|^s |g|^s) \log^\gamma \left(e + \frac{|f|^s |g|^s}{\||f|^s |g|^s\|_1} \right) dx$$

$$\leq C_2 \left(\int_\Omega (|f|^s)^{m/s} \log^\alpha \left(e + \frac{|f|^s}{\||f|^s\|_{m/s}} \right) dx \right)^{s/m}$$

$$\times \left(\int_\Omega (|g|^s)^{n/s} \log^\beta \left(e + \frac{|g|^s}{\||g|^s\|_{n/s}} \right) dx \right)^{s/n} \tag{8.2.13}$$

$$= C_2 \left(\int_\Omega |f|^m \log^\alpha \left(e + \left(\frac{|f|}{\|f\|_m} \right)^s \right) dx \right)^{s/m}$$

$$\times \left(\int_\Omega |g|^n \log^\beta \left(e + \left(\frac{|g|}{\|g\|_n} \right)^s \right) dx \right)^{s/n}$$

$$\leq C_3 \left(\int_\Omega |f|^m \log^\alpha \left(e + \frac{|f|}{\|f\|_m} \right) dx \right)^{s/m}$$

$$\times \left(\int_\Omega |g|^n \log^\beta \left(e + \frac{|g|}{\|g\|_n} \right) dx \right)^{s/n} .$$

Hence, it follows that

$$\left(\int_\Omega |fg|^s \log^\gamma \left(e + \frac{|fg|}{\|fg\|_s}\right) dx\right)^{1/s}$$

$$\leq C_4 \left(\int_\Omega |f|^m \log^\alpha \left(e + \frac{|f|}{\|f\|_m}\right) dx\right)^{1/m} \tag{8.2.14}$$

$$\times \left(\int_\Omega |g|^n \log^\beta \left(e + \frac{|g|}{\|g\|_n}\right) dx\right)^{1/n},$$

which shows that inequality (8.2.8) is true. ∎

From Theorem 8.2.5, we have the following general result immediately.

Corollary 8.2.6. *Let $p_i > 0$, $\alpha_i > 0$ for $i = 1, 2, \ldots, k$,*

$$\frac{1}{p_1} + \frac{1}{p_2} + \cdots + \frac{1}{p_k} = \frac{1}{p} \quad \text{and} \quad \frac{\alpha_1}{p_1} + \frac{\alpha_2}{p_2} + \cdots + \frac{\alpha_k}{p_k} = \frac{\alpha}{p}.$$

Assume that $f_i \in L^{p_i}(\log L)^{\alpha_i}(\Omega)$ for $i = 1, 2, \ldots, k$. Then, $f_1 f_2 \cdots f_k \in L^p(\log L)^\alpha(\Omega)$ and

$$[f_1 f_2 \cdots f_k]_{L^p(\log L)^\alpha(\Omega)}$$

$$\leq C [f_1]_{L^{p_1}(\log L)^{\alpha_1}(\Omega)} [f_2]_{L^{p_2}(\log L)^{\alpha_2}(\Omega)} \cdots [f_k]_{L^{p_k}(\log L)^{\alpha_k}(\Omega)}, \tag{8.2.15}$$

where C is a positive constant and the norms $[f_1 f_2 \cdots f_k]_{L^p(\log L)^\alpha(\Omega)}$ and $[f_i]_{L^{p_i}(\log L)^{\alpha_i}(\Omega)}$, $i = 1, 2, \ldots, k$, are defined in Section 1.9.

8.2.3 Applications

Next, we explore some applications of the new version of the Hölder inequality established above. Specifically, we study the integrability of the Jacobian of the composition of mappings $f : \Omega \to \mathbf{R}^n$ defined by

$$f = (f^1(u_1, u_2, \ldots, u_n), f^2(u_1, u_2, \ldots, u_n), \ldots, f^n(u_1, u_2, \ldots, u_n))$$

of Sobolev class $W^{1,p}_{loc}(\Omega, \mathbf{R}^n)$, where

$$u_i = u_i(x_1, x_2, \ldots, x_n), \quad i = 1, 2, \ldots, n,$$

are functions of $x = (x_1, x_2, \ldots, x_n) \in \Omega$ with continuous partial derivatives $\frac{\partial u_i}{\partial x_j}$, $j = 1, 2, \ldots, n$. For this, we assume that distributional differentials $Df(u) = [\partial f^i/\partial u_j]$ and $Du(x) = [\partial u_i/\partial x_j]$ are locally integrable functions with values in the space $GL(n)$ of all $n \times n$-matrices. As usual, we will write

$$J(x, f) = \det Df(u(x)) = \frac{\partial(f^1 \cdots f^n)}{\partial(x_1 \cdots x_n)}, \qquad (8.2.16)$$

$$J(u, f) = \det Df(u) = \frac{\partial(f^1 \cdots f^n)}{\partial(u_1 \cdots u_n)}, \qquad (8.2.17)$$

and

$$J(x, u) = \det Du(x) = \frac{\partial(u_1 \cdots u_n)}{\partial(x_1 \cdots x_n)}, \qquad (8.2.18)$$

respectively. From Theorem 8.2.5, we have the following integrability result for the Jacobian of the composition of mappings.

Theorem 8.2.7. *Let $s, t, \beta, \gamma > 0$, with $\frac{1}{p} = \frac{1}{s} + \frac{1}{t}$ and $\frac{\beta}{s} + \frac{\gamma}{t} = \frac{\alpha}{p}$. Assume that $J(x, f)$, $J(u, f)$, and $J(x, u)$ are Jacobians defined in (8.2.16), (8.2.17), and (8.2.18), respectively. If $J(u(x), f) \in L^s(\log L)^\beta(\Omega)$ and $J(x, u) \in L^t(\log L)^\gamma(\Omega)$, then $J(x, f) \in L^p(\log L)^\alpha(\Omega)$ and*

$$\left(\int_\Omega |J(x, f)|^p \log^\alpha \left(e + \frac{|J(x,f)|}{\|J(x,f)\|_p} \right) dx \right)^{1/p}$$

$$\leq C \left(\int_\Omega |J(u, f)|^s \log^\beta \left(e + \frac{|J(u,f)|}{\|J(u,f)\|_s} \right) dx \right)^{1/s} \qquad (8.2.19)$$

$$\times \left(\int_\Omega |J(x, u)|^t \log^\gamma \left(e + \frac{|J(x,u)|}{\|J(x,u)\|_t} \right) dx \right)^{1/t},$$

where C is a positive constant.

Proof. Note that the Jacobian of the composition of f and u can be expressed as

$$J(x, f) = \frac{\partial(f^1 \cdots f^n)}{\partial(x_1 \cdots x_n)}$$

$$= \frac{\partial(f^1 \cdots f^n)}{\partial(u_1 \cdots u_n)} \cdot \frac{\partial(u_1 \cdots u_n)}{\partial(x_1 \cdots x_n)} \qquad (8.2.20)$$

$$= J(u, f) \cdot J(x, u).$$

From Theorem 8.2.5 and (8.2.20), we find that

$$\left(\int_\Omega |J(x,f)|^p \log^\alpha \left(e + \frac{|J(x,f)|}{||J(x,f)||_p} \right) dx \right)^{1/p}$$

$$= \left(\int_\Omega |J(u,f) \cdot J(x,u)|^p \log^\alpha \left(e + \frac{|J(u,f) \cdot J(x,u)|}{||J(u,f) \cdot J(x,u)||_p} \right) dx \right)^{1/p}$$

$$\leq C \left(\int_\Omega |J(u,f)|^s \log^\beta \left(e + \frac{|J(u,f)|}{||J(u,f)||_s} \right) dx \right)^{1/s} \tag{8.2.21}$$

$$\times \left(\int_\Omega |J(x,u)|^t \log^\gamma \left(e + \frac{|J(x,u)|}{||J(x,u)||_t} \right) dx \right)^{1/t}$$

$$< \infty$$

since $J(u(x),f) \in L^s(\log L)^\beta(\Omega)$ and $J(x,u) \in L^t(\log L)^\gamma(\Omega)$. Thus, $J(x,f) \in L^p(\log L)^\alpha(\Omega)$.   ∎

Now, as an application of the Hölder inequality with L^p-norm

$$|| fg ||_{s,E} \leq || f ||_{\alpha,E} \cdot || g ||_{\beta,E}, \tag{8.2.22}$$

where $0 < \alpha, \beta < \infty$, $s^{-1} = \alpha^{-1} + \beta^{-1}$, and f and g are any measurable functions on a measurable set $E \subset \mathbf{R}^n$, we have the following L^p-integrability theorem for the Jacobian of a composite mapping.

Theorem 8.2.8.   *Let $J(x,f)$, $J(u,f)$, and $J(x,u)$ be Jacobians defined in* $(8.2.16), (8.2.17),$ *and* $(8.2.18)$, *respectively. If $J(u(x),f) \in L^s(\Omega)$ and $J(x,u) \in L^t(\Omega)$, $s,t > 0$, then $J(x,f) \in L^p(\Omega)$ and*

$$||J(x,f)||_{L^p(\Omega)} \leq C||J(u(x),f)||_{L^s(\Omega)}||J(x,u)||_{L^t(\Omega)},$$

where C is a positive constant and the integrability exponent p of $J(x,f)$ determined by $\frac{1}{p} = \frac{1}{s} + \frac{1}{t}$ is the best possible.

8.2.4 Examples

The following example shows that the integrability exponent p of $J(x,f)$ cannot be improved further.

Example 8.2.9. For any $(x, y) \in D = \{(x, y) : 0 < x^2 + y^2 \leq \rho^2\}$, we consider the mappings defined by

$$f(x, y) = (f^1, f^2) = \left(\frac{x}{(x^2 + y^2)^\sigma}, \frac{y}{(x^2 + y^2)^\sigma} \right),$$

and

$$x = r^{-k} \cos \theta, \quad y = r^{-k} \sin \theta, \quad (r, \theta) \in \Omega = \{(r, \theta) : 0 < r < \rho, \, 0 < \theta \leq 2\pi\},$$

where σ and ρ are positive constants. After a simple calculation, we obtain the following Jacobians:

$$J_1 = \frac{\partial(f^1, f^2)}{\partial(r, \theta)} = \frac{k(2\sigma - 1)}{r^{4\sigma + 2k + 1}},$$

$$J_2 = \frac{\partial(f^1, f^2)}{\partial(x, y)} = \frac{1 - 2\sigma}{r^{4\sigma}}, \tag{8.2.23}$$

$$J_3 = \frac{\partial(x, y)}{\partial(r, \theta)} = \frac{-k}{r^{2k+1}}, \quad 0 < r < \rho.$$

It is easy to see that $J_1 \in L^{1/(4\sigma + 2k + 1)}(\Omega)$, but $J_1 \notin L^p(\Omega)$ for any $p > \frac{1}{4\sigma + 2k + 1}$. Similarly, $J_2 \in L^{1/4\sigma}(\Omega)$, but $J_2 \notin L^s(\Omega)$ for any $s > \frac{1}{4\sigma}$; and $J_3 \in L^{1/(2k+1)}(\Omega)$, but $J_3 \notin L^t(\Omega)$ for any $t > \frac{1}{2k+1}$. Here, the integrability exponent $p = 1/(4\sigma + 2k + 1)$ of $\partial(f^1, f^2)/\partial(r, \theta)$ is determined by

$$\frac{1}{p} = 4\sigma + 2k + 1 = \frac{1}{s} + \frac{1}{t},$$

where $s = 1/4\sigma$ and $t = 1/(2k + 1)$ are the integrability exponents of the Jacobians $\partial(f^1, f^2)/\partial(x, y)$ and $\partial(x, y)/\partial(r, \theta)$, respectively.

Thus, the integrability exponent p of $J(x, f)$ is the best possible.

Example 8.2.10. Let $J_1 = \partial(f^1, f^2)/\partial(r, \theta)$, $J_2 = \partial(f^1, f^2)/\partial(x, y)$, and $J_3 = \partial(x, y)/\partial(r, \theta)$ be the Jacobians obtained in Example 8.2.9. For any $\varepsilon > 0$, there exists a constant $C_1 > 0$ such that

$$|J_1|^p \log \left(e + \frac{|J_1|}{\|J_1\|_p} \right) \leq C_1 |J_1|^{p + \varepsilon/(4\sigma + 2k + 1)}. \tag{8.2.24}$$

From (8.2.23) and (8.2.24), we have

$$\int_\Omega |J_1|^p \log\left(e + \frac{|J_1|}{\|J_1\|_p}\right) dr d\theta$$

$$= 2\pi \int_0^\rho |J_1|^p \log\left(e + \frac{|J_1|}{\|J_1\|_p}\right) dr$$

$$= C_2 \int_0^\rho \left(\frac{1}{r^{4\sigma+2k+1}}\right)^p \log\left(e + \frac{\left|\frac{(2\sigma-1)k}{r^{4\sigma+2k+1}}\right|}{\left\|\frac{(2\sigma-1)k}{r^{4\sigma+2k+1}}\right\|_p}\right) dr$$

$$\leq C_3 \int_0^\rho r^{-(4\sigma+2k+1)p} \left(r^{-(4\sigma+2k+1)}\right)^{\varepsilon/4\sigma+2k+1} dr$$

$$\leq C_4 \int_0^\rho r^{-(4\sigma+2k+1)p-\varepsilon} dr$$

$$= C_5 < \infty$$

for any p satisfying

$$0 < p \leq \frac{1}{4\sigma+2k+1} - \frac{\varepsilon}{4\sigma+2k+1}.$$

Since $\varepsilon > 0$ is arbitrary, it follows that $J_1 \in L^p \log L(\Omega)$ for any p with $0 < p < \frac{1}{4\sigma+2k+1}$. Similarly, we have $J_2 \in L^s \log L(\Omega)$ for any s with $0 < s < \frac{1}{4\sigma}$ and $J_3 \in L^t \log L(\Omega)$ for any t with $0 < t < \frac{1}{2k+1}$. This example shows that the integrability exponent p of $\frac{\partial(f^1,f^2)}{\partial(r,\theta)}$ determined by $1/p = 1/s + 1/t$ is the best possible when $\alpha = \beta = \gamma = 1$ in Theorem 8.2.7.

8.2.5 The norm comparison

In this section, we shall discuss the relationship between norms $\|f\|_{L^p \log^\alpha L}$ and $[f]_{L^p (\log L)^\alpha(\Omega)}$, which will provide an alternative method to prove Theorems 8.2.5 and 8.2.7. For this, we shall need the following general inequality which first appeared in [312, Theorem A.1].

Theorem 8.2.11. *Suppose $A, B, C : [0,\infty) \rightarrow [0,\infty)$ are continuous, monotone-increasing functions for which there exist positive constants c and d such that*

(i) $B^{-1}(t)C^{-1}(t) \leq cA^{-1}(t)$ *for all $t > 0$ and*

(ii) $A(\frac{t}{d}) \leq \frac{1}{2}A(t)$ *for all $t > 0$.*

Suppose that G is an open subset of $\mathbf{R}^n$, and $f \in L_B(G)$ and $g \in L_C(G)$. Then, $fg \in L_A(G)$ and

$$\|fg\|_A \leq cd\|f\|_B\|g\|_C.$$

Proof. Note that if $f \in L_B(G)$, the monotonicity of A and an application of the monotone convergence theorem gives

$$\int_G A\left(\frac{|f(x)|}{\|f\|_A}\right) dx \leq 1.$$

Hence, from the definition of the Luxemburg norm, we obtain

$$\int_G B\left(\frac{f(x)g(x)}{c\|f\|_B\|g\|_C}\right) dx$$
$$\leq \int_G B\left(\frac{|f(x)|}{\|f\|_B}\right) dx + \int_G C\left(\frac{|g(x)|}{\|g\|_C}\right) dx$$
$$\leq 2.$$

We therefore have

$$\int_G A\left(\frac{f(x)g(x)}{cd\|f\|_B\|g\|_C}\right) dx \leq 1$$

and, again by the definition of the Luxemburg norm, we have the result. ∎

Thus, the norm $\|f\|_{L^p \log^\alpha L}$ is equivalent to the norm $[f]_{L^p(\log L)^\alpha(\Omega)}$ for $1 < p < \infty$. We have also proved in Theorem 1.9.1 that the norm $\|f\|_{L^p \log^\alpha L}$ is equivalent to the norm $[f]_{L^p(\log L)^\alpha(\Omega)}$, that is, for each $f \in L^p(\log L)^\alpha(\Omega)$, $0 < p < \infty$ and $\alpha \geq 0$,

$$\|f\|_p \leq \|f\|_{L^p \log^\alpha L} \leq [f]_{L^p(\log L)^\alpha(\Omega)} \leq C\|f\|_{L^p \log^\alpha L}, \qquad (8.2.25)$$

where $C = 2^{\alpha/p}\left(1 + \left(\frac{\alpha}{ep}\right)^\alpha\right)^{1/p}$ is a constant independent of f.

Hence, for any $0 < p < \infty$ and $\alpha \geq 0$, the Luxemburg norm $\|f\|_{L^p \log^\alpha L}$ is equivalent to the norm $[f]_{L^p(\log L)^\alpha(\Omega)}$ defined in Section 1.9. Therefore, Theorems 8.2.5 and 8.2.7 can also be proved by employing Theorem 8.2.11 with suitable choices of functions $A(t), B(t)$, and $C(t)$.

The following norm comparison theorem appeared in [312] whose proof is available in [204, p. 23].

Theorem 8.2.12. *Let f be supported in a ball $B \subset \mathbf{R}^n$ and let $\mathbb{M}$ be the Hardy–Littlewood maximal operator. Then, $f \in L \log L(B)$ if and only if*

$\mathbf{M}f \in L^1(B)$. *Furthermore, there exist constants C_1 and C_2 independent of f for which*

$$C_1\|\mathbf{M}f\|_{L^1(B)} \leq \|f\|_{L\log L(B)} \leq C_2\|\mathbf{M}f\|_{L^1(B)}.$$

Now, for $k = 0, 1, \ldots, n - 1$, consider the subdeterminant of the Jacobian $J(x, f)$ which is obtained by deleting the k rows and k columns from $J(x, f)$, i.e.,

$$J(x_{j_1}, x_{j_2}, \ldots, x_{j_{n-k}}; f^{i_1}, f^{i_2}, \ldots, f^{i_{n-k}})$$

$$= \begin{vmatrix} f^{i_1}_{x_{j_1}} & f^{i_1}_{x_{j_2}} & f^{i_1}_{x_{j_3}} & \cdots & f^{i_1}_{x_{j_{n-k}}} \\ f^{i_2}_{x_{j_1}} & f^{i_2}_{x_{j_2}} & f^{i_2}_{x_{j_3}} & \cdots & f^{i_2}_{x_{j_{n-k}}} \\ \vdots & \vdots & \vdots & \ddots & \vdots \\ f^{i_{n-k}}_{x_{j_1}} & f^{i_{n-k}}_{x_{j_2}} & f^{i_{n-k}}_{x_{j_3}} & \cdots & f^{i_{n-k}}_{x_{j_{n-k}}} \end{vmatrix}$$

which is an $(n - k) \times (n - k)$ subdeterminant of $J(x, f)$, $\{i_1, i_2, \ldots, i_{n-k}\} \subset \{1, 2, \ldots, n\}$ and $\{j_1, j_2, \ldots, j_{n-k}\} \subset \{1, 2, \ldots, n\}$. Let

$$u = J(x_{j_1}, x_{j_2}, \ldots, x_{j_{n-k}}; f^{i_1}, f^{i_2}, \ldots, f^{i_{n-k}})dx_{j_1} \wedge dx_{j_2} \wedge \cdots \wedge dx_{j_{n-k}}.$$
$$(8.2.26)$$

It is clear that u is an $(n - k)$-form. From Corollary 8.2.3, the following estimate for a subdeterminant of the Jacobian $J(x, f)$ follows immediately.

Theorem 8.2.13. *Let u be an $(n - k) \times (n - k)$ subdeterminant of $J(x, f)$ defined in (8.2.26), $k = 0, 1, \ldots, n - 1$, and φ be a strictly increasing convex function on $[0, \infty)$ with $\varphi(0) = 0$, and D be a domain in $\mathbf{R}^n$. If $\varphi(|u|) \in L^1(D; \mu)$ and $\mu(\{x \in D : |u - c| > 0\}) > 0$ for any constant c, then for any positive constant a, it follows that*

$$\int_D \varphi(\frac{a}{2}|u - u_{D,\mu}|)d\mu \leq \int_D \varphi(a|u|)d\mu \leq C \int_D \varphi(2a|u - u_{D,\mu}|)d\mu, \quad (8.2.27)$$

where C is a positive constant.

If we choose $k = 0$ in Theorem 8.2.13, we have the following estimate for the Jacobian $J(x, f)$.

Corollary 8.2.14. *Let φ be a strictly increasing convex function on $[0, \infty)$ with $\varphi(0) = 0$, and D be a domain in $\mathbf{R}^n$. If $\varphi(|J(x, f)|) \in L^1(D; \mu)$ and $\mu(\{x \in D : |J(x, f) - c| > 0\}) > 0$ for any constant c, then for any positive constant a, it follows that*

$$\int_D \varphi(\tfrac{a}{2}|J(x,f) - J(x,f)_{D,\mu}|)d\mu$$

$$\leq \int_D \varphi(a|J(x,f)|)d\mu$$

$$\leq C \int_D \varphi(2a|J(x,f) - J(x,f)_{D,\mu}|)d\mu,$$

where C is a positive constant.

Choosing $\varphi(t) = t^p \log^\alpha(e+t)$, $1 < p < \infty$ and $\alpha \geq 0$, in Corollary 8.2.14, we obtain the following estimate for the Jacobian:

$$C_1 \int_D |J(x,f) - J(x,f)_{D,\mu}|^p \log^\alpha \left(e + \tfrac{a}{2}|J(x,f) - J(x,f)_{D,\mu}|\right) d\mu$$

$$\leq \int_D |J(x,f)|^p \log^\alpha \left(e + a|J(x,f)|\right) d\mu$$

$$\leq C_2 \int_D |J(x,f) - J(x,f)_{D,\mu}|^p \log^\alpha \left(e + 2a|J(x,f) - J(x,f)_{D,\mu}|\right) d\mu,$$

where $a \geq 0$ is any constant and C_1 and C_2 are some positive constants.

Similarly, choosing $\varphi(t) = t^p$, $1 < p < \infty$, in Corollary 8.2.14, we obtain the following estimate for the Jacobian:

$$C_1 \int_D |J(x,f) - J(x,f)_{D,\mu}|^p d\mu$$

$$\leq \int_D |J(x,f)|^p d\mu$$

$$\leq C_2 \int_D |J(x,f) - J(x,f)_{D,\mu}|^p d\mu,$$

where C_1 and C_2 are some positive constants.

In our next result we shall provide an estimate for the sharp maximal operator $\mathbb{M}_s^\sharp$.

Theorem 8.2.15. *Let $u \in L^s(\log L)^\alpha(\Omega, \wedge^l)$, $l = 1, 2, \ldots, n$, $1 < s < \infty$, and $\alpha \geq 0$, be a differential form satisfying (1.2.10) in a bounded, convex domain Ω, $\mathbb{M}_s^\sharp$ be the sharp maximal operator defined in (7.1.2), and T be the homotopy operator defined in (1.5.1), then*

$$c\|\mathbb{M}_s^\sharp(Tu)\|_{s,\Omega} \leq \|u\|_{s,\Omega} \leq \|u\|_{L^s \log^\alpha L(\Omega)} \leq [u]_{L^s(\log L)^\alpha(\Omega)} \qquad (8.2.28)$$

for some constant c, independent of u.

Proof. From Theorem 7.8.1, there exists a constant $C_1 > 0$ such that

$$\|\mathbb{M}_s^\sharp(Tu)\|_{s,\Omega} \leq C_1 \|u\|_{s,\Omega}, \qquad (8.2.29)$$

which is equivalent to

$$c\|\mathbf{M}_s^\sharp(Tu)\|_{s,\Omega} \le \|u\|_{s,\Omega}, \tag{8.2.30}$$

where $c = 1/C_1$. On the other hand, from Theorem 1.9.1, we obtain

$$\|u\|_{s,\Omega} \le \|u\|_{L^s \log^\alpha L(\Omega)} \le [u]_{L^s(\log L)^\alpha(\Omega)}. \tag{8.2.31}$$

Combining (8.2.30) and (8.2.31), we have (8.2.28) immediately. ∎

Finally, we state the following corollary, which follows from Theorem 1.9.1.

Corollary 8.2.16. *Let* $u \in L^s(\log L)^\alpha(\Omega, \wedge^l)$, $l = 1, 2, \ldots, n$, $1 < s < \infty$, *and* $\alpha \ge 0$, *be an* $(n - k) \times (n - k)$ *subdeterminant of* $J(x, f)$ *defined in* (8.2.26), $k = 0, 1, \ldots, n - 1$. *Then,*

$$\|u\|_{s,\Omega} \le \|u\|_{L^s \log^\alpha L(\Omega)} \le [u]_{L^s(\log L)^\alpha(\Omega)}, \tag{8.2.32}$$

where Ω *is a bounded domain.*

If $k = 0$ in Corollary 8.2.16, we obtain the following estimate:

$$\|J(x, f)\|_{s,\Omega} \le \|J(x, f)\|_{L^s \log^\alpha L(\Omega)} \le [J(x, f)]_{L^s(\log L)^\alpha(\Omega)},$$

where Ω is a bounded domain.

Chapter 9
Lipschitz and BMO norms

In this chapter we provide some norm comparison theorems related to the BMO norms and the Lipschitz norms. We prove that the integrability exponents described in the Lipschitz norm comparison theorem (Theorem 9.2.1) are the best possible. We also develop some norm comparison theorems for the operators.

9.1 Introduction

The bounded mean oscillation (BMO) space was introduced by John and Nirenberg in 1961. "Bounded mean oscillation" soon became one of the main concepts in many fields, such as harmonic analysis, complex analysis, and partial differential equations. A function $f \in L^1_{loc}(\Omega, \mu)$ is said to be in $BMO(\Omega, \mu)$ if there is a constant C such that

$$\frac{1}{\mu(B)} \int_B |f - f_B| d\mu \leq C \tag{9.1.1}$$

for all balls B with $\sigma B \subset \Omega$, where $\sigma > 1$ is a constant. The least C for which (9.1.1) holds is denoted by $\|f\|_\star = \|f\|_{\star,\Omega}$ and called the BMO norm of f. Equivalently,

$$\|f\|_{\star,\Omega} = \sup_{\sigma B \subset \Omega} \frac{1}{\mu(B)} \int_B |f - f_B| d\mu, \tag{9.1.2}$$

where $\sigma > 1$ is a constant.

One of the most useful results for BMO is the John–Nirenberg lemma which was first proved by Muckenhoupt and Wheeden in [146]. Abstract versions of the John–Nirenberg lemma are also available, see [313], for example.

R.P. Agarwal et al., *Inequalities for Differential Forms*,
DOI 10.1007/978-0-387-68417-8_9, © Springer Science+Business Media, LLC 2009

For relationship between BMO and quasiconformal mappings, see [70, 71]. Also, see [123] for properties of BMO spaces.

Theorem 9.1.1. (John–Nirenberg lemma for doubling weights). *Let μ be defined by $d\mu = w(x)dx$ and $w(x)$ be a doubling weight. Then, a function f is in $BMO(\Omega, \mu)$ if and only if*

$$\mu(\{x \in B : |f(x) - f_B| > t\}) \le c_1 e^{-c_2 t}\mu(B) \qquad (9.1.3)$$

for each ball $B \subset \Omega$ and $t > 0$. Here c_1 and c_2 are positive constants.

The proof of John–Nirenberg lemma is also available in [59] where the Calderón–Zygmund decomposition technique is adopted. We state the following corollary from [59] which provides a useful tool to study the averaging domains.

Corollary 9.1.2. *A function f is in $BMO(\Omega, \mu)$ if and only if there exist positive constants k and C such that*

$$\frac{1}{\mu(B)} \int_B e^{k|f - f_B|} d\mu \le C \qquad (9.1.4)$$

for any ball B which is a compact subset of Ω. If (9.1.4) holds, then $\|f\|_\star \le C/k$. Conversely, if f is in $BMO(\Omega, \mu)$, then (9.1.4) holds with $C = 3$ and $k = (\log 2)/(8c_0 N\|f\|_\star)$, where N and c_0 are constants appearing in the Calderón–Zygmund decomposition for doubling weights.

9.2 BMO spaces and Lipschitz classes

In this section, we will first present some interesting results for the BMO spaces and the Lipschitz spaces of differential forms, and, then provide an example to show that the integrability exponents in Lipschitz conditions for conjugate A-harmonic tensors are the best possible.

9.2.1 Some recent results

Let $\omega \in L^1_{loc}(\Omega, \wedge^l)$, $l = 0, 1, \dots, n$. We write $\omega \in \mathrm{locLip}_k(\Omega, \wedge^l)$, $0 \le k \le 1$, if

$$\|\omega\|_{locLip_k, \Omega} = \sup_{\sigma Q \subset \Omega} |Q|^{-(n+k)/n} \|\omega - \omega_Q\|_{1,Q} < \infty \qquad (9.2.1)$$

for some $\sigma \geq 1$. Further, we write $\mathrm{Lip}_k(\Omega, \wedge^l)$ for those forms whose coefficients are in the usual Lipschitz space with exponent k and write $\|\omega\|_{\mathrm{Lip}_k,\Omega}$ for this norm. Similarly, for $\omega \in L^1_{loc}(\Omega, \wedge^l)$, $l = 0,1,\ldots,n$, we write $\omega \in \mathrm{BMO}(\Omega, \wedge^l)$ if

$$\|u\|_{\star,\Omega} = \sup_{\sigma Q \subset \Omega} |Q|^{-1} \|\omega - \omega_Q\|_{1,Q} < \infty \qquad (9.2.2)$$

for some $\sigma \geq 1$. When ω is a 0-form, (9.2.2) reduces to the classical definition of $\mathrm{BMO}(\Omega)$.

In 1999, C. Nolder [71] proved the following result.

Theorem 9.2.1. *If $0 \leq k, l \leq 1$ satisfy $p(k-1) = q(l-1)$, then there exists a constant C such that*

$$C^{-1}\|u\|^p_{locLip_k,\Omega} \leq \|\star v\|^q_{locLip_l,\Omega} \leq C\|u\|^p_{locLip_k,\Omega} \qquad (9.2.3)$$

for all conjugate A-harmonic tensors u and v in Ω.

Note that a pair of conjugate A-harmonic tensors are solutions of the conjugate A-harmonic equation (1.2.6).

Proof. We shall only prove the left inequality. The right inequality follows similarly. From the definition of $\|\cdot\|_{locLip_k,\Omega}$-norm, we have

$$\|u\|_{locLip_k,\Omega} = \sup_{2Q \subset \Omega} |Q|^{-(n+k)/n}\|u - u_Q\|_{1,Q}. \qquad (9.2.4)$$

Now, using the condition $p(k-1) = q(l-1)$ and Theorem 1.5.2, we obtain

$$|Q|^{-(n+k)/n}\|u - u_Q\|_{1,Q} \leq C_1(|Q|^{-(n+l)/n}\|v - c\|_{1,2Q})^{q/p} \qquad (9.2.5)$$

for all cubes Q with $2Q \subset \Omega$. Next, we choose c such that $\star c = (\star v)_{2Q}$. The left inequality now follows from (9.2.4) and (9.2.5). $\blacksquare$

The following result which appeared in [71] describes the relationship between a *BMO* space and a local Lipschitz space.

Theorem 9.2.2. *Let ω be a solution to the homogeneous A-harmonic equation (1.2.4). Then, the following statements are equivalent:*

(a) $\omega \in \mathrm{BMO}(\Omega, \wedge)$;

(b) $\sup\{|Q|^{(p-n)/pn}\|d\omega\|_{p,Q}|\sigma Q \subset \Omega\} < \infty$ *for some $\sigma > 1$.*

Similarly, the following statements are equivalent:

(c) $\omega \in \mathrm{locLip}_k(\Omega, \wedge)$;

(d) $\sup\{|Q|^{(p-pk-n)/pn}\|d\omega\|_{p,Q}|\sigma Q \subset \Omega\} < \infty$ *for some* $\sigma > 1$.

We conclude this section with the following global result established in [71].

Theorem 9.2.3. *There exists a constant C such that*

$$C^{-1}\|u - u_Q\|_{Lip_k,Q} \leq \| \star v - c\|_{Lip_k,Q} \leq C\|u - u_Q\|_{Lip_k,Q}$$

for all conjugate A-harmonic tensors u and v in a cube $Q \subset \mathbf{R}^n$. Here $0 < k \leq 1$ and $\star c = \star(\star v)_Q$.

9.2.2 Sharpness of integrability exponents

From Section 1.8, we know that

$$f(x) = (f^1, f^2, f^3) = x|x|^\beta = (x_1|x|^\beta, x_2|x|^\beta, x_3|x|^\beta)$$

is a K-quasiregular mapping in $\mathbf{R}^3$. Here $\beta \neq -1$ is a real number. Applying Example 1.2.4 with $l = 1$, we find that

$$u = f^1 = x_1|x|^\beta$$

and

$$v = \star f^2 df^3 = \star x_2|x|^\beta d(x_3|x|^\beta)$$

form a pair of conjugate A-harmonic tensors. Now since $p = \frac{n}{l} = 3$, $q = \frac{n}{n-l} = \frac{3}{2}$, by simple calculation, it follows that

$$v = \beta x_1 x_2 x_3|x|^{2\beta-2}dx_2 \wedge dx_3 - \beta x_2^2 x_3|x|^{2\beta-2}dx_1 \wedge dx_3$$

$$+|x|^{2\beta-2}x_2(|x|^2 + \beta x_3^2)dx_1 \wedge dx_2.$$

We also know that

$$|u| = |x_1||x|^\beta$$

and

$$|v| = |x|^{2\beta-1}|x_2|(|x|^2 + (\beta^2 + 2\beta)x_3^2)^{1/2} .$$

Thus, if $\beta \geq 0$ or $\beta \leq -2$, then

$$(|x|^2 + (\beta^2 + 2\beta)x_3^2)^{1/2} \leq (|x|^2 + (\beta^2 + 2\beta)|x|^2)^{1/2}$$
$$= |x|(\beta^2 + 2\beta + 1)^{1/2}$$
$$= |\beta + 1||x|.$$

Hence, we have

$$|x_2||x|^{2\beta} \leq |v| \leq |\beta + 1||x_2||x|^{2\beta}.$$

Now, we select $\Omega \subset \mathbf{R}^3$ so that for any $x \in \Omega$,

$$C_1|x_1| \leq |x| \leq C_2|x_1|$$

and

$$C_1|x_2| \leq |x| \leq C_2|x_2|,$$

where C_1 and C_2 are constants. For example, when

$$\Omega = \{(x_1, x_2, x_3) \mid 1 \leq x_1 \leq 2,\ 1 \leq x_2 \leq 2,\ 0 \leq x_3 \leq 1\},$$

then for any $x \in \Omega$,

$$|x| \leq (x_1^2 + 4x_1^2 + 1)^{1/2}$$
$$\leq \left(1 + 4 + \tfrac{1}{|x_1|^2}\right)^{1/2}$$
$$\leq |x_1|(1 + 4 + 1)^{1/2}$$
$$= \sqrt{6}|x_1|.$$

Thus,

$$|x_1| \leq |x| \leq \sqrt{6}|x_1|$$

holds for any $x \in \Omega$. Similarly,

$$|x_2| \leq |x| \leq \sqrt{6}|x_2|$$

holds for any $x \in \Omega$. Therefore, $|x_1| \sim |x|$ and $|x_2| \sim |x|$ in Ω. So, we have

$$|u| \sim |x_1||x|^\beta \sim |x|^{\beta+1} = |x|^{k'} \tag{9.2.6}$$

and

$$|v| \sim |x_2||x|^{2\beta} \sim |x|^{2\beta+1} = |x|^{l'} \tag{9.2.7}$$

in Ω. Here $k' = \beta + 1$ and $l' = 2\beta + 1$. Now, since $p = n/l = 3$ and $q = n/(n - l) = 3/2$, it follows that

$$p(k'-1) = 3((\beta+1)-1) = 3\beta = \frac{3}{2}((2\beta+1)-1) = q(l'-1),$$

that is

$$p(k'-1) = q(l'-1).$$

Thus, the integrability exponents in Theorem 9.2.1 are sharp.

9.3 Global integrability

In recent years new interest has developed in the study of the global integrability for solutions of the A-harmonic equation. T. Kilpeläinen and P. Koskela have proved the following global integrability theorem in [314].

Theorem 9.3.1. *If u is a bounded A-harmonic function (solution to equation (1.2.2)) in a ball $B \subset \mathbf{R}^n$, then*

$$\int_B |\nabla u|(\log(2+|\nabla u|))^{-1-\varepsilon} < \infty \qquad (9.3.1)$$

for all $\varepsilon > 0$.

We say that Ω satisfies a Whitney cube number condition with exponent λ if there is a constant M such that

$$N(j) \le M 2^{\lambda j},$$

where $N(j)$ is the number of Whitney cubes with the side length 2^{-j}. Balls and cubes satisfy a Whitney cube number condition with $\lambda = n-1$; see [204] for more properties of Whitney cubes.

9.3.1 Estimates for du

In [72], C. Nolder obtained the following result for more general domains.

Theorem 9.3.2. *Suppose that Ω satisfies a Whitney cube number condition with exponent $\lambda < n$. If u is A-harmonic in Ω and $u \in \mathrm{locLip}_k(\Omega)$, $0 < k \le 1$, then*

$$\int_\Omega |\nabla u|^q < \infty \qquad (9.3.2)$$

for $q < (n-\lambda)/(1-k)$.

He also proved the following global integrability theorem for the solutions of the nonhomogeneous A-harmonic equation (1.2.10).

Theorem 9.3.3. *Suppose that the differential form $u \in W_{loc}^{1,p}(\Omega, \wedge)$ satisfies (1.2.10) and Ω satisfies a Whitney cube number condition with exponent λ, $0 \leq \lambda < n$. If $u \in \mathrm{BMO}(\Omega, \wedge)$ and $n - \lambda \leq p$, then*

$$\int_\Omega |du|^{n-\lambda} (\log(2 + |du|))^{-1-\varepsilon} < \infty \qquad (9.3.3)$$

for all $\varepsilon > 0$.

If $u \in \mathrm{locLip}_k(\Omega, \wedge)$ and $(n - \lambda)/(1 - k) \leq p$, then

$$\int_\Omega |du|^{(n-\lambda)/(1-k)} (\log(2 + |du|))^{-1-\varepsilon} < \infty \qquad (9.3.4)$$

for all $\varepsilon > 0$.

9.3.2 Estimates for du^+

Theorem 9.3.3 is a far-reaching extension of Theorem 9.3.1. The following global integrability theorems also appear in [72].

Theorem 9.3.4. *Let u be a solution to the nonhomogeneous A-harmonic equation (1.2.10) in Ω and $\eta \in C_0^\infty(\Omega)$ with $\eta \geq 0$. Then, there exists a constant C, depending only on a, b, and p, such that*

$$\int_\Omega |du^+|^p \eta^p \geq C \left(\int_\Omega |u^+|^p |\nabla \eta|^p + \int_\Omega |u^+|^p \eta^p \right). \qquad (9.3.5)$$

The same is true for u^-.

Theorem 9.3.5. *Let u be a solution to the nonhomogeneous A-harmonic equation (1.2.10) in Ω and $\eta \in C_0^\infty(\Omega)$ with $\eta \geq 0$. For each $q \geq 0$, there exists a constant C, depending only on a, b, p, q, and n, such that*

$$\int_\Omega |u^+|^q |du|^p \eta^p \geq C \int_\Omega |u^+|^{p+q} (|\nabla \eta|^p + \eta^p). \qquad (9.3.6)$$

The inequality also holds for u^-.

Similar to Theorem 9.2.2 for the case of homogeneous A-harmonic equation, the equivalent statements hold for the solutions of the nonhomogeneous A-harmonic equation (1.2.10).

Theorem 9.3.6. *Let ω be a solution to the nonhomogeneous A-harmonic equation (1.2.10). Then, the following statements are equivalent:*

(a) $\omega \in \mathrm{BMO}(\Omega, \wedge)$;

(b) $\sup\{|Q|^{(p-n)/pn}\|d\omega\|_{p,Q}|\sigma Q \subset \Omega\} < \infty$ *for some* $\sigma > 1$.

Similarly, the following statements are equivalent:

(c) $\omega \in \mathrm{locLip}_k(\Omega, \wedge)$;

(d) $\sup\{|Q|^{(p-pk-n)/pn}\|d\omega\|_{p,Q}|\sigma Q \subset \Omega\} < \infty$ *for some* $\sigma > 1$.

9.4 Lipschitz and BMO norms

In this section, we will establish some estimates for the Lipschitz and BMO norms of the operators applied to some differential forms. We will also obtain some norm comparison theorems and discuss weighted cases.

9.4.1 Estimates for Lipschitz norms

First, we will develop some Lipschitz norm estimates for the operators. We begin with the following results for the homotopy operator T defined in (1.5.1).

Theorem 9.4.1. *Let $u \in L^s(\Omega, \wedge^l)$, $l = 1, 2, \ldots, n$, $1 < s < \infty$, be a solution of the A-harmonic equation (1.2.10) in a bounded, convex domain Ω and $T : C^\infty(\Omega, \wedge^l) \to C^\infty(\Omega, \wedge^{l-1})$ be the homotopy operator defined in (1.5.1). Then, there exists a constant C, independent of u, such that*

$$\|T(u)\|_{locLip_k, \Omega} \le C\|u\|_{s,\Omega}, \tag{9.4.1}$$

where k is a constant with $0 \le k \le 1$.

Proof. From Theorem 7.5.1, we have

$$\|T(u) - (T(u))_B\|_{s,B} \le C_1|B|diam(B)\|u\|_{s,\sigma B} \tag{9.4.2}$$

for all balls B with $\sigma B \subset \Omega$, where $\sigma > 1$ is a constant. Using the Hölder inequality with $1 = 1/s + (s-1)/s$, we find that

$$
\begin{aligned}
&\|T(u) - (T(u))_B\|_{1,B} \\
&= \int_B |T(u) - (T(u))_B| dx \\
&\leq \left(\int_B |T(u) - (T(u))_B|^s dx \right)^{1/s} \left(\int_B 1^{s/(s-1)} dx \right)^{(s-1)/s} \\
&= |B|^{(s-1)/s} \|T(u) - (T(u))_B\|_{s,B} \\
&= |B|^{1-1/s} \|T(u) - (T(u))_B\|_{s,B} \\
&\leq |B|^{1-1/s} (C_1 |B| diam(B) \|u\|_{s,\sigma B}) \\
&\leq C_2 |B|^{2-1/s+1/n} \|u\|_{s,\sigma B}.
\end{aligned}
$$
(9.4.3)

Now, from the definition of the Lipschitz norm, (9.4.3), and

$$2 - 1/s + 1/n - 1 - k/n = 1 - 1/s + 1/n - k/n > 0,$$

we obtain

$$
\begin{aligned}
&\|T(u)\|_{locLip_k,\Omega} \\
&= \sup_{\sigma B \subset \Omega} |B|^{-(n+k)/n} \|T(u) - (T(u))_B\|_{1,B} \\
&= \sup_{\sigma B \subset \Omega} |B|^{-1-k/n} \|T(u) - (T(u))_B\|_{1,B} \\
&\leq \sup_{\sigma B \subset \Omega} |B|^{-1-k/n} C_2 |B|^{2-1/s+1/n} \|u\|_{s,\sigma B} \\
&= \sup_{\sigma B \subset \Omega} C_2 |B|^{1-1/s+1/n-k/n} \|u\|_{s,\sigma B} \\
&\leq \sup_{\sigma B \subset \Omega} C_2 |\Omega|^{1-1/s+1/n-k/n} \|u\|_{s,\sigma B} \\
&\leq C_3 \sup_{\sigma B \subset \Omega} \|u\|_{s,\sigma B} \\
&\leq C_3 \|u\|_{s,\Omega},
\end{aligned}
$$
(9.4.4)

that is,

$$\|T(u)\|_{locLip_k,\Omega} \leq C_3 \|u\|_{s,\Omega}. \quad \blacksquare$$

Note that for a bounded domain D, both

$$\|T(u)\|_{s,B} \leq C|B| diam(B) \|u\|_{s,B}$$
(9.4.5)

and

$$\|T(u)\|_{s,B} \leq C diam(B)\|u\|_{s,B} \qquad (9.4.6)$$

hold for any ball B with $B \subset D$. Thus, if we use (9.4.5) in the proof of Theorem 3.3.7, instead of (9.4.6), we obtain the following version of the Poincaré-type estimate

$$\|G(u) - (G(u))_B\|_{s,B} \leq C|B|diam(B)\|du\|_{s,B}. \qquad (9.4.7)$$

Further, from the above inequality we have the following Poincaré-type estimate for the projection operator

$$\|H(u) - (H(u))_B\|_{s,B} \leq C|B|diam(B)\|du\|_{s,B}. \qquad (9.4.8)$$

From (9.4.7) and (9.4.8) and the method similar to the proof of Theorem 9.4.1, noticing that both of G and H commute with d, we can establish the following Lipschitz norm inequalities for Green's operator G and the projection operator H.

Theorem 9.4.2. *Let $u \in L^s(\Omega, \wedge^l)$, $l = 1, 2, \ldots, n-1$, $1 < s < \infty$, be a smooth differential form in a bounded domain Ω and G be Green's operator, and H be the projection operator. Then, there exists a constant C, independent of u, such that*

$$\|G(u)\|_{locLip_k,\Omega} \leq C\|du\|_{s,\Omega} \qquad (9.4.9)$$

and

$$\|H(u)\|_{locLip_k,\Omega} \leq C\|du\|_{s,\Omega}, \qquad (9.4.10)$$

where k is a constant with $0 \leq k \leq 1$.

Theorem 9.4.3. *Let $u \in L^s_{loc}(\Omega, \wedge^l)$, $l = 1, 2, \ldots, n$, $1 < s < \infty$, be a solution of the A-harmonic equation (1.2.10) in a bounded domain Ω and G be Green's operator. Then, there exists a constant C, independent of u, such that*

$$\|G(u)\|_{locLip_k,\Omega} \leq C\|u\|_{\star,\Omega}, \qquad (9.4.11)$$

where k is a constant with $0 \leq k \leq 1$.

Proof. Using the Hölder inequality with $1 = 1/s + (s-1)/s$ and (9.4.7), we find that

$$\|G(u) - (G(u))_B\|_{1,B}$$

$$= \int_B |G(u) - (G(u))_B|\, dx$$

$$\leq \left(\int_B |G(u) - (G(u))_B|^s dx\right)^{1/s} \left(\int_B 1^{s/(s-1)} dx\right)^{(s-1)/s}$$

$$= |B|^{(s-1)/s}\|G(u) - (G(u))_B\|_{s,B} \tag{9.4.12}$$

$$= |B|^{1-1/s}\|G(u) - (G(u))_B\|_{s,B}$$

$$\leq |B|^{1-1/s}(C_1|B|diam(B)\|du\|_{s,B})$$

$$\leq C_2|B|^{2-1/s+1/n}\|du\|_{s,B},$$

where we have used $diam(B) = C|B|^{1/n}$. Now, from the Caccioppoli inequality (Theorem 4.2.1), we have

$$\|du\|_{s,B} \leq C_3|B|^{-1/n}\|u - c\|_{s,\sigma_1 B} \tag{9.4.13}$$

for any ball B and some constant $\sigma_1 > 1$, where c is any closed form. Next, choosing $c = u_B$ in (9.4.13), we find that

$$\|du\|_{s,B} \leq C_3|B|^{-1/n}\|u - u_B\|_{s,\sigma_1 B}. \tag{9.4.14}$$

Combining (9.4.12) and (9.4.14), it follows that

$$\|G(u) - (G(u))_B\|_{1,B}$$

$$\leq C_2|B|^{2-1/s+1/n}\|du\|_{s,B} \tag{9.4.15}$$

$$\leq C_4|B|^{2-1/s}\|u - u_B\|_{s,\sigma_1 B}.$$

Applying the weak reverse Hölder inequality for the solutions of the nonhomogeneous A-harmonic equation, we obtain

$$\|u - u_B\|_{s,\sigma_1 B} \leq C_5|B|^{(1-s)/s}\|u - u_B\|_{1,\sigma_2 B}, \tag{9.4.16}$$

where $\sigma_2 > \sigma_1 > 1$ is a constant. Substituting (9.4.16) into (9.4.15), we have

$$\|G(u) - (G(u))_B\|_{1,B} \leq C_6|B|\|u - u_B\|_{1,\sigma_2 B}, \tag{9.4.17}$$

which is equivalent to

$$|B|^{-(n+k)/n}\|G(u) - (G(u))_B\|_{1,B}$$

$$\leq C_6 |B|^{1-k/n}|B|^{-1}\|u - u_B\|_{1,\sigma_2 B} \qquad (9.4.18)$$

$$\leq C_7 |B|^{1-k/n}|\sigma_2 B|^{-1}\|u - u_B\|_{1,\sigma_2 B}.$$

Finally, taking the supremum over all balls $\sigma_3 B \subset \Omega$ with $\sigma_3 > \sigma_2$ and using the definitions of the Lipschitz and BMO norms, we obtain

$$\|G(u)\|_{locLip_k,\Omega}$$

$$= \sup_{\sigma_3 B \subset \Omega} |B|^{-(n+k)/n}\|G(u) - (G(u))_B\|_{1,B}$$

$$\leq \sup_{\sigma_3 B \subset \Omega} C_6 |B|^{1-k/n}|B|^{-1}\|u - u_B\|_{1,\sigma_2 B} \qquad (9.4.19)$$

$$\leq C_7 \sup_{\sigma_3 B \subset \Omega} |B|^{-1}\|u - u_B\|_{1,\sigma_2 B}$$

$$\leq C_8 \|u\|_{\star,\Omega}.$$

Thus, inequality (9.4.11) follows.   ∎

9.4.2 Lipschitz norms of $T \circ G$

Now, using Theorem 7.4.1 and the method developed in the proof of Theorem 9.4.1, we have the following estimate for the Lipschitz norm of the composition $T \circ G$ of the homotopy operator T and Green's operator G.

Theorem 9.4.4. *Let $u \in L^s(\Omega, \wedge^l)$, $l = 1, 2, \ldots, n$, $1 < s < \infty$, be a smooth differential form in a bounded, convex domain Ω, G be Green's operator, and $T : C^\infty(\Omega, \wedge^l) \to C^\infty(\Omega, \wedge^{l-1})$ be the homotopy operator defined in (1.5.1). Then, there exists a constant C, independent of u, such that*

$$\|T(G(u))\|_{locLip_k,\Omega} \leq C\|u\|_{s,\Omega}, \qquad (9.4.20)$$

where k is a constant with $0 \leq k \leq 1$.

Note that in Theorem 9.4.4, u need not be a solution of the A-harmonic equation. Further, if $u \in L^s(\Omega, \wedge^1)$, then $T(G(u)) \in L^s(\Omega, \wedge^0)$.

Using Lemma 8.2.2 with $\varphi(t) = t^s$ and $c = u_B$ over the ball B, we have

$$\|u\|_{s,B} \leq C\|u - u_B\|_{s,B}, \tag{9.4.21}$$

where C is a constant.

Theorem 9.4.5. *Let $u \in L^s_{loc}(\Omega, \wedge^1)$, $1 < s < \infty$, be a solution of the A-harmonic equation (1.2.10) in a bounded, convex domain Ω and G be Green's operator. Then, there exists a constant C, independent of u, such that*

$$\|T(G(u))\|_{locLip_k, \Omega} \leq C\|u\|_{\star, \Omega}, \tag{9.4.22}$$

where k is a constant with $0 \leq k \leq 1$.

Proof. From the definition of the Lipschitz norm, the Hölder inequality with $1 = 1/s + (s-1)/s$, (7.4.1), and (9.4.21), for any ball B with $B \subset \Omega$, we find that

$$
\begin{aligned}
&\|T(G(u)) - (T(G(u)))_B\|_{1,B} \\
&= \int_B |T(G(u)) - (T(G(u)))_B| dx \\
&\leq \left(\int_B |T(G(u)) - (T(G(u)))_B|^s dx\right)^{1/s} \left(\int_B 1^{\frac{s}{s-1}} dx\right)^{(s-1)/s} \\
&= |B|^{(s-1)/s} \|T(G(u)) - (T(G(u)))_B\|_{s,B} \\
&= |B|^{1-1/s} \|T(G(u)) - (T(G(u)))_B\|_{s,B} \\
&\leq |B|^{1-1/s} (C_1 |B| diam(B) \|u\|_{s,B}) \\
&\leq C_2 |B|^{2-1/s+1/n} \|u\|_{s,B} \\
&\leq C_3 |B|^{2-1/s+1/n} \|u - u_B\|_{s,B}.
\end{aligned}
\tag{9.4.23}
$$

Next, from the weak reverse Hölder inequality for solutions of the nonhomogeneous A-harmonic equation, we have

$$\|u - u_B\|_{s,B} \leq C_4 |B|^{(1-s)/s} \|u - u_B\|_{1,\sigma_1 B} \tag{9.4.24}$$

for some constant $\sigma_1 > 1$. Combination of (9.4.23) and (9.4.24) gives

$$
\begin{aligned}
&\|T(G(u)) - (T(G(u)))_B\|_{1,B} \\
&\leq C_3 |B|^{2-1/s+1/n} \|u - u_B\|_{s,B} \\
&\leq C_5 |B|^{1+1/n} \|u - u_B\|_{1,\sigma_1 B}.
\end{aligned}
\tag{9.4.25}
$$

Hence, we obtain

$$
\begin{aligned}
|B|^{-(n+k)/n}&\|T(G(u)) - (T(G(u)))_B\|_{1,B} \\
&\le C_5 |B|^{1/n-k/n}\|u - u_B\|_{1,\sigma_1 B} \\
&= C_5 |B|^{1+1/n-k/n}|B|^{-1}\|u - u_B\|_{1,\sigma_1 B} \\
&\le C_6 |B|^{1+1/n-k/n}|\sigma_1 B|^{-1}\|u - u_B\|_{1,\sigma_1 B} \\
&\le C_6 |\Omega|^{1+1/n-k/n}|\sigma_1 B|^{-1}\|u - u_B\|_{1,\sigma_1 B} \\
&\le C_7 |\sigma_1 B|^{-1}\|u - u_B\|_{1,\sigma_1 B}.
\end{aligned}
\tag{9.4.26}
$$

Thus, taking the supremum on both sides of (9.4.26) over all balls $\sigma_2 B \subset \Omega$ with $\sigma_2 > \sigma_1$ and using the definitions of the Lipschitz and BMO norms, we find that

$$
\begin{aligned}
\|T(G(u))\|_{locLip_k,\Omega} \\
= \sup\nolimits_{\sigma_2 B\subset\Omega} |B|^{-(n+k)/n}&\|T(G(u)) - (T(G(u)))_B\|_{1,B} \\
\le C_7 \sup\nolimits_{\sigma_2 B\subset\Omega} |\sigma_1 B|^{-1}&\|u - u_B\|_{1,\sigma_1 B} \\
\le C_7 \|u\|_{\star,\Omega},
\end{aligned}
$$

that is,

$$
\|T(G(u))\|_{locLip_k,\Omega} \le C\|u\|_{\star,\Omega}. \qquad\blacksquare
$$

Remark. (1) The above global norm comparison inequality has been proved on a bounded, convex domain Ω. We encourage readers to explore similar estimates in other kind of domains, such as L^s-averaging domains and $L^s(\mu)$-averaging domains.

(2) Let $0 < p < \infty$, $0 \le k \le 1$, and $u \in L^p_{loc}(\Omega, \wedge^l, \mu)$, $l = 0, 1, \ldots, n$, where the measure μ is defined by $d\mu = w(x)dx$ for some weight $w(x)$. We may define the generalized Lipschitz norm by

$$
\|u\|_{locLip_k,p,\Omega,w} = \sup_{\sigma B\subset\Omega} \left(\frac{1}{(\mu(B))^{(n+k)/n}} \int_B |u - u_{B,\mu}|^p d\mu \right)^{1/p}
$$

for some $\sigma > 1$, where the measure μ is defined by $d\mu = w(x)dx$ for some weight $w(x)$.

(3) For $0 < p < \infty$ and $u \in L^p_{loc}(\Omega, \wedge^l, \mu)$, $l = 0, 1, \ldots, n$, we define the generalized BMO norm by

$$\|u\|_{\star,p,\Omega,w} = \sup_{\sigma B \subset \Omega} \left(\frac{1}{\mu(B)} \int_B |u - u_{B,\mu}|^p d\mu \right)^{1/p}$$

for some $\sigma > 1$, where the measure μ is defined by $d\mu = w(x)dx$ for some weight $w(x)$. When u is a 0-form and $p = 1$, the above BMO norm reduces to the classical BMO norm for functions.

(4) Similarly, we can study the relationship between $\|u\|_{locLip_k,p,\Omega,w}$ and $\|u\|_{\star,p,\Omega,w}$ as we discussed in Theorems 9.4.1, 9.4.2, 9.4.3, 9.4.4, and 9.4.5.

9.4.3 Lipschitz norms of $G \circ T$

Similarly, using Theorem 7.4.15 and the method developed in the proof of Theorem 9.4.1, we have the following estimate for the Lipschitz norm of the composition $G \circ T$.

Theorem 9.4.6. *Let $u \in L^s(\Omega, \wedge^l)$, $l = 1, 2, \ldots, n$, $1 < s < \infty$, be a smooth differential form in a bounded, convex domain Ω, G be Green's operator, and $T : C^\infty(\Omega, \wedge^l) \to C^\infty(\Omega, \wedge^{l-1})$ be the homotopy operator defined in (1.5.1). Then, there exists a constant C, independent of u, such that*

$$\|G(T(u))\|_{locLip_k,\Omega} \le C\|u\|_{s,\Omega}, \tag{9.4.27}$$

where k is a constant with $0 \le k \le 1$.

Applying the same method used in the proof of Theorem 9.4.5, we find the following inequality for the Lipschitz and BMO norms.

Theorem 9.4.7. *Let $u \in L^s_{loc}(\Omega, \wedge^1)$, $1 < s < \infty$, be a solution of the A-harmonic equation (1.2.10) in a bounded, convex domain Ω and G be Green's operator. Then, there exists a constant C, independent of u, such that*

$$\|G(T(u))\|_{locLip_k,\Omega} \le C\|u\|_{\star,\Omega}, \tag{9.4.28}$$

where k is a constant with $0 \le k \le 1$.

Now, if in the proof of Lemma 3.7.2, we use

$$\|T(u)\|_{s,B} \le C|B|diam(B)\|u\|_{s,B},$$

instead of (3.7.7), then (3.7.11) will appear as

$$\|\Delta G(u) - (\Delta G(u))_B\|_{s,B} \leq C|B| diam(B) \|du\|_{s,B}. \qquad (9.4.29)$$

As an application of (9.4.29), we have the following result, which is analogous to Theorem 9.4.2.

Theorem 9.4.8. *Let* $u \in L^s(\Omega, \wedge^l)$, $l = 1, 2, \ldots, n - 1$, $1 < s < \infty$, *be a solution of the A-harmonic equation* (1.2.10) *in a bounded domain* Ω, Δ *be the Laplace–Beltrami operator, and G be Green's operator. Then, there exists a constant C, independent of u, such that*

$$\|\Delta G(u)\|_{locLip_k, \Omega} \leq C\|du\|_{s,\Omega},$$

where k is a constant with $0 \leq k \leq 1$.

Finally, following the proof of Theorem 9.4.3, we obtain the following norm inequality.

Theorem 9.4.9. *Let* $u \in L^s_{loc}(\Omega, \wedge^l)$, $l = 1, 2, \ldots, n$, $1 < s < \infty$, *be a solution of the A-harmonic equation* (1.2.10) *in a bounded domain* Ω, Δ *be the Laplace–Beltrami operator, and G be Green's operator. Then, there exists a constant C, independent of u, such that*

$$\|\Delta G(u)\|_{locLip_k, \Omega} \leq C\|u\|_{\star,\Omega}, \qquad (9.4.30)$$

where k is a constant with $0 \leq k \leq 1$.

9.4.4 Lipschitz norms of $T \circ H$ and $H \circ T$

From Theorem 7.6.1 and the method developed in the proof of Theorem 9.4.1, we have the following results.

Theorem 9.4.10. *Let* $u \in L^s(\Omega, \wedge^l)$, $l = 1, 2, \ldots, n$, $1 < s < \infty$, *be a solution of the A-harmonic equation* (1.2.10) *in a bounded, convex domain* Ω, *H be the projection operator, and* $T : C^\infty(\Omega, \wedge^l) \to C^\infty(\Omega, \wedge^{l-1})$ *be the homotopy operator defined in* (1.5.1). *Then, there exists a constant C, independent of u, such that*

$$\|T(H(u))\|_{locLip_k, \Omega} \leq C\|u\|_{s,\Omega}, \qquad (9.4.31)$$

where k is a constant with $0 \leq k \leq 1$.

Theorem 9.4.11. *Let* $u \in L^s_{loc}(\Omega, \wedge^1)$, $1 < s < \infty$, *be a solution of the A-harmonic equation* (1.2.10) *in a bounded, convex domain* Ω, *H be the* projection operator, and $T : C^\infty(\Omega, \wedge^l) \to C^\infty(\Omega, \wedge^{l-1})$ be the homotopy operator defined in (1.5.1). Then, there exists a constant C, independent of u, such that

$$\|T(H(u))\|_{locLip_k,\Omega} \leq C\|u\|_{\star,\Omega}, \qquad (9.4.32)$$

where k is a constant with $0 \leq k \leq 1$.

Note that inequality (9.4.32) implies that the norm $\|T(H(u))\|_{locLip_k,\Omega}$ of $T(H(u))$ can be controlled by the norm $\|u\|_{\star,\Omega}$ when u is a 1-form.

Now, using Theorem 7.6.15 and the same method as in Theorem 9.4.1, we obtain the following Theorems 9.4.12 and 9.4.13.

Theorem 9.4.12. *Let $u \in L^s(\Omega, \wedge^l)$, $l = 1, 2, \ldots, n$, $1 < s < \infty$, be a solution of the A-harmonic equation (1.2.10) in a bounded, convex domain Ω, H be the projection operator, and $T : C^\infty(\Omega, \wedge^l) \to C^\infty(\Omega, \wedge^{l-1})$ be the homotopy operator defined in (1.5.1). Then, there exists a constant C, independent of u, such that*

$$\|H(T(u))\|_{locLip_k,\Omega} \leq C\|u\|_{s,\Omega}, \qquad (9.4.33)$$

where k is a constant with $0 \leq k \leq 1$.

Theorem 9.4.13. *Let $u \in L^s_{loc}(\Omega, \wedge^1)$, $1 < s < \infty$, be a solution of the A-harmonic equation (1.2.10) in a bounded, convex domain Ω, H be the projection operator, and $T : C^\infty(\Omega, \wedge^l) \to C^\infty(\Omega, \wedge^{l-1})$ be the homotopy operator defined in (1.5.1). Then, there exists a constant C, independent of u, such that*

$$\|H(T(u))\|_{locLip_k,\Omega} \leq C\|u\|_{\star,\Omega}, \qquad (9.4.34)$$

where k is a constant with $0 \leq k \leq 1$.

9.4.5 Estimates for BMO norms

In Section 9.4.1, we have developed some estimates for the Lipschitz norm $\|\cdot\|_{locLip_k,\Omega}$. Now, we will focus on the estimates for the *BMO* norm $\|\cdot\|_{\star,\Omega}$. For this, let $u \in locLip_k(\Omega, \wedge^l)$, $l = 0, 1, \ldots, n$, $0 \leq k \leq 1$, and Ω be a bounded domain. Then, from the definitions of the Lipschitz and *BMO* norms, we have

$$\|u\|_{\star,\Omega} = \sup_{\sigma B \subset \Omega} |B|^{-1} \|u - u_B\|_{1,B}$$

$$= \sup_{\sigma B \subset \Omega} |B|^{k/n} |B|^{-(n+k)/n} \|u - u_B\|_{1,B}$$

$$\leq \sup_{\sigma B \subset \Omega} |\Omega|^{k/n} |B|^{-(n+k)/n} \|u - u_B\|_{1,B}$$

$$\leq |\Omega|^{k/n} \sup_{\sigma B \subset \Omega} |B|^{-(n+k)/n} \|u - u_B\|_{1,B} \qquad (9.4.35)$$

$$\leq C_1 \sup_{\sigma B \subset \Omega} |B|^{-(n+k)/n} \|u - u_B\|_{1,B}$$

$$\leq C_1 \|u\|_{locLip_k,\Omega},$$

where C_1 is a positive constant. Hence, we have proved the following result.

Theorem 9.4.14. *If a differential form* $u \in \text{locLip}_k(\Omega, \wedge^l)$, $l = 0, 1, \ldots, n$, $0 \leq k \leq 1$, *in a bounded domain* Ω, *then* $u \in \text{BMO}(\Omega, \wedge^l)$ *and*

$$\|u\|_{\star,\Omega} \leq C \|u\|_{locLip_k,\Omega}, \qquad (9.4.36)$$

where C *is a constant.*

Theorem 9.4.15. *Let* $u \in L^s(\Omega, \wedge^l)$, $l = 1, 2, \ldots, n$, $1 < s < \infty$, *be a solution of the A-harmonic equation* (1.2.10) *in a bounded, convex domain* Ω *and* $T : C^\infty(\Omega, \wedge^l) \to C^\infty(\Omega, \wedge^{l-1})$ *be the homotopy operator defined in* (1.5.1). *Then, there exists a constant* C, *independent of* u, *such that*

$$\|Tu\|_{\star,\Omega} \leq C \|u\|_{s,\Omega}. \qquad (9.4.37)$$

Similar to the proof of Theorem 9.4.1, we can prove Theorem 9.4.15 directly by using Theorem 7.5.1 and the Hölder inequality. Alternatively, using Theorems 9.4.1 and 9.4.14, we have the following simple proof.

Proof of Theorem 9.4.15. Since inequality (9.4.36) holds for any differential form, we may replace u by Tu in inequality (9.4.36). Thus, it follows that

$$\|Tu\|_{\star,\Omega} \leq C_1 \|Tu\|_{locLip_k,\Omega}, \qquad (9.4.38)$$

where k is a constant with $0 \leq k \leq 1$. On the other hand, from Theorem 9.4.1, we have

$$\|T(u)\|_{locLip_k,\Omega} \leq C_2 \|u\|_{s,\Omega}. \qquad (9.4.39)$$

Combination of (9.4.38) and (9.4.39) yields

$$\|Tu\|_{\star,\Omega} \leq C_3 \|u\|_{s,\Omega}. \qquad \blacksquare$$

As in the proof of Theorem 9.4.15, using inequality (9.4.36) and Theorem 9.4.2, we obtain the following result immediately.

Theorem 9.4.16. *Let $u \in L^s(\Omega, \wedge^l)$, $l = 1, 2, \ldots, n-1$, $1 < s < \infty$, be a solution of the A-harmonic equation (1.2.10) in a bounded domain Ω, G be Green's operator, and H be the projection operator. Then, there exists a constant C, independent of u, such that*

$$\|G(u)\|_{\star,\Omega} \leq C\|du\|_{s,\Omega} \qquad (9.4.40)$$

and

$$\|H(u)\|_{\star,\Omega} \leq C\|du\|_{s,\Omega}. \qquad (9.4.41)$$

For the compositions of two or three operators, including $T \circ G$, $G \circ T$, $H \circ T$, $T \circ H$, and $\Delta \circ G$ which have been well studied in the previous sections, we have similar estimates. Some of these we list as follows:

$$\|T(G(u))\|_{\star,\Omega} \leq C\|u\|_{s,\Omega}, \qquad (9.4.42)$$

$$\|G(T(u))\|_{\star,\Omega} \leq C\|u\|_{s,\Omega}, \qquad (9.4.43)$$

$$\|H(T(u))\|_{\star,\Omega} \leq C\|u\|_{s,\Omega}, \qquad (9.4.44)$$

$$\|T(H(u))\|_{\star,\Omega} \leq C\|u\|_{s,\Omega}, \qquad (9.4.45)$$

$$\|\Delta(G(u))\|_{\star,\Omega} \leq C\|u\|_{s,\Omega}. \qquad (9.4.46)$$

9.4.6 Weighted norm inequalities

So far, we have developed various basic estimates for the Lipschitz and BMO norms of differential forms and the operators applied to these forms. In this section, we discuss the weighted Lipschitz and BMO norms. For $\omega \in L^1_{loc}(\Omega, \wedge^l, w^\alpha)$, $l = 0, 1, \ldots, n$, we write $\omega \in \mathrm{locLip}_k(\Omega, \wedge^l, w^\alpha)$, $0 \leq k \leq 1$, if

$$\|\omega\|_{locLip_k,\Omega,w^\alpha} = \sup_{\sigma Q \subset \Omega} (\mu(Q))^{-(n+k)/n} \|\omega - \omega_Q\|_{1,Q,w^\alpha} < \infty \qquad (9.4.47)$$

for some $\sigma > 1$, where the measure μ is defined by $d\mu = w(x)^\alpha dx$, w is a weight and α is a real number. For convenience, we shall write the simple notation $\mathrm{locLip}_k(\Omega, \wedge^l)$ for $\mathrm{locLip}_k(\Omega, \wedge^l, w^\alpha)$. Similarly, for $\omega \in L^1_{loc}(\Omega, \wedge^l, w^\alpha)$, $l = 0, 1, \ldots, n$, we will write $\omega \in \mathrm{BMO}(\Omega, \wedge^l, w^\alpha)$ if

$$\|\omega\|_{\star,\Omega,w^\alpha} = \sup_{\sigma Q \subset \Omega} (\mu(Q))^{-1} \|\omega - \omega_Q\|_{1,Q,w^\alpha} < \infty \qquad (9.4.48)$$

for some $\sigma > 1$, where the measure μ is defined by $d\mu = w(x)^\alpha dx$, w is a weight, and α is a real number. Again, we shall write $\text{BMO}(\Omega, \wedge^l)$ to replace $\text{BMO}(\Omega, \wedge^l, w^\alpha)$ when it is clear that the integral is weighted.

Theorem 9.4.17. *Let $u \in L^s(\Omega, \wedge^l, \mu)$, $l = 1, 2, \ldots, n$, $1 < s < \infty$, be a solution of the nonhomogeneous A-harmonic equation (1.2.10) in a bounded, convex domain Ω and T be the homotopy operator defined in (1.5.1), where the measure μ is defined by $d\mu = w^\alpha dx$ and $w \in A_r(\Omega)$ for some $r > 1$ with $w(x) \geq \varepsilon > 0$ for any $x \in \Omega$. Then, there exists a constant C, independent of u, such that*

$$\|T(u)\|_{locLip_k, \Omega, w^\alpha} \leq C\|u\|_{s, \Omega, w^\alpha}, \tag{9.4.49}$$

where k and α are constants with $0 \leq k \leq 1$ and $0 < \alpha \leq 1$.

Proof. First, we note that

$$\mu(B) = \int_B w^\alpha dx \geq \int_B \varepsilon^\alpha dx = C_1|B|, \tag{9.4.50}$$

which implies that

$$\frac{1}{\mu(B)} \leq \frac{C_2}{|B|} \tag{9.4.51}$$

for any ball B. From Theorem 7.5.2, we have

$$\|T(u) - (T(u))_B\|_{s, B, w^\alpha} \leq C_3|B|diam(B)\|u\|_{s, \sigma B, w^\alpha} \tag{9.4.52}$$

for all balls B with $\sigma B \subset \Omega$, where $\sigma > 1$ is a constant. Using the Hölder inequality with $1 = 1/s + (s-1)/s$, we find that

$$\|T(u) - (T(u))_B\|_{1, B, w^\alpha}$$
$$= \int_B |T(u) - (T(u))_B| d\mu$$
$$\leq \left(\int_B |T(u) - (T(u))_B|^s d\mu\right)^{1/s} \left(\int_B 1^{s/(s-1)} d\mu\right)^{(s-1)/s}$$
$$= (\mu(B))^{(s-1)/s}\|T(u) - (T(u))_B\|_{s, B, w^\alpha} \tag{9.4.53}$$
$$= (\mu(B))^{1-1/s}\|T(u) - (T(u))_B\|_{s, B, w^\alpha}$$
$$\leq (\mu(B))^{1-1/s}(C_3|B|diam(B)\|u\|_{s, \sigma B, w^\alpha})$$
$$\leq C_4(\mu(B))^{1-1/s}|B|^{1+1/n}\|u\|_{s, \sigma B, w^\alpha}.$$

Now, using the definition of the weighted Lipschitz norm, (9.4.53), and (9.4.51), we obtain

$$\|T(u)\|_{locLip_k,\Omega,w^\alpha}$$

$$= \sup\nolimits_{\sigma B \subset \Omega}(\mu(B))^{-(n+k)/n}\|T(u) - (T(u))_B\|_{1,B,w^\alpha}$$

$$= \sup\nolimits_{\sigma B \subset \Omega}(\mu(B))^{-1-k/n}\|T(u) - (T(u))_B\|_{1,B,w^\alpha}$$

$$\le C_5 \sup\nolimits_{\sigma B \subset \Omega}(\mu(B))^{-1/s-k/n}|B|^{1+1/n}\|u\|_{s,\sigma B,w^\alpha}$$

$$\le C_6 \sup\nolimits_{\sigma B \subset \Omega}|B|^{-1/s-k/n+1+1/n}\|u\|_{s,\sigma B,w^\alpha} \qquad (9.4.54)$$

$$\le C_6 \sup\nolimits_{\sigma B \subset \Omega}|\Omega|^{-1/s-k/n+1+1/n}\|u\|_{s,\sigma B,w^\alpha}$$

$$\le C_6 |\Omega|^{-1/s-k/n+1+1/n}\sup\nolimits_{\sigma B \subset \Omega}\|u\|_{s,\sigma B,w^\alpha}$$

$$\le C_7\|u\|_{s,\Omega,w^\alpha}$$

since $-1/s - k/n + 1 + 1/n > 0$ and $|\Omega| < \infty$. Inequality (9.4.54) is the same as (9.4.49). ∎

As we have done in the previous section, we now develop the $\|\cdot\|_{\star,\Omega,w^\alpha}$ norm estimate. Let $u \in locLip_k(\Omega, \wedge^l)$, $l = 0, 1, \ldots, n$, $0 \le k \le 1$, in a bounded domain Ω. From the definitions of the weighted Lipschitz and the weighted BMO norms, we have

$$\|u\|_{\star,\Omega,w^\alpha}$$

$$= \sup\nolimits_{\sigma B \subset \Omega}(\mu(B))^{-1}\|u - u_B\|_{1,B,w^\alpha}$$

$$= \sup\nolimits_{\sigma B \subset \Omega}(\mu(B))^{k/n}(\mu(B))^{-(n+k)/n}\|u - u_B\|_{1,B,w^\alpha}$$

$$\le \sup\nolimits_{\sigma B \subset \Omega}(\mu(\Omega))^{k/n}(\mu(B))^{-(n+k)/n}\|u - u_B\|_{1,B,w^\alpha} \qquad (9.4.55)$$

$$\le (\mu(\Omega))^{k/n}\sup\nolimits_{\sigma B \subset \Omega}(\mu(B))^{-(n+k)/n}\|u - u_B\|_{1,B,w^\alpha}$$

$$\le C_1 \sup\nolimits_{\sigma B \subset \Omega}(\mu(B))^{-(n+k)/n}\|u - u_B\|_{1,B,w^\alpha}$$

$$\le C_1\|u\|_{locLip_k,\Omega,w^\alpha},$$

where C_1 is a positive constant. Thus, we have proved the following result.

Theorem 9.4.18. *Let $u \in locLip_k(\Omega, \wedge^l, \mu)$, $l = 0, 1, \ldots, n$, $0 \le k \le 1$, be any differential form in a bounded domain Ω, where $w \in A_r(\Omega)$ is a weight for some $r > 1$. Then, $u \in BMO(\Omega, \wedge^l, w^\alpha)$ and*

$$\|u\|_{\star,\Omega,w^\alpha} \le C\|u\|_{locLip_k,\Omega,w^\alpha}, \qquad (9.4.56)$$

where C and α are constants with $0 < \alpha \le 1$.

Theorem 9.4.19. *Let $u \in L^s(\Omega, \wedge^l, \mu)$, $l = 1, 2, \ldots, n$, $1 < s < \infty$, be a solution of the nonhomogeneous A-harmonic equation (1.2.10) in a bounded, convex domain Ω and T be the homotopy operator defined in (1.5.1), where the measure μ is defined by $d\mu = w^\alpha dx$ and $w \in A_r(\Omega)$ for some $r > 1$ with $w(x) \geq \varepsilon > 0$ for any $x \in \Omega$. Then, there exists a constant C, independent of u, such that*

$$\|Tu\|_{\star,\Omega,w^\alpha} \leq C\|u\|_{s,\Omega,w^\alpha}, \tag{9.4.57}$$

where α is a constant with $0 < \alpha \leq 1$.

Proof. Replacing u by Tu in Theorem 9.4.18, we have

$$\|Tu\|_{\star,\Omega,w^\alpha} \leq C_1\|Tu\|_{locLip_k,\Omega,w^\alpha}, \tag{9.4.58}$$

where k is a constant with $0 \leq k \leq 1$. Now, from Theorem 9.4.17, we find that

$$\|T(u)\|_{locLip_k,\Omega,w^\alpha} \leq C_2\|u\|_{s,\Omega,w^\alpha}. \tag{9.4.59}$$

Substituting (9.4.59) into (9.4.58), we obtain

$$\|Tu\|_{\star,\Omega,w^\alpha} \leq C_3\|u\|_{s,\Omega,w^\alpha}. \qquad \blacksquare$$

9.4.7 Estimates in averaging domains

Next, we discuss the weighted norm inequalities for the solutions of the A-harmonic equation (1.2.2) in the $L^1(\mu)$-averaging domain. From Definition 1.6.2, we know that a proper subdomain $\Omega \subset \mathbf{R}^n$ is called an $L^1(\mu)$-averaging domain if there exists a constant C such that

$$\frac{1}{\mu(B_0)} \int_\Omega |u - u_{B_0,\mu}| d\mu \leq C \sup_{4B \subset \Omega} \left(\frac{1}{\mu(B)} \int_B |u - u_{B,\mu}| d\mu \right) \tag{9.4.60}$$

for some ball $B_0 \subset \Omega$ and all $u \in L^1_{loc}(\Omega; \mu)$. Here the supremum is taken over all balls $B \subset \Omega$ with $4B \subset \Omega$. The factor 4 here is for convenience and these domains are independent of this expansion factor, see [183]. Now, recall that in Lemmas 8.2.1 and 8.2.2, we have proved that for any positive constant a, the inequalities

$$\int_D \varphi(\frac{a}{2}|u - u_{D,\mu}|) d\mu \leq \int_D \varphi(a|u|) d\mu \tag{9.4.61}$$

and

$$\int_D \varphi(a|u|)d\mu \le C \int_D \varphi(2a|u-c|)d\mu \qquad (9.4.62)$$

hold for any strictly increasing convex function φ on $[0,\infty)$ with $\varphi(0) = 0$ and a function u in the domain $D \subset \mathbf{R}^n$ with $\varphi(|u|) \in L^1(D;\mu)$, and

$$\mu(\{x \in D : |u-c| > 0\}) > 0$$

for any constant c. The choice $\varphi(t) = t^s$, $s \ge 1$, in (9.4.61) and (9.4.62) is very useful in estimating the L^s-norms. Also, we stress the importance of the arbitrary of c on the right-hand side of (9.4.62). In particular, c may or may not be zero, and we do not need to concern ourselves with the difference between u_B and $u_{B,\mu}$ in the integrand.

Theorem 9.4.20. *Let $u \in L^1(\Omega, \wedge^0, \mu)$ be a solution of the A-harmonic equation (1.2.2), where the measure μ is defined by $d\mu = wdx$ and $w \in A_r(\Omega)$ for some $r > 1$. Then, there exists a constant C, independent of u, such that*

$$\|u\|_{1,\Omega,w} \le C\|u\|_{\star,\Omega,w} \qquad (9.4.63)$$

for any bounded $L^1(\mu)$-averaging domain $\Omega \subset \mathbf{R}^n$.

Proof. Choosing $\varphi(t) = t$ and $c = u_{B_0}$ in (9.4.62), where B_0 is a ball in Ω, we have

$$\begin{aligned}
\|u\|_{1,\Omega,w} &= \int_\Omega |u|wdx \\
&= \int_\Omega |u|d\mu \qquad (9.4.64)\\
&\le C_1 \int_\Omega |u - u_{B_0}|d\mu.
\end{aligned}$$

Thus, it follows that

$$\frac{1}{\mu(B_0)}\int_\Omega |u|d\mu \le \frac{C_1}{\mu(B_0)}\int_\Omega |u - u_{B_0}|d\mu. \qquad (9.4.65)$$

From (9.4.60), (9.4.64), (9.4.65), and (9.4.48), we find that

$$\begin{aligned}
&\mu(B_0)^{-1}\|u\|_{1,\Omega,w} \\
&\le \frac{C_1}{\mu(B_0)}\int_\Omega |u - u_{B_0}|d\mu \\
&\le C_2 \sup_{4B \subset \Omega}\left(\frac{1}{\mu(B)}\int_B |u - u_B|d\mu\right) \qquad (9.4.66)\\
&= C_2 \sup_{4B \subset \Omega}(\mu(B))^{-1}\|u - u_B\|_{1,B,w} \\
&\le C_3\|u\|_{\star,\Omega,w},
\end{aligned}$$

which is equivalent to

$$\|u\|_{1,\Omega,w} \leq C_3\mu(B_0)\|u\|_{\star,\Omega,w} \leq C_4\|u\|_{\star,\Omega,w}. \qquad \blacksquare$$

Note that if $w \in A_r(\Omega)$ and α is a constant with $0 < \alpha \leq 1$, then w^α is also an $A_r(\Omega)$ weight. Hence, we may replace w by w^α in Theorem 9.4.20, to obtain the following version of weighted norm inequality:

$$\|u\|_{1,\Omega,w^\alpha} \leq C\|u\|_{\star,\Omega,w^\alpha}. \qquad (9.4.67)$$

Now, from inequality (9.4.67) and Theorem 9.4.18, we have the following norm comparison theorem for harmonic functions in an $L^1(\mu)$-averaging domain.

Theorem 9.4.21. *Let* $u \in L^1(\Omega, \wedge^0, \mu) \cap \mathrm{locLip}_k(\Omega, \wedge^0, w^\alpha)$, $0 \leq k \leq 1$, *be a solution of the A-harmonic equation* (1.2.2), *where the measure* μ *is defined by* $d\mu = wdx$ *and* $w \in A_r(\Omega)$ *for some* $r > 1$. *Then, there exist constants* C_1 *and* C_2, *independent of* u, *such that*

$$\|u\|_{1,\Omega,w^\alpha} \leq C_1\|u\|_{\star,\Omega,w^\alpha} \leq C_2\|u\|_{\mathrm{locLip}_k,\Omega,w^\alpha} \qquad (9.4.68)$$

for any bounded $L^1(\mu)$-*averaging domain* $\Omega \subset \mathbf{R}^n$.

Similarly, as in Theorem 9.4.17, we have the following weighted norm comparison inequalities for Green's operator G and the projection operator H.

Theorem 9.4.22. *Let* $u \in L^s(\Omega, \wedge^l, \mu)$, $l = 0, 1, 2, \ldots, n-1$, $1 < s < \infty$, *be a solution of the nonhomogeneous A-harmonic equation* (1.2.10) *in a bounded domain* Ω, G *be Green's operator, and* H *be the projection operator, where the measure* μ *is defined by* $d\mu = w^\alpha dx$ *and* $w \in A_r(\Omega)$ *for some* $r > 1$ *with* $w(x) \geq \varepsilon > 0$ *for any* $x \in \Omega$. *Then, there exists a constant* C, *independent of* u, *such that*

$$\|G(u)\|_{\mathrm{locLip}_k,\Omega,w^\alpha} \leq C\|du\|_{s,\Omega,w^\alpha} \qquad (9.4.69)$$

and

$$\|H(u)\|_{\mathrm{locLip}_k,\Omega,w^\alpha} \leq C\|du\|_{s,\Omega,w^\alpha}, \qquad (9.4.70)$$

where k *and* α *are constants with* $0 \leq k \leq 1$ *and* $0 < \alpha \leq 1$.

Applying Theorems 9.4.18 and 9.4.22, we obtain the following estimates for the BMO norms for Green's operator G and the projection operator H, respectively.

Theorem 9.4.23. *Let* $u \in L^s(\Omega, \wedge^l, \mu)$, $l = 0, 1, 2, \ldots, n-1$, $1 < s < \infty$, *be a solution of the nonhomogeneous A-harmonic equation* (1.2.10) *in a bounded domain* Ω, G *be Green's operator, and* H *be the projection operator, where the measure* μ *is defined by* $d\mu = w^\alpha dx$ *and* $w \in A_r(\Omega)$ *for some* $r > 1$ *with* $w(x) \geq \varepsilon > 0$ *for any* $x \in \Omega$. *Then, there exists a constant* C, *independent of* u, *such that*

$$\|G(u)\|_{\star,\Omega,w^\alpha} \leq C\|du\|_{s,\Omega,w^\alpha} \qquad (9.4.71)$$

and

$$\|H(u)\|_{\star,\Omega,w^\alpha} \leq C\|du\|_{s,\Omega,w^\alpha}, \qquad (9.4.72)$$

where α *is a constant with* $0 < \alpha \leq 1$.

Now, we make the following observation. For any differential form u, we have $\|u_B\|_{s,B} \leq C\|u\|_{s,B}$. Also, if c is a closed form, then $c = c_B$. Thus, it follows that

$$\|G(u) - (G(u))_B\|_{\star,\Omega}$$

$$= \sup_{\sigma B \subset \Omega} \left(|B|^{-1}\|G(u) - (G(u))_B - (G(u) - (G(u))_B)_B\|_{1,B}\right)$$

$$= \sup_{\sigma B \subset \Omega} \left(|B|^{-1}\|G(u) - (G(u))_B - (G(u))_B + (G(u))_B\|_{1,B}\right)$$

$$\leq \sup_{\sigma B \subset \Omega} \left(|B|^{-1}\|G(u) - (G(u))_B\|_{1,B}\right)$$

$$\leq \sup_{\sigma B \subset \Omega} \left(|B|^{-1}\|G(u) - c - ((G(u))_B - c_B)\|_{1,B}\right)$$

$$\leq \sup_{\sigma B \subset B} \left(|B|^{-1}\|G(u) - c - (G(u) - c)_B\|_{1,B}\right)$$

$$= \|G(u) - c\|_{\star,\Omega}.$$

Therefore, the following result holds.

Theorem 9.4.24. *Let* $u \in BMO(\Omega, \wedge^l, \mu)$, $l = 0, 1, 2, \ldots, n$, $0 \leq k \leq 1$, *be a smooth differential form and* G *be Green's operator, where the measure* μ *is defined by* $d\mu = wdx$ *and* $w \in A_r(\Omega)$ *for some* $r > 1$. *Then,*

$$\|G(u) - (G(u))_B\|_{\star,\Omega} \leq \|G(u) - c\|_{\star,\Omega} \qquad (9.4.73)$$

holds for any bounded domain Ω *and any closed form* c.

It should be noticed that in Theorem 9.4.24 the differential form u need not be a solution of any version of the A-harmonic equation. Further, similar to the case of Green's operator, similar estimates hold for the homotopy operator and the projection operator, and also for the compositions of these operators.

Theorem 9.4.25. *Let $u \in \mathrm{BMO}(\Omega, \wedge^l, \mu)$, $l = 0, 1, 2, \ldots, n$, $0 \leq k \leq 1$, be a smooth differential form, T be the homotopy operator, and H be the projection operator, where the measure μ is defined by $d\mu = wdx$ and $w \in A_r(\Omega)$ for some $r > 1$. Then,*

$$\|H(u) - (H(u))_B\|_{\star,\Omega} \leq \|H(u) - c\|_{\star,\Omega} \qquad (9.4.74)$$

and

$$\|T(u) - (T(u))_B\|_{\star,\Omega} \leq \|T(u) - c\|_{\star,\Omega} \qquad (9.4.75)$$

hold for any bounded, convex domain Ω and any closed form c.

We have presented the *BMO* norm estimates above. By the same method, we can develop a similar estimate for the Lipschitz norm as follows:

$$\|G(u) - (G(u))_B\|_{locLip_k,\Omega}$$

$$= \sup_{\sigma B \subset \Omega} \left(|B|^{-(n+k)/n} \|G(u) - (G(u))_B - (G(u) - (G(u))_B)_B\|_{1,B}\right)$$

$$= \sup_{\sigma B \subset \Omega} \left(|B|^{-(n+k)/n} \|G(u) - (G(u))_B - (G(u))_B + (G(u))_B\|_{1,B}\right)$$

$$\leq \sup_{\sigma B \subset \Omega} \left(|B|^{-(n+k)/n} \|G(u) - (G(u))_B\|_{1,B}\right)$$

$$\leq \sup_{\sigma B \subset \Omega} \left(|B|^{-(n+k)/n} \|G(u) - c - ((G(u))_B - c_B)\|_{1,B}\right)$$

$$\leq \sup_{\sigma B \subset B} \left(|B|^{-(n+k)/n} \|G(u) - c - (G(u) - c)_B\|_{1,B}\right)$$

$$= \|G(u) - c\|_{locLip_k,\Omega}.$$

Hence, we have established the following Lipschitz norm inequality for smooth differential forms.

Theorem 9.4.26. *Let $u \in locLip_k(\Omega, \wedge^l, \mu)$, $l = 0, 1, 2, \ldots, n$, $0 \leq k \leq 1$, be a smooth differential form and G be Green's operator, where the measure μ is defined by $d\mu = wdx$ and $w \in A_r(\Omega)$ for some $r > 1$. Then,*

$$\|G(u) - (G(u))_B\|_{locLip_k,\Omega} \leq \|G(u) - c\|_{locLip_k,\Omega} \qquad (9.4.76)$$

holds for any bounded domain Ω and any closed form c.

Now we state a result which is analogous to Theorem 9.4.25.

Theorem 9.4.27. *Let $u \in locLip_k(\Omega, \wedge^l, \mu)$, $l = 1, 2, \ldots, n$, $0 \leq k \leq 1$, be a smooth differential form, T be the homotopy operator, and H be the projection operator, where the measure μ is defined by $d\mu = wdx$ and $w(x) \in A_r(\Omega)$ for* *some $r > 1$. Then,*

$$\|H(u) - (H(u))_B\|_{locLip_k,\Omega} \leq \|H(u) - c\|_{locLip_k,\Omega} \tag{9.4.77}$$

and

$$\|T(u) - (T(u))_B\|_{locLip_k,\Omega} \leq \|T(u) - c\|_{locLip_k,\Omega} \tag{9.4.78}$$

hold for any bounded, convex domain Ω and any closed form c.

If we choose μ to be the Lebesgue measure in the definition of the $L^s(\mu)$-averaging domain, we have

$$\left(\frac{1}{|B_0|}\int_\Omega |u - u_{B_0}|^s dx\right)^{1/s} \leq C \sup_{4B \subset \Omega} \left(\frac{1}{|B|}\int_B |u - u_B|^s dx\right)^{1/s}, \tag{9.4.79}$$

which can be considered as the definition of the L^s-averaging domain. We also note that when $\varphi(t) = t^s$ and $c = u_{B_0}$ in (9.4.62), we obtain

$$\int_\Omega |u|^s dx \leq C \int_\Omega |u - u_{B_0}|^s dx. \tag{9.4.80}$$

We are now ready to prove the following inequality in the L^s-averaging domain.

Theorem 9.4.28. *Let $u \in BMO(\Omega, \wedge^0)$ be a solution of the nonhomogeneous A-harmonic equation (1.2.10) in an L^s-averaging domain Ω, $1 \leq s < \infty$. Then, there exists a constant C, independent of u, such that*

$$\|u\|_{s,\Omega} \leq C\|u\|_{\star,\Omega}. \tag{9.4.81}$$

Proof. From the weak reverse Hölder inequality, we have

$$\|u - u_B\|_{s,B} \leq C_3|B|^{(1-s)/s}\|u - u_B\|_{1,\sigma B} \tag{9.4.82}$$

for any ball B and some $\sigma > 1$. Now, applying (9.4.79) and (9.4.80), we find that

$$\begin{aligned}
\|u\|_{s,\Omega} &= \left(\int_\Omega |u|^s dx\right)^{1/s} \\
&\leq C_1|B_0|^{1/s}\left(\frac{1}{|B_0|}\int_\Omega |u - u_{B_0}|^s dx\right)^{1/s} \\
&\leq C_2 \sup_{\sigma B \subset \Omega}\left(\frac{1}{|B|}\int_B |u - u_B|^s dx\right)^{1/s} \\
&\leq C_2 \sup_{\sigma B \subset \Omega} |B|^{-1/s}\|u - u_B\|_{s,B}.
\end{aligned} \tag{9.4.83}$$

Finally, substituting (9.4.82) into (9.4.83), we obtain

$$\|u\|_{s,\Omega} \leq C_2 \sup_{\sigma B \subset \Omega} |B|^{-1/s} \|u - u_B\|_{s,B}$$

$$\leq C_2 \sup_{\sigma B \subset \Omega} |B|^{-1/s} (C_3 |B|^{(1-s)/s} \|u - u_B\|_{1,\sigma B})$$

$$\leq C_4 \sup_{\sigma B \subset \Omega} |B|^{-1} \|u - u_B\|_{1,\sigma B} \tag{9.4.84}$$

$$\leq C_5 \|u\|_{\star,\Omega}. \qquad \blacksquare$$

9.4.8 Applications

Let u be an $(n-k) \times (n-k)$ subdeterminant of Jacobian $J(x,f)$ of a mapping $f : \Omega \to \mathbf{R}^n$, which is obtained by deleting the k rows and k columns, $k = 0, 1, \ldots, n-1$, say,

$$u = J(x_{j_1}, x_{j_2}, \ldots, x_{j_{n-k}}; f^{i_1}, f^{i_2}, \ldots, f^{i_{n-k}})$$

$$= \begin{vmatrix} f^{i_1}_{x_{j_1}} & f^{i_1}_{x_{j_2}} & f^{i_1}_{x_{j_3}} & \cdots & f^{i_1}_{x_{j_{n-k}}} \\ f^{i_2}_{x_{j_1}} & f^{i_2}_{x_{j_2}} & f^{i_2}_{x_{j_3}} & \cdots & f^{i_2}_{x_{j_{n-k}}} \\ \vdots & \vdots & \vdots & \ddots & \vdots \\ f^{i_{n-k}}_{x_{j_1}} & f^{i_{n-k}}_{x_{j_2}} & f^{i_{n-k}}_{x_{j_3}} & \cdots & f^{i_{n-k}}_{x_{j_{n-k}}} \end{vmatrix},$$

where $\{i_1, i_2, \ldots, i_{n-k}\} \subset \{1, 2, \ldots, n\}$ and $\{j_1, j_2, \ldots, j_{n-k}\} \subset \{1, 2, \ldots, n\}$. Also, it is easy to see that $J(x_{j_1}, x_{j_2}, \ldots, x_{j_{n-k}}; f^{i_1}, f^{i_2}, \ldots, f^{i_{n-k}}) dx_{j_1} \wedge dx_{j_2} \wedge \cdots \wedge dx_{j_{n-k}}$ is an $(n-k)$-form. Thus, all estimates for differential forms are applicable to the $(n-k)$-form $J(x_{j_1}, x_{j_2}, \ldots, x_{j_{n-k}}; f^{i_1}, f^{i_2}, \ldots, f^{i_{n-k}}) dx_{j_1} \wedge dx_{j_2} \wedge \cdots \wedge dx_{j_{n-k}}$. For example, choosing $u = J(x,f) dx$ and applying Theorems 9.4.14 and 9.4.18 to u, respectively, we have the following theorems.

Theorem 9.4.29. *Let $J(x,f) dx \in \text{locLip}_k(\Omega, \wedge^n)$, $0 \leq k \leq 1$, where $J(x,f)$ is the Jacobian of the mapping $f = (f^1, \ldots, f^n) : \Omega \to \mathbf{R}^n$ and Ω is a bounded domain in $\mathbf{R}^n$. Then, $J(x,f) dx \in \text{BMO}(\Omega, \wedge^n)$ and*

$$\|J(x,f)\|_{\star,\Omega} \leq C \|J(x,f)\|_{\text{locLip}_k,\Omega}, \tag{9.4.85}$$

where C is a constant.

Theorem 9.4.30. *Let* $J(x,f)dx \in \text{locLip}_k(\Omega, \wedge^n, \mu)$, $0 \le k \le 1$, *where* $J(x,f)$ *is the Jacobian of the mapping* $f = (f^1, \ldots, f^n) : \Omega \to \mathbf{R}^n$, $w \in A_r(\Omega)$ *is a weight for some* $r > 1$, *and* Ω *is a bounded domain in* $\mathbf{R}^n$. *Then,* $J(x,f)dx \in \text{BMO}(\Omega, \wedge^n, w^\alpha)$ *and*

$$\|J(x,f)\|_{\star, \Omega, w^\alpha} \le C\|J(x,f)\|_{locLip_k, \Omega, w^\alpha}, \tag{9.4.86}$$

where C *and* α *are constants with* $0 < \alpha \le 1$.

Notes to Chapter 9. Considering the size of the monograph, we will not discuss the norm comparison theorems for the compositions of three operators here. For recent results on *BMO* spaces, Lipschitz classes and domains, and related domains, see [315–324, 137–143, 154, 167–169, 309]. We also encourage readers to see [325–333] for different versions of the Poincaré inequality.

References

1. **J. M. Ball**, Convexity conditions and existence theorems in nonlinear elasticity, *Arch. Rational Mech. Anal.*, **63** (1977), 337–403.
2. **J. M. Ball and F. Murat**, $W^{1,p}$-quasi-convexity and variational problems for multiple integrals, *J. Funct. Anal.*, **58** (1984), 225–253.
3. **G. F. D. Duff**, Differential forms in manifolds with boundary, *Ann. Math.*, **56** (1952), 115–127.
4. **G. F. D. Duff and D. C. Spencer**, Harmonic tensors on Riemannian manifolds with boundary, *Ann. Math.*, **56** (1952), 128–156.
5. **F. W. Gehring**, Symmetrization of rings in space, *Trans. Amer. Math. Soc.*, **101** (1961), 499–519.
6. **F. W. Gehring**, The L^p-integrability of partial derivatives of a quasiconformal mapping, *Acta Math.*, **130** (1973), 265–277.
7. **F. W. Gehring and J. Väisälä**, Hausdorff dimension and quasiconformal mappings, *J. London Math. Soc.*, **6** (1973), 504–512.
8. **F. W. Gehring**, The L^p-integrability of the partial derivatives of quasiconformal mapping, *Bull. Amer. Math. Soc.*, **79** (1973), 465–466.
9. **F. W. Gehring**, Topics in quasiconformal mappings. Quasiconformal space mappings, 20–38, *Lect. Notes Math.*, **1508**, Springer, Berlin, 1992.
10. **F. W. Gehring and B. G. Osgood**, Lipschitz classes and quasi-conformal extension domains, *Complex Variables*, **9** (1987), 175–188.
11. **F. W. Gehring and B. G. Osgood**, Uniform domains and the quasihyperbolic metric, *J. Anal. Math.*, **36** (1979), 50–74.
12. **F. W. Gehring and B. Palka**, Quasiconformally homogeneous domains, *Journal d'analyse mathématique*, **30** (1976), 172–199.
13. **F. W. Gehring**, Topics in quasiconformal mappings, *Proceedings of ICM, Berkeley*, 1986, 62–82.
14. **F. W. Gehring**, Some metric properties of quasi-conformal mappings, *Proceedings of the International Congress of Mathematicians* (Vancouver, B. C., 1974), Vol. 2, 203–206. Canad. Math. Congress, Montreal, Que., 1975.
15. **F. W. Gehring**, Quasiconformal mappings in R^n, Lectures on quasiconformal mappings, 44–110, *Lect. Note*, **14**, Dept. Math., Univ. Maryland, College Park, Md., 1975.
16. **F. W. Gehring**, Lower dimensional absolute continuity properties of quasiconformal mappings, *Math. Proc. Cambridge Philos. Soc.*, **78** (1975), 81–93.
17. **F. W. Gehring, W. K. Hayman and A. Hinkkanen**, Analytic functions satisfying Hölder conditions on the boundary, *J. Approx. Theory*, **35** (1982), 243–249.

18. **F. W. Gehring and O. Martio**, Quasidisks and the Hardy-Littlewood property, *Complex Variables Theory Appl.*, **2** (1983), 67–78.

19. **F. W. Gehring and O. Martio**, Lipschitz classes and quasi-conformal mappings. *Ann. Acad. Sci. Fenn. Ser.* A.I. Math., **10** (1985), 203–219.

20. **R. P. Agarwal and S. Ding**, Advances in differential forms and the A-harmonic equations, *Math. Comput. Model.*, **37** (2003), 1393–1426.

21. **C. Scott**, L^p-theory of differential forms on manifolds, *Trans. Amer. Math. Soc.*, **347** (1995), 2075–2096.

22. **M. P. do Carmo**, Differential forms and applications, Springer-Verlag, Berlin, 1994.

23. **F. W. Warner**, Foundations of differentiable manifolds and Lie groups, Springer-Verlag, New York, 1983.

24. **S. Morita**, Geometry of differential forms. Translated from the two-volume Japanese original (1997, 1998) by Teruko Nagase and Katsumi Nomizu. Translations of Mathematical Monographs, 201. Iwanami Series in Modern Mathematics. American Mathematical Society, Providence, RI, 2001.

25. **S. Lang**, Differential and Riemannian manifolds, Springer-Verlag, New York, 1995.

26. **J. W. Robbin, R. C. Rogers and B. Temple**, On weak continuity and the Hodge decomposition, *Trans. Amer. Math. Soc.*, **303** (1987), 609–618.

27. **S. M. Buckley, J. J. Manfredi and E. Villamor**, Regularity theory and traces of A-harmonic functions, *Trans. Amer. Math. Soc.*, **348** (1996), 1–12.

28. **J. Cheng and A. Fang**, (G, H)-quasiregular mappings and B-harmonic equations (Chinese), *Acta Math. Sinica*, **42** (1999), 883–888.

29. **M. G. Crandall and J. Zhang**, Another way to say harmonic, *Trans. Amer. Math. Soc.*, **355** (2003), 241–263.

30. **B. E. J. Dahlberg**, Estimates of harmonic measure, *Arch. Rational Mech. Anal.*, **65** (1977), 272–288.

31. **S. Ding**, Some weighted integral inequalities for solutions of the A-harmonic equation, invited lecture at the ICM2002 Complex Analysis Satellite Conference in Shanghai, August 15–17, 2002.

32. **S. Ding, G. Bao and Y. Xing**, Sobolev-Poincaré embeddings for operators on harmonic forms on manifolds, *Comput. Math. Appl.*, **47** (2004), 259–270.

33. **S. Ding, Y. Xing and G. Bao**, $A_r(\Omega)$-weighted inequalities for A-harmonic tensors and related operators, *J. Math. Anal. Appl.*, **322** (2006), 219–232.

34. **S. Ding and Y. Gai**, A_r-weighted Poincaré-type inequalities for differential forms in some domains, *Acta Math. Sinica*, **46** (2003), 23–28.

35. **S. Ding**, Local and global estimates for solutions to the A-harmonic equation and related operators, *Proceedings of Conference on Differential & Difference Equations and Applications*, Florida Institute of technology, Hindawi, 2006, 351–361.

36. **S. Ding and R. P. Agarwal**, Integrability of the solutions to conjugate A-harmonic equation in $L^s(\mu)$-averaging domains, *Arch. Inequal. Appl.*, **2** (2004), 517–526.

37. **P. Duren**, F. W. Gehring: a biographical sketch, Quasiconformal mappings and analysis (Ann Arbor, MI, 1995), 1–4, Springer, New York, 1998.

38. **P. Duren, J. Heinonen, B. Osgood and B. Palka**, Quasiconformal mappings and analysis, a collection of papers honoring F. W. Gehring (papers from the International Symposium held in Ann Arbor, MI, August 1995), Springer-Verlag, New York, 1998.

39. **A. Fang**, On the compactness of homeomorphisms with integrable dilatations, *Acta Math. Sinica (N.S.)*, **14** (1998), 161–168.

40. **A. Fang**, Quasiconformal mappings and the theory of functions for systems of nonlinear elliptic partial differential equations of first order (Chinese), *Acta Math. Sinica*, **23** (1980), 280–292.

41. **A. Fang**, Riemann's existence theorem for nonlinear quasiconformal mappings (Chinese), *Acta Math. Sinica*, **23** (1980), 341–353.

42. **A. Fang**, Two kinds of boundary value problems for first-order systems of nonlinear elliptic partial differential equations (Chinese), *Chinese Ann. Math.*, **3** (1982), 587–594.

43. **R. Fehlmann and M. Vuorinen**, Mori's theorem for n-dimensional quasiconformal mappings, *Ann. Acad. Sci. Fenn. Ser. A I Math.*, **13** (1988), 111–124.

44. **R. Finn**, On partial differential equations (whose solutions admit no isolated singularities), *Scripta Math.*, **26** (1963), 107–115.

45. **R. Finn and J. Serrin**, On the Hölder continuity of quasi-conformal and elliptic mappings. *Trans. Amer. Math. Soc.*, **89** (1958), 1–15.

46. **R. Finn and V. Solonnikov**, Gradient estimates for solutions of the Navier-Stokes equations, *Topol. Methods Nonlinear Anal.*, **9** (1997), 29–39.

47. **O. Lehto**, Fred Gehring and Finnish mathematics, *Quasiconformal mappings and analysis* (Ann Arbor, MI, 1995), 19–32, Springer, New York, 1998.

48. **A. K. Lerner**, Weighted rearrangement inequalities for local sharp maximal functions, *Trans. Amer. Math. Soc.*, **357** (2005), 2445–2465.

49. **P. Li and J Wang**, Hölder estimates and regularity for holomorphic and harmonic functions, *J. Differ. Geom.*, **58** (2001), 309–329.

50. **Q. Ran**, Fusion problems of very weak solutions for nonhomogeneous A-harmonic equations (Chinese), *Math. Appl (Wuhan)*, **15** (2002), 113–117.

51. **J. Serrin**, Local behavior of solutions of quasi-linear equations, *Acta Math.*, **111** (1964), 247–302.

52. **B. Stroffolini**, On weakly A-harmonic tensors, *Studia Math.*, **114** (1995), 289–301.

53. **H. Grötzsch**, Über die Verzerrung bei schlichten nichtkonformen Abbildungen und über eine damit zusammenhängende Erweiterung des Picardschen Satzes, *Ber. Verh. Sachs. Akad. Wiss. Leipzig*, **80** (1928), 503–507.

54. **M. A. Lavrent'ev**, Sur un critére différentiel des transformations homeomorphes des domaines a trois dimensions, *Dokl. Akad. Nauk SSSR*, **22** (1938), 241–242.

55. **O. Teichmüller**, Untersuchungen über konforme und quasikonforme Abbildung, *Deutsche Math.*, **3** (1938), 621–678.

56. **L. V. Ahlfors**, On quasiconformal mappings, *J. Analyse Math.*, **3** (1954), 1–58, 207–208.

57. **J. Väisälä**, On quasiconformal mappings in space, *Ann. Acad. Sci. Fenn. Ser. A I*, **298** (1961), 1–36.

58. **Yu. G. Reshetnyak**, Estimates of the modulus of continuity for certain mappings (Russian), *Sibirsk. Mat. Zh.*, **7** (1966), 1106–1114.

59. **J. Heinonen, T. Kilpelainen and O. Martio**, Nonlinear potential theory of degenerate elliptic equations, Oxford University Press Inc., New York, 1993.

60. **S. Ding**, Integral estimates for the Laplace-Beltrami and Green's operators applied to differential forms, *Z. Anal. Anwendungen*, **22** (2003), 939–957.

61. **H. Gao**, Regularity for very weak solutions of A-harmonic equations (Chinese), *Acta Math. Sinica*, **44** (2001), 605–610.

62. **H. Gao**, Weighted integral inequalities for conjugate A-harmonic tensors, *J. Math. Anal. Appl.*, **281** (2003), 253–263.

63. **H. Gao**, Some properties of weakly quasiregular mappings (Chinese), *Acta Math. Sinica*, **45** (2002), 191–196.

64. **C. E. Gutierrez and R. L. Wheeden**, Sobolev interpolation inequalities with weights, *Trans. Amer. Math. Soc.*, **323** (1991), 263–281.

65. **V. Gutlyanskii, O. Martio, T. Sugawa and M. Vuorinen**, On the degenerate Beltrami equation, *Trans. Amer. Math. Soc.*, **357** (2005), 875–900.

66. **A. Iovita and M. Spiess**, Logarithmic differential forms on p-adic symmetric spaces, *Duke Math. J.*, **110** (2001), 253–278.

67. **C. A. Nolder**, Conjugate harmonic functions and Clifford algebras. *J. Math. Anal. Appl.*, **302** (2005), 137–142.
68. **C. A. Nolder**, The H^p-norm of a quasiconformal mapping, *J. Math. Anal. Appl.*, **275** (2002), 557–561.
69. **C. A. Nolder**, A characterization of certain measures using quasiconformal mappings, *Proc. Amer. Math. Soc.*, **109** (1990), 349–456.
70. **C. A. Nolder**, A quasiregular analogue of theorem of Hardy and Littlewood, *Trans. Amer. Math. Soc.*, **331**(1992), 215–226.
71. **C. A. Nolder**, Hardy-Littlewood theorems for A-harmonic tensors, *Illinois J. Math.*, **43** (1999), 613–631.
72. **C. A. Nolder**, Global integrability theorems for A-harmonic tensors, *J. Math. Anal. Appl.*, **247**(2000), 236–245.
73. **C. A. Nolder**, An L^p definition of interpolating Blaschke products, *Proc. Amer. Math. Soc.*, **128** (2000), 1799–1806.
74. **C. A. Nolder**, Carleson's inequality and quasiconformal mappings, *Rocky Mountain J. Math.*, **24** (1994), 1055–1063.
75. **S. Xie and A. Fang**, Global higher integrability for the gradient of very weak solutions of a class of nonlinear elliptic systems, *Nonlinear Anal.*, **53** (2003), 1127–1147.
76. **S. Xie and B. Dai**, Local regularity of very weak solutions of a class of nonlinear elliptic systems (Chinese), *Math. Appl. (Wuhan)*, **14** (2001), 93–97.
77. **S. Xie and J. Zhang**, Boundary higher integrability for the gradient of very weak solutions of a nonhomogeneous harmonic-type equation (Chinese), *Natur. Sci. J. Xiangtan Univ.*, **23** (2001), 6–10.
78. **S. Xie**, Regularity of very weak solutions for a nonhomogeneous A-harmonic type equation (Chinese), *J. Shanghai Jiaotong Univ. (Chin. Ed.)*, **35** (2001), 1105–1108.
79. **S. Xie**, Weakly Quasi-regular mappings and very weak solutions, invited lecture at the International Congress of Mathematicians 2002 Satellite Conference for Analysis, Shanghai, August, 2002.
80. **Y. Xing**, Weighted integral inequalities for solutions of the A-harmonic equation, *J. Math. Anal. Appl.*, **279** (2003), 350–363.
81. **Y. Xing**, Two-weight imbedding inequalities for solutions to the A-harmonic equation, *J. Math. Anal. Appl.*, **307** (2005), 555–564.
82. **Y. Xing**, Weighted Poincaré-type estimates for conjugate A-harmonic tensors, *J. Inequal. Appl.*, **1** (2005), 1–6.
83. **Y. Xing, B. Wang and S. Ding**, Global estimates for compositions of operators applied to differential forms, *Comput. Math. Appl.*, **48** (2004), 1905–1913.
84. **Y. Xing and G. Bao**, Two-weight weak reverse Hölder inequality for differential forms, *J. Nat. Sci. Heilongjiang Univ.* (Heilongjiang Daxue Ziran Kexue Xuebao), **22** (2005), 138–140.
85. **Y. Xing**, Weighted integral inequalities for solutions of the A-harmonic equation, *J. Math. Anal. Appl.*, **279** (2003), 350–363.
86. **Y. Xing and C. Wu**, Global weighted inequalities for operators and harmonic forms on manifolds, *J. Math. Anal. Appl.*, **294** (2004), 294–309.
87. **O. Martio, S. Rickman and J. Väisälä**, Definitions for quasiregular mappings, *Ann. Acad. Sci. Fenn. Ser. A I*, **448** (1969), 1–40.
88. **O. Martio, S. Rickman and J. Väisälä**, Distortion and singularities of quasiregular mappings, *Ann. Acad. Sci. Fenn. Ser. A I*, **465** (1970), 1–13.
89. **O. Martio, S. Rickman and J. Väisälä**, Topological and metric properties of quasiregular mappings, *Ann. Acad. Sci. Fenn. Ser. A I*, **488** (1971), 1–31.
90. **J. Väisälä**, Modulus and capacity inequalities for quasiregular mappings, *Ann. Acad. Sci. Fenn. Ser. A I*, **509** (1972), 1–14.
91. **L. V. Ahlfors and A. Beurling**, Conformal invariants and function-theoretic null-sets, *Acta Math.*, **83** (1950), 101–129.

92. **M. Vuorinen**, Conformal geometry and quasiregular mappings, **1319**, Springer-Verlag, Berlin, 1988.

93. **M. Vuorinen**, Conformal invariants and quasiregular mappings, *J. Anal. Math.*, **45** (1985), 69–115.

94. **M. Vuorinen**, On the distortion of n-dimensional quasiconformal mappings, *Proc. Amer. Math. Soc.*, **96** (1986), 275–283.

95. **M. Vuorinen**, On quasiconformal mappings and domains with a complete conformal metric, *Math. Z.*, **194** (1987), 459–470.

96. **G. D. Anderson, M. Vamanamurthy and M. Vuorinen**, Dimension-free quasiconformal distortion in n-space, *Trnas. Amer. Math. Soc.*, **297** (1986), 687–706.

97. **G. D. Anderson, M. Vamanamurthy and M. Vuorinen**, Sharp distortion theorems for quasiconformal mappings, *Trnas. Amer. Math. Soc.*, **305** (1988), 95–111.

98. **M. Lehtinen and M. Vuorinen**, On Teichmüller's modulus problem in the plane, *Rev. Roumaine Math. Pures Appl.*, **33** (1988), 97–106.

99. **T. Iwaniec and A. Lutoborski**, Integral estimates for null Lagrangians, *Arch. Rational Mech. Anal.*, **125** (1993), 25–79.

100. **G. H. Hardy and J. E. Littlewood**, Some properties of conjugate functions, *J. Reine Angew. Math.*, **167** (1932), 405–423.

101. **G. H. Hardy and J. E. Littlewood**, Some properties of fractional Integrals II, *Math. Z.*, **34** (1932), 403–439.

102. **S. Ding**, Two-weight Caccioppoli inequalities for solutions of nonhomogeneous A-harmonic equations on Riemannian manifolds, *Proc. Amer. Math. Soc.*, **132** (2004), 2367–2375.

103. **Y. Wang and C. Wu**, Sobolev imbedding theorems and Poincaré inequalities for Green's operator on solutions of the nonhomogeneous A-harmonic equation, *Comput. and Math. Appl.*, **47** (2004), 1545–1554.

104. **S. Ding**, Local and global norm comparison theorems for solutions to the nonhomogeneous A-harmonic equation, *J. Math. Anal. Appl.*, **335** (2007), 1274–1293.

105. **T. Iwaniec**, p-harmonic tensors and quasiregular mappings, *Ann. Math.*, **136** (1992), 589–624.

106. **D. Franke, O. Martio, V. M. Miklyukov, M. Vuorinen and R. Wisk**, Quasiregular mappings and WT-classes of differential forms on Riemannian manifolds, *Pacific J. Math.*, **202** (2002), 73–92.

107. **G. Aronsson and P. Lindqvist**, On p-harmonic functions in the plane and their stream functions, *J. Differ. Equ.*, **74** (1988), 157–178.

108. **G. Aronsson**, Construction of singular solutions to the p-harmonic equation and its limit equation for $p = \infty$, *Manuscripta Math.*, **56** (1986), 135–158.

109. **G. Aronsson**, On certain p-harmonic functions in the plane, *Manuscripta Math.*, **61** (1988), 79–101.

110. **M. Carozza and A. Passarelli Di Napoli**, A regularity result for p-harmonic equations with measure data, *Collect. Math.*, **55** (2004), 11–19.

111. **S. Ding**, Some examples of conjugate p-harmonic differential forms, *J. Math. Anal. Appl.*, **227** (1998), 251–270.

112. **P. E. Herman, R. Peirone and R. S. Strichartz**, p-energy and p-harmonic functions on Sierpinski Gasket type fractals, *Potential Anal.*, **20** (2004), 125–148.

113. **S. Kim**, Harnack inequality for nondivergent elliptic operators on Riemannian manifolds, *Pacific J. Math.*, **213** (2004), 281–293.

114. **K. Kodaira**, Harmonic fields in Riemannian manifolds (generalized potential theory), *Ann. Math.*, **50** (1949), 587–665.

115. **P. Koskela and J. Onninen**, Mappings of finite distortion: the sharp modulus of continuity, *Trans. Amer. Math. Soc.*, **355** (2003), 1905–1920.

116. **P. Lindqvist**, On p-harmonic functions in the complex plane and curvature, *Israel J. Math.*, **63** (1988), 257–269.

117. **G. Moscariello**, On the regularity of solutions to a class of p-harmonic equations. Nonlinear elliptic and parabolic equations and systems (Pisa, 2002), *Comm. Appl. Nonlinear Anal.*, **10** (2003), 1–8.

118. **K. Rogovin**, Local compactness for families of A-harmonic functions, *Illinois J. Math.*, **48** (2004), 71–87.

119. **T. Iwaniec and C. A. Nolder**, Hardy-Littlewood inequality for quasiregular mappings in certain domains in R^n, *Ann. Acad. Sci. Fenn. Ser. A.I. Math.* **10** (1985), 267–282.

120. **S. M. Buckley**, Pointwise multipliers for reverse Hölder spaces, *Studia Math.*, **109** (1994), 23–39.

121. **S. Ding and P. Shi**, Weighted Poincaré-type inequalities for differential forms in $L^s(\mu)$-averaging domains, *J. Math. Anal. Appl.*, **227** (1998), 200–215.

122. **M. Frazier and S. Roudenko**, Matrix-weighted Besov spaces and conditions of A_p type for $0 < p < 1$, *Indiana Univ. Math. J.*, **5** (2004), 1225–1254.

123. **J. B. Garnett**, Bounded analytic functions, Academic Press, New York, 1970.

124. **S. Ding**, New weighted integral inequalities for differential forms in some domains, *Pacific J. Math.*, **149** (2000), 43–56.

125. **C. J. Neugebauer**, Inserting A_p-weights, *Proc. Amer. Math. Soc.*, **87** (1983), 644–648.

126. **D. Cruz-Uribe and C. Peter**, Two-weight, weak-type norm inequalities for fractional integrals, Calderón-Zygmund operators and commutators, *Indiana Univ. Math. J.*, **49** (2000), 697–721.

127. **S. Ding and B. Liu**, Singular integral of the composite operator, *Applied Mathematics Letters*, **22** (2009), 1271–1275.

128. **Y. Ding and C. Lin**, Two-weight norm inequalities for the rough fractional integrals, *Int. J. Math. Math. Sci.*, **25** (2001), 517–524.

129. **J. García-Cuerva and J. M. Martell**, Two-weight norm inequalities for maximal operators and fractional integrals on non-homogeneous spaces, *Indiana Univ. Math. J.*, **50** (2001), 1241–1279.

130. **E. T. Sawyer**, A characterization of two weight norm inequalities for fractional and Poisson integrals, *Trans. Amer. Math. Soc.*, **308** (1988), 533–545.

131. **L. Wang**, Generalized abstract mean values with two weight functions and related inequalities (Chinese), *Sichuan Daxue Xuebao*, **40** (2003), 618–621.

132. **C. Wu and L. Shu**, Two-weight norm inequalities for a class of rough multilinear operators, *Anal. Theory Appl.*, **19** (2003), 209–219.

133. **Y. Xing and S. Ding**, Norm comparison inequalities for the composite operator, *Journal of Inequalities and Applications*, **2009** (2009), Article ID 212915, 13 pages, doi:10.1155/2009/212915.

134. **M. Riesz**, Sur les fonctions conjugues, *Math. Z.*, **27** (1927), 218–277.

135. **K. Michalik and M. Ryznar**, Relative Fatou theorem for α-harmonic functions in Lipschitz domains, *Illinois J. Math.*, **48** (2004), 977–998.

136. **M. Miranda**, Renato Caccioppoli and geometric measure theory (Italian), International Symposium in honor of Renato Caccioppoli (Naples, 1989) *Ricerche Mat.*, **40** (1991), suppl., 111–118.

137. **D. Mitrea and M. Mitrea**, Boundary integral methods for harmonic differential forms in Lipschitz domains, *Electron. Res. Announc. Amer. Math. Soc.*, **2** (1996), 92–97.

138. **M. Mitrea**, Sharp Hodge decompositions, Maxwell's equations, and vector Poisson problems on nonsmooth, three-dimensional Riemannian manifolds, *Duke Math. J.*, **125** (2004), 467–547.

139. **M. Mitrea and M. Taylor**, Boundary layer methods for Lipschitz domains in Riemannian manifolds, *J. Funct. Anal.*, **163** (1999), 181–251.

140. **M. Mitrea and M. Taylor**, Potential theory on Lipschitz domains in Riemannian manifolds: Hölder continuous metric tensors, *Comm. Partial Differ. Equ.*, **25** (2000), 1487–1536.

141. **M. Mitrea and M. Taylor**, Potential theory on Lipschitz domains in Riemannian manifolds: Sobolev-Besov space results and the Poisson problem, *J. Funct. Anal.*, **176** (2000), 1–79.

142. **M. Mitrea and M. Taylor**, Potential theory on Lipschitz domains in Riemannian manifolds: L^p Hardy, and Hölder space results, *Comm. Anal. Geom.*, **9** (2001), 369–421.

143. **T. Mitsis**, Embedding B_∞ into Muckenhoupt classes, *Proc. Amer. Math. Soc.*, **133** (2005), 1057–1061.

144. **C. B. Morrey**, Multiple integrals in the calculus of variations, Springer-Verlag, Berlin, 1966.

145. **B. Muckenhoupt**, Weighted norm inequalities for the Hardy maximal function, *Trans. Amer. Math. Soc.*, **165** (1972), 207–226.

146. **B. Muckenhoupt an R. L. Wheeden**, Weighted bounded mean oscillation and the Hilbert transform, *Studia Math.*, **54** (1976), 221–237.

147. **S. Müller**, Higher integrability of determinants and weak convergence in L^1, *J. Reine Angew. Math.*, **412** (1990), 20–34.

148. **J. Nmeth**, Generalizations of the Hardy-Littlewood inequality, *Acta Sci. Math.(Szeged)*, **32** (1971), 295–299.

149. **J. Nmeth**, Generalizations of the Hardy-Littlewood inequality II, *Acta Sci. Math.(Szeged)*, **35** (1973), 127–134.

150. **L. Ni, Y. Shi and L. Tam**, Poisson equation, Poincaré-Lelong equation and curvature decay on complete Köhler manifolds, *J. Differ. Geom.*, **57** (2001), 339–388.

151. **R. Nibbi**, Some generalized Poincaré inequalities and applications to problems arising in electromagnetism, *J. Inequal. Appl.*, **4** (1999), 283–299.

152. **F. Nier**, Quelques critéres pour l'inégalité de Poincaré dans $R^d, d \geq 2$. (French) [Some criteria for the Poincare inequality in $R^d, d \geq 2$] Seminaire: Équations aux Dérivées Partielles, 2002–2003, Exp. No. V, 16 pp., Sémin. Équ. Driv. Partielles, École Polytech., Palaiseau, 2003.

153. **C. A. Nolder**, Hardy-Littlewood inequality for quasiregular maps on Carnot groups, *Nonlinear Analysis*, **63** (2005), 407–415.

154. **C. A. Nolder**, Lipschitz classes of solutions to certain elliptic equations, *Ann. Acad. Sci. Fenn. Ser. A I Math.*, **17** (1992), 211–219.

155. **C. A. Nolder**, Hardy-Littlewood theorems for solutions of elliptic equations in divergence form, *Indiana Univ. Math. J.*, **40** (1991), 149–160.

156. **C. A. Nolder**, Conjugate functions and moduli of continuity, *Illinois J. Math.*, **31** (1987), 699–709.

157. **C. A. Nolder and D. M. Oberlin**, Moduli of continuity and a Hardy-Littlewood theorem. Complex analysis, Joensuu 1987, 265–272, *Lect. Notes in Math.*, **1351**, Springer, Berlin, 1988.

158. **B. Osikiewicz and A. Tonge**, An interpolation approach to Hardy-Littlewood inequalities for norms of operators on sequence spaces, *Linear Algebra Appl.*, **331** (2001), 1–9.

159. **L. E. Payne and H. F. Weinberger**, An optimal Poincaré inequality for convex domains, *Arch. Rational Mech. Anal.*, **5** (1960), 286–292.

160. **M. Pearson**, The Poincaré inequality and entire functions, *Proc. Amer. Math. Soc.*, **118** (1993), 1193–1197.

161. **I. Perić and D. Žubrinić**, Caccioppoli's inequality for quasilinear elliptic operators, *Math. Inequal. Appl.*, **2** (1999), 251–261.

162. **A. Ponce**, A variant of Poincaré inequality, *C. R. Math. Acad. Sci. Paris*, **337** (2003), 253–257.

163. **Y. Rakotondratsimba**, Weighted vector-valued inequalities for fractional maximal operators, *Southeast Asian Bull. Math.*, **23** (1999), 471–495.

164. **Y. Rakotondratsimba**, Fractional maximal and integral operators on weighted amalgam spaces, *J. Korean Math. Soc.*, **36** (1999), 855–890.

165. **Y. Rakotondratsimba**, Two-weight inequality for fractional integral operators and Adams inequality, *Vietnam J. Math.*, **29** (2001), 269–280.

166. **A. Ranjbar-Motlagh**, A note on the Poincaré inequality, *Studia Math.*, **154** (2003), 1–11.

167. **M. M. Rao and Z. D. Ren**, Theory of Orlicz spaces (Pure and Applied Mathematics), Marcel Dekker, Newyork 1991.

168. **H. M. Reimann**, Functions of bounded mean oscillation and quasiconformal mappings, *Commentarii Helveetici*, **49** (1974), 260–276.

169. **H. M. Reimann and T. Rychener**, Funktionen beschränkter mittlerer Oszillation, *Lect. Notes Math.*, **487** (1975), Springer-Verlag, Berlin.

170. **D. C. Sanyal and M. Dutta**, On extensions of Caccioppoli's fixed point theorem, *J. Indian Acad. Math.*, **24** (2002), 371–379.

171. **C. Sbordone**, Rearrangement of functions and reverse Hölder inequalities. Ennio De Giorgi colloquium (Paris, 1983), 139–148, *Res. Notes in Math.*, **125** (1983), Pitman, Boston, MA, 1985.

172. **G. Scheffer**, Local Poincaré inequalities in non-negative curvature and finite dimension, *J. Funct. Anal.*, **198** (2003), 197–228.

173. **G. A. Serëgin**, A local estimate of the Caccioppoli inequality type for extremal variational problems of Hencky plasticity. (Russian) Some applications of functional analysis to problems of mathematical physics (Russian), 127–138, **145**, Akad. Nauk SSSR Sibirsk. Otdel., Inst. Mat., Novosibirsk, 1988.

174. **A. Seeger, S. Wainger, J. Wright and S. Ziesler**, Singular integral and maximal operators associated to hypersurfaces: L^p theory, *J. Geom. Anal.*, **15** (2005), 477–498.

175. **Z. Shen**, Weighted estimates in L^2 for Laplace's equation on Lipschitz domains, *Trans. Amer. Math. Soc.*, **357** (2005), 2483–2870.

176. **Y. Shen and X. Guo**, Weighted Poincaré inequalities on unbounded domains and nonlinear elliptic boundary value problems, *Acta Math. Sci.* (English Ed.), **4** (1984), 277–286.

177. **J. Strömberg and A. Torchinsky**, Weighted Hardy spaces, *Lect. Notes Math.*, **1381** (1989), Springer-Verlag, Berlin.

178. **F. Weisz**, Two-parameter Hardy-Littlewood inequalities, *Studia Math.*, **118** (1996), 175–184.

179. **S. Ding**, Weighted Hardy-Littlewood inequality for A-harmonic tensors, *Proc. Amer. Math. Soc.*, **125** (1997), 1727–1735.

180. **S. Ding**, Parametric weighted integral inequalities for A-harmonic tensors, *Z. Anal. Anwendungen*, **20** (2001), 691–708.

181. **B. Liu**, $A_r(\lambda)$-weighted Caccioppoli-type and Poincaré-type inequalities for A-harmonic tensors, *Int. J. Math. Math. Sci.*, **31** (2002), 115–122.

182. **B. Liu**, $A_r^\lambda(\Omega)$-weighted imbedding inequalities for A-harmonic tensors, *J. Math. Anal. Appl.*, **273** (2002), 667–676.

183. **S. G. Staples**, L^p-averaging domains and the Poincaré inequality, *Ann. Acad. Sci. Fenn, Ser. AI. Math.*, **14** (1989), 103–127.

184. **S. Ding and C. A. Nolder**, $L^s(\mu)$-averaging domains, *J. Math. Anal. Appl.*, **283** (2003), 85–99.

185. **S. Ding and B. Liu**, Whitney covers and quasi-isometry of $L^s(\mu)$-averaging domains, *J. Inequal. Appl.*, **6** (2001), 435–449.

186. **G. Bao**, $A_r(\lambda)$-weighted integral inequalities for A-harmonic tensors, *J. Math. Anal. Appl.*, **247** (2000), 466–477.

187. **G. Bao, T. Li and Y. Xing**, Two-weighted Hardy-Littlewood inequality for A-harmonic tensors, *Chinese Ann. Math. Ser. A*, **26** (2005), 113–120.

188. **G. Bao and H. Zhu**, $A_r^\lambda(\Omega)$-weighted Hardy-Littlewood inequality for differential forms satisfying the A-harmonic equation, *Heilongjiang Daxue Ziran Kexue Xuebao*, **23** (2006), 26–30.

189. **S. M. Buckley and P. Koskela**, Orlicz-Hardy Inequalities, *Illinois J. Math.*, **48** (2004), 782–802.

190. **R. R. Coifman, P. L. Lion, Y. Meyer and S. Semmes**, Compensated compactness and Hardy spaces, *J. Math. Pures Appl.*, **72** (1993), 247–286.

191. **Z. Liu and S. Lu**, Two-weight weak-type norm inequalities for the commutators of fractional integrals, *Integr. Equ. Oper. Theory*, **48** (2004), 397–409.

192. **Z. Lou**, Extensions of Hardy-Littlewood inequalities, *Int. J. Math. Math. Sci.*, **17** (1994), 193–195.

193. **Z. Lou and A. McIntosh**, Hardy space of exact forms on R^N, *Trans. Amer. Math. Soc.*, **357** (2005), 1469–1496.

194. **S. Ding and B. Liu**, Global integrability of the Jacobians of a composite mapping, *J. Inequal. Appl.*, **2006** (2006).

195. **K. Astala and F. W. Gehring**, Quasiconformal analogues of theorems of Koebe and Hardy-Littlewood, *Michigan Math. J.*, **32** (1985), 99–107.

196. **F. G. Avkhadiev and K. J. Wirths**, Asymptotically sharp bounds in the Hardy-Littlewood inequalities on mean values of analytic functions, *Bull. London Math. Soc.*, **33** (2001), 695–700.

197. **S. Ding and H. Xie**, Norm comparison theorems for conjugate A-harmonic tensors (Chinese), *Chinese Ann. Math. Ser. A* , **18** (1997), 205–212; translation in *Chinese J. Contemp. Math.*, **18** (1997), 151–159.

198. **W. D. Evans, D. J. Harris and L. Pick**, Weighted Hardy and Poincaré inequalities on trees, *J. London Math. Soc.*, **52** (1995), 121–136.

199. **M. K. Kwong and A. Zettl**, An extension of the Hardy-Littlewood inequality, *Proc. Amer. Math. Soc.*, **77** (1979), 117–118.

200. **J. McMullen and B. Sharp**, A remark on the Hardy-Littlewood inequality, *Proc. Amer. Math. Soc.*, **90** (1984), 95–96.

201. **S. Ding and C. A. Nolder**, Weighted Poincaré-type inequalities for solutions to the A-harmonic equation, *Illinois J. Math.*, **2** (2002), 199–205.

202. **S. G. Staples**, Averaging domains: from Euclidean spaces to homogeneous spaces, *Proceedings of Conference on Differential & Difference Equations and Applications*, Florida Institute of technology, Hindawi Publishing, 2006, 1041–1048.

203. **S. Ding and B. Liu**, Generalized Poincaré inequalities for solutions to the A-harmonic equation in certain domains, *J. Math. Anal. Appl.*, **252** (2000), 538–548.

204. **E. M. Stein**, Singular integrals and differentiability properties of functions, Princeton University Press, Princeton, 1970.

205. **R. P. Agarwal and S. Ding**, Global Caccioppoli-type and Poincaré inequalities with Orlicz norms, *J. Inequal. Appl.*, submitted.

206. **Y. Wang**, Two-weight Poincaré-type inequalities for differential forms in $L^s(\mu)$-averaging domains, *Applied Mathematics Letters*, **20** (2007), 1161–1166.

207. **P. A. Zharov**, On a two-weight inequality. Generalization of Hardy and Poincaré inequalities (Russian), *Trudy Mat. Inst. Steklov.*, **194** (1992), Issled. po Teor. Differ. Funktsii Mnogikh Peremen. i ee Prilozh. 14, 97–110; translation in *Proc. Steklov Inst. Math.*, **4** (1993), 101–114.

208. **J. Heinonen and P. Koskela**, Weighted Sobolev and Poincaré inequalities and quasiconregular mappings of polynomial type, *Math. Scand.*, **77** (1995), 251–271.

209. **D. E. Edmunds and R. Hurri-Syrjänen**, Weighted Poincaré inequalities and Minkowski content, *Proc. Roy. Soc. Edinburgh Sect. A*, **125** (1995), 817–825.

210. **A. Gavrilov, S. Nicaise and O. Penkin**, Poincaré inequality on stratified sets and applications. Evolution equations: applications to physics, industry, life sciences and economics (Levico Terme, 2000), 195–213, Progr. Nonlinear Differ. Equ. Appl., **55**, Birkhäuser, Basel, 2003.

211. **S. M. Buckley and P. Koskela**, Sobolev-Poincaré inequalities for $p < 1$, *Indiana Univ. Math. J.*, **43** (1994), 221–240.

212. **E. Hebey**, Sharp Sobolev-Poincaré inequalities on compact Riemannian manifolds, *Trans. Amer. Math. Soc.*, **354** (2002), 1193–1213.

213. **P. Koskela and A. Stanoyevitch**, Poincaré inequalities and Steiner symmetrization, *Illinois J. Math.*, **40** (1996), 365–389.

214. **P. Koskela, N. Shanmugalingam and J. Tyson**, Dirichlet forms, Poincaré inequalities, and the Sobolev spaces of Korevaar and Schoen, *Potential Anal.*, **21** (2004), 241–262.

215. **P. Koskela and N. Shanmugalingam and H. Tuominen**, Removable sets for the Poincaré inequality on metric spaces, *Indiana Univ. Math. J.*, **49** (2000), 333–352.

216. **O. Martio**, John domains, bi-Lipschitz balls and Poincaré inequality, *Rev. Roumaine Math. Pures Appl.*, **33** (1988), 107–112.

217. **C. Wang**, A weighted Poincaré inequality and removable singularities for harmonic maps, *Manuscripta Math.*, **112** (2003), 259–270.

218. **G. Acosta and R. Durán**, An optimal Poincaré inequality in L^1 for convex domains, *Proc. Amer. Math. Soc.*, **132** (2004), 195–202.

219. **M. Bebendorf**, A note on the Poincaré inequality for convex domains, *Z. Anal. Anwendungen*, **22** (2003), 751–756.

220. **I. Chavel and E. A. Feldman**, An optimal Poincaré inequality for convex domains of non-negative curvature, *Arch. Rational Mech. Anal.*, **65** (1977), 263–273.

221. **A. Stanoyevitch and D. Stegenga**, Equivalence of analytic and Sobolev Poincaré inequalities for planar domains, *Pacific J. Math.*, **178** (1997), 363–375.

222. **A. Elcrat and A. MacLean**, Weighted Wirtinger and Poincaré inequalities on unbounded domains, *Indiana Univ. Math. J.*, **29** (1980), 321–332.

223. **W. D. Evans, D. J. Harris and L. Pick**, Ridged domains, embedding theorems and Poincaré inequalities, *Math. Nachr.*, **221** (2001), 41–74.

224. **S. Chanillo and R. L. Wheeden**, Poincaré inequalities for a class of non-A_p weights, *Indiana Univ. Math. J.*, **41** (1992), 605–623.

225. **R. Hurri-Syrjänen**, An improved Poincaré inequality, *Proc. Amer. Math. Soc.*, **120** (1994), 213–222.

226. **R. Hurri-Syrjänen**, A Weighted Poincaré inequality with a doubling weight, *Proc. Amer. Math. Soc.*, **126** (1998), 545–552.

227. **R. Hurri-Syrjänen**, The Poincaré inequality and reverse doubling weights, *Canad. Math. Bull.*, **47** (2004), 206–214.

228. **J. Fleckinger-Pellé and P. Takáč**, An improved Poincaré inequality and the p-Laplacian at resonance for $p > 2$, *Adv. Differ. Equ.*, **7** (2002), 951–971.

229. **B. Franchi**, Weighted Sobolev-Poincaré inequalities and pointwise estimates for a class of degenerate elliptic equations, *Trans. Amer. Math. Soc.*, **327** (1991), 125–158.

230. **B. Franchi, P. Carlos and R. L. Wheeden**, A sum operator with applications to self-improving properties of Poincaré inequalities in metric spaces, *J. Fourier Anal. Appl.*, **9** (2003), 511–540.

231. **B. Franchi, P. Carlos and R. L. Wheeden**, Self-improving properties of John-Nirenberg and Poincaré inequalities on spaces of homogeneous type, *J. Funct. Anal.*, **153** (1998), 108–146.

232. **B. Franchi, C. E. Gutiérrez and R. L. Wheeden**, Weighted Sobolev-Poincaré inequalities for Grushin type operators, *Comm. Partial Differ. Equ.*, **19** (1994), 523–604.

233. **B. Franchi, G. Lu and R. L. Wheeden**, Weighted Poincaré inequalities for Hörmander vector fields and local regularity for a class of degenerate elliptic equations. Potential theory and degenerate partial differential operators (Parma), *Potential Anal.*, **4** (1995), 361–375.

234. **B. Franchi, C. Pérez and R. L. Wheeden**, Sharp geometric Poincaré inequalities for vector fields and non-doubling measures, *Proc. London Math. Soc.*, **80** (2000), 665–689.

235. **B. Helffer and F. Nier**, Criteria to the Poincaré inequality associated with Dirichlet forms in R^d, $d \geq 2$, *Int. Math. Res. Not.*, **22** (2003), 1199–1223.

236. **C. Amick**, Some remarks on Rellich's theorem and the Poincaré inequality, *J. London Math. Soc.*, **18** (1978), 81–93.

237. **M. Bourdon and H. Pajot**, Poincaré inequalities and quasiconformal structure on the boundary of some hyperbolic buildings, *Proc. Amer. Math. Soc.*, **127** (1999), 2315–2324.

238. **S. Chua and R. Wheeden**, Sharp conditions for weighted 1-dimensional Poincaré inequalities, *Indiana Univ. Math. J.*, **49** (2000), 143–175.

239. **D. H. Hamilton**, On the Poincaré inequality, *Complex Variables Theory Appl.*, **5** (1986), 265–270.

240. **B. Hanson and J. Heinonen**, An n-dimensional space that admits a Poincaré inequality but has no manifold points, *Proc. Amer. Math. Soc.*, **128** (2000), 3379–3390.

241. **S. Keith**, Modulus and the Poincaré inequality on metric measure spaces, *Math. Z.*, **245** (2003), 255–292.

242. **S. Keith**, Measurable differentiable structures and the Poincaré inequality, *Indiana Univ. Math. J.*, **53** (2004), 1127–1150.

243. **T. J. Laakso**, Ahlfors Q-regular spaces with arbitrary $Q > 1$ admitting weak Poincaré inequality, *Geom. Funct. Anal.*, **10** (2000), 111–123.

244. **E. H. Lieb, R. Seiringer and J. Yngvason**, Poincaré inequalities in punctured domains, *Ann. Math.*, **158** (2003), 1067–1080.

245. **G. Lu and C Pérez**, $L^1 \to L^q$ Poincaré inequalities for $0 < q < 1$ imply representation formulas, *Acta Math. Sin. (Engl. Ser.)*, **18** (2002), 1–20.

246. **G. Lu and R. L. Wheeden**, Poincaré inequalities, isoperimetric estimates, and representation formulas on product spaces, *Indiana Univ. Math. J.*, **47** (1998), 123–151.

247. **P. MacManus**, Poincaré inequalities and Sobolev spaces, *Proceedings of the 6th International Conference on Harmonic Analysis and Partial Differential Equations* (El Escorial, 2000), *Publ. Mat.* 2002, Vol. Extra, 181–197.

248. **P. MacManus and C. Pérez**, Generalized Poincaré inequalities: sharp self-improving properties, *Int. Math. Res. Not.*, **2** (1998), 101–116.

249. **P. Souplet**, Geometry of unbounded domains, Poincaré inequalities and stability in semilinear parabolic equations, *Comm. Partial Differ. Equ.*, **24** (1999), 951–973.

250. **A. Stanoyevitch**, Products of Poincaré domains, *Proc. Amer. Math. Soc.*, **117** (1993), 79–87.

251. **P. Sternberg and K. Zumbrun**, A Poincaré inequality with applications to volume-constrained area-minimizing surfaces, *J. Reine Angew. Math.*, **503** (1998), 63–85.

252. **F. Wang**, Coupling, convergence rates of Markov processes and weak Poincaré inequalities, *Sci. China Ser. A*, **45** (2002), 975–983.

253. **W. Wang, Z. Zheng and J. Sun**, Weighted Poincaré inequalities on one-dimensional unbounded domains, *Appl. Math. Lett.*, **16** (2003), 1143–1149.

254. **M. Zhu**, On the extremal functions of Sobolev-Poincaré inequality, *Pacific J. Math.*, **214** (2004), 185–199.

255. **S. Chua**, Weighted inequalities on John domains, *J. Math. Anal. Appl.*, **258** (2001), 763–776.

256. **S. Chua**, Weighted Sobolev interpolation inequalities on certain domains, *J. London Math. Soc.*, **51** (1995), 532–544.

257. **S. Chua**, Some remarks on extension theorems for weighted Sobolev spaces, *Illinois J. Math.*, **38** (1994), 95–126.

258. **S. Chua**, On weighted Sobolev interpolation inequalities, *Proc. Amer. Math. Soc.*, **121** (1994), 441–449.

259. **S. Chua**, Weighted Sobolev inequalities on domains satisfying the chain condition, *Proc. Amer. Math. Soc.*, **117** (1993), 449–457.

260. **S. Chua**, Extension theorems on weighted Sobolev spaces, *Indiana Univ. Math. J.*, **41** (1992), 1027–1076.

261. **S. Ding and Y. Xing**, Estimates for Lipschitz and BMO norms of the composite operator, preprint.

262. **E. Lutwak, D. Yang and G. Zhang**, Sharp affine L_p Sobolev inequalities, *J. Differ. Geom.*, **62** (2002), 17–38.

263. **J. Maly, D. Swanson and W. Ziemer**, The co-area formula for Sobolev mappings, *Trans. Amer. Math. Soc.*, **355** (2003), 477–492.

264. **J. Meng**, Principal frequency and existence of solutions for quasilinear elliptic obstacle problems, *J. Math. Anal. Appl.*, **286** (2003), 271–280.

265. **J. Meng and H. Gao**, Boundary regularity of weak solutions to some degenerate elliptic obstacle problems (Chinese), *Chinese Ann. Math. Ser. A*, **24** (2003), 407–414; translation in *Chinese J. Contemp. Math.*, **24** (2003), 205–214.

266. **J. Meng and H. Gao**, Existence of positive solutions for a class of nonlinear elliptic obstacle problems (Chinese), *J. Shanghai Jiaotong Univ. (Chin. Ed.)*, **37** (2003), 308–312.

267. **S. Ding**, Weighted Caccioppoli-type estimates and weak reverse Hölder inequalities for A-harmonic tensors, *Proc. Amer. Math. Soc.*, **127** (1999), 2657–2664.

268. **S. Ding and D. Sylvester**, Weak reverse Hölder inequalities and imbedding inequalities for solutions to the A-harmonic equation, *Nonlinear Anal.*, **51** (2002), 783–800.

269. **M. Giaquinta and J. Souček**, Caccioppoli's inequality and Legendre-Hadamard condition, *Math. Ann.*, **270** (1985), 105–107.

270. **R. F. Gariepy**, A Caccioppoli inequality and partial regularity in the calculus of variations, *Proc. Roy. Soc. Edinburgh Sect. A*, **112** (1989), 249–255.

271. **J. K. Brooks and D. Candeloro**, On the Caccioppoli integral, *Atti Sem. Mat. Fis. Univ. Modena*, **51** (2003), 415–431.

272. **G. M. Troianiello**, Estimates of the Caccioppoli-Schauder type in weighted function spaces, *Trans. Amer. Math. Soc.*, **334** (1992), 551–573.

273. **E. Vesentini**, Renato Caccioppoli and complex analysis (Italian), International Symposium in honor of Renato Caccioppoli (Naples, 1989), *Ricerche Mat.*, **40** (1991), suppl., 119–125.

274. **D. Gilbarg and N. Trudinger**, Elliptic partial differential equations of second order, Springer-Verlag, New York, 1983.

275. **B. Bojarski**, Remarks on Sobolev imbedding inequalities, Complex analysis, Joensuu 1987, 52–68, *Lec. Notes Math.*, **1351**, Springer, Berlin, 1988.

276. **S. Ding**, $A_r(\Omega)$-Weighted imbedding A-harmonic tensors, *Potential Anal.*, **18** (2003), 25–34.

277. **S. Ding**, Weighted imbedding theorems in the space of differential forms, *J. Math. Anal. Appl.*, **262** (2001), 435–445.

278. **S. Ding**, Weighted imbedding theorems in the space of differential forms, *J. Math. Anal. Appl.*, **262** (2001), 435–445.

279. **X. Wu**, Some boundary properties and imbedding inequalities of analytic functions I (Chinese), *J. Huazhong Inst. Tech.*, **7** (1979), 3, 41–49.

280. **X. Wu**, Some boundary properties and imbedding inequalities of analytic functions II (Chinese), *J. Huazhong Inst. Tech.*, **7** (1979), 49–61.

281. **G. Chiti**, A reverse Hölder inequality for the eigenfunctions of linear second order elliptic operators, *Z. Angew. Math. Phys.*, **33** (1982), 143–148.

282. **K. Hoshino and N. Kikuchi**, Reverse Hölder inequality for functions with discrete-time variable, *Dynam. Contin. Discrete Impuls. Syst.*, **6** (1999), 503–517.

283. **A. A. Arkhipova**, Reverse Hölder inequalities with boundary integrals and L_p-estimates for solutions of nonlinear elliptic and parabolic boundary-value problems, *Nonlinear evolution equations*, 15–42, Amer. Math. Soc. Transl. Ser., **164** (1995), Amer. Math. Soc., Providence, RI.

284. **A. A. Arkhipova**, Reverse Hölder inequalities in parabolic problems with anisotropic data spaces (translation of theoretical and numerical problems of mathematical physics (Russian), 3–19, "Nauka", Sibirsk. Otdel., Novosibirsk, 1994), *Siberian Adv. Math.*, **6** (1996), 1–18.

285. **B. Bojarski**, Remarks on the stability of reverse Hölder inequalities and quasiconformal mappings, *Ann. Acad. Sci. Fenn. Ser. A I Math.*, **10** (1985), 89–94.

286. **D. Cruz-Uribe**, The class $A_\infty^+(g)$ and the one-sided reverse Hölder inequality, *Canad. Math. Bull.*, **40** (1997), 169–173.

287. **D. Cruz-Uribe and C. J. Neugebauer**, The structure of the reverse Hölder classes, *Trans. Amer. Math. Soc.*, **347** (1995), 2941–2960.

288. **D. Cruz-Uribe, C. J. Neugebauer and V. Olesen**, The one-sided minimal operator and the one-sided reverse Hölder inequality, *Studia Math.*, **116** (1995), 255–270.

289. **L. D'Apuzzo and C. Sbordone**, Reverse Hölder inequalities: a sharp result, *Rend. Mat. Appl.*, **10** (1990), 357–366.

290. **S. Ding**, The weak reverse Hölder inequality for conjugate A-harmonic tensors, preprint.

291. **J. Duoandikoetxea**, Reverse Hölder inequalities for spherical harmonics, *Proc. Amer. Math. Soc.*, **101** (1987), 487–491.

292. **B. Liu and S. Ding**, The monotonic property of the $L^s(\mu)$-averaging domains and weighted weak reverse Hölder inequality, *J. Math. Anal. Appl.*, **237** (1999), 730–739.

293. **J. Martín and M. Milman**, Reverse Hölder inequalities and approximation spaces, *J. Approx. Theory*, **109** (2001), 82–109.

294. **E. W. Stredulinsky**, Higher integrability from reverse Hölder inequalities, *Indiana Univ. Math. J.*, **29** (1980), 407–413.

295. **V. I. Vasyunin**, The exact constant in the inverse Hölder inequality for Muckenhoupt weights(Russian), *Algebra i Analiz*, **15** (2003), 73–117; translation in *St. Petersburg Math. J.*, **15** (2004), 49–79.

296. **X. Yang**, Refinement of Hölder inequality and application to Ostrowski inequality, *Appl. Math. Comput.*, **138** (2003), 455–461.

297. **B. Yan and Z. Zhou**, A theorem on improving regularity of minimizing sequences by reverse Hölder inequalities, *Michigan Math. J.*, **44** (1997), 543–553.

298. **B. Bojarski**, Sharp maximal operator of fractional order and Sobolev imbedding inequalities, *Bull. Polish Acad. Sci. Math.*, **33** (1985), 7–16.

299. **Y. Xing and S. Ding**, Norm of the composition of the maximal and projection operators, preprint.

300. **S. Ding**, Norm estimates for the maximal operators and Green's operator, Proceedings of the 6th International Conference on Differential Equations in Baltimore, Maryland, May, 2008.

301. **L. D'Onofrio and L. Greco**, A counter-example in G-convergence of nondivergence elliptic operators, *Proc. Roy. Soc. Edinburgh Sect. A*, **133** (2003), 1299–1310.

302. **M. Goldberg**, Matrix A_p weights via maximal functions, *Pacific J. Math.*, **211** (2003), 201–220.

303. **J. M. Martell, C. Pérez and R. Trujillo-González**, Lack of natural weighted estimates for some singular integral operators, *Trans. Amer. Math. Soc.*, **357** (2005), 385–396.

304. **Y. Terasawa**, Outer measures and weak type $(1,1)$ estimates of Hardy-Littlewood maximal operators, *J. Inequal. Appl.*, **2006** (2006), 1–13.

305. **C. Wang**, The Calderón-Zygmund inequality on a compact Riemannian manifold, *Pacific J. Math.*, **217** (2004), 181–200.
306. **Y. Xing and Y. Wang**, BMO and Lipschitz norm estimates for composite operators, *Potential Analysis*, DOI 10.1007/s11118-009-9137-5.
307. **Y. Xing and S. Ding**, Global estimates for the maximal operator and homotopy operator, preprint.
308. **Y. Xing and S. Ding**, Norms of the composition of the homotopy and Green's operators, preprint.
309. **Y. Xing and S. Ding**, Inequalities for Green's operator with Lipschitz and BMO norms, *Computers and Mathematics with Applications*, **58** (2009), 273–280.
310. **H. Brezis, N. Fusco and C. Sbordone**, Integrability for the Jacobian of orientation preserving mappings, *J. Funct. Anal.*, **115** (1993), 425–431.
311. **S. Ding**, Global estimates for the Jacobians of orientation preserving mappings, *Comput. Math. Appl.*, **50** (2005), 707–718.
312. **J. Hogan, C. Li, A. McInton and K. Zhang**, Global higher integrability of Jacobians on bounded domains, *Ann. Inst. H. Poincaré Anal.*, **17** (2000), 193–217.
313. **J. Strömberg and A. Torchinsky**, Weights, sharp maximal functions and Hardy spaces, *Bull. Amer. Math. Soc. (N.S.)*, **3** (1980), 1053–1056.
314. **T. Kilpeläinen and P. Koskela**, Global integrability of gradients of solutions to partial differential equations, *Nonlinear Anal.*, **23** (1994), 899–909.
315. **G. Bao and S. Ding**, Invariance properties of $L^\varphi(\mu)$-averaging domains under some mappings, *J. Math. Anal. Appl.*, **259** (2001), 241–252.
316. **S. Byun**, Elliptic equations with BMO coefficients in Lipschitz domains, *Trans. Amer. Math. Soc.*, **357** (2005), 1025–1046.
317. **B. E. J. Dahlberg**, On the Poisson integral for Lipschitz and C^1-domains, *Studia Math.*, **66** (1979), 13–24.
318. **B. E. J. Dahlberg**, L^q-Estimates for Green potentials in Lipschitz domains, *Math. Scand.*, **44** (1979), 149–170.
319. **B. E. J. Dahlberg and C. Kenig**, Hardy spaces and the Neumann problem in L^p for Laplace's equation in Lipschitz domains, *Ann. Math.*, **125** (1987), 437–465.
320. **B. E. J. Dahlberg and C. Kenig**, L^q-estimates for the three-dimensional system of elastostatics on Lipschitz domains, in Analysis and partial differential equations, ed. C. Sadosky, 621–634*Lect. Notes in Pure Appl. Math.*, **122**, Dekker, New York, 1990.
321. **S. Ding**, $L^\varphi(\mu)$-averaging domains and the quasihyperbolic metric, *Comput. Math. Appl.*, **47** (2004), 1611–1618.
322. **S. Ding**, Lipschitz and BMO norm inequalities for operator, presented at the 5th World Congress of Nonlinear Analysts in Orlando, Florida, in July of 2008 (WCNA-2008).
323. **J. Heinonen and P. Koskela**, A note on Lipschitz functions, upper gradients, and the Poincaré inequality, *New Zealand J. Math.*, **28** (1999), 37–42.
324. **J. Väisälä**, Domains and maps, *Lect. Notes Math., Springer-Verlag*, **1508** (1992), 119–131.
325. **G. Bao, C. Wu and Y. Xing**, Two-weight Poincaré inequalities for the projection operator and A-harmonic tensors on Riemannian manifolds, *Illinois J. Math.*, **51**, (2007), 831–842.
326. **W. Beckner**, A generalized Poincaré inequality for Gaussian measures, *Proc. Amer. Math. Soc.*, **105** (1989), 397–400.
327. **M. Belloni and B. Kawohl**, A symmetry problem related to Wirtinger's and Poincaré's inequality, *J. Differ. Equ.*, **156** (1999), 211–218.
328. **R. D. Benguria and C. Depassier**, A reversed Poincaré inequality for monotone functions, *J. Inequal. Appl.*, **5** (2000), 91–96.

329. **T. Bhattacharya and F. Leonetti**, A new Poincaré inequality and its application to the regularity of minimizers of integral functionals with nonstandard growth, *Nonlinear Anal.*, **17** (1991), 833–839.

330. **J. Björn**, Poincaré inequalities for powers and products of admissible weights, *Ann. Acad. Sci. Fenn. Math.*, **26** (2001), 175–188.

331. **J. Bruna**, L^p-estimates for Riesz transforms on forms in the Poincaré space, *Indiana Univ. Math. J.*, **54** (2005), 153–186.

332. **S. M. Buckley and P. Koskela**, New Poincaré inequalities from old, *Ann. Acad. Sci. Fenn. Math.*, **23** (1998), 251–260.

333. **F. Cavallini and F. Crisciani**, A generalized 2-D Poincaré inequality, *J. Inequal. Appl.*, **5** (2000), 343–349.

Index